Springer Complexity

Springer Complexity is an interdisciplinary program publishing the best research and academic-level teaching on both fundamental and applied aspects of complex systems—cutting across all traditional disciplines of the natural and life sciences, engineering, economics, medicine, neuroscience, social and computer science.

Complex Systems are systems that comprise many interacting parts with the ability to generate a new quality of macroscopic collective behavior the manifestations of which are the spontaneous formation of distinctive temporal, spatial or functional structures. Models of such systems can be successfully mapped onto quite diverse "real-life" situations like the climate, the coherent emission of light from lasers, chemical reaction-diffusion systems, biological cellular networks, the dynamics of stock markets and of the internet, earthquake statistics and prediction, freeway traffic, the human brain, or the formation of opinions in social systems, to name just some of the popular applications.

Although their scope and methodologies overlap somewhat, one can distinguish the following main concepts and tools: self-organization, nonlinear dynamics, synergetics, turbulence, dynamical systems, catastrophes, instabilities, stochastic processes, chaos, graphs and networks, cellular automata, adaptive systems, genetic algorithms and computational intelligence.

The three major book publication platforms of the Springer Complexity program are the monograph series "Understanding Complex Systems" focusing on the various applications of complexity, the "Springer Series in Synergetics", which is devoted to the quantitative theoretical and methodological foundations, and the "Springer Briefs in Complexity" which are concise and topical working reports, case studies, surveys, essays and lecture notes of relevance to the field. In addition to the books in these two core series, the program also incorporates individual titles ranging from textbooks to major reference works.

Editorial and Programme Advisory Board

Understanding Complex Systems

Founding Editor: S. Kelso

Future scientific and technological developments in many fields will necessarily depend upon coming to grips with complex systems. Such systems are complex in both their composition—typically many different kinds of components interacting simultaneously and nonlinearly with each other and their environments on multiple levels—and in the rich diversity of behavior of which they are capable.

The Springer Series in Understanding Complex Systems series (UCS) promotes new strategies and paradigms for understanding and realizing applications of complex systems research in a wide variety of fields and endeavors. UCS is explicitly transdisciplinary. It has three main goals: First, to elaborate the concepts, methods and tools of complex systems at all levels of description and in all scientific fields, especially newly emerging areas within the life, social, behavioral, economic, neuro and cognitive sciences (and derivatives thereof); second, to encourage novel applications of these ideas in various fields of engineering and computation such as robotics, nano-technology and informatics; third, to provide a single forum within which commonalities and differences in the workings of complex systems may be discerned, hence leading to deeper insight and understanding.

UCS will publish monographs, lecture notes and selected edited contributions aimed at communicating new findings to a large multidisciplinary audience.

More information about this series at http://www.springer.com/series/5394

Victoriano Carmona · Jesús Cuevas-Maraver
Fernando Fernández-Sánchez
Elisabeth García-Medina
Editors

Nonlinear Systems, Vol. 1

Mathematical Theory and Computational Methods

Springer

Editors
Victoriano Carmona
Departamento de Matemática Aplicada II
Universidad de Sevilla
Sevilla, Spain

Fernando Fernández-Sánchez
Departamento de Matemática Aplicada II
Universidad de Sevilla
Sevilla, Spain

Jesús Cuevas-Maraver
Departamento de Física Aplicada I
Universidad de Sevilla
Sevilla, Spain

Elisabeth García-Medina
Departamento de Matemática Aplicada II
Universidad de Sevilla
Sevilla, Spain

ISSN 1860-0832 ISSN 1860-0840 (electronic)
Understanding Complex Systems
ISBN 978-3-030-09781-3 ISBN 978-3-319-66766-9 (eBook)
https://doi.org/10.1007/978-3-319-66766-9

Printed on acid-free paper

This Springer imprint is published by the registered company Springer Nature Switzerland AG
The registered company address is: Gewerbestrasse 11, 6330 Cham, Switzerland

To Prof. Antonio Castellanos Mata, founder and head of the Group of Electrohydrodynamics and Cohesive Granular Media of the University of Seville. Antonio directed research projects for more than 30 years, and this made it possible for him to organize two laboratories at the University. Dedicating a lot of efforts to pure science, Antonio was also interested in practical problems and collaborated with industry (Xerox Corporation, Novartis, Dow Corning, IFPRI).

Antonio belonged to a generation that played an important role in the revival of physics in Spain. In 2013, he was awarded the Prize FAMA for the research career by the University of Seville. In his last years, Antonio worked on thermodynamics in relativity and on triboelectricity in fine powders As a researcher, Antonio combined a strong theoretical mind, experimental intuition, profound understanding of physics of phenomena, and passionate love for science.

Preface

This book commemorates the conference Nolineal2016: International Conference on Nonlinear Mathematics and Physics, that took place in Sevilla, Spain, from 7 to 10 July 2016. There were delegates from many different countries in Europe and also three other continents.

At the end of the conference, the decision was taken of writing a book to provide the readers with a landscape of the many different fields in which nonlinear science is being developed with great success. Contributions would not be proceedings but present an introduction to the different subjects, provide context, present the state of art and certainly the own research of the authors in the field.

Although the mathematics and physics of nonlinear systems are closely intertwined, it has been considered convenient to divide the matter in two volumes:

- *Nonlinear Systems, Vol. 1. Mathematical Theory and Computational Methods in Nonlinear Systems*, edited by Victoriano Carmona, Jesús Cuevas-Maraver, Fernando Fernández-Sánchez and Elisabeth García-Medina
- *Nonlinear Systems, Vol 2. Nonlinear Phenomena in Biology, Optics and Condensed Matter*, edited by Juan F. R. Archilla, Faustino Palmero, M. Carmen Lemos, Bernardo Sánchez-Rey and Jesús Casado-Pascual

The present book is the first volume, it deals with the theory of nonlinear systems, especially from mathematical and computational approaches, and it is divided into four large areas of knowledge, namely bifurcation analysis, wave equations, differential or difference equations and computational methods. From Lorenz system and its bifurcations to applications of cellular automata to laser dynamics, from fast–slow systems to dark matter, from integrability and normal forms to gravitational waves and solitons, this book oscillates between theoretical analysis of nonlinear phenomena and the most current topics in science today.

The outline of the book is as follows:

1. *Bifurcation Analysis*

    - "A Review on Some Bifurcations in the Lorenz System" by A. Algaba, M. C. Domínguez-Moreno, M. Merino and A. J. Rodríguez-Luis
    - "Normal Form for a Class of Three-Dimensional Systems with Free-Divergence Principal Part" by A. Algaba, N. Fuentes, E. Gamero and C. García
    - "Piecewise-Linear (PWL) Canard Dynamics: Simplifying Singular Perturbation Theory in the Canard Regime Using Piecewise-linear Systems" by M. Desroches, S. Fernández-García, M. Krupa, R. Prohens and A. E. Teruel

2. *Wave Equations*

    - "Solitary Waves in the Nonlinear Dirac Equation" by J. Cuevas-Maraver, N. Boussaïd, A. Comech, R. Lan, P. G. Kevrekidis and A. Saxena
    - "On Nonlinear Schrödinger Equation as a Model for Dark Matter: Comments on Galactic Collisions, Supermassive Black Holes and Analogue Laboratory Implementations" by A. Paredes and H. Michinel
    - "Adiabatic Invariants of Second Order Korteweg-de Vries Type Equation" by P. Rozmej and A. Karczewska
    - "Nonlinear Gravitational Waves and Solitons" by F. R. Villatoro

3. *Differential and Difference Equations*

    - "Local Integrability for Some Degenerate Nilpotent Vector Fields" by A. Algaba, I. Checa and C. García
    - "A Logistic Non-linear Difference Equation with Two Delays" by F. Balibrea
    - "Diffusive Limits of the Master Equation in Inhomogeneous Media" by L. Salasnich, A. Bonato and F. Sattin

4. *Computational Methods*

    - "Anticipating Abrupt Changes in Complex Networks: Significant Falls in the Price of a Stock Index" by A. Cordoba, C. Castillejo, J. J. García-Machado and A. M. Lara
    - "On the Numerical Approximation to Generalized Ostrovsky Equations: I" by Ángel Durán
    - "On the Numerical Approximation to Generalized Ostrovsky Equations: II" by Ángel Durán
    - "Simulating Laser Dynamics with Cellular Automata" by F. Jiménez-Morales, J. L. Guisado and J. M. Guerra

Chapters will provide an opportunity for the readers to understand subjects which are normally dispersed in different journals for specialists. We expect them to feel the fascination of nonlinear physics and its broad applicability, stimulating their curiosity and perhaps extending their own research to unexpected territory.

The editors of this book do not want to miss the opportunity to acknowledge the University of Seville and the *Instituto de Matemáticas de la Universidad de Sevilla Antonio Castro Brzezicki* for the financial and administrative support.

Sevilla, Spain
July 2017

Victoriano Carmona
Jesús Cuevas-Maraver
Fernando Fernández-Sánchez
Elisabeth García-Medina

In memoriam of Prof. Antonio Castellanos Mata (07.03.1947–27.01.2016), Full Professor of Electromagnetism at the University of Seville, and Director of the group of Electrohydrodynamics and Cohesive Granular Media

Contents

Contributors

Antonio Algaba Departamento de Ciencias Integradas, Centro de Investigación de Física Teórica y Matemática FIMAT, Universidad de Huelva, Huelva, Spain

Francisco Balibrea Group of Dynamical Systems, Facultad de Matemáticas, Universidad de Murcia, Murcia, Spain

Andrea Bonato Dipartimento di Fisica e Astronomia Galileo Galilei, Universitá di Padova, Padova, Italy

Nabile Boussaïd Université Bourgogne Franche-Comté, Besançon CEDEX, France

Christian Castillejo Instituto IBT, Parque Empresarial Nuevo Torneo, Sevilla, Spain

Isabel Checa Departamento de Matemáticas, Centro de Investigación de Física Teórica y Matemática FIMAT, Universidad de Huelva, Huelva, Spain

Andrew Comech Department of Mathematics, Texas A&M University, College Station, TX, USA; IITP, Moscow, Russia

Antonio Cordoba Departamento de Física de la Materia Condensada, Universidad de Sevilla, Sevilla, Spain

Jesús Cuevas-Maraver Departamento de Física Aplicada I, Escuela Politécnica Superior, Grupo de Física No Lineal, Universidad de Sevilla, Sevilla, Spain; Instituto de Matemáticas de la Universidad de Sevilla (IMUS), Sevilla, Spain

Mathieu Desroches MathNeuro Team, Inria Sophia Antipolis Research Centre, Sophia Antipolis Cedex, France

M. Cinta Domínguez-Moreno Departamento de Ciencias Integradas, Centro de Investigación de Física Teórica y Matemática FIMAT, Universidad de Huelva, Huelva, Spain

Ángel Durán Applied Mathematics Department, University of Valladolid, Valladolid, Spain

Soledad Fernández-García Departamento EDAN, Facultad de Matemáticas, University of Sevilla, Sevilla, Spain

Natalia Fuentes Department of Integrated Sciences, Investigation Center of Theoretical Physics and Mathematic FIMAT, Huelva University, Huelva, Spain

Estanislao Gamero Department of Applied Mathematic II, E.T.S.I. Sevilla University, Sevilla, Spain

Juan J. García-Machado Departamento de Economía Financiera, Contabilidad y Dirección de Operaciones, Universidad de Huelva, Huelva, Spain

José Manuel Guerra Departamento de Óptica, Facultad de CC. Físicas, Universidad Complutense de Madrid, Madrid, Spain

Cristóbal García Departamento de Matemáticas, Centro de Investigación de Física Teórica y Matemática FIMAT, Universidad de Huelva, Huelva, Spain

José Luis Guisado Departamento de Arquitectura y Tecnología de Computadores, Universidad de Sevilla, Sevilla, Spain

Francisco Jiménez-Morales Departamento de Física de la Materia Condensada, Universidad de Sevilla, Sevilla, Spain

Anna Karczewska Faculty of Mathematics, Computer Science and Econometrics, University of Zielona Góra, Zielona Góra, Poland

Panayotis G. Kevrekidis Department of Mathematics and Statistics, University of Massachusetts, Amherst, MA, USA

Martin Krupa MathNeuro Team, Inria Sophia Antipolis Research Centre, Sophia Antipolis Cedex, France; Université Côte d'Azur (UCA), Nice, France; Laboratoire J. A. Dieudonné, Université de Nice Sophia Antipolis, Nice Cedex 02, France

Ruomeng Lan Department of Mathematics, Texas A&M University, College Station, TX, USA

Ana M. Lara Instituto IBT, Parque Empresarial Nuevo Torneo, Sevilla, Spain

Manuel Merino Departamento de Ciencias Integradas, Centro de Investigación de Física Teórica y Matemática FIMAT, Universidad de Huelva, Huelva, Spain

Humberto Michinel Área de óptica, Escola de Enxeñaría Aeroespacial, Ourense, Spain

Angel Paredes Área de óptica, Departamento de Física Aplicada, Ourense, Spain

Rafel Prohens Departament de Matemàtiques i Informàtica, Universitat de les Illes Balears, Palma de Mallorca, Spain

Alejandro J. Rodríguez-Luis Departamento de Matemática Aplicada II, E.S. Ingenieros, Universidad de Sevilla, Sevilla, Spain

Piotr Rozmej Faculty of Physics and Astronomy, Institute of Physics, University of Zielona Góra, Zielona Góra, Poland

Luca Salasnich Dipartimento di Fisica e Astronomia Galileo Galilei, Universitá di Padova, Padova, Italy

Fabio Sattin Consorzio RFX (CNR, ENEA, INFN, Universitá di Padova, Acciaierie Venete SPA), Padova, Italy

Avadh Saxena Los Alamos National Laboratory, Center for Nonlinear Studies and Theoretical Division, Los Alamos, NM, USA

Antonio E. Teruel Departament de Matemàtiques i Informàtica, Universitat de les Illes Balears, Palma de Mallorca, Spain

Francisco R. Villatoro Department Lenguajes y Ciencias de la Computación, Escuela de Ingenierías Industriales, Ampliación del Campus de Teatinos, Universidad de Málaga, Málaga, Spain

Part I
Bifurcation Analysis

A Review on Some Bifurcations in the Lorenz System

Antonio Algaba, M. Cinta Domínguez-Moreno, Manuel Merino and Alejandro J. Rodríguez-Luis

Abstract In this chapter, we review some bifurcations exhibited by the classical Lorenz system, where the parameters can have any real value. Analytical results on the pitchfork, Hopf and Takens–Bogdanov bifurcations of the origin, as well as the Hopf bifurcation of the nontrivial equilibria, are summarized. These results serve as a guide for the numerical study that reveals other important organizing centers of the dynamics: Takens–Bogdanov bifurcations of periodic orbits, torus bifurcations and the resonances associated, homoclinic and heteroclinic connections with several degeneracies, etc. We also point out that the analysis of the Hopf-pitchfork and the triple-zero bifurcations of the origin cannot be performed with the usual tools and propose a way to carry out this study avoiding the structural singularities exhibited by the Lorenz system.

Keywords Lorenz · Bifurcation · Pitchfork · Hopf · Takens–Bogdanov · Torus Resonances · Hopf-pitchfork · Triple-zero

1 Introduction

The famous Lorenz system was derived from a simplified model of convection in the atmosphere: a two-dimensional fluid cell is warmed from below and cooled from above and the resulting convective motion is modeled by a partial differential

A. Algaba · M. C. Domínguez-Moreno (✉) · M. Merino
Departamento de Ciencias Integradas, Centro de Investigación de Física Teórica y Matemática FIMAT, Universidad de Huelva, 21071 Huelva, Spain
e-mail: mcinta.dominguez@dmat.uhu.es

A. Algaba
e-mail: algaba@uhu.es

M. Merino
e-mail: merino@uhu.es

A. J. Rodríguez-Luis
Departamento de Matemática Aplicada II, E.S. Ingenieros, Universidad de Sevilla, Camino de los Descubrimientos s/n, 41092 Sevilla, Spain
e-mail: ajrluis@us.es

V. Carmona et al. (eds.), *Nonlinear Systems, Vol. 1*,
Understanding Complex Systems, https://doi.org/10.1007/978-3-319-66766-9_1

equation. This system is obtained after a Galerkin approximation, that is, the variables are expanded into an infinite number of modes and all except three of them are put to zero [86, 97]:

$$\begin{cases} \dot{x} = \sigma(y - x), \\ \dot{y} = \rho x - y - xz, \\ \dot{z} = -bz + xy. \end{cases} \tag{1}$$

Thus, the variable x is proportional to the intensity of convective motion, y is proportional to the temperature difference between ascending and descending currents and z is proportional to the distortion from linearity of the vertical temperature profile. The dimensionless parameters have a physical interpretation: σ is a Prandtl number (the ratio of kinematic viscosity and thermal diffusivity), ρ is a Rayleigh number (proportional to the temperature difference across the fluid layer and the gravitational acceleration acting on the fluid) and b is a ratio of the height and width of the fluid layer. Consequently, the three parameters considered by Lorenz were positive.

The Lorenz equations also arise in simplified models in a variety of fields, for instance, lasers [73], dynamos [77], thermosyphons [71], chemical reactions [93], electric circuits [49] and brushless DC motors [75]. They have even been proposed very recently in the thermodynamic modelling of leukaemia malignancy [1].

The complex dynamics exhibited by the Lorenz system has fascinated to a large number of scientists in such a way that, in the last 50 years, hundreds of studies have examined this emblematic dynamical system. To illustrate this fact without pretending to be too exhaustive, we cite several of them, indicating briefly the topic considered: Lorenz chaotic attractor [39–41, 100, 103], T-point heteroclinic cycle and some degeneracies [16, 48, 69], global invariant manifolds and chaos visualization [54–56, 88], dimension of the global attractor [83], its relation with Kolmogorov systems [90], numerical modeling of its dynamics [91] and its preeminence over other proposed Lorenz-like systems [11, 13]. However, among so many papers, there are some that present incorrect results. For example, in Ref. [104, Appendices A and B] the authors claim to have proved the presence of a Shilnikov heteroclinic orbit in the Lorenz system, via the undetermined coefficient method. Unfortunately, this method (used for the first time in Refs. [105, 106]) is wrong as it is demonstrated in Refs. [7, 9, 10, 12, 15] (see also references therein).

On the other hand, even if this fact is less known, the Lorenz system also appears, when $\sigma < 0$, in the study of traveling-wave solutions in the Maxwell-Bloch equations [57] and in the analysis of a thermosolutal convection model [78]. This is one of the reasons why the Lorenz system is also considered for negative values of the parameters. Moreover, from a dynamical point of view, it is also stimulating to analyze the behavior of this iconic system when the parameters can take any real value. In this context, several aspects on the dynamics of the Lorenz system have been investigated, for instance, in the following works: invariant algebraic surfaces [43, 84, 85, 98], degenerate heteroclinic cycles [79, 87], Hopf bifurcation [2, 14, 101], Takens–Bogdanov and global bifurcations [3], resonances of periodic orbits [24] and superluminal periodic orbits [36].

The aim of this survey is to comment some results on the Lorenz system obtained by means of the Local Bifurcation Theory (see, for instance, [72, 82, 102]), in the way we briefly summarize below. When an autonomous system is analyzed an usual target is the knowledge of its dynamics in certain zones of the parameter space. In practice, the parameter space is divided in regions, bounded by bifurcation loci, and the goal is to determine the qualitative behaviour in each of such regions. This can be done in several steps. First, the detection of the equilibrium points and the analysis of the linearization around such equilibria, allow to show the presence of linear degeneracies (nonhyperbolicities) for some values of the control parameters. Second, the computation of approximations of the center manifold (and also of the reduced system on the center manifold) enables to reduce the dimension of the problem, transforming the reduced system into the corresponding normal form by means of changes of variables (sometimes a reparametrization of the time is also needed). Symbolic computation algorithms greatly facilitate this task. Third, the analysis of the unfolding of the normal form in the nondegenerate cases, provides local information on the bifurcation sets. Furthermore, possible nonlinear degeneracies giving rise to a higher codimension bifurcation problem can be detected at this step. Finally, from the information achieved in the study of local bifurcations, good starting points for the application of adequate numerical techniques can be obtained. This will provide a global picture of the dynamics of the system in the parameter space (see, for instance, [64]).

To illustrate how the method described above allows to obtain a deep knowledge of the dynamical system under consideration, we mention now two three-dimensional systems with a very rich dynamics. On the one hand, for a modified van der Pol-Duffing electronic oscillator, interesting information can be found in the following references on some local and global bifurcations: Hopf and Takens–Bogdanov [17], Hopf-pitchfork [18, 19, 21], triple-zero [67], periodic orbits bifurcations [4, 32, 59], homoclinic connections and some degeneracies [45, 65], T-points and some degeneracies [5, 8, 58, 60, 62]. Secondly, in the case of the widely studied Chua's equation, the following references clarify how to deal with the corresponding local and global bifurcations: Hopf [27], Takens–Bogdanov [20], Hopf-pitchfork [28], triple-zero [29], homoclinic connections and some degeneracies [33–35], T-points and some degeneracies [6, 25, 26, 61].

This work is organized as follows. In Sect. 2 we enumerate the linear degeneracies that the equilibrium at the origin of the Lorenz system can exhibit. The analysis of the pitchfork bifurcation is considered in Sect. 3. In Sect. 4 we present results on Hopf bifurcations. Section 5 is devoted to Takens–Bogdanov bifurcations of equilibria as well as of periodic orbits. Section 6 is dedicated to the study of resonances, whose presence is motivated by the existence of torus bifurcations of periodic orbits. Finally, some conclusions are reported in Sect. 7.

2 Linear Degeneracies

Along this work we consider Lorenz system (1) where σ, ρ and b are real parameters. We exclude two degenerate situations: the system is linear if $\sigma = 0$ and non-isolated equilibria on the z-axis exist for $b = 0$.

The Lorenz system (1) is invariant to the change $(x, y, z) \to (-x, -y, z)$. The origin $E_0 = (0, 0, 0)$ is always one equilibrium point and, for $b(\rho - 1) > 0$, two symmetric nontrivial equilibria, $E_{\pm} = (\pm\sqrt{b(\rho - 1)}, \pm\sqrt{b(\rho - 1)}, \rho - 1)$, exist.

The linearization matrix of system (1) at the origin is

$$\begin{pmatrix} -\sigma & \sigma & 0 \\ \rho & -1 & 0 \\ 0 & 0 & -b \end{pmatrix}, \tag{2}$$

whose characteristic polynomial is

$$p = \lambda^3 + p_1\lambda^2 + p_2\lambda + p_3,$$

where

$$p_1 = b + 1 + \sigma, \quad p_2 = \sigma(1 + b - \rho) + b, \quad p_3 = -b\sigma(\rho - 1).$$

It is easy to check that the following nonhyperbolic situations may arise:

- A simple zero eigenvalue, and the other two with nonvanishing real part. This case comes up if $p_1 \neq 0$, $p_2 \neq 0$, $p_3 = 0$, that, in terms of the parameters, occurs into the set $\rho = 1, \sigma \neq 0, -1, b \neq 0$.
  The corresponding codimension-one pitchfork bifurcation gives rise to the appearance of the two nontrivial equilibria $E_{\pm}$ in the region $b(\rho - 1) > 0$.
- A pair of imaginary eigenvalues and the third one nonzero. This degeneration occurs when $p_1 p_2 = p_3$, $p_2 > 0$, $p_3 \neq 0$, i.e., for $\sigma = -1, \rho > 1, b \neq 0$.
  The analysis of the corresponding Hopf bifurcation of the origin was carried out in Ref. [2], where the Hopf bifurcation exhibited by the nontrivial equilibria $E_{\pm}$ is also studied.
- A double-zero eigenvalue and the third one nonzero. This case appears when $p_2 = p_3 = 0$, $p_1 \neq 0$, or, in terms of the σ, ρ, b parameters, in the set $\sigma = -1$, $\rho = 1, b \neq 0$.
  The associated Takens–Bogdanov bifurcation, both in the homoclinic and in the heteroclinic case, has been analyzed in Ref. [3].
- A pair of imaginary eigenvalues and the third one zero. This situation corresponds to $p_1 = p_3 = 0$, $p_2 > 0$, that is, it occurs when $\sigma = -1, b = 0, \rho > 1$. A Hopf-pitchfork bifurcation is exhibited but, as non-isolated equilibria appear when $b = 0$, this bifurcation cannot be analyzed by the standard procedures.
- A triple-zero eigenvalue. This situation arises if $p_1 = p_2 = p_3 = 0$, which, in the parameter space, corresponds to the point $\sigma = -1, \rho = 1, b = 0$.
  This bifurcation cannot be studied by the standard methods because, for $b = 0$, the origin is a non-isolated equilibrium.

3 Pitchfork Bifurcation

For $\rho = 1$, the linearization matrix (2) has the eigenvalues $0, -(\sigma+1), -b$. Therefore, as a consequence of its symmetry, the Lorenz system (1) exhibits a pitchfork bifurcation. To study this bifurcation, we examine the Lorenz system at the critical values of the parameters and use the linear change of variables given by

$$\begin{pmatrix} x \\ y \\ z \end{pmatrix} = \begin{pmatrix} 1 & \sigma & 0 \\ 1 & -1 & 0 \\ 0 & 0 & 1 \end{pmatrix} \begin{pmatrix} u \\ v \\ w \end{pmatrix} \tag{3}$$

in order to obtain the linearization matrix in canonical form.

Thus, the Lorenz system is transformed into

$$\begin{pmatrix} \dot{u} \\ \dot{v} \\ \dot{w} \end{pmatrix} = \begin{pmatrix} 0 & 0 & 0 \\ 0 & -(\sigma+1) & 0 \\ 0 & 0 & -b \end{pmatrix} \begin{pmatrix} u \\ v \\ w \end{pmatrix} + \begin{pmatrix} -\frac{\sigma}{\sigma+1} w(u+\sigma v) \\ \frac{1}{\sigma+1} w(u+\sigma v) \\ (u-v)(u+\sigma v) \end{pmatrix}. \tag{4}$$

Assuming that $\sigma + 1 \neq 0$, $b \neq 0$, the second-order approximation to the center manifold is given by

$$v = 0 + O(u^3), \quad w = \frac{1}{b}u^2 + O(u^3),$$

and the third-order reduced system on the center manifold is

$$\dot{u} = -\frac{\sigma}{b(\sigma+1)} u^3.$$

Consequently, the bifurcation is supercritical when the coefficient of u^3 is negative and subcritical if it is positive.

The above results can be summarized in the following theorem.

Theorem 1 *The locus in the (σ, ρ, b)-parameter space where the origin of the Lorenz system undergoes a nondegenerate pitchfork bifurcation is defined by*

$$\rho = 1, \quad \sigma \neq 0, -1, \quad b \neq 0.$$

This bifurcation is supercritical if:

(i) $\sigma \in (-1, 0)$, $b < 0$;
(ii) $\sigma \in (-\infty, -1) \cup (0, \infty)$, $b > 0$.

It is subcritical when:

(i) $\sigma \in (-\infty, -1) \cup (0, \infty)$, $b < 0$;
(ii) $\sigma \in (-1, 0)$, $b > 0$.

4 Hopf Bifurcations

In this section we precis the principal results obtained in Ref. [2], devoted to the analysis of Hopf bifurcations and their degeneracies in the Lorenz system (to do that, the computation of some Lyapunov coefficients of the corresponding normal form is needed [70, 72, 82, 102]). First, we consider the Hopf bifurcation of the origin and later, the Hopf bifurcation exhibited by the nontrivial equilibria $E_{\pm}$.

As was mentioned in Sect. 2, the origin E_0 undergoes a Hopf bifurcation if $\sigma = -1,\ \ \rho > 1,\ \ b \neq 0$. The corresponding normal form to third order, obtained with the recursive algorithm developed in Ref. [66] is (see the details in Ref. [2, Sect. 2])

$$\begin{cases} \dot{r} = a_1 r^3 + \cdots, \\ \dot{\theta} = 1 + b_1 r^2 + \cdots, \end{cases} \tag{5}$$

where the first Lyapunov coefficient is given by

$$a_1 = \frac{-b-2}{8\sqrt{\rho-1}\left(4(\rho-1)+b^2\right)}.$$

A degeneracy occurs if $a_1 = 0$, i.e., if $b = -2$.

In this case, as the center manifold is an algebraic invariant surface, the Hopf bifurcation has infinite codimension: the Lorenz equations have a center at the origin (see Fig. 1). In fact, it can be easily proved that $x^2 + 2z = 0$ is the only polynomial center manifold for the Hopf bifurcation of the origin in the Lorenz system. To prove this fact it is enough to consider the six invariant algebraic surfaces the Lorenz system has [85, 98] and apply the conditions for the Hopf bifurcation ($b = -2$, $\sigma = -1$, $\rho > 1$). Thus, it is easily obtained that the only solution is $x^2 + 2z = 0$.

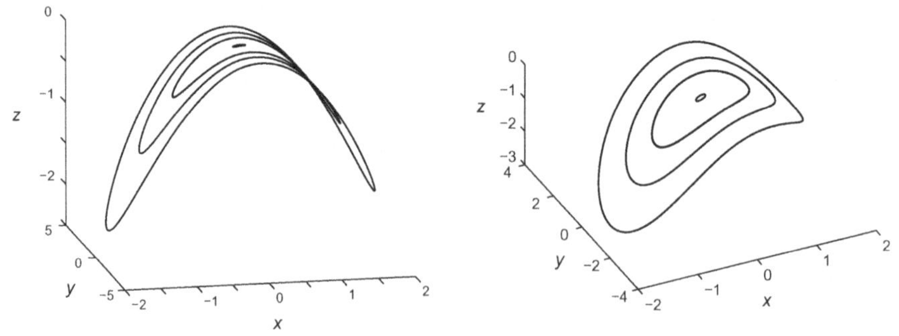

Fig. 1 Two different perspectives of the phase space of the Lorenz system (1) for $b = -2, \sigma = -1$, $\rho = 2$ where the origin undergoes a degenerate Hopf bifurcation of infinite codimension. Some periodic orbits on the center manifold are drawn. Reproduced with permission from [2]. Copyright (2015) by Springer

The following statement sum up all the results on the Hopf bifurcation of the origin.

Theorem 2 ([2, Theorem 1]) *The locus in the (σ, ρ, b)-parameter space where the origin of the Lorenz system undergoes a Hopf bifurcation is defined by*

$$\sigma = -1, \quad \rho > 1, \quad b \neq 0.$$

This bifurcation is supercritical when $b > -2$ and subcritical if $b < -2$. A degenerate Hopf bifurcation of infinite codimension occurs if $b = -2$.

The rest of this section is devoted to the Hopf bifurcation of the nontrivial equilibria (all the details can be found in Ref. [2, Sect. 3]). The standard techniques used in the study of a Hopf bifurcation allow to determine, in a first step, the locus where it occurs and to compute, in a second step, the Lyapunov coefficients that lead to the detection of all the degeneracies this bifurcation can have. The results obtained are summarized below.

Proposition 1 ([2, Proposition 2]) *The nontrivial equilibria of the Lorenz system experiment a Hopf bifurcation in the surface parameterized in explicit form by*

$$S_{hnt} = \left\{ \left(\sigma, \rho, b = \frac{-\sigma^2 - (3-\rho)\sigma - \rho}{\sigma + \rho} \right) : (\sigma, \rho) \in \Omega \right\},$$

with $\Omega = \Omega_1 \cup \Omega_2 \cup \Omega_3 \cup \Omega_4 \cup \Omega_5$, where

$$\begin{aligned}
\Omega_1 &= \left\{ (\sigma, \rho) \in \mathbb{R} : \sigma < -1, \rho < \frac{\sigma^2 + 3\sigma}{\sigma - 1} \right\}, \\
\Omega_2 &= \left\{ (\sigma, \rho) \in \mathbb{R} : \sigma = -1, \rho < 1 \right\}, \\
\Omega_3 &= \left\{ (\sigma, \rho) \in \mathbb{R} : -1 < \sigma < 0, \rho < -\sigma \right\}, \\
\Omega_4 &= \left\{ (\sigma, \rho) \in \mathbb{R} : 0 < \sigma < 1, \rho < \frac{\sigma^2 + 3\sigma}{\sigma - 1} \right\}, \\
\Omega_5 &= \left\{ (\sigma, \rho) \in \mathbb{R} : \sigma > 1, \rho > \frac{\sigma^2 + 3\sigma}{\sigma - 1} \right\}.
\end{aligned}$$

The region Ω is drawn in Fig. 2.

The study of the first Lyapunov coefficient of the third-order normal form for the reduced system [see Eq. (5)], given by

$$a_1 = \frac{(\sigma + 1)\sqrt{-\Delta}\,(\sigma + \rho)\, N1(\sigma, \rho)}{D1(\sigma, \rho)}, \tag{6}$$

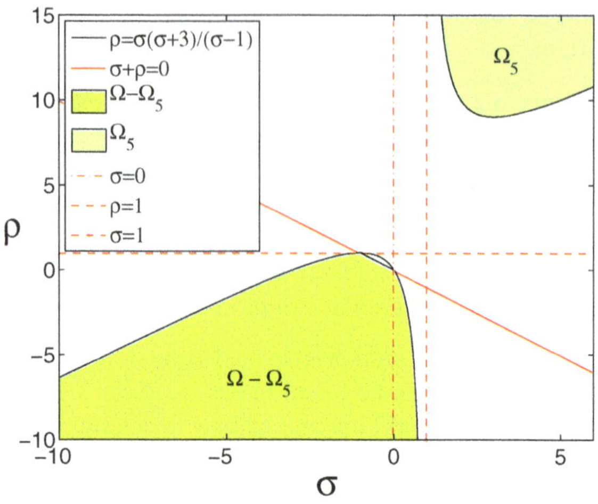

Fig. 2 Region Ω which corresponds to the projection onto the (σ, ρ)-plane of the surface S_{hnt} where the Hopf bifurcation of the nontrivial equilibria occurs. To get that surface recall that $b = \frac{-\sigma^2-(3-\rho)\sigma-\rho}{\sigma+\rho}$. Reproduced with permission from [2]. Copyright (2015) by Springer

where

$$\begin{aligned}
\Delta &= \sigma^2 + (3-\rho)\sigma + \rho, \\
N1(\sigma, \rho) &= 6\sigma^4 + 15\sigma^3 + 9\sigma^3\rho + 35\sigma^2\rho + \sigma^2\rho^2 \\
&\quad +21\sigma\rho^2 + 2\sigma\rho + \sigma\rho^3 + 5\rho^3 + \rho^4, \\
D1(\sigma, \rho) &= 8(\rho-1)\left[\sigma^2(\rho-1)^2 - (\sigma+\rho)^2\Delta\right] \\
&\quad \times \left[4\sigma^2(\rho-1)^2 - (\sigma+\rho)^2\Delta\right],
\end{aligned}$$

and of the second Lyapunov coefficient a_2 of the fifth-order normal form for the reduced system (whose expression appears in Ref. [2, Appendix A]) allow to find all the degeneracies this Hopf bifurcation may experiment. This information is summarized below.

Theorem 3 *The nontrivial equilibria of the Lorenz system undergo a degenerate Hopf bifurcation in the following cases:*

1. *The first Lyapunov coefficient a_1 vanishes for all the values $(\sigma, \rho) \in \Omega$ where the polynomial $N1(\sigma, \rho)$ is zero. A codimension-two bifurcation occurs in this case when the second Lyapunov coefficient a_2 is nonzero.*
2. *On the two points $(\sigma, \rho, b) \in S_{hnt}$ given by*

$$P_1 \approx (-0.646547, -6.605871, -1.709567)$$

and

$$P_2 \approx (-0.0100012, -0.0396965, -1.408456)$$

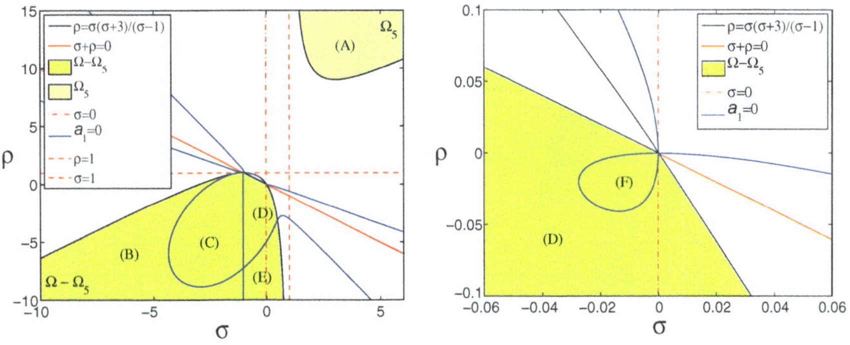

Fig. 3 (Left) Projection onto the (σ, ρ)-plane of the locus where the first Lyapunov coefficient a_1 is zero. When this curve is inside the region Ω, it corresponds to a degenerate Hopf bifurcation. (Right) Zoom in a neighborhood of the origin. Reproduced with permission from [2]. Copyright (2015) by Springer

a codimension-three Hopf bifurcation occurs because the first and the second Lyapunov coefficient vanish simultaneously and the third one a_3 is nonzero. These are the unique codimension-three Hopf bifurcation points.

3. *On the half-line given by $\sigma = -1$, $b = -2$, $\rho < 1$ a Hopf bifurcation of codimension infinite occurs because the reduced system on the center manifold is Hamiltonian (centers on center manifolds).*

We would like to do several remarks on the above result. First, the region Ω is split in six zones (see Fig. 3). A subcritical Hopf bifurcation occurs in the zones (A), (C), (E) and (F), while on the contrary it is supercritical in the zones (B) and (D). Remark that it is well-known that the Hopf bifurcation in the region where the three parameters are positive (our region (A) that corresponds to Ω_5) is always subcritical [89, 95, 99].

Second, to guarantee the existence of the two codimension-three points P_1 and P_2 the Poincaré-Miranda theorem was used [81]. A detailed analysis of the roots of a polynomial of degree 104 (it appears in the computation of the resultant of $N1(\sigma, \rho)$ and the numerator of a_2) is also needed.

Finally, as it occurs for the origin, the Hopf bifurcation of the nontrivial equilibria has infinite codimension because the center manifold is an algebraic invariant surface, namely $x^2 + 2z = 0$ (see Fig. 4). Moreover, the Hopf bifurcation of the nontrivial equilibria only has this polynomial center manifold in the Lorenz system.

In the following we provide the results of some numerical continuations, obtained with AUTO [52], in order to illustrate the dynamical consequences of the presence of a Hopf bifurcation of codimension-three (that occurs at P_1 and at P_2). From this degeneracy, a curve of cusp bifurcations of periodic orbits appears [70]. Thus, two curves of saddle-node bifurcation contact tangentially at the cusp point, giving rise to a semicubic parabola. Three periodic orbits exists in the system for proximate param-

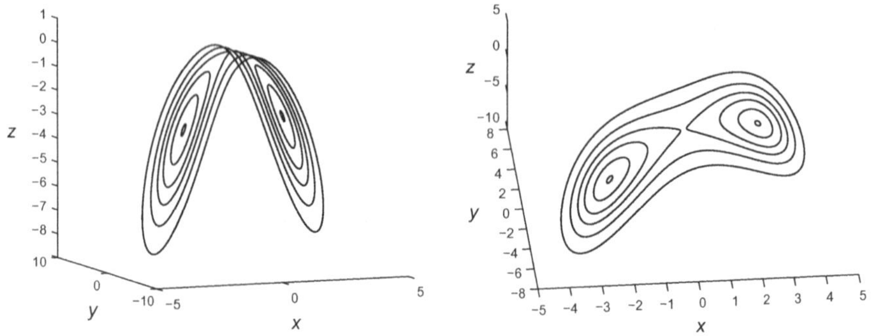

Fig. 4 Two different perspectives of the phase space of the Lorenz system (1) for $b = -2$, $\sigma = -1$, $\rho = -2$, where the nontrivial equilibria undergo a degenerate Hopf bifurcation of infinite codimension. Some periodic orbits on the center manifold appear. Reproduced with permission from [2]. Copyright (2015) by Springer

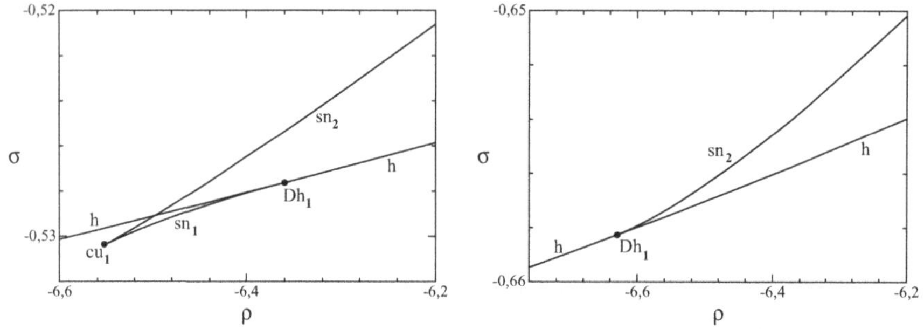

Fig. 5 Two partial bifurcation sets in the (ρ, σ)-plane in a neighborhood of the point P_1: (Left) for $b = -1.6$; (Right) for $b = -1.72$. Reproduced with permission from [2]. Copyright (2015) by Springer

eter values. These orbits disappear in pairs by means of saddle-node bifurcations. The existence of hysteresis phenomenon is associated with the cusp bifurcation.

As similar results are obtained in the vicinity of P_1 and P_2, we only consider here the point P_1. Thus, we show for $b = -1.6$ and $b = -1.72$ (values at both sides of $b_1 \approx -1.709567$) the corresponding partial bifurcation sets in the (ρ, σ)-plane.

A degeneracy in the Hopf curve of the nontrivial equilibria h occurs at the point $\mathtt{Dh_1}$ when $b = -1.6 > b_1$ (see Fig. 5 (Left)). Thus, if we move along h from left to right, the Hopf bifurcation changes from subcritical to supercritical. A curve of saddle-node of periodic orbits $\mathtt{sn_1}$ arises from $\mathtt{Dh_1}$ and remains below h. It collapses with the saddle-node curve $\mathtt{sn_2}$ into the cusp $\mathtt{cu_1}$.

To be placed on the other part of the point P_1, a value $b < b_1$ has to be taken, for instance $b = -1.72$ (see Fig. 5 (Right)). In this situation, as a consequence of the degenerate point Dh$_1$ on the curve h, the Hopf bifurcation varies from subcritical (on the left) to supercritical (on the right). Now the saddle-node curve sn$_2$ emerges from Dh$_1$ and remains above h.

5 Takens–Bogdanov Bifurcations

In this section we summarize the results obtained in Ref. [3], devoted to the analysis of Takens–Bogdanov bifurcations in the Lorenz system. In the first part we mention the analytical results in the case of the Takens–Bogdanov bifurcation of the equilibrium at the origin. Secondly, we precis some numerical results on the existence of Takens–Bogdanov bifurcations exhibited by periodic orbits.

As was stated in Sect. 2, the origin E_0 exhibits a Takens–Bogdanov bifurcation when

$$\sigma = -1, \quad \rho = 1, \quad b \neq 0. \tag{7}$$

The corresponding normal form to third order for the reduced system on the center manifold, obtained with the recursive algorithm developed in Ref. [66], is (see the details in Ref. [3, Sect. 2])

$$\begin{cases} \dot{u} = v, \\ \dot{v} = a_3 u^3 + b_3 u^2 v, \end{cases} \tag{8}$$

with

$$a_3 = \frac{1}{b}, \quad b_3 = \frac{-2-b}{b^2}.$$

Whereas the coefficient a_3 cannot vanish, a degenerate Takens–Bogdanov bifurcation occurs when $b_3 = 0$, i.e. when $b = -2$.

As it is well known (see, for instance, [47, 72]), when $b_3 \neq 0$, the nondegenerate Takens–Bogdanov bifurcation is of heteroclinic type if $a_3 > 0$ and of homoclinic type for $a_3 < 0$. Therefore, a nondegenerate Takens–Bogdanov of heteroclinic type exists in the Lorenz system if $b > 0$ and of homoclinic type for $b < 0$ ($b \neq -2$).

In symmetric systems, the Takens–Bogdanov point TB appears when a curve of pitchfork bifurcations of the origin Pi collapses with a curve of Hopf bifurcations of the same equilibrium H. In the heteroclinic case (see Fig. 7 (Left)), a curve of heteroclinic connections of the nontrivial equilibria He emerges from TB. In the homoclinic case (see Fig. 10 (Left)), three curves emerge from TB: h (of Hopf bifurcations of the nontrivial equilibria), Ho (of homoclinic connections to the origin) and SN (of saddle-node bifurcations of symmetric periodic orbits).

When the coefficient b_3 of the normal form (8) vanishes (if $b = -2$) a nonlinear degeneracy appears. Specifically, as the center manifold is an algebraic invariant surface, the Takens–Bogdanov bifurcation has infinite codimension: the origin is a center in the Lorenz system when $b = -2, \sigma = -1$ and $\rho = 1$. This fact is illustrated in Fig. 6.

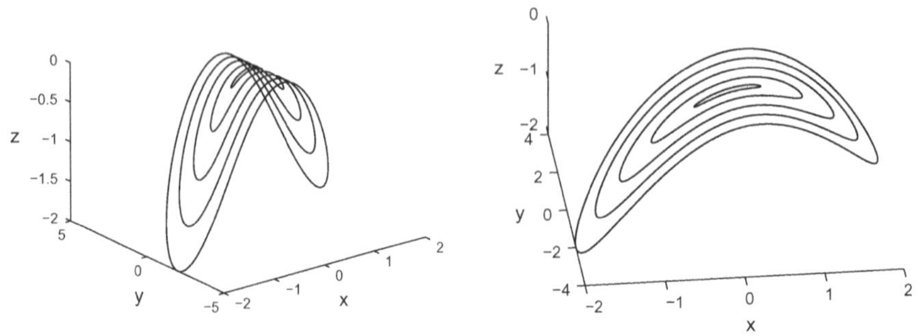

Fig. 6 Two different perspectives of the phase space of the Lorenz system (1) for $b = -2, \sigma = -1$, $\rho = 1$, where the origin undergoes a degenerate Takens–Bogdanov bifurcation of infinite codimension. Some periodic orbits on the center manifold $x^2 + 2z = 0$ appear. Reproduced with permission from [3]. Copyright (2016) by Elsevier

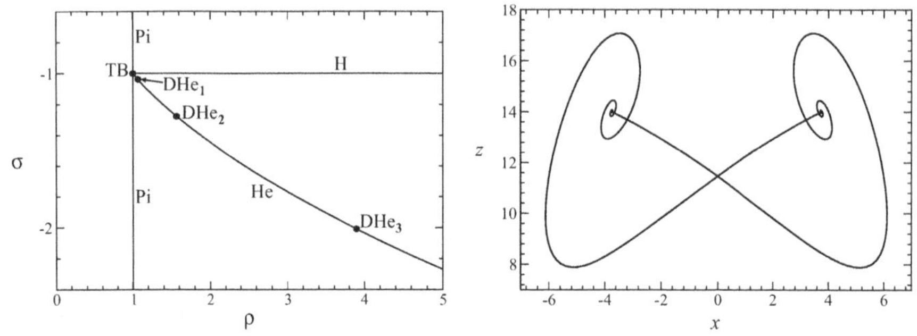

Fig. 7 (Left) For $b = 1$, partial bifurcation set on the (ρ, σ)-plane in a neighborhood of the Takens–Bogdanov point TB (heteroclinic case). Three degeneracies $\mathtt{DHe}_1$, $\mathtt{DHe}_2$ and $\mathtt{DHe}_3$ are present on the curve of heteroclinic connections He. (Right) For $b = 1$ and $\rho = 15$, projection onto the (x, z)-plane of the Shilnikov heteroclinic loop He existing for $\sigma \approx -3.874338$. Reproduced with permission from [3]. Copyright (2016) by Elsevier

All the information on the Takens–Bogdanov bifurcation is condensed in the following result.

Theorem 4 *The locus in the (σ, ρ, b)-parameter space where the origin of the Lorenz system undergoes a Takens–Bogdanov bifurcation is defined by*

$$\sigma = -1, \quad \rho = 1, \quad b \neq 0.$$

This bifurcation is nondegenerate if $b \neq -2$: of heteroclinic type when $b > 0$ and of homoclinic type when $b \in (-\infty, -2) \cup (-2, 0)$. A degenerate Takens–Bogdanov bifurcation of infinite codimension occurs if $b = -2$.

We summarize now the results mentioned in Sects. 4 and 5 on bifurcations of infinite codimension. All three cases appear on the straight line $b = -2, \sigma = -1$:

a Hopf of the origin if $\rho > 1$, a Hopf of the nontrivial equilibria when $\rho < 1$ and a Takens–Bogdanov of the origin for $\rho = 1$.

On the other hand, it is interesting to comment that, in these three situations, it is possible to find analytical expressions for the period of the orbits existing in the center manifold $x^2 + 2z = 0$ (see Figs. 1, 4 and 6). Thus, by taking limit in the corresponding expressions, for finite values of the parameter ρ, the existence of superluminal periodic orbits (periodic orbits with unbounded amplitude and whose period tends to zero) is demonstrated. All the details can be found in Ref. [36]. In this work, it is also numerically shown that superluminal periodic orbits also exists in other situations of physical interest when the parameter ρ tends to infinity.

In the rest of this section, we highlight the most important numerical results on the Takens–Bogdanov bifurcation of the origin presented in Ref. [3, Sect. 3], which have been obtained with AUTO [52]. Specifically, we present two partial bifurcation sets in the (ρ, σ)-parameter plane. The first one, for $b = 1$, illustrates the heteroclinic case ($b = 1$) whereas the second one, for $b = -1.6$, corresponds to the homoclinic case. Note that, as for $b = 0$ a triple-zero degeneracy occurs, we obtain information on both sides of such rich bifurcation.

In Fig. 7 (Left), for $b = 1$, a partial bifurcation set is drawn in a neighborhood of the Takens–Bogdanov point on the (ρ, σ)-plane. According to the well-known results in the heteroclinic case [47, 72], the Takens–Bogdanov point TB is placed on the curve where the origin exhibits a pitchfork bifurcation Pi. From that point a curve of Hopf bifurcation of the origin H emerges. As it is a supercritical Hopf bifurcation, a stable symmetric periodic orbit arises at H. This periodic orbit disappears in the curve He, where a heteroclinic orbit to the nontrivial equilibria occurs. In Fig. 7 (Right) a heteroclinic loop is drawn for $\rho = 15$.

Three degeneracies are numerically detected on the curve He. For their description, the eigenvalues of the nontrivial equilibria are denoted by $\alpha \pm \beta i$, λ, and the saddle-quantity $\delta = |\alpha/\lambda|$ is considered. The first degeneracy $\mathtt{He}_1$ appears when the nontrivial equilibria change from real saddle to saddle-focus. As $\delta > 1$, this global connection remains *tame* [44] and, in this way, a symmetric stable periodic orbit is born from the curve He. A second degeneracy, $\mathtt{He}_2$, is present when $\delta = 1$, namely the eigenvalues are resonant. At this point the heteroclinic orbit changes from *tame* to *chaotic* Shilnikov [82, 102].

The third degeneracy $\mathtt{He}_3$ occurs because $\delta = 1/2$ (null divergence). Since the expression of the divergence in the Lorenz system (1) is $divF = -(b + \sigma + 1)$ and we have fixed $b = 1$, then $divF = 0$ along the straight line $\sigma = -2$. Observe that $divF$ has no dependence on the spatial variables but only on the system parameters. This fact has important consequences on the bifurcations of the periodic orbits as we briefly explain in the following. If $\gamma(t)$ denotes a periodic orbit in the autonomous system $\dot{x} = F(x)$, then the variational equation is defined by the linear system

$$\dot{y} = DF(\gamma(t))y = A(t)y.$$

As it is known (see, for example, [74, Lemma III.7.3]), in a three-dimensional autonomous system, the product of the three Floquet multipliers of a periodic orbit fulfills:

$$m_1 m_2 m_3 = e^{\int_0^T tr(A(s))ds} = e^{\int_0^T divF(\gamma(s))ds}.$$

Therefore, on the assumption $divF = 0$ we can infer that the two Floquet multipliers of all the periodic orbits must verify $m_1 m_2 = 1$ (note that in a 3D autonomous system the third Floquet multiplier is at any time $m_3 = +1$). Consequently, one of the following three scenarios can take place:

(a) $0 < m_1 < 1 < m_2$;
(b) $m_2 < -1 < m_1 < 0$;
(c) $m_{1,2} = \alpha \pm i\beta$, with $\alpha^2 + \beta^2 = 1$.

In this last case, when $\beta = 0$, a double Floquet multiplier ($+1$ or -1) appears and, in symmetric systems, a Takens–Bogdanov bifurcation of periodic orbits is present if the double Floquet multiplier is nondiagonalizable (see, for instance, [42, 47, 50]). If $\beta \neq 0$, a torus bifurcation of periodic orbits is present. Note that, in a 3D continuous parameterized autonomous system, a periodic orbit undergoes a torus bifurcation (named secondary Hopf bifurcation too, or more appropriately, secondary Poincaré–Andronov–Hopf bifurcation) if its two Floquet multipliers traverse the unit circle subject to generical assumptions on the bifurcation parameter (see, for instance, [72, 82]).

In the dynamics of the Lorenz system we consider in the rest of this survey, Takens–Bogdanov and torus bifurcations of periodic orbits are important organizing centers. We have just seen that the only locus where these bifurcations can occur corresponds to the parameter plane defined by $\sigma + b + 1 = 0$.

Takens–Bogdanov bifurcations of periodic orbits are placed on curves of saddle-node, symmetry-breaking and period-doubling bifurcations and permit the change of stability of the periodic orbits involved. These codimension-two points have been detected in several systems (see, for instance, [18, 32, 35, 68]). Remark that interesting dynamical behavior can also appear if a diagonalizable double $+1$ Floquet multiplier occurs (see, for instance, [19, 21, 80]).

Our goal now is to acquire numerically new information on the bifurcation set near the degenerations $\mathtt{He}_2$ and $\mathtt{He}_3$. To achieve this aim (see the details in Ref. [3, Sect. 3.1]), the continuation for certain values of the parameters of the symmetric periodic orbit born at the Hopf bifurcation of the origin H shows that it exhibits symmetry-breaking PPO, saddle-node SN and torus bifurcations $\mathtt{HH}^{\mathtt{s}}$ before it disappears in a heteroclinic loop He. For its part, the asymmetric periodic orbit emanated in the first PPO bifurcation undergoes period-doubling PD, saddle-node sn and torus bifurcations $\mathtt{HH}^{\mathtt{a}}$ before it disappears in a homoclinic orbit to the nontrivial equilibria Hnt.

The loci where some of the bifurcations just mentioned above exist are drawn in Fig. 8 (Left). Specifically, in comparison with Fig. 7 (Left), five new curves appear: the first two symmetry-breaking bifurcations give rise to a single curve whose branches are labelled $\mathtt{PPO}_\mathtt{A}$ and $\mathtt{PPO}_\mathtt{B}$; the first two period-doubling bifurcations give rise to

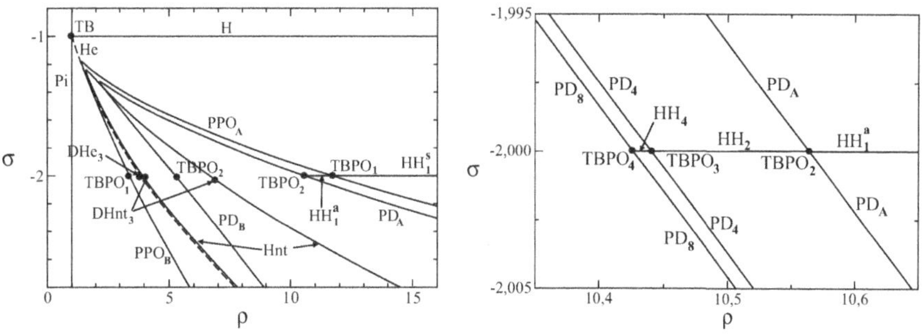

Fig. 8 For $b = 1$: (Left) For $b = 1$, partial bifurcation set on the (ρ, σ)-plane in a vicinity of $\sigma = -2$. (Right) Zoom of the bifurcation set in a neighborhood of the point $\mathtt{TBPO_2}$ situated on the branch $\mathtt{PD_A}$. Reproduced with permission from [3]. Copyright (2016) by Elsevier

a single curve whose branches are labelled $\mathtt{PD_A}$ and $\mathtt{PD_B}$; a curve of the homoclinic connections of the nontrivial equilibria $\mathtt{Hnt}$; the torus bifurcation of the symmetric periodic orbits occurs on the curve $\mathtt{HH_1^s}$; the torus bifurcation of the asymmetric periodic orbits occurs on the curve $\mathtt{HH_1^a}$. These two curves of torus bifurcations are located, as was predicted above, on the line $\sigma = -2$ where the divergence vanishes.

On the other hand, as was also anticipated above, the Takens–Bogdanov bifurcations of periodic orbits also occurs when $\sigma = -2$. Thus, the points marked $\mathtt{TBPO_1}$ (on the branches $\mathtt{PPO_A}$ and $\mathtt{PPO_B}$) correspond to a nondiagonalizable double $+1$ Floquet multiplier (1:1 resonance) whereas the points marked $\mathtt{TBPO_2}$ (on the branches $\mathtt{PD_A}$ and $\mathtt{PD_B}$) correspond to a nondiagonalizable double -1 Floquet multiplier (1:2 resonance). Moreover, the global connections to the nontrivial equilibria are also degenerate on $\sigma = -2$. Thus, two points $\mathtt{DHnt_3}$ appears on the curve $\mathtt{Hnt}$ because $\delta = 1/2$. Note that both curves $\mathtt{HH_1^s}$ and $\mathtt{HH_1^a}$ starts at the Takens–Bogdanov point $\mathtt{TBPO_1}$, placed on the curve $\mathtt{PPO_A}$. Specifically, the curve $\mathtt{HH_1^s}$ is unbounded whereas the curve $\mathtt{HH_1^a}$ ends at the Takens–Bogdanov point $\mathtt{TBPO_2}$, situated on the curve $\mathtt{PD_A}$.

In Fig. 8 (Right), a zoom in the vicinity of the point $\mathtt{TBPO_2}$ marked on the branch $\mathtt{PD_A}$ is shown. As we can see, this point is essential since it is the birth of a Feigenbaum cascade of flip bifurcations ($\mathtt{PD_A}$, $\mathtt{PD_4}$, $\mathtt{PD_8}$, ...), in addition to infinitely many torus bifurcation curves (of double pulse $\mathtt{HH_2}$, quadruple pulse $\mathtt{HH_4}$, ...). The accompanying Takens–Bogdanov bifurcations of periodic orbits that originate the curves of torus bifurcations $\mathtt{TBPO_3}$, $\mathtt{TBPO_4}$, ... are all situated on the half-line $\sigma = -2$, $\rho > 0$ (see [37, 82]).

An important final comment on the bifurcations existing when $b = 1$ refers to the significant role as organizing centers of the points $\mathtt{DHe_3}$ and $\mathtt{DHnt_3}$. In Fig. 9 (Left) the first curves of saddle-node bifurcations of symmetric periodic orbits ($\mathtt{SN_1}$, $\mathtt{SN_2}$, $\mathtt{SN_3}$ and $\mathtt{SN_4}$) are drawn. We notice in Fig. 9 (Left) that the points of Takens–Bogdanov bifurcation of periodic orbits $\mathtt{TBPO_1^i}$, situated on the curves $\mathtt{SN_i}$, collect to the degenerate point $\mathtt{DHe_3}$. A similar situation occurs with the Takens–Bogdanov points $\mathtt{TBPO_1^i}$, placed on the curves of saddle-node bifurcations of asymmetric peri-

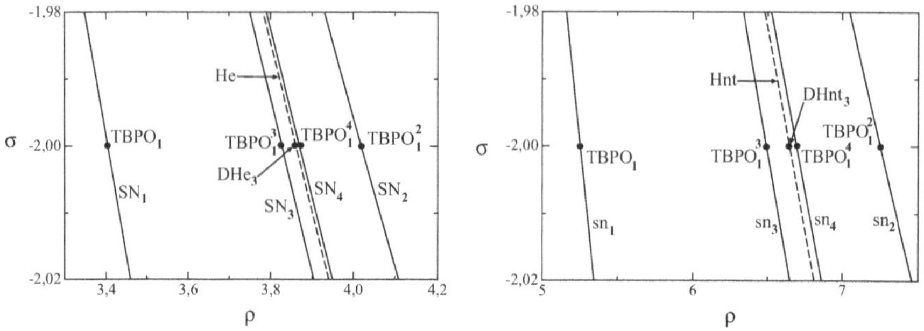

Fig. 9 For $b = 1$, partial bifurcation set on the (ρ, σ)-plane in a neighborhood of the point: (Left) $\mathtt{DHe}_3$. (Right) $\mathtt{DHnt}_3$ situated on the right branch of Hnt. Reproduced with permission from [3]. Copyright (2016) by Elsevier

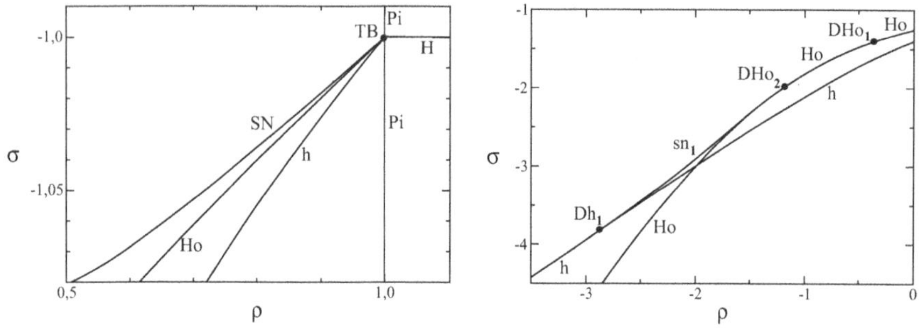

Fig. 10 For $b = -1.6$, partial bifurcation set in a neighborhood of: (Left) the Takens–Bogdanov point TB (homoclinic case). (Right) the degeneracies $\mathtt{DHo}_2$ and $\mathtt{Dh}_1$ situated, respectively, on the curves Ho and h. Reproduced with permission from [3]. Copyright (2016) by Elsevier

odic orbits $\mathtt{sn}_i$. As can be seen in Fig. 9 (Right), they accumulate to the degenerate point $\mathtt{DHnt}_3$.

Another remarkable fact that occurs outside the windows shown in Fig. 9 is that the curves of saddle-node bifurcations of periodic orbits disappear in pairs ($\mathtt{SN}_1$ with $\mathtt{SN}_2$, $\mathtt{SN}_3$ with $\mathtt{SN}_4$, $\mathtt{sn}_1$ with $\mathtt{sn}_2$, $\mathtt{sn}_3$ with $\mathtt{sn}_4$) in cusp bifurcations that accumulate to the degenerate point $\mathtt{DHe}_2$.

It is worthy to note that a family of invariant tori, existing on the locus where the divergence of the system is null (a degenerate situation present in the Lorenz equations), exists too in the numerical simulation presented in Ref. [96], in a vicinity of the triple-zero degeneracy of a truncated normal form for $\mathbb{Z}_2$-symmetric systems considered in the paper [29].

In the following we comment the results obtained for $b = -1.6$, when the Takens–Bogdanov bifurcation of the origin is of homoclinic type (see the details in Ref. [3, Sect. 3.2]). In the vicinity of this point TB (see Fig. 10 (Left)) five curves are present [72]: Pi (pitchfork bifurcation of equilibria), H (supercritical Hopf bifurcation of the origin from which a saddle symmetric periodic orbit emerges), h (subcritical Hopf bifurcation of the nontrivial equilibria from which a pair of repulsive asymmetric periodic orbits is born), Ho (homoclinic orbit to the origin where, in a neighborhood of the point TB, the two periodic orbits risen from h disappear) and SN (saddle-node bifurcation of two symmetric periodic orbits, the saddle one is born in H and the unstable one that is present after Ho).

The numerical continuation of the curve Ho reveals the existence of two degenerate points (see Fig. 10 (Right)). In an analogous way as we did above for the study of the degeneracies of the global connections related to the nontrivial equilibria, let us denominate the eigenvalues of the origin by $\lambda_1 < 0 < \lambda_2 < \lambda_3$ and take into account the saddle quantity $\delta^* = |\lambda_2/\lambda_1|$. The origin E_0 is a saddle equilibrium, with $\delta^* > 1$, when the curve Ho emanates from TB. A first degeneracy occurs on the point $\mathtt{DHo}_1$ due to the presence of a double eigenvalue $\lambda_2 = \lambda_3 = -1.6$. In this case, the homoclinic bifurcation remains *tame* because this double eigenvalue is not determining [44]. Thus, a single symmetric unstable periodic orbit arises from the curve Ho to the left whereas two asymmetric unstable periodic orbits are present on the right part of Ho.

The second degeneration occurs at the point $\mathtt{DHo}_2$ when $\delta^* = 1$ (homoclinic orbit at resonance $\lambda_1 = -\lambda_2 = -1.6$). Due to the fact that the homoclinic orbit is non-twisted close to $\mathtt{DHo}_2$, a curve of saddle-node bifurcation of asymmetric periodic orbits $\mathtt{sn}_1$ emerges from $\mathtt{DHo}_2$ (see [46]) and ends at the point $\mathtt{Dh}_1$ situated on the Hopf curve h.

To recapitulate the bifurcations that act as principal organizing centers in this area of the parameter space in the Lorenz equations, we draw in Fig. 11 the projection onto the (b, σ) and the (b, ρ) planes of the curves where codimension-two bifurcations exist. Thus, we find the following curves: TB Takens–Bogdanov bifurcation of

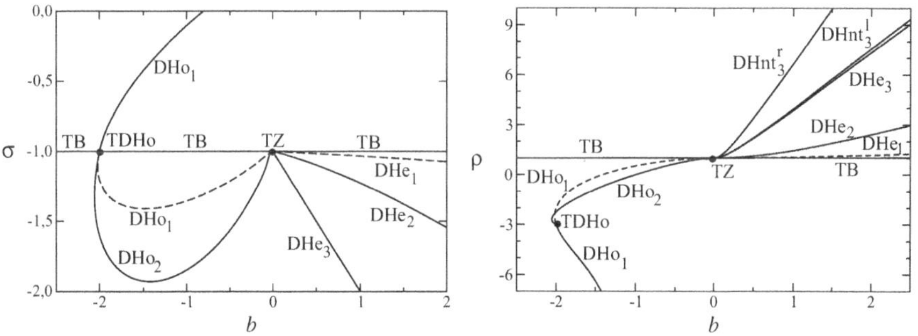

Fig. 11 Projection of the loci where the Takens–Bogdanov bifurcation of the origin, TB, degenerate homoclinic orbits of the origin ($\mathtt{DHo}_1$ and $\mathtt{DHo}_2$) and degenerate heteroclinic connections of the nontrivial equilibria ($\mathtt{DHe}_1$, $\mathtt{DHe}_2$ and $\mathtt{DHe}_3$) exist. These curves are organized by a triple-zero degeneracy TZ and by a codimension-three homoclinic connection of the origin TDHo. (Left) On the (b, σ) parameter plane. (Right) On the (b, ρ) parameter plane. In this case, we have also drawn the curves of degenerate homoclinic orbits of the nontrivial equilibria ($\mathtt{DHnt}_3^r$ and $\mathtt{DHnt}_3^l$) that do not appear in panel (Left) for the sake of clarity. Reproduced with permission from [3]. Copyright (2016) by Elsevier

the origin; $\mathtt{DHe}_1$ and $\mathtt{DHo}_1$ where the heteroclinic orbits to the nontrivial equilibria and the homoclinic orbits to the origin, respectively, exhibit a double-real leading eigenvalue degeneration (solid line) and a double-real nonleading eigenvalue singularity (dashed line); $\mathtt{DHe}_2$ and $\mathtt{DHo}_2$ where the heteroclinic connections and the homoclinic of the origin undergo, respectively, a $\delta = 1$ and $\delta^* = 1$ degeneracy (resonant eigenvalues); $\mathtt{DHe}_3$ where the heteroclinic orbits exhibit a $\delta = 1/2$ degeneracy; $\mathtt{DHnt}_3^{\mathtt{r}}$ and $\mathtt{DHnt}_3^{\mathtt{l}}$ where the homoclinic connections of the nontrivial equilibria experiment the degeneracy $\delta = 1/2$ (these connections appear in Fig. 7 (Left) and superscripts r and l refer to the right and to the left branch, respectively, of that figure).

A remarkable fact is that all the curves of degenerate homoclinic and heteroclinic orbits likely are born from the triple-zero bifurcation point $\mathtt{TZ} = (1, 0, -1)$. Moreover, another codimension-three point, $\mathtt{TDHo} = (-3, -2, -1)$, where a double degeneration of the homoclinic orbits to the origin exists (resonant eigenvalues and double real eigenvalue since $\lambda_1 = -\lambda_2 = -\lambda_3 = b = -2$) is marked. Whereas the double eigenvalues are not determining in the portion of the curve $\mathtt{DHo}_1$ between TZ and TDHo, they become determining on the other side of TDHo (in this situation, curves of saddle-node and flip bifurcations arise from $\mathtt{DHo}_1$). Note that the point where TB undergoes a degeneration of infinite codimension ($\sigma = -1$, $\rho = 1$, $b = -2$) is not indicated.

Notice that analogous arrangements of curves of codimension-two global bifurcations close to a triple-zero degeneracy TZ have been detected in other systems [29, 30].

We terminate this section with some comments on Shilnikov chaos (see more details in Ref. [3, Sect. 3.3]). As it is widely known (see, for example, [76, Sect. 5.1.2]) infinitely many saddle periodic orbits exist in every vicinity of a Shilnikov homoclinic/heteroclinic orbit. The existence of chaotic behavior is assured because these periodic orbits are contained in suspended horseshoes that cumulate onto the homoclinic/heteroclinic connection.

Consequently, according to the results commented in this review, Shilnikov homoclinic and heteroclinic connections exist in the Lorenz system. However, for negative values of the parameters, further studies are needed on the global dynamics of the Lorenz system because, as far as we know, there is no result on the existence of a compact invariant set. For instance, unlimited orbits in both onward and reversed time are obtained in most numerical simulations when $\sigma < 0$. To exemplify the cases where bounded dynamics occurs, we draw in Fig. 12 two chaotic attractors (whose basin of attraction is somewhat little) emerged from successions of flip bifurcations structured by Shilnikov global orbits (see a similar situation, for instance, in Ref. [53]). The divergence is negative in the region where these two chaotic attractors exist. The attractor presented in Fig. 12 (Left) is associated to a Takens–Bogdanov bifurcation of heteroclinic type while on the contrary the attractor drawn in Fig. 12 (Right) corresponds to a Takens–Bogdanov of homoclinic type.

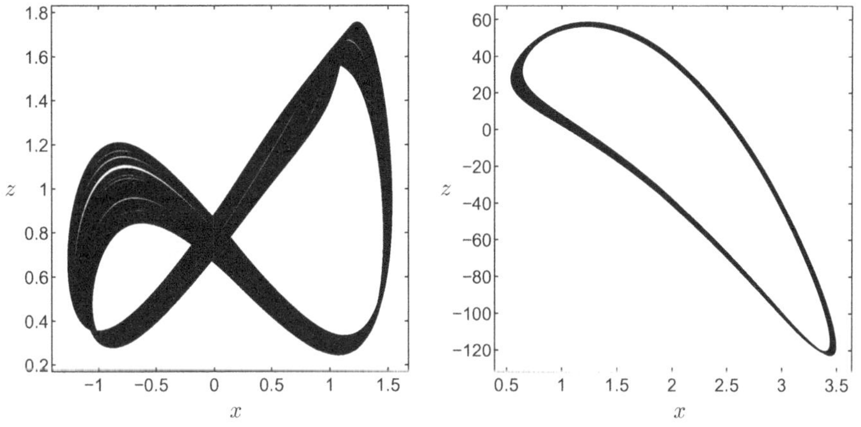

Fig. 12 Projection onto the (x, z)-plane of two chaotic attractors obtained integrating in a time interval of 2000 s, for the parameter values and initial conditions given: (Left) $(\rho, \sigma, b) = (2.54, -1.4, 1)$, $(x, y, z) = (1.4, 1.68, 1.25)$. (Right) $(\rho, \sigma, b) = (-30.83, -0.03, -0.957)$, $(x, y, z) = (2, 87.4, -39.7)$. Reproduced with permission from [3]. Copyright (2016) by Elsevier

6 Resonances of Periodic Orbits

As it was shown in Ref. [3] (see Sect. 5 of this survey) when $\sigma < 0$, the periodic orbit emanated from the Hopf bifurcation of the origin exhibits a torus bifurcation. The locus where this bifurcation occurs is a curve in a bidimensional parameter plane. On this curve, the pair of conjugate complex Floquet multipliers of the periodic orbit moves on the unit circle. Each time they meet a root of unity, a couplet of curves of saddle-node bifurcation of periodic orbits arises from the corresponding point on the torus curve. This originates a resonance region (Arnold's tongue) locally bounded by those curves of saddle-node bifurcation [47, 51, 72, 82]. The dynamical effects of the presence of torus bifurcations in some relevant autonomous systems has been considered in the literature (see, for instance, [22, 23, 31, 32]).

The aim of this section is to highlight important features on the resonances of periodic orbits that appear in the Lorenz system (all the details can be found in Ref. [24]). In the following, we will focus on the torus bifurcation curve $\mathtt{HH}_1^{\mathtt{s}}$ drawn in Fig. 8 (Left) for $b = 1$. This unbounded curve exists for $\sigma = -2$ on the right of the Takens–Bogdanov point $\mathtt{TBPO}_1$ ($\rho_{\mathtt{TBPO}_1} \approx 11.6960$), placed on the curve $\mathtt{PPO}_\mathtt{A}$.

To know what resonances will appear we analyze the evolution of the Floquet multipliers along the unit circle on the torus curve $\mathtt{HH}_1^{\mathtt{s}}$ (see Fig. 13 (Left)). The argument Arg (angle between the horizontal axis and the Floquet multiplier of the complex conjugate pair with positive imaginary part) is represented versus parameter ρ. Note that a maximum, 180°, is reached for $\rho_m \approx 21.4623$. Consequently, when we move along the torus curve, the Floquet multipliers change from $+1$ (Arg $= 0°$) to -1 (Arg $= 180°$) on the unit circle in the interval $I_A \equiv [\rho_{\mathtt{TBPO}_1}, \rho_m]$ and from

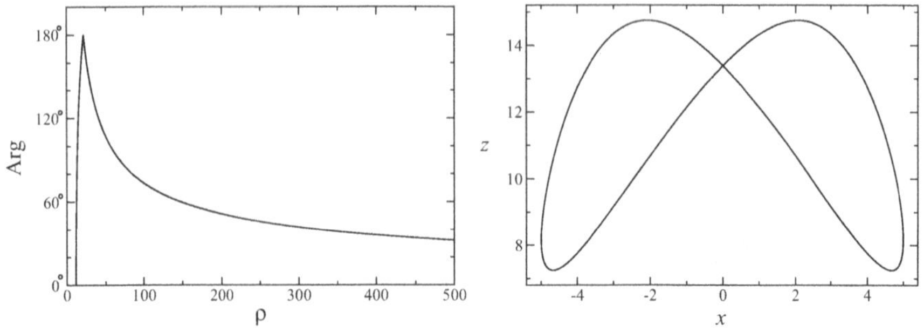

Fig. 13 For $b = 1$: (Left) Argument of the Floquet multipliers of the principal periodic orbit versus ρ along the torus bifurcation curve $\mathrm{HH}_1^{\mathrm{s}}$. It reaches its maximum, 180°, at $\rho_m \approx 21.4623$. (Right) Projection onto the (x, z)-plane of the principal periodic orbit exhibiting the torus bifurcation when $\rho = 15, \sigma = -2$. Reproduced with permission from [24]. Copyright (2016) by Springer

-1 to $+1$ in the interval $I_B \equiv [\rho_m, \infty)$. It seems numerically that $\lim_{\rho\to\infty} Arg = 0°$, that is, a Takens–Bogdanov bifurcation of periodic orbits (with a double $+1$ Floquet multiplier) is present at infinity. Therefore, all the resonances occur in the two intervals I_A and I_B.

A projection of the principal periodic orbit undergoing the torus bifurcation when $(\rho, \sigma) = (15, -2)$ is depicted in Fig. 13 (Right). Remark that, because the geometry this periodic orbit has and the $\mathbb{Z}_2$-symmetry exhibited by the Lorenz system, it is not possible that symmetric periodic orbits of period 2n-T bifurcate from this principal periodic orbit. According to this argument, symmetric periodic orbits of period 3T, 5T, 7T, … will emerge in the resonances of the principal periodic orbit, but not of period 2T, 4T, 6T, ….

In this section we employ the following notation for the corresponding curves. In the case of the torus bifurcations, the superscript s is used when the periodic orbit involucrated is symmetric and the superscript a if it is asymmetric. In the case of the 1:p resonances, a subscript n signifies that it corresponds to an nT-periodic orbit. In the case of curves of period-doubling bifurcations, curves of pitchfork bifurcations and Takens–Bogdanov points, so as not to unduly complicate the notation, superscripts to indicate the resonance concerned are not included (even though identical names are used in different figures, there is no place for error).

We start considering the periodic orbit emanated from the 1:2 resonance that occurs in a vicinity of $\rho_m \approx 21.4623$ (see all the details in Ref. [24, Sect. 3.1]).

The bifurcation diagram of the asymmetric 2T periodic orbit born in the 1:2 resonance is represented in Fig. 14a. The first two saddle-node bifurcations are labelled $\mathrm{sn2_A}$ and $\mathrm{sn2_B}$ (we use lowercase letters for saddle-node bifurcations of asymmetric periodic orbits and uppercase letters if they are symmetric) and the first two period-doubling bifurcations are marked as $\mathrm{PD_A}$ and $\mathrm{PD_B}$. The wiggles of this diagram are organized by a pair of Shilnikov homoclinic orbits to the nontrivial equilibria, $\mathrm{Hnt_A}$ and $\mathrm{Hnt_B}$.

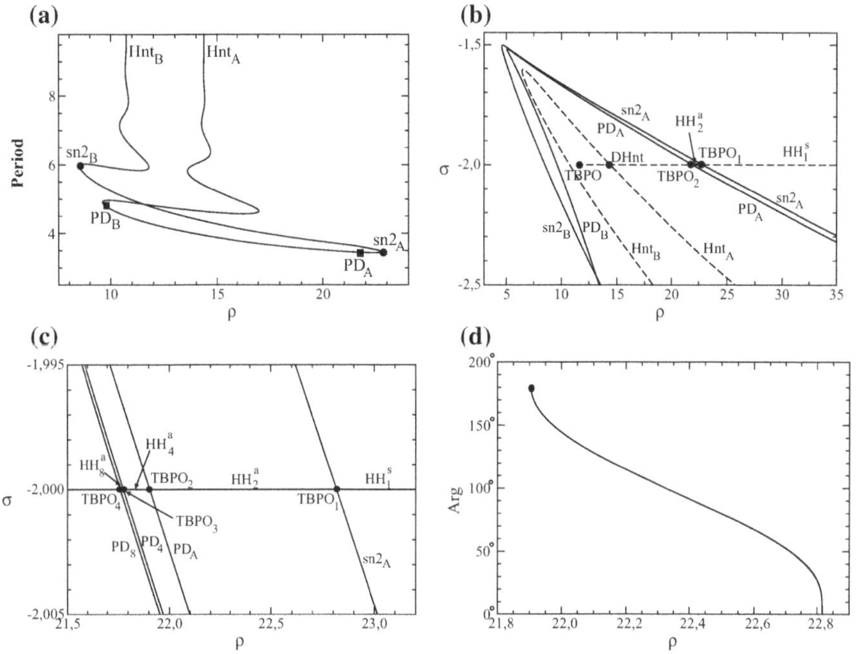

Fig. 14 For $b = 1$: **a** Bifurcation diagram of the 2T asymmetric periodic orbit of the resonance 1:2 when $\sigma = -1.999$. **b** Partial bifurcation set of the 1:2 open resonance zone $\mathtt{sn2}$ in a vicinity of $\sigma = -2$. **c** Partial bifurcation set for the 2T periodic orbit emerged at the resonance 1:2. **d** Argument of the Floquet multipliers versus ρ along the torus bifurcation curve $\mathtt{HH}_2^{\mathtt{a}}$. It reaches its maximum, 180°, at $\rho_m \approx 21.9$. Reproduced with permission from [24]. Copyright (2016) by Springer

The numerical continuation of these bifurcations allows obtaining the partial bifurcation set depicted in Fig. 14b for $b = 1$. The curves $\mathtt{sn2}_{\mathtt{A}}$ and $\mathtt{sn2}_{\mathtt{B}}$ collapse in a turning point in a similar way as the curves of period-doublings ($\mathtt{PD}_{\mathtt{A}}$ and $\mathtt{PD}_{\mathtt{B}}$) and the curves of homoclinic connections of the nontrivial equilibria ($\mathtt{Hnt}_{\mathtt{A}}$ and $\mathtt{Hnt}_{\mathtt{B}}$) do. Remark that all these curves present a degeneracy when $\sigma = -2$ (they are only marked on the right branches of the corresponding curves): the saddle-node curves have a non-diagonalizable double $+1$ Floquet multiplier $\mathtt{TBPO}_1$, the flip curves undergo a non-diagonalizable double -1 Floquet multiplier $\mathtt{TBPO}_2$ and the homoclinic connections curves exhibit a $\delta = 1/2$ degeneracy $\mathtt{DHnt}$. The unbounded torus curve $\mathtt{HH}_1^{\mathtt{s}}$ is drawn as a dashed line emanating from the point called $\mathtt{TBPO}$ for easiness (it correspond to the point $\mathtt{TBPO}_1$ of Fig. 8 (Left)). Observe that on the curve $\mathtt{HH}_2^{\mathtt{a}}$, that exists between the points $\mathtt{TBPO}_1$ and $\mathtt{TBPO}_2$, a 2T periodic orbit emerged in the 1:2 resonance undergoes a torus bifurcation.

A zoom in the vicinity of the curve $\mathtt{HH}_2^{\mathtt{a}}$ appears in Fig. 14c. The succession of flip bifurcations $\mathtt{PD}_{\mathtt{A}}$, $\mathtt{PD}_4$, $\mathtt{PD}_8, \ldots$ originates a sequence of Takens–Bogdanov points $\mathtt{TBPO}_2, \mathtt{TBPO}_3, \mathtt{TBPO}_4, \ldots$ (all correspond to non-diagonalizable double -1 Floquet multiplier). A torus bifurcation of a 2×2T periodic orbit occurs on the

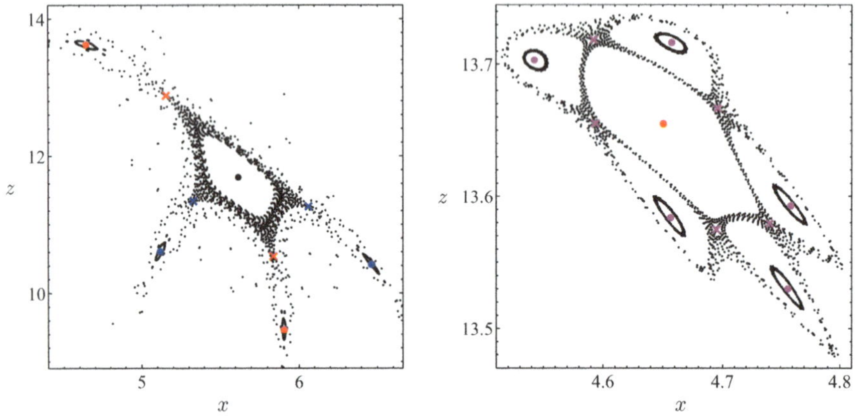

Fig. 15 Poincaré sections in the plane $y = 0$ of the Lorenz system with $b = 1$: (Left) resonance 1:2 of the principal periodic orbit $\sigma = -1.999$ and $\rho = 22.5$. (Right) resonance 1:5 of the 2T periodic orbit when $\rho = 22.54$, $\sigma = -1.9999$. Reproduced with permission from [24]. Copyright (2016) by Springer

curve $\mathrm{HH}_4^{\mathrm{a}}$ between the points TBPO_2 and TBPO_3. Similarly, a torus bifurcation of a 4×2T periodic orbit appears between the points TBPO_3 and TBPO_4 on the curve $\mathrm{HH}_8^{\mathrm{a}}$. Deserves to be highlighted that all the torus bifurcation curves $\mathrm{HH}_2^{\mathrm{a}}$, $\mathrm{HH}_4^{\mathrm{a}}$, $\mathrm{HH}_8^{\mathrm{a}}, \ldots$ overlap with the *principal* torus curve $\mathrm{HH}_1^{\mathrm{s}}$ because all are situated on the straight-line $\sigma = -2$.

In Fig. 14d, the evolution of the Floquet multipliers on the curve $\mathrm{HH}_2^{\mathrm{a}}$ is shown. As they evolve from $+1$ (Arg $= 0°$ at TBPO_1) to -1 (Arg $= 180°$ at TBPO_2), we infer that the 2T periodic orbit also exhibits all the resonances.

In Fig. 15 a pair of Poincaré sections on the plane $y = 0$ are represented. In all the Poincaré sections considered in this review, found with DsTool [38], a value of σ a bit greater than -2 is considered. Thus, the periodic orbits are stable (the modulus of the complex Floquet multipliers is nearby one) or saddle. A filled circle indicates a stable periodic orbit of focus type whereas a cross corresponds to a saddle periodic orbit. In Fig. 15 (Left) the black circle identifies the principal periodic orbit. The two asymmetric 2T stable periodic orbits are indicated by red and blue circles, severally, whereas the corresponding 2T saddle orbits are identified with red and blue crosses.

In Fig. 15 (Right) a zoom of the left-up part of the region shown in Fig. 15 (Left) is presented. Magenta circles and crosses correspond to periodic orbits emerged in the resonance 1:5 of the 2T periodic orbits.

In the following we consider the 1:3 resonance of the principal symmetric periodic orbit that takes place in the interval I_A. In this case, the 3T periodic orbit emanated from this resonance is symmetric. Its numerical continuation when $\sigma = -1.9995$ affords the bifurcation diagram drawn in Fig. 16a, organized by a couple of Shilnikov heteroclinic loops of the nontrivial equilibria, He_{A} and He_{B}. The first two saddle-

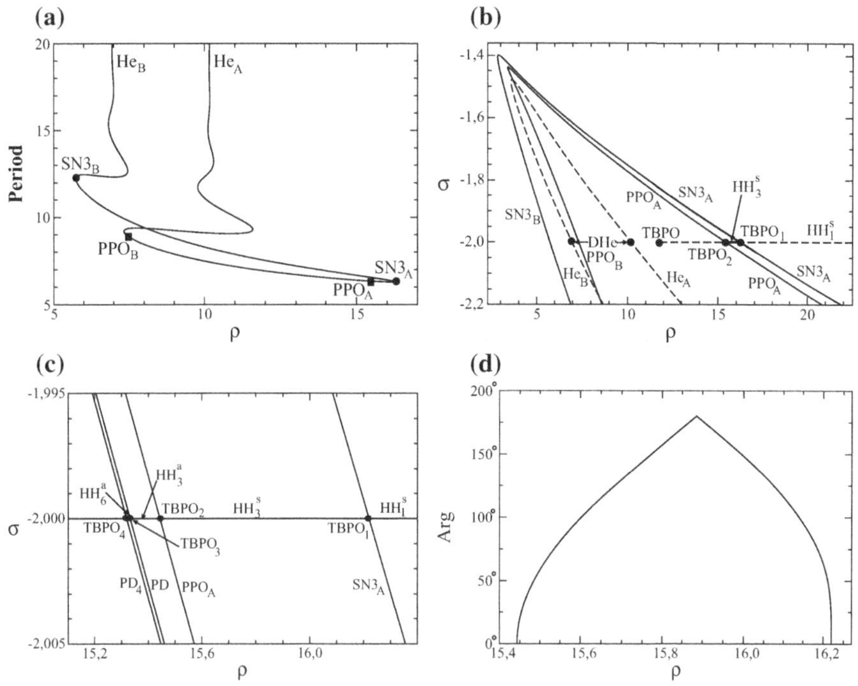

Fig. 16 For $b = 1$: **a** Bifurcation diagram of the 3T symmetric periodic orbit born in the resonance 1:3, for $\sigma = -1.9995$. **b** Partial bifurcation set of the 1:3 open resonance zone SN3 in a vicinity of $\sigma = -2$. **c** Partial bifurcation set for the 3T periodic orbit emerged from the resonance 1:3 placed on the interval I_A (on the left of the resonance 1:2). **d** Argument of the Floquet multipliers versus ρ along the torus bifurcation curve $\mathtt{HH}_3^{\mathtt{S}}$ of the 3T periodic orbit. It reaches its maximum, 180°, at $\rho_m \approx 15.89$. Reproduced with permission from [24]. Copyright (2016) by Springer

node bifurcations are marked as $\mathtt{SN3}_\mathtt{A}$ and $\mathtt{SN3}_\mathtt{B}$ and the first two symmetry-breaking bifurcations are labelled $\mathtt{PPO}_\mathtt{A}$ and $\mathtt{PPO}_\mathtt{B}$.

The corresponding loci where these bifurcations occur, obtained by numerical continuation, are represented in Fig. 16b.

The curves $\mathtt{SN3}_\mathtt{A}$ and $\mathtt{SN3}_\mathtt{B}$ collapse in a turning point in a similar way as the curves of symmetry-breakings ($\mathtt{PPO}_\mathtt{A}$ and $\mathtt{PPO}_\mathtt{B}$) and the curves of heteroclinic connections of the nontrivial equilibria ($\mathtt{He}_\mathtt{A}$ and $\mathtt{He}_\mathtt{B}$) do. Remark that all these curves present a degeneracy when $\sigma = -2$: the saddle-node curves and the symmetry-breaking curves have a non-diagonalizable double $+1$ Floquet multiplier (they are only marked on the right branches of the corresponding curves, named $\mathtt{TBPO}_1$ and $\mathtt{TBPO}_2$, respectively) and the heteroclinic loop curves exhibit a $\delta = 1/2$ degeneracy DHe. The unbounded torus curve $\mathtt{HH}_1^{\mathtt{S}}$ is represented by the dashed line emanating from the point TBPO. Observe that the curve $\mathtt{HH}_3^{\mathtt{S}}$, that exists between the points $\mathtt{TBPO}_1$ and $\mathtt{TBPO}_2$, is a torus bifurcation of a 3T symmetric periodic orbit emerged in the 1:3 resonance.

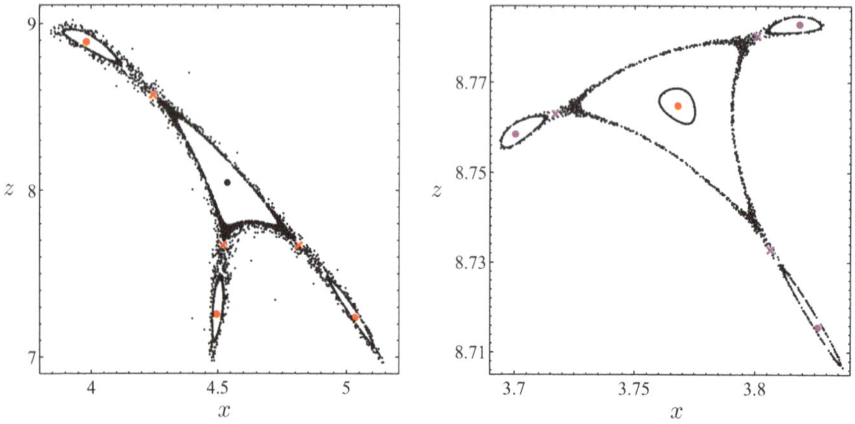

Fig. 17 Poincaré sections in the plane $y = 0$ of the Lorenz system with $b = 1$: (Left) Resonance 1:3 of the principal periodic orbit for $\rho = 16.1$, $\sigma = -1.9999$. (Right) Resonance 1:3 of the 3T periodic orbit for $\rho = 15.707$, $\sigma = -1.99995$. Reproduced with permission from [24]. Copyright (2016) by Springer

A zoom in the vicinity of the curve $\mathtt{HH}_3^{\mathtt{s}}$ appears in Fig. 16c. The concatenation of flip bifurcations $\mathtt{PD}$, $\mathtt{PD}_4, \ldots$ originates a sequence of Takens–Bogdanov points $\mathtt{TBPO}_3, \mathtt{TBPO}_4, \ldots$ (all correspond to non-diagonalizable double -1 Floquet multiplier). A torus bifurcation of a 2×3T asymmetric periodic orbit occurs on the curve $\mathtt{HH}_6^{\mathtt{a}}$ between the points $\mathtt{TBPO}_3$ and $\mathtt{TBPO}_4$. Remark again that all the torus bifurcation curves $\mathtt{HH}_3^{\mathtt{s}}$, $\mathtt{HH}_3^{\mathtt{a}}$, $\mathtt{HH}_6^{\mathtt{a}}, \ldots$ cooccur with the *principal* torus curve $\mathtt{HH}_1^{\mathtt{s}}$ because all of them are situated on the line $\sigma = -2$.

In Fig. 16d, the evolution along the curve $\mathtt{HH}_3^{\mathtt{s}}$ of the Floquet multipliers is shown. As they move from $+1$ to $+1$, attaining its maximum, 180°, at $\rho_{m3} \approx 15.89$, all the resonances of the 3T periodic orbit also occur twice, once when $\rho < \rho_{m3}$ and again for $\rho > \rho_{m3}$. The situation is analogous to that drawn in Fig. 13 (Left) for the principal periodic orbit although now the torus curve is finite.

In Fig. 17 (Left) a Poincaré section taken on the plane $y = 0$ is drawn for $\rho = 16.1$, $\sigma = -1.9999$. A black circle indicates the principal periodic orbit whereas the pair of 3T symmetric periodic orbits born in the 1:3 resonance are tagged by red circles (stable focus-type periodic orbit) and by red crosses (saddle periodic orbit).

In Fig. 17 (Right) the symmetric 3×3T periodic orbits, emerged in the 1:3 resonance of the 3T periodic orbit, are represented when $\rho = 15.707$. The drawn region is a zoom of the left-up part of Fig. 17 (Left). The magenta circles and crosses correspond to these periodic orbits (stable and saddle, respectively). The red circle stands for the 3T symmetric periodic orbit that undergoes this secondary resonance.

Now we briefly comment on the resonance 1:4 of the principal symmetric periodic orbit that takes place in the interval I_A. A partial bifurcation set for this 4T asymmetric periodic orbit appears in Fig. 18 (Left). Remark that one of the saddle-node curves $\mathtt{sn41}$ remains over $\mathtt{HH}_1^{\mathtt{s}}$ whereas the other one is under it. Normally, the two curves

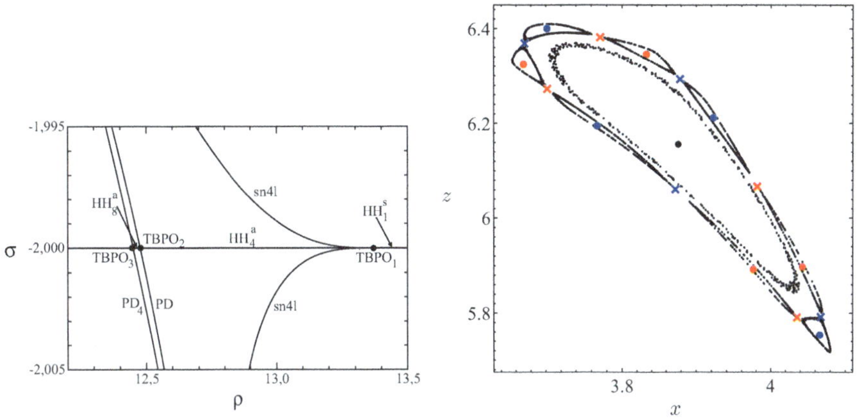

Fig. 18 For $b = 1$: (Left) Partial bifurcation set for the 4T asymmetric periodic orbit born in the resonance 1:4 of the principal periodic orbit, exhibited in the interval I_A. (Right) Poincaré section in the plane $y = 0$ of the Lorenz system for $\rho = 12.78, \sigma = -1.9999$. Reproduced with permission from [24]. Copyright (2016) by Springer

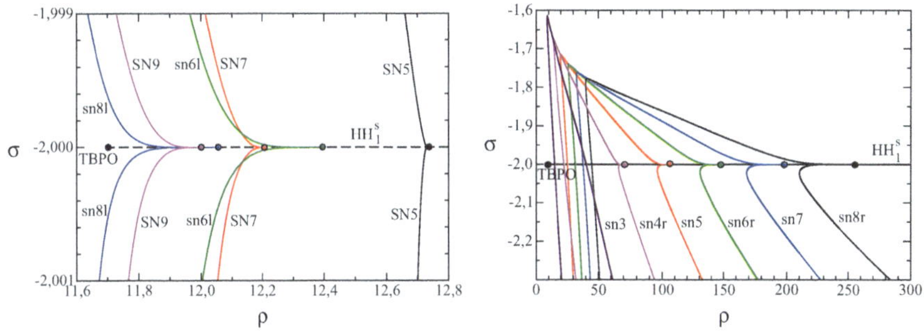

Fig. 19 For $b = 1$: (Left) Partial bifurcation set for the periodic orbits born at the resonances $1{:}p$ $(5 \leq p \leq 9)$ on the interval I_A. (Right) Partial bifurcation set for the periodic orbits born at the resonances $1{:}p$, with $3 \leq p \leq 8$, on the interval I_B. Reproduced with permission from [24]. Copyright (2016) by Springer

of saddle-node bifurcations emanated from a tip stay situated on the same side of the torus curve [47, 82]. A cascade of period-doublings $\mathtt{PD}$, $\mathtt{PD}_4, \ldots$, and another one of Takens–Bogdanov bifurcations (non-diagonalizable double -1 Floquet multiplier) $\mathtt{TBPO}_2$, $\mathtt{TBPO}_3, \ldots$, are present. The torus bifurcation curves $\mathtt{HH}_4^{\mathtt{a}}$, $\mathtt{HH}_8^{\mathtt{a}}, \ldots$ coexist with $\mathtt{HH}_1^{\mathtt{s}}$. In Fig. 18 (Right) a Poincaré section on the plane $y = 0$ for $\rho = 12.78$ and $\sigma = -1.9999$ is drawn. The principal periodic orbit (stable and symmetric) is indicated with a black circle and the two asymmetric 4T stable/saddle periodic orbits are marked with red and blue circles/crosses.

Now we represent in Fig. 19 (Left) the curves of saddle-node bifurcations emanated from the first weak resonances $1{:}p$ $(5 \leq p \leq 9)$ undergone by the principal periodic orbit in the interval I_A. Observe that the curves of symmetric periodic orbits (capital

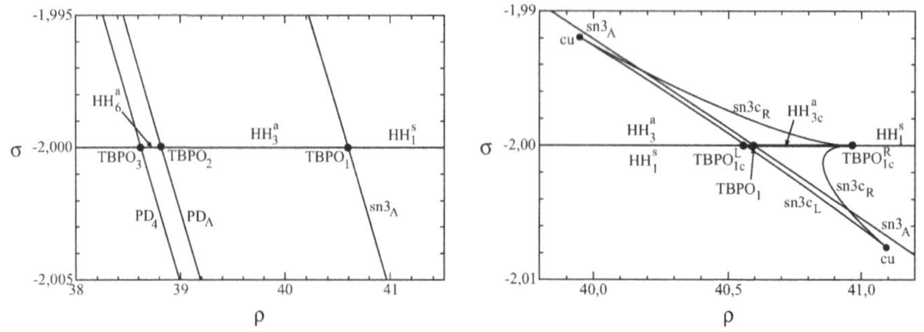

Fig. 20 For $b = 1$: (Left) Partial bifurcation set of the 1:3 open resonance zone sn3 corresponding to the 3T asymmetric periodic orbit of the resonance 1:3 of the interval I_B. (Right) Zoom of panel (Left) in a neighborhood of the Takens–Bogdanov bifurcation $\mathtt{TBPO}_1$. A small closed resonance zone organized by two cusps exists. Reproduced with permission from [24]. Copyright (2016) by Springer

letters) separate faster from the torus curve $\mathtt{HH}_1^{\mathtt{s}}$ than those of the asymmetric periodic orbits (lowercase letters). On the other hand, near the related tip, always one of the saddle-node curves emanates and continues above the torus curve $\mathtt{HH}_1^{\mathtt{s}}$ and the other one comes out and stays below it.

In the following, we briefly focus on resonances in the interval $I_B \equiv [\rho_m, \infty)$, where $\rho_m \approx 21.4623$ [24, Sects. 3.4 and 3.5]. First, in Fig. 19 (Right) the curves of saddle-node bifurcations arisen from the resonances $1{:}p$ $(3 \leq p \leq 8)$ of the principal periodic orbit on the curve $\mathtt{HH}_1^{\mathtt{s}}$ in the interval I_B are superimposed (note that all the periodic orbits born are asymmetric). In this situation, the comportment evidenced for all the curves is identical for $4 \leq p \leq 8$: the curves of saddle-node bifurcations appear from a tip; in every single case, in the neighborhood of the tip, one of the saddle-node curve comes out and stays over the curve $\mathtt{HH}_1^{\mathtt{s}}$ whereas the other one emanates and continues under it; the *velocity of separation* from the curve $\mathtt{HH}_1^{\mathtt{s}}$ is always *alike* (recall that in I_A it depends on the parity of p); ulteriorly the curves have turning points giving rise to new Takens–Bogdanov points when they intersect the line $\sigma = -2$. On the other hand, the saddle-node curve sn3 does not emanate from a tip on the curve $\mathtt{HH}_1^{\mathtt{s}}$. It crosses the torus curve, as it is habitual in the case of the strong resonance 1:3 (see, for example, [47, 82]).

Now we specifically consider the resonance 1:3. In Fig. 20 (Left) the window of Fig. 19 (Right) is enlarged in the neighborhood of the intersection of the right branch of curve sn3 with $\sigma = -2$ (labelled here as $\mathtt{sn3}_{\mathtt{A}}$). The 3T asymmetric periodic orbits live on the left side of $\mathtt{sn3}_{\mathtt{A}}$. The Takens–Bogdanov points $\mathtt{TBPO}_1$ (non-diagonalizable double $+1$ Floquet multiplier) and $\mathtt{TBPO}_2$ (non-diagonalizable double -1 Floquet multiplier) are the limits of the curve of torus bifurcation exhibited by these 3T periodic orbits, $\mathtt{HH}_3^{\mathtt{a}}$. The succession of flip bifurcations $\mathtt{PD}_{\mathtt{A}}$, $\mathtt{PD}_4, \ldots$ originates a chain of Takens–Bogdanov points $\mathtt{TBPO}_2$, $\mathtt{TBPO}_3, \ldots$. A torus bifurcation of 2×3T periodic orbits occurs on the curve $\mathtt{HH}_6^{\mathtt{a}}$. Recall that the torus curves $\mathtt{HH}_3^{\mathtt{a}}$, $\mathtt{HH}_6^{\mathtt{a}}, \ldots$ coexist with the curve $\mathtt{HH}_1^{\mathtt{s}}$.

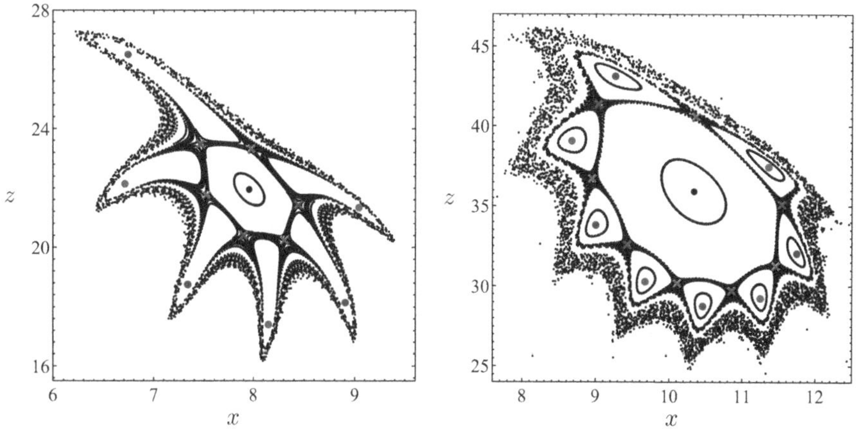

Fig. 21 For $b = 1$, $\sigma = -1.9999$, Poincaré section in the plane $y = 0$, in a neighborhood of the pT asymmetric periodic orbits emerged in the resonance $1{:}p$ of the principal periodic orbit in the interval I_B: (Left) $p = 3$ when $\rho = 40.5$. (Right) $p = 4$ when $\rho = 65$. Reproduced with permission from [24]. Copyright (2016) by Springer

But a more detailed study in the vicinity of the point $\mathtt{TBPO}_1$ (see [24, Fig. 13]) reveals the striking partial bifurcation set shown in Fig. 20 (Right). On the one hand, two interconnected cusps appear when the curves $\mathtt{sn3c}_\mathrm{L}$ and $\mathtt{sn3c}_\mathrm{R}$ collapse. This originates the existence of a relatively small closed resonance zone (the letter c means 'closed' in the labels of this figure). Note that the saddle-node bifurcation curves $\mathtt{sn3c}_\mathrm{R}$ appear from a tip on the curve $\mathtt{HH}_1^\mathtt{s}$, exactly at the point $\mathtt{TBPO}_{1\mathtt{c}}^\mathrm{R}$ where the argument of the Floquet multipliers of the principal periodic orbit is 120°. This is an abnormal situation for a strong resonance (1:3). On the other hand, a new torus curve $\mathtt{HH}_{3\mathtt{c}}^\mathtt{a}$ is limited by the Takens–Bogdanov points $\mathtt{TBPO}_{1\mathtt{c}}^\mathrm{R}$ and $\mathtt{TBPO}_{1\mathtt{c}}^\mathrm{L}$. It coexists with $\mathtt{HH}_3^\mathtt{a}$ between the points $\mathtt{TBPO}_{1\mathtt{c}}^\mathrm{L}$ and $\mathtt{TBPO}_1$. As the maximum of the argument along the curve $\mathtt{HH}_{3\mathtt{c}}^\mathtt{a}$ is more or less 3.3155°, these 3T periodic orbits will merely undergo very high resonances (namely $1{:}p$ where $p \geq 109$). This is a strange fact in the Lorenz system since, in almost all the cases reported in Ref. [24], the maximum of the argument is 180°.

A Poincaré section obtained with the plane $y = 0$, when $\sigma = -1.9999$ and $\rho = 40.5$, is drawn in Fig. 21 (Left). The black circle stands for the principal periodic orbit. Blue and red circles indicate, respectively, the two asymmetric 3T stable periodic orbits whereas the corresponding 3T saddle periodic orbits are marked by blue and red crosses. A similar Poincaré section appears in Fig. 21 (Right), to illustrate the 4T asymmetric periodic orbits emerged in the 1:4 resonance of the interval I_B, for $\sigma = -1.9999$ and $\rho = 65$.

A new remarkable fact is that the resonances $2{:}(2n + 3)$, for $n \geq 1$, in the interval I_B originates symmetric periodic orbits. This is exemplified in Fig. 22 where two Poincaré sections, related to resonances 2:5 and 2:7, are drawn: a 5T symmetric

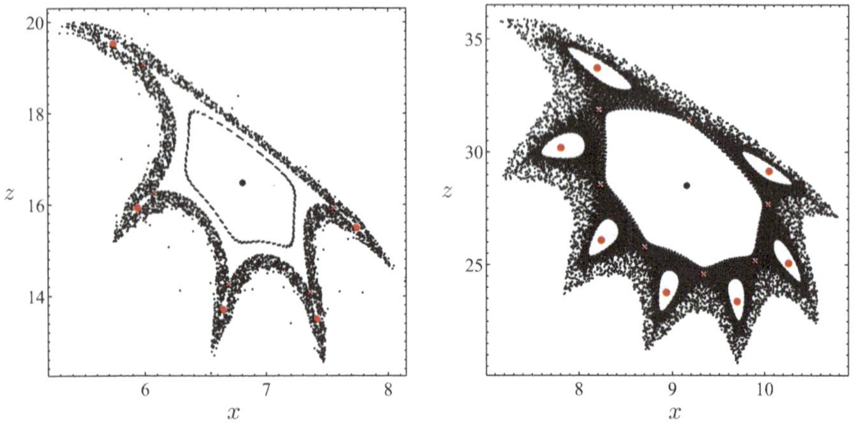

Fig. 22 Poincaré sections in the plane $y = 0$ of the Lorenz system with $b = 1$, $\sigma = -1.9999$: (Left) Resonance 2:5 in the interval I_B when $\rho = 30.9$. (Right) Resonance 2:7 in the interval I_B when $\rho = 52$. Reproduced with permission from [24]. Copyright (2016) by Springer

periodic orbit appears in the first case (Left) and a 7T symmetric periodic orbit in the second one (Right).

To finish this section we summarize the main results on resonances of periodic orbits in the Lorenz system, obtained for $b = 1$ (see [24, Conclusions]): (i) Various curves of torus bifurcations related to different periodic orbits coexist because all the torus curves are placed at $\sigma = -2$ and the *principal* torus curve $\mathrm{HH}_1^{\mathrm{S}}$ is unbounded. (ii) When the torus curves exist between curves of saddle-node and symmetry-breaking bifurcations (both Takens–Bogdanov points correspond to nondegenerate double $+1$ Floquet multiplier) mostly all the resonances occur twice because the maximum of the argument is 180°. If one of the limit curves corresponds to period-doublings (a nondegenerate double -1 Floquet multiplier occurs) all the resonances occur one time in every case found. (iii) The presence of a concatenation of flip bifurcations implies the existence of a sequence of torus curves. (iv) Most of the time the resonance regions (limited by the curves of saddle-node bifurcations) are open. However, in some cases, two cusp points organize small closed regions and, thus, an angular degeneracy on the torus curve might lead to the existence of closed Arnold's tongues [32, 92]. This possibility should be investigated in the future. (v) For the strong resonances (1:3, 1:4 and 2:5) there are cases where the curves of saddle-node intersect transversely with the related torus curve as well as situations where they come out from a tip on the torus curve. (vi) The ordinary situation in a tip is that the two saddle-node curves emanate towards the same part of the torus curve. However, in the Lorenz system, in all the tips found, one of the curve arises over the torus curve and the other one emerges under it. Moreover, in almost all the cases, both curves emanate towards the left side. (vii) Symmetric periodic orbits emerge from resonances $1{:}p$ (with p odd) for the interval I_A and from resonances $2{:}q$ (with q odd) for the interval I_B.

7 Conclusions

The goal of this chapter is to review some bifurcations exhibited by the classical Lorenz system, when the parameters can have any real value. On the one hand, we have described analytical and numerical results recently obtained. The theoretical study of the pitchfork, Hopf and Takens–Bogdanov bifurcations of the origin, as well as the Hopf bifurcation of the nontrivial equilibria, has been successfully completed. Moreover, from the information achieved in the study of the above local bifurcations, other important organizing centers of the dynamics have been found with the help of the adequate numerical techniques: Takens–Bogdanov bifurcations of periodic orbits, torus bifurcations and the resonances associated, homoclinic and heteroclinic connections with several degeneracies, etc.

However, as pointed out in Sect. 2, the analysis with the usual tools of the Hopf-pitchfork and the triple-zero bifurcations of the origin cannot be performed because it is not an isolated equilibrium when $b = 0$. Furthermore, other symptoms of the singularity of the Lorenz system have also been mentioned in this work. For instance, the presence of bifurcations of codimension infinite (Hopf and Takens–Bogdanov) and the coexistence of torus bifurcation curves of different periodic orbits (a direct consequence of the fact that the divergence in the Lorenz system does not depend on the spatial variables but only on the parameters).

In the next future our objective is to analyze the Hopf-pitchfork bifurcation in the Lorenz system. A way to avoid the degeneration present when $b = 0$ is the introduction of new nonlinear terms so that the Lorenz system is embedded in the structurally stable system obtained. After analyzing the Hopf-pitchfork bifurcation in this new Lorenz-like system, it will suffice to take the appropriate limit to obtain valuable information for the Lorenz system. This same idea should be valid to study the triple-zero bifurcation (a partial study of this bifurcation in the famous Rössler system can be found in Ref. [63]). Remark that, in this new system, all the bifurcation of codimension infinite will disappear. For instance, a codimension-three degeneracy will occur in the case of the Takens–Bogdanov bifurcation, as the one considered in Ref. [94]. At the same time, the simultaneous existence of torus bifurcation curves of distinct periodic orbits will also disappear.

Acknowledgements This work has been partially supported by the *Ministerio de Economía y Competitividad, Plan Nacional I+D+I* co-financed with FEDER funds, in the frame of the project MTM2014-56272-C2, and by the *Consejería de Economía, Innovación, Ciencia y Empleo de la Junta de Andalucía* (FQM-276, TIC-0130 and P12-FQM-1658).

References

1. Alexeev, I.: Lorenz system in the thermodynamic modelling of leukaemia malignancy. Med. Hypotheses **102**, 150–155 (2017)
2. Algaba, A., Domínguez-Moreno, M.C., Merino, M., Rodríguez-Luis, A.J.: Study of the Hopf bifurcation in the Lorenz, Chen and Lü systems. Nonlinear Dyn. **79**, 885–902 (2015)

3. Algaba, A., Domínguez-Moreno, M.C., Merino, M., Rodríguez-Luis, A.J.: Takens–Bogdanov bifurcations of equilibria and periodic orbits in the Lorenz system. Commun. Nonlinear Sci. Numer. Simul. **30**, 328–343 (2016)
4. Algaba, A., Fernández-Sánchez, F., Freire, E., Gamero, E., Rodríguez-Luis, A.J.: Oscillation-sliding in a modified van der Pol-Duffing electronic oscillator. J. Sound Vib. **249**, 899–907 (2003)
5. Algaba, A., Fernández-Sánchez, F., Freire, E., Merino, M., Rodríguez-Luis, A.J.: Nontransversal curves of T-points: a source of closed curves of global bifurcations. Phys. Lett. A **303**, 204–211 (2002)
6. Algaba, A., Fernández-Sánchez, F., Merino, M., Rodríguez-Luis, A.J.: Analysis of the T-point-Hopf bifurcation with $\mathbb{Z}_2$-symmetry. Application to Chua's equation. Int. J. Bifurc. Chaos **20**, 979–993 (2010)
7. Algaba, A., Fernández-Sánchez, F., Merino, M., Rodríguez-Luis, A.J.: Comment on "Sil'nikov chaos of the Liu system" [Chaos **18**, 013113 (2008)]. Chaos **21**, 048101 (2011)
8. Algaba, A., Fernández-Sánchez, F., Merino, M., Rodríguez-Luis, A.J.: Structure of saddle-node and cusp bifurcations of periodic orbits near a non-transversal T-point. Nonlinear Dyn. **63**, 455–476 (2011)
9. Algaba, A., Fernández-Sánchez, F., Merino, M., Rodríguez-Luis, A.J.: Comment on "Existence of heteroclinic orbits of the Shil'nikov type in a 3D quadratic autonomous chaotic system" [J. Math. Anal. Appl. **315**, 106–119 (2006)]. J. Math. Anal. Appl. **392**, 99–101 (2012)
10. Algaba, A., Fernández-Sánchez, F., Merino, M., Rodríguez-Luis, A.J.: Comment on "Heteroclinic orbits in Chen circuit with time delay" [Commun. Nonlinear Sci. Numer. Simulat. **15**, 3058–3066 (2010)]. Commun. Nonlinear Sci. Numer. Simulat. **17**, 2708–2710 (2012)
11. Algaba, A., Fernández-Sánchez, F., Merino, M., Rodríguez-Luis, A.J.: Chen's attractor exists if Lorenz repulsor exists: the Chen system is a special case of the Lorenz system. Chaos **23**, 033108 (2013)
12. Algaba, A., Fernández-Sánchez, F., Merino, M., Rodríguez-Luis, A.J.: Comment on 'Šilnikov-type orbits of Lorenz-family systems' [Physica A **375**, 438–446 (2007)]. Physica A **392**, 4252–4257 (2013)
13. Algaba, A., Fernández-Sánchez, F., Merino, M., Rodríguez-Luis, A.J.: The Lü system is a particular case of the Lorenz system. Phys. Lett. A **377**, 2771–2776 (2013)
14. Algaba, A., Fernández-Sánchez, F., Merino, M., Rodríguez-Luis, A.J.: Centers on center manifolds in the Lorenz, Chen and Lü systems. Commun. Nonlinear Sci. Numer. Simul. **19**, 772–775 (2014)
15. Algaba, A., Fernández-Sánchez, F., Merino, M., Rodríguez-Luis, A.J.: Comment on "Existence of heteroclinic and homoclinic orbits in two different chaotic dynamical systems" [Appl. Math. Comput. **218**, 11859–11870 (2012)]. Appl. Math. Comput. **244**, 49–56 (2014)
16. Algaba, A., Fernández-Sánchez, F., Merino, M., Rodríguez-Luis, A.J.: Analysis of the T-point-Hopf bifurcation in the Lorenz system. Commun. Nonlinear Sci. Numer. Simul. **22**, 676–691 (2015)
17. Algaba, A., Freire, E., Gamero, E., Rodríguez-Luis, A.J.: Analysis of Hopf and Takens–Bogdanov bifurcations in a modified van der Pol-Duffing oscillator. Nonlinear Dyn. **16**, 369–404 (1998)
18. Algaba, A., Freire, E., Gamero, E., Rodríguez-Luis, A.J.: A three-parameter study of a degenerate case of the Hopf-pitchfork bifurcation. Nonlinearity **12**, 1177–1206 (1999)
19. Algaba, A., Freire, E., Gamero, E., Rodríguez-Luis, A.J.: On a codimension-three unfolding of the interaction of degenerate Hopf and pitchfork bifurcations. Int. J. Bifurc. Chaos **9**, 1333–1362 (1999)
20. Algaba, A., Freire, E., Gamero, E., Rodríguez-Luis, A.J.: On the Takens–Bogdanov bifurcation in the Chua's equation. IEICE T. Fund. Electr. **E82-A**, 1722–1728 (1999)
21. Algaba, A., Freire, E., Gamero, E., Rodríguez-Luis, A.J.: A tame degenerate Hopf-pitchfork bifurcation in a modified van der Pol-Duffing oscillator. Nonlinear Dyn. **22**, 249–269 (2000)

22. Algaba, A., Freire, E., Gamero, E., Rodríguez-Luis, A.J.: Resonances of periodic orbits in Rössler system in presence of a triple-zero bifurcation. Int. J. Bifurc. Chaos **17**, 1997–2008 (2007)
23. Algaba, A., Gamero, E., García, C., Merino, M.: A degenerate Hopf-saddle-node bifurcation analysis in a family of electronic circuits. Nonlinear Dyn. **48**, 55–76 (2007)
24. Algaba, A., Gamero, E., Merino, M., Rodríguez-Luis, A.J.: Resonances of periodic orbits in the Lorenz system. Nonlinear Dyn. **84**, 2111–2136 (2016)
25. Algaba, A., Merino, M., Fernández-Sánchez, F., Rodríguez-Luis, A.J.: Closed curves of global bifurcations in Chua's equation: a mechanism for their formation. Int. J. Bifurc. Chaos **13**, 609–616 (2003)
26. Algaba, A., Merino, M., Fernández-Sánchez, F., Rodríguez-Luis, A.J.: Open-to-closed curves of saddle-node bifurcations of periodic orbits near a nontransversal T-point in Chua's equation. Int. J. Bifurc. Chaos **16**, 2637–2647 (2006)
27. Algaba, A., Merino, M., Fernández-Sánchez, F., Rodríguez-Luis, A.J.: Hopf bifurcations and their degeneracies in Chua's equation. Int. J. Bifurc. Chaos **21**, 2749–2763 (2011)
28. Algaba, A., Merino, M., Freire, E., Gamero, E., Rodríguez-Luis, A.J.: On the Hopf-pitchfork bifurcation in the Chua's equation. Int. J. Bifurc. Chaos **10**, 291–305 (2000)
29. Algaba, A., Merino, M., Freire, E., Gamero, E., Rodríguez-Luis, A.J.: Some results on Chua's equation near a triple-zero linear degeneracy. Int. J. Bifurc. Chaos **13**, 583–608 (2003)
30. Algaba, A., Merino, M., García, C., Reyes, M.: Degenerate global bifurcations in a simple circuit. Int. J. Pure Appl. Math. **57**, 265–278 (2009)
31. Algaba, A., Merino, M., Rodríguez-Luis, A.J.: Evolution of Arnold's Tongues in a $\mathbb{Z}_2$-symmetric electronic circuit. IEICE T. Fund. Electr. **E82-A**, 1714–1721 (1999)
32. Algaba, A., Merino, M., Rodríguez-Luis, A.J.: Takens–Bogdanov bifurcations of periodic orbits and Arnold's tongues in a three-dimensional electronic model. Int. J. Bifurc. Chaos **11**, 513–531 (2001)
33. Algaba, A., Merino, M., Rodríguez-Luis, A.J.: Homoclinic connections near a Belyakov point in Chua's equation. Int. J. Bifurc. Chaos **15**, 1239–1252 (2005)
34. Algaba, A., Merino, M., Rodríguez-Luis, A.J.: Analysis of a Belyakov homoclinic connection with $\mathbb{Z}_2$-symmetry. Nonlinear Dyn. **69**, 519–529 (2012)
35. Algaba, A., Merino, M., Rodríguez-Luis, A.J.: Homoclinic interactions near a triple-zero degeneracy in Chua's equation. Int. J. Bifurc. Chaos **22**, 1250,129 (2012)
36. Algaba, A., Merino, M., Rodríguez-Luis, A.J.: Superluminal periodic orbits in the Lorenz system. Commun. Nonlinear Sci. Numer. Simulat. **39**, 220–232 (2016)
37. Arnold, V.I.: Geometrical Methods in the Theory of Ordinary Differential Equations. Springer, New York (1983)
38. Back, A., Guckenheimer, J., Myers, M.R., Wicklin, F.J., Worfolk, P.A.: DsTool: computer assisted exploration of dynamical systems. Notices Am. Math. Soc. **39**, 303–309 (1992)
39. Barrio, R., Blesa, F., Serrano, S.: Global organization of spiral structures in biparameter space of dissipative systems with Shilnikov saddle-foci. Phys. Rev. E **84**, 035,201 (2011)
40. Barrio, R., Serrano, S.: Bounds for the chaotic region in the Lorenz model. Physica D **238**, 1615–1624 (2009)
41. Barrio, R., Shilnikov, A.L., Shilnikov, L.P.: Kneadings, symbolic dynamics and painting Lorenz chaos. Int. J. Bifurc. Chaos **22**, 1230016 (2012)
42. Broer, H., Roussarie, R., Simó, C.: Invariant circles in the Bogdanov–Takens bifurcation for diffeomorphisms. Ergod. Theory Dyn. Syst. **16**, 1147–1172 (1996)
43. Cao, J., Zhang, X.: Dynamics of the Lorenz system having an invariant algebraic surface. J. Math. Phys. **48**, 1–13 (2007)
44. Champneys, A.R., Kuznetsov, Y.A.: Numerical detection and continuation of codimension-two homoclinic bifurcations. Int. J. Bifurc. Chaos **4**, 795–822 (1994)
45. Champneys, A.R., Rodríguez-Luis, A.J.: The non-transverse Shil'nikov-Hopf bifurcation: uncoupling of homoclinic orbits and homoclinic tangencies. Physica D **128**, 130–158 (1999)
46. Chow, S., Deng, B., Fiedler, B.: Homoclinic bifurcation at resonant eigenvalues. J. Dyn. Differ. Equ. **2**, 177–244 (1990)

47. Chow, S., Li, C., Wang, D.: Normal Forms and Bifurcation of Planar Vector Fields. Cambridge University Press, Cambridge (1994)
48. Creaser, J.L., Krauskopf, B., Osinga, H.M.: α-flips and T-points in the Lorenz system. Nonlinearity **28**, R39–R65 (2015)
49. Cuomo, K.M., Oppenheim, A.V.: Circuit implementation of synchronized chaos with applications to communications. Phys. Rev. Lett. **71**, 65–68 (1993)
50. De Witte, V., Della Rossa, F., Govaerts, W., Kuznetsov, Y.A.: Numerical periodic normalization for codim 2 bifurcations of limit cycles: computational formulas, numerical implementation, and examples. SIAM J. Appl. Dyn. Syst. **12**, 722–788 (2013)
51. Devaney, R.L.: An Introduction to Chaotic Dynamics. Benjamin/Cummings, Menlo Park (1986)
52. Doedel, E.J., Champneys, A.R., Dercole, F., Fairgrieve, T., Kuznetsov, Y.A., Oldeman, B.E., Paffenroth, R., Sandstede, B., Wang, X., Zhang, C.: AUTO-07P: continuation and bifurcation software for ordinary differential equations (with HomCont). Technical report, Concordia University (2010)
53. Doedel, E.J., Freire, E., Gamero, E., Rodríguez-Luis, A.J.: An analytical and numerical study of a modified van der Pol oscillator. J. Sound Vib. **256**, 755–771 (2002)
54. Doedel, E.J., Krauskopf, B., Osinga, H.M.: Global bifurcations of the Lorenz manifold. Nonlinearity **19**, 2947–2972 (2006)
55. Doedel, E.J., Krauskopf, B., Osinga, H.M.: Global invariant manifolds in the transition to preturbulence in the Lorenz system. Indag. Math. **22**, 222–240 (2011)
56. Doedel, E.J., Krauskopf, B., Osinga, H.M.: Global organization of phase space in the transition to chaos in the Lorenz system. Nonlinearity **28**, R113R139 (2015)
57. Elgin, J.N., Molina-Garza, J.B.: Traveling wave solutions of the Maxwell-Bloch equations. Phys. Rev. A **35**, 3986–3988 (1987)
58. Fernández-Sánchez, F., Freire, E., Pizarro, L., Rodríguez-Luis, A.J.: A model for the analysis of the dynamical consequences of a nontransversal intersection of the two-dimensional manifolds involved in a T-point. Phys. Lett. A **320**, 169–179 (2003)
59. Fernández-Sánchez, F., Freire, E., Rodríguez-Luis, A.J.: Isolas, cusps and global bifurcations in an electronic oscillator. Dyn. Syst. **12**, 319–336 (1997)
60. Fernández-Sánchez, F., Freire, E., Rodríguez-Luis, A.J.: T-points in a $\mathbb{Z}_2$-symmetric electronic oscillator. (I) Analysis. Nonlinear Dyn. **28**, 53–69 (2002)
61. Fernández-Sánchez, F., Freire, E., Rodríguez-Luis, A.J.: Bi-spiraling homoclinic curves around a T-point in Chua's equation. Int. J. Bifurc. Chaos **14**, 1789–1793 (2004)
62. Fernández-Sánchez, F., Freire, E., Rodríguez-Luis, A.J.: Analysis of the T-point-Hopf bifurcation. Physica D **237**, 292–305 (2008)
63. Freire, E., Gamero, E., Rodríguez-Luis, A.J., Algaba, A.: A note on the triple-zero linear degeneracy: normal forms, dynamical and bifurcation behaviors of an unfolding. Int. J. Bifurc. Chaos **12**, 2799–2820 (2002)
64. Freire, E., Rodríguez-Luis, A.J.: Numerical bifurcation analysis of electronic circuits. In: Krauskopf, B., et al. (eds.) Numerical Continuation Methods for Dynamical Systems: Path Following and Boundary Value Problems, pp. 221–251. Springer, Dordrecht (2007)
65. Freire, E., Rodríguez-Luis, A.J., Gamero, E., Ponce, E.: A case study for homoclinic chaos in an autonomous electronic circuit. A trip from Takens–Bogdanov to Hopf-Šil'nikov. Physica D **62**, 230–253 (1993)
66. Gamero, E., Freire, E., Ponce, E.: Normal forms for planar systems with nilpotent linear part. In: Seydel, R., et al. (eds.) Bifurcation and Chaos: Analysis, Algorithms, Applications, International Series of Numerical Mathematics, vol. 97, pp. 123–127. Birkhäuser, Basel (1991)
67. Gamero, E., Freire, E., Rodríguez-Luis, A.J., Ponce, E., Algaba, A.: Hypernormal form calculation for triple-zero degeneracies. B. Belg. Math. Soc. Sim. **6**, 357–368 (1999)
68. Gelfreich, V.: Chaotic zone in the Bogdanov–Takens bifurcation for diffeomorphisms. In: Begehr, H.G.W., Gilbert, R.P., Wong, M.W. (eds.) International Society for Analysis, Applications and Computation, Analysis and Applications-ISAAC 2001, vol. 10, pp. 187–197. Kluwer Acad. Publ, Dordrecht (2003)

69. Glendinning, P., Sparrow, C.: T-points: a codimension two heteroclinic bifurcation. J. Stat. Phys. **43**, 479–488 (1986)
70. Golubitsky, M., Langford, W.F.: Classification and unfoldings of degenerate Hopf bifurcations. J. Differ. Equ. **41**, 375–415 (1981)
71. Gorman, M., Widmann, P.J., Robbins, K.A.: Nonlinear dynamics of a convection loop: a quantitative comparison of experiment with theory. Physica D **19**, 255–267 (1986)
72. Guckenheimer, J., Holmes, P.J.: Nonlinear Oscillations, Dynamical Systems, and Bifurcations of Vector Fields. Springer, New York (1983)
73. Haken, H.: Analogy between higher instabilities in fluids and lasers. Phys. Lett. A **53**, 77–78 (1975)
74. Hale, J.K.: Ordinary Differential Equations. Krieger Publishing Company, Malabar (1980)
75. Hemati, N.: Strange attractors in brushless DC motors. IEEE T. Circuits-I **41**, 40–45 (1994)
76. Homburg, A.J., Sandstede, B.: Homoclinic and heteroclinic bifurcations in vector fields. In: Broer, H., et al. (eds.) Handbook of Dynamical Systems, vol. 3, pp. 379–524. Elsevier, Amsterdam (2010)
77. Knobloch, E.: Chaos in the segmented disc dynamo. Phys. Lett. A **82**, 439–440 (1981)
78. Knobloch, E., Proctor, M.R.E., Weiss, N.O.: Heteroclinic bifurcations in a simple model of double-diffusive convection. J. Fluid Mech. **239**, 273–292 (1992)
79. Kokubu, H., Roussarie, R.: Existence of a singularly degenerate heteroclinic cycle in the Lorenz system and its dynamical consequences: Part 1. J. Dyn. Differ. Equ. **16**, 513–557 (2004)
80. Krauskopf, B., Rousseau, C.: Codimension-three unfoldings of reflectionally symmetric planar vector fields. Nonlinearity **10**, 1115–1150 (1997)
81. Kulpa, W.: The Poincaré-Miranda theorem. Am. Math. Mon. **104**, 545–550 (1997)
82. Kuznetsov, Y.A.: Elements of Applied Bifurcation Theory. Springer, New York (2004)
83. Leonov, G.A., Kuznetsov, N.V., Korzhemanova, N.A., Kusakin, D.V.: Lyapunov dimension formula for the global attractor of the Lorenz system. Commun. Nonlinear Sci. Numer. Simul. **41**, 84–103 (2016)
84. Llibre, J., Messias, M., da Silva, P.R.: Global dynamics of the Lorenz system with invariant algebraic surfaces. Int. J. Bifurc. Chaos **20**, 3137–3155 (2010)
85. Llibre, J., Zhang, X.: Invariant algebraic surfaces of the Lorenz system. J. Math. Phys. **43**, 1622–1645 (2002)
86. Lorenz, E.N.: Deterministic nonperiodic flow. J. Atmos. Sci. **20**, 130–141 (1963)
87. Messias, M.: Dynamics at infinity and the existence of singularly degenerate heteroclinic cycles in the Lorenz system. J. Phys. A **42**, 115,101 (2009)
88. Osinga, H.M., Krauskopf, B.: Visualizing the structure of chaos in the Lorenz system. Comput. Graph. **26**, 815–823 (2002)
89. Pade, J., Rauh, A., Tsarouhas, G.: Analytical investigation of the Hopf bifurcation in the Lorenz model. Phys. Lett. A **115**, 93–96 (1986)
90. Pasini, A., Pelino, V.: A unified view of Kolmogorov and Lorenz systems. Phys. Lett. A **275**, 435–446 (2000)
91. Pchelintsev, A.N.: Numerical and physical modeling of the dynamics of the Lorenz system. Numer. Anal. Appl. **7**, 159–167 (2014)
92. Peckman, B.B., Frouzakis, C.E., Kevrekidis, I.: Bananas and bananas splits: a parametric degeneracy in the Hopf bifurcation for maps. SIAM J. Math. Anal. **26**, 190–217 (1995)
93. Poland, D.: Cooperative catalysis and chemical chaos: a chemical model for the Lorenz equations. Physica D **65**, 86–99 (1993)
94. Rodríguez-Luis, A.J., Freire, E., Ponce, E.: On a codimension 3 bifurcation arising in an autonomous electronic circuit. In: Seydel, R., et al. (eds.) Bifurcation and Chaos: Analysis, Algorithms, Applications, International Series of Numerical Mathematics, vol. 97, pp. 301–306. Birkhäuser, Basel (1991)
95. Roschin, M.: Dangerous stability boundaries in the Lorenz model. Prikl. Mat. Mekh. **42**, 950–952 (1978)

96. Sieber, J., Krauskopf, B.: Bifurcation analysis of an inverted pendulum with delayed feedback control near a triple-zero eigenvalue singularity. Nonlinearity **17**, 85–103 (2004)
97. Sparrow, C.: The Lorenz Equations. Springer, New York (1982)
98. Swinnerton-Dyer, P.: The invariant algebraic surfaces of the Lorenz system. Math. Proc. Camb. Philos. Soc. **132**, 385–393 (2002)
99. Tsarouhas, G., Pade, J.: The Hopf bifurcation in the Lorenz by the 2-timing method model. Physica A **138**, 505–517 (1986)
100. Tucker, W.: The Lorenz attractor exists. C. R. Acad. Sci. **328**, 1197–1202 (1999)
101. Wang, Q., Huang, W., Feng, J.: Multiple limit cycles and centers on center manifolds for Lorenz system. Appl. Math. Comput. **238**, 281–288 (2014)
102. Wiggins, S.: Introduction to Applied Dynamical Systems and Chaos. Springer, New York (2003)
103. Yajima, T., Nagahama, H.: Tangent bundle viewpoint of the Lorenz system and its chaotic behavior. Phys. Lett. A **374**, 1315–1319 (2010)
104. Zhou, T., Chen, G.: Classification of chaos in 3-D auto nomous quadratic systems-I. Basic framework and methods. Int. J. Bifurc. Chaos **16**, 2459–2479 (2006)
105. Zhou, T., Chen, G., Čelikovský, S.: Ši'lnikov chaos in the generalized Lorenz canonical form of dynamical systems. Nonlinear Dyn. **39**, 319–334 (2005)
106. Zhou, T., Tang, Y., Chen, G.: Chen's attractor exists. Int. J. Bifurc. Chaos **14**, 3167–3178 (2004)

Normal Form for a Class of Three-Dimensional Systems with Free-Divergence Principal Part

Antonio Algaba, Natalia Fuentes, Estanislao Gamero and Cristóbal García

Abstract We present the basic ideas of the *Normal Form Theory* by using quasi-homogeneous expansions of the vector field, where the structure of the normal form is determined by the principal part of the vector field. We focus on a class of tridimensional systems whose principal part is the coupling of a Hamiltonian planar system and an unidimensional system, in such a way that the quoted principal part does not depend on the last variable and has free divergence. Our study is based on several decompositions of quasi-homogeneous vector fields. An application, corresponding to the coupling of a Takens-Bogdanov and a saddle-node singularities, (in fact, it is a triple-zero singularity with geometric multiplicity two), that falls into the class considered, is analyzed.

Keywords Normal forms · Conservative-disipative splitting · Hamiltonian Homological operator · Lie operator · Quasi-homogeneous

A. Algaba · N. Fuentes · C. García
Department of Integrated Sciences, Investigation Center of Theoretical Physics and Mathematic FIMAT, Huelva University, Av. Fuerzas Armadas s/n, 21071 Huelva, Spain
e-mail: algaba@uhu.es

N. Fuentes
e-mail: natalia.fuentes@dmat.uhu.es

C. García
e-mail: cristoba@uhu.es

E. Gamero (✉)
Department of Applied Mathematic II, E.T.S.I. Sevilla University, Avd. Camino de los Descubridores s/n, Sevilla, Spain
e-mail: estanis@us.es

V. Carmona et al. (eds.), *Nonlinear Systems, Vol. 1*,
Understanding Complex Systems, https://doi.org/10.1007/978-3-319-66766-9_2

1 Introduction

A wide class of disciplines, among which we can mention electronics, mechanics, chemistry,... use magnitudes for describing certain type of phenomena which obey laws that can be expressed by means of dynamical systems.

In this chapter we focus on autonomous dynamical systems, that arise as models of systems whose laws do not change in time. Mathematically, they consist in a system of ordinary differential equations of the form

$$\dot{\mathbf{x}}(t) = \frac{d\mathbf{x}(t)}{dt} = \mathbf{F}(\mathbf{x}(t)), \tag{1}$$

where $\mathbf{x} = (x_1, x_2, \ldots, x_n) \in \mathbb{R}^n$ and the right-hand side does not explicitly depends on the independent variable t (usually called time).

Roughly, the aim of dynamical systems theory is determine the structure of the solutions set of these models. This study usually starts by simplifying the system (i.e,. by finding a simplification in the analytical expression of the vector field $\mathbf{F}(\mathbf{x})$). The most important of these simplifications is the reduction to *normal form*. The Normal Form Theory (also called *classic normal form*) was introduced by Poincare and was later developed by Dulac, Lyapunov, Birkhoff...

The basic idea of the Normal Form Theory is to use changes of variables to simplify the analytical expressions of a given vector field degree by degree, by removing the nonlinear terms which are non-essential in the dynamic behavior of the system. This procedure, called smooth conjugation, is the main subject of the present chapter.

There is another possibility of simplifying the analytical expression of the vector field by considering not only changes in the state variables but also transformations in time (this procedure is called smooth equivalence) that is considered in Algaba et al. [5, 7, 8].

We must mention three significant features concerning to the simplification procedure. Firstly, it is a local method. This means that the coordinate transformations are valid in a neighborhood of some solution, which we assume that is an *equilibrium point* (which is determined by the equation $\frac{d\mathbf{x}(t)}{dt} = \mathbf{F}(\mathbf{x}) = \mathbf{0}$).

Secondly, the coordinate transformations are usually nonlinear functions of the dependent variables. However, these coordinate transformations are obtained by solving a sequence of linear problems.

Finally, the structure of the normal form is determined by the principal part of the vector field (in the classical theory, the principal part reduces to the linear part).

Let us consider a quasi-homogeneous tridimensional system of some type $\mathbf{t} = (t_1, t_2, t_3)$ and degree r of the form

$$\begin{pmatrix} \dot{x} \\ \dot{y} \\ \dot{z} \end{pmatrix} = \begin{pmatrix} -\frac{\partial h}{\partial y}(x, y) \\ \frac{\partial h}{\partial x}(x, y) \\ f(x, y) \end{pmatrix}$$

where h and f are quasi-homogeneous scalar functions of type $\hat{\mathbf{t}} = (t_1, t_2)$ and degrees $r + t_1 + t_2$ and $r + t_3$, respectively. We observe that the vector field of this system is independent on z and has zero divergence. In this chapter we analyze normal forms for quasi-homogeneous higher-order perturbations of the above system:

$$\begin{pmatrix} \dot{x} \\ \dot{y} \\ \dot{z} \end{pmatrix} = \begin{pmatrix} -\frac{\partial h}{\partial y}(x, y) + F(x, y, z) \\ \frac{\partial h}{\partial x}(x, y) + G(x, y, z) \\ f(x, y) + H(x, y, z) \end{pmatrix}, \tag{2}$$

where $(F, G, H)^T$ contains the higher-order quasi-homogeneous terms of type $\mathbf{t} = (t_1, t_2, t_3)$.

This kind of systems embeds some interesting situations.

For instance, the normal form for the non-degenerate Hopf-zero singularity can be obtained following our approach if we use the type $\mathbf{t} = (1, 1, 2)$, and write the singularity as

$$\begin{pmatrix} \dot{x} \\ \dot{y} \\ \dot{z} \end{pmatrix} = \begin{pmatrix} -y + F(x, y, z) \\ x + G(x, y, z) \\ x^2 + y^2 + H(x, y, z) \end{pmatrix},$$

where $(F, G, H)^T$ contains higher-oder quasi-homogeneous terms of the quoted type. Notice that it corresponds to $2h = f = x^2 + y^2$ (which is quasi-homogeneous of type $\hat{\mathbf{t}} = (1, 1)$ and degree $r = 2$). This case has been analyzed in Algaba et al. [1], Chen et al. [16, 17] and Gazor and Mokhtari [22].

Also, the normal form for a triple-zero singularity with geometric multiplicity two (that corresponds to the coupling of a Takens-Bogdanov and a saddle-node singularities) can be obtained following our approach. It is enough to use the type $\mathbf{t} = (2, 3, 5)$ and write the singularity as

$$\begin{pmatrix} \dot{x} \\ \dot{y} \\ \dot{z} \end{pmatrix} = \begin{pmatrix} y + F(x, y, z) \\ x^2 + G(x, y, z) \\ x^3/3 - y^2/2 + H(x, y, z) \end{pmatrix}.$$

Now, we have $h = f = x^3/3 - y^2/2$, that is quasi-homogeneous of type $\hat{\mathbf{t}} = (2, 3)$ and degree 6. This case will be analyzed in Sect. 6.

We have structured this chapter as follows. In the next section, we summarize the main concepts of the classical Normal Form Theory. Later, in Sect. 3, we extend the ideas for vector fields developed in quasi-homogeneous terms. Section 4 is devoted to describe several decompositions of quasi-homogeneous vector fields, that we use which allow to calculate normal forms for planar and tridimensional systems. The main goal of this chapter is considered in the Sect. 5, where we determine normal forms for tridimensional systems whose principal part does not depend on the last variable and has free divergence. Finally, as commented before, in Sect. 6 we consider a case that falls in this situation, corresponding to a triple-zero singularity.

2 Classical Normal Forms

The properties and concepts that are presented below are known and can be seen, for more details, in Chua and Kokubu [20] and Golubitsky and Shaeffer [23].

Definition 1 Let $\mathscr{H}_k^n$ be the vectorial space of n-dimensional homogeneous polynomial vector fields n variables of degree k. We define the *Lie Bracket* of two differentiable vector fields $\mathbf{F}$ and $\mathbf{G}$ as

$$[\mathbf{F}, \mathbf{G}](\mathbf{x}) = D\mathbf{F}(\mathbf{x})\,\mathbf{G}(\mathbf{x}) - D\mathbf{G}(\mathbf{x})\,\mathbf{F}(\mathbf{x}), \quad \text{for all } x \in \mathbb{R}^n.$$

Moreover, it is a simple matter to show that, given $\mathbf{F} \in \mathscr{H}_i^n$ and $\mathbf{G} \in \mathscr{H}_j^n$, then $[\mathbf{F}, \mathbf{G}] \in \mathscr{H}_{i+j}^n$.

The Lie product has the following properties.

1. **Bilinearity**. If $a_1, a_2, b_1, b_2 \in \mathbb{R}$ and $\mathbf{F}_1, \mathbf{F}_2, \mathbf{G}_1, \mathbf{G}_2 \in \mathscr{H}_j^n$, then

$$\begin{aligned}[a_1\mathbf{F}_1 + a_2\mathbf{F}_2, b_1\mathbf{G}_1 + b_2\mathbf{G}_2] =& a_1b_1[\mathbf{F}_1, \mathbf{G}_1] + a_1b_2[\mathbf{F}_1, \mathbf{G}_2] + a_2b_1[\mathbf{F}_2, \mathbf{G}_1] \\ &+ a_2b_2[\mathbf{F}_2, \mathbf{G}_2].\end{aligned}$$

2. **Antisymmetry**.

$$[\mathbf{F}, \mathbf{G}] = -[\mathbf{G}, \mathbf{F}], \quad \text{with } \mathbf{F},\ \mathbf{G} \in \mathscr{H}_j^n.$$

3. **Jacobi Identity**.

$$[[\mathbf{F}, \mathbf{G}], \mathbf{H}] + [[\mathbf{G}, \mathbf{H}], \mathbf{F}] + [[\mathbf{H}, \mathbf{F}], \mathbf{G}] = 0, \quad \text{with } \mathbf{F},\ \mathbf{G},\ \mathbf{H} \in \mathscr{H}_j^n.$$

Next, we present the basic ideas of the classical theory of normal forms (see Takens [30], Chow and Hale [18], Guckenheimer and Holmes [24], Elphick et al. [21], Iooss and Adelmeyer [25] and Chow et al. [19]). Roughly, it consist into simplifying the analytical expression of a vector field degree by degree through changes of variables. Hence, let us assume that the vector field $\mathbf{F}$ of system (1) is written in homogeneous components, by its Taylor expansion. Then, the quoted system is expressed as

$$\dot{\mathbf{x}} = A\mathbf{x} + \mathbf{F}_2(\mathbf{x}) + \mathbf{F}_3(\mathbf{x}) + \cdots, \tag{3}$$

where $A = D\mathbf{F}(0)$ is the Jacobian matrix at the origin of $\mathbf{F}(\mathbf{x})$ and $\mathbf{F}_k(\mathbf{x})$ represents the k-degree terms of the Taylor expansion of $\mathbf{F}(\mathbf{x})$.

The starting point in the simplification procedure consist into simplifying the lowest degree term, using linear transformations. Let T be the matrix that transforms $A = D\mathbf{F}(\mathbf{0})$ into real Jordan canonical form J. Then, the linear transformation $\mathbf{x} = T\tilde{\mathbf{x}}$ brings system (3) into

$$\dot{\tilde{\mathbf{x}}} = T^{-1}AT\tilde{\mathbf{x}} + T^{-1}\mathbf{F}_2(T\tilde{\mathbf{x}}) + T^{-1}\mathbf{F}_3(T\tilde{\mathbf{x}}) + \cdots = J\tilde{\mathbf{x}} + \tilde{\mathbf{F}}_2(\tilde{\mathbf{x}}) + \tilde{\mathbf{F}}_3(\tilde{\mathbf{x}}) + \cdots, \tag{4}$$

where we have defined $\tilde{\mathbf{F}}_k(\tilde{\mathbf{x}}) \equiv T^{-1}\mathbf{F}_k(T\tilde{\mathbf{x}})$. Dropping the tildes, the above system can be written as

$$\dot{\mathbf{x}} = J\mathbf{x} + \mathbf{F}_2(\mathbf{x}) + \mathbf{F}_3(\mathbf{x}) + \cdots. \tag{5}$$

This first step in the simplification procedure is not essential. In fact, the further simplification procedure is applicable for other canonical forms for the matrix A (for instance, the Frobenius normal form could be selected) and even the initial matrix A could be chosen as it is.

Next, we fix $k \geq 2$ and use a near-identity transformation $\mathbf{x} = \mathbf{y} + \mathbf{P}_k(\mathbf{y})$, with $\mathbf{P}_k \in \mathcal{H}_k^n$, in order to simplify the terms of degree k. The transformed system is

$$\begin{aligned}\dot{\mathbf{y}} &= (I + D_\mathbf{y}\mathbf{P}_k(\mathbf{y}))^{-1}J(\mathbf{y} + \mathbf{P}_k(\mathbf{y})) + \sum_{k\geq 2}(I + D_\mathbf{y}\mathbf{P}_k(\mathbf{y}))^{-1}\mathbf{F}_k(\mathbf{x})(\mathbf{y} + \mathbf{P}_k(\mathbf{y}))\\ &= J\mathbf{y} + \sum_{k\geq 2}\mathbf{G}_k(\mathbf{y}),\end{aligned} \tag{6}$$

where $\mathbf{G}_k$ are the k-degree terms of the transformed system. It can be shown (see Guckenheimer and Holmes [24]), that this transformed vector field does not change up to order $k-1$, i.e.:

$$\mathbf{G}_2(\mathbf{y}) = \mathbf{F}_2(\mathbf{y}),\ \mathbf{G}_3(\mathbf{y}) = \mathbf{F}_3(\mathbf{y}),\ \ldots, \mathbf{G}_{k-1}(\mathbf{y}) = \mathbf{F}_{k-1}(\mathbf{y}). \tag{7}$$

Moreover, k-degree terms of the transformed vector field are:

$$\mathbf{G}_k(\mathbf{y}) = \mathbf{F}_k(\mathbf{y}) - (D_\mathbf{y}\mathbf{P}_k(\mathbf{y})J\mathbf{y} - J\mathbf{P}_k(\mathbf{y})). \tag{8}$$

This expression suggest to define the *homological operator*:

$$\begin{aligned}\mathbf{L}_k^J :\mathcal{H}_k^n &\longrightarrow \mathcal{H}_k^n\\ \mathbf{P}_k &\longrightarrow \mathbf{L}_k^J(\mathbf{P}_k) = D_\mathbf{x}\mathbf{P}_k(\mathbf{x})J\mathbf{x} - J\mathbf{P}_k(\mathbf{x}), = [\mathbf{P}_k,\ J](\mathbf{x}).\end{aligned} \tag{9}$$

It is easy to prove that $\mathbf{L}_k^J$ is linear. Therefore, we can write the k-degree terms of the transformed vector field (8) as

$$\mathbf{G}_k(\mathbf{y}) = \mathbf{F}_k(\mathbf{y}) - \mathbf{L}_k^J(\mathbf{P}_k). \tag{10}$$

We observe that we could achieve $\mathbf{G}_k = \mathbf{0}$ by selecting $\mathbf{P}_k$ such that $\mathbf{L}_k^J(\mathbf{P}_k) = \mathbf{F}_k$. Nevertheless, in general we can not eliminate all the k-degree terms since the above equation can be incompatible, but we can proceed as follows:

- We consider a complementary subspace $\mathrm{Cor}(\mathbf{L}_k^J)$ to the range of the *homological operator* in (9), i.e.: $\mathcal{H}_k^n = \mathrm{Range}(\mathbf{L}_k^J) \oplus \mathrm{Cor}(\mathbf{L}_k^J)$.

- We decompose $\mathbf{F}_k = \mathbf{F}_k^r + \mathbf{F}_k^c$, where $\mathbf{F}_k^r \in \text{Range}(\mathbf{L}_k^J)$ and $\mathbf{F}_k^c \in \text{Cor}(\mathbf{L}_k^J)$.
- We select $\mathbf{P}_k \in \mathscr{H}_k^n$ verifying the homological equation

$$\mathbf{L}_k^J(\mathbf{P}_k) = \mathbf{F}_k^r. \tag{11}$$

In this way, we obtain $\mathbf{G}_k = \mathbf{F}_k - \mathbf{L}_k^J(\mathbf{P}_k) = \mathbf{F}_k^c$. In other words, we have simplified $\mathbf{F}_k$ by eliminating the part belonging to the image of the homological operator.

Repeating this procedure for $k = 2, 3, 4, \ldots$ and using a version of *Borel's Theorem* (see Vanderbauwhede [32]), we obtain the normal form Theorem .

Theorem 1 *There exists a $\mathscr{C}^\infty$-diffeomorphism Φ verifying $\Phi(\mathbf{0}) = \mathbf{0}$ and $D\Phi(\mathbf{0}) = I$ such that the change of variables $\mathbf{x} = \Phi(\mathbf{y})$ transforms system* (5) *into* (6) *where $\mathbf{G}_k \in \text{Cor}(\mathbf{L}_k^J)$, for all $k \geq 2$. In this case, we say that* (6) *is a normal form for system* (3).

The homological equation (11) and, consequently, the corresponding normal form are based on the linear part of the vector field. This equation does not have, in general, an unique solution. Then, its solution will depend on arbitrary terms belonging to the kernel of the homological operator. These terms can be used later to make simplifications in the normal form terms of order higher than k (see Takens [30], Ushiki [31], Chua and Kokubu [20], Baider [13], Algaba et al. [1–4]) giving rise to simplified normal forms. We notice that these simplified normal forms are not only determined by the linear part, but they are also influenced by the nonlinear terms.

In the next subsection, we present a new approach that sometimes provides the simplified normal forms directly because the homological equations have unique solutions. Moreover, the new approach allows to work with systems having null linearization matrix, where the classical normal form theory does not provides any advantage.

3 Quasi-homogeneous Normal Forms

The use of quasi-homogeneous expansions of vector field instead Taylor expansions has several benefits. For instance, it allows to manage linear and nonlinear terms at once (because monomials with different degrees may have the same quasi-homogeneous degree).

Moreover, the linear part does not play a predominant role because instead we use the principal part of the vector field.

In this section, we extend the ideas of the Normal Form Theory for vector fields expanded in quasi-homogeneous terms. These ideas have been used in the case of the Bogdanov-Takens singularity (see Baider and Sanders [14], Kokubu et al. [26], Wang et al. [33]), Lombardi and Stolovich [27] and Strozyna and Zoladek [29]) and in the case of degenerate vector field (see Algaba et al. [6, 7, 10, 11], Basov and Slutskaya [15] and Strozyna [28]).

We start with some definitions and properties about quasi-homogeneity that we will use later. For more details, see Algaba et al. [5].

3.1 Some Properties About Quasi-homogeneity

Let consider a *type* $\mathbf{t} = (t_1, t_2, \ldots, t_n) \in \mathbb{N}^n$ (here, $\mathbb{N}$ is the set of natural numbers not including zero, whereas $\mathbb{N}_0$ will denote the set of natural numbers including zero). We define its module as $|\mathbf{t}| = t_1 + t_2 + \cdots + t_n$. We use standard multi-index notations: a *multi-index* is an element $\mathbf{a} = (a_1, a_2, \ldots, a_n) \in \mathbb{N}_0^n$. We write the monomials as $\mathbf{x}^{\mathbf{a}} = x_1^{a_1}, \ldots, x_n^{a_n}$, the canonical basis of $\mathbb{R}^n$ is denoted by $\{\mathbf{e}_1, \ldots, \mathbf{e}_n\}$, and the canonical basis of polynomial vector fields by

$$\mathscr{B} = \left\{\mathbf{x}^{\mathbf{a}}\mathbf{e}_j : \mathbf{a} \in \mathbb{N}_0^n, 1 \leq j \leq n\right\}.$$

Here we perform a *formal* analysis of normal forms, which means that we will not address any question about the convergence of the expansions.

Definition 2 A scalar function f is quasi-homogeneous of type $\mathbf{t}$ and degree k if its monomials $\mathbf{x}^{\mathbf{a}}$ satisfy

$$\mathbf{a} \cdot \mathbf{t} = a_1 t_1 + a_2 t_2 + \cdots + a_n t_n = k. \tag{12}$$

The vector space of quasi-homogeneous polynomials in n variables of type $\mathbf{t}$ and degree k is denoted by $\mathcal{P}_k^{\mathbf{t}}$.

A vector field $\mathbf{F} = (F_1, F_2, \ldots, F_n)$ is quasi-homogeneous of type $\mathbf{t}$ and degree k if its components $F_j \in \mathcal{P}_{k+t_j}^{\mathbf{t}}$, for all $j = 1, 2, \ldots, n$. We denote $\mathcal{Q}_k^{\mathbf{t}}$ the vector space of quasi-homogeneous vector fields of type $\mathbf{t}$ and degree k.

Next result present an alternative characterization for quasi-homogeneous functions and vector fields. Let us consider the diagonal matrix

$$E = \operatorname{diag}\left(\varepsilon^{t_1}, \varepsilon^{t_2}, \ldots, \varepsilon^{t_n}\right).$$

Proposition 1 *(a) The function f is quasi-homogeneous of type $\mathbf{t}$ and degree k if and only if*

$$f(E\mathbf{x}) = \varepsilon^k f(\mathbf{x}). \tag{13}$$

(b) The vector field $\mathbf{F}$ is quasi-homogeneous of type $\mathbf{t}$ and degree k if and only if

$$\mathbf{F}(E\mathbf{x}) = \varepsilon^k E\mathbf{F}(\mathbf{x}). \tag{14}$$

Proof (a) If $f \in \mathcal{P}_k^{\mathbf{t}}$, then

$$f(\mathbf{x}) = \sum_{\substack{\mathbf{a}\in\mathbb{N}_0^n \\ \mathbf{a}\cdot\mathbf{t}=k}} \alpha_{\mathbf{a}} \mathbf{x}^{\mathbf{a}} = \sum_{\substack{\mathbf{a}\in\mathbb{N}_0^n \\ \mathbf{a}\cdot\mathbf{t}=k}} \alpha_{\mathbf{a}} x_1^{a_1} x_2^{a_2} \dots x_n a_n.$$

In consequence:

$$\begin{aligned} f(E\mathbf{x}) &= f\left(\varepsilon^{t_1}x_1, \varepsilon^{t_2}x_2, \dots, \varepsilon^{t_n}x_n\right) = \sum_{\substack{\mathbf{a}\in\mathbb{N}_0^n \\ \mathbf{a}\cdot\mathbf{t}=k}} \alpha_{\mathbf{a}} (\varepsilon^{t_1}x_1)^{a_1} (\varepsilon^{t_2}x_2)^{a_2} \dots (\varepsilon^{t_n}x_n)^{a_n} \\ &= \sum_{\substack{\mathbf{a}\in\mathbb{N}_0^n \\ \mathbf{a}\cdot\mathbf{t}=k}} \alpha_{\mathbf{a}} \varepsilon^{t_1 a_1} \varepsilon^{t_2 a_2} \dots \varepsilon^{t_n a_n} x_1^{a_1} x_2^{a_2} \dots x_n^{a_n} = \varepsilon^k \sum_{\substack{\mathbf{a}\in\mathbb{N}_0^n \\ \mathbf{a}\cdot\mathbf{t}=k}} \alpha_{\mathbf{a}} \mathbf{x}^{\mathbf{a}} = \varepsilon^k f(\mathbf{x}). \end{aligned}$$

The converse can be proven analogously.

(b) Note that $\mathbf{F} = (F_1, F_2, \dots, F_n)^T \in \mathcal{Q}_k^{\mathbf{t}}$ if, and only if, $F_j \in \mathcal{P}_{k+t_j}$. Using item **(a)**, the proof can be easily completed.

Next, we show the behavior of the Lie product for quasi-homogeneous vector fields. The quoted result requires a previous lemma:

Lemma 1 *Let us consider* $\mathbf{F} \in \mathcal{Q}_k^{\mathbf{t}}$. *Then:* $D\mathbf{F}(E\mathbf{x}) = \varepsilon^k E D\mathbf{F}(\mathbf{x}) E^{-1}$.

Moreover, the j*-th column of the matrix* $D\mathbf{F}$ *is a quasi-homogeneous vector field of degree* $k - t_j$ *with respect to the type* $\mathbf{t}$.

Proof Differentiating (14) with respect to x, we obtain the equality.

In addition, the column j of $D\mathbf{F}(\mathbf{x})$ is given by $D\mathbf{F}(\mathbf{x})\mathbf{e}_j$. Then:

$$D\mathbf{F}(E\mathbf{x})\mathbf{e}_j = \varepsilon^k E D\mathbf{F}(\mathbf{x}) E^{-1}\mathbf{e}_j = \varepsilon^k E D\mathbf{F}(\mathbf{x}) \varepsilon^{-t_j}\mathbf{e}_j = \varepsilon^{k-t_j} E D\mathbf{F}(\mathbf{x})\mathbf{e}_j.$$

The result follows from Proposition 1(b).

Proposition 2 *Let us consider* $\mathbf{F} \in \mathcal{Q}_k^{\mathbf{t}}$ *and* $\mathbf{G} \in \mathcal{Q}_l^{\mathbf{t}}$. *Then,* $[\mathbf{F}, \mathbf{G}] \in \mathcal{Q}_{k+l}^{\mathbf{t}}$.

Proof From Lemma 1, we obtain $D\mathbf{F}(E\mathbf{x}) = \varepsilon^k E D\mathbf{F}(\mathbf{x}) E^{-1}$ and $D\mathbf{G}(E\mathbf{x}) = \varepsilon^l E D\mathbf{G}(\mathbf{x}) E^{-1}$. Then:

$$\begin{aligned} [\mathbf{F}, \mathbf{G}](E\mathbf{x}) &= D\mathbf{F}(E\mathbf{x})\mathbf{G}(E\mathbf{x}) - D\mathbf{G}(E\mathbf{x})\mathbf{F}(E\mathbf{x}) \\ &= \varepsilon^l E D\mathbf{F}(\mathbf{x}) E^{-1} \varepsilon^k E\mathbf{G}(\mathbf{x}) - \varepsilon^k E D\mathbf{G}(\mathbf{x}) E^{-1} \varepsilon^l E\mathbf{F}(\mathbf{x}) \\ &= \varepsilon^{k+l} E\left(D\mathbf{F}(\mathbf{x})\mathbf{G}(\mathbf{x}) - D\mathbf{G}(\mathbf{x})\mathbf{F}(\mathbf{x})\right) = \varepsilon^{k+l} E[\mathbf{F}, \mathbf{G}](\mathbf{x}). \end{aligned}$$

Using Proposition 1(b), we complete the proof.

Next result is a version of *Euler's theorem* in the quasi-homogeneous case. Let us denote $\mathbf{D}_0 = (t_1x_1, t_2x_2, \dots, t_nx_n)^T \in \mathcal{Q}_0^{\mathbf{t}}$.

Lemma 2 *Let us consider* $f \in \mathcal{P}_k^{\mathbf{t}}$. *Then:* $\nabla f \cdot \mathbf{D}_0 = kf$.

Proof As $f \in \mathcal{P}_k^{\mathbf{t}}$, we have $f(E\mathbf{x}) = \varepsilon^k f(\mathbf{x})$. Differentiating with respect to ε,we obtain

$$\nabla f(E\mathbf{x}) \cdot (D_\varepsilon E)\mathbf{x} = k\varepsilon^{k-1} f(\mathbf{x}).$$

Taking $\varepsilon = 1$, we obtain the result.

Definition 3 The divergence of the vector field $\mathbf{F}$ is

$$\operatorname{div}(\mathbf{F}) := \frac{\partial F_1}{\partial x_1} + \frac{\partial F_2}{\partial x_2} + \cdots + \frac{\partial F_n}{\partial x_n},$$

i.e., it is the trace of matrix $D\mathbf{F}(\mathbf{x})$: $\operatorname{tr}(D\mathbf{F}(\mathbf{x}))$.

Lemma 3 *Let us consider* $\mathbf{F} \in \mathcal{Q}_k^{\mathbf{t}}$. *Then:* $\operatorname{div}(\mathbf{F}) \in \mathcal{P}_k^{\mathbf{t}}$.

Proof From Lemma 1, we have:

$$\operatorname{div}(\mathbf{F})(E\mathbf{x}) = \operatorname{tr}(D\mathbf{F}(E\mathbf{x})) = \operatorname{tr}\left(\varepsilon^k E D\mathbf{F}(\mathbf{x})E^{-1}\right) = \varepsilon^k \operatorname{tr}(D\mathbf{F}(\mathbf{x})) = \varepsilon^k \operatorname{div}(\mathbf{F})(\mathbf{x}).$$

Using Proposition 1(b), we complete the proof.

Lemma 4 *Let us consider* $\mathbf{F} \in \mathcal{Q}_k^{\mathbf{t}}$. *Then:* $[\mathbf{F}, \mathbf{D}_0] = k\mathbf{F}$.
In particular, if $\mathbf{F} \in \mathcal{Q}_0^{\mathbf{t}}$, *then* $[\mathbf{F}, \mathbf{D}_0] = \mathbf{0}$.

Proof Consider $j = 1, \ldots, n$. Applying Lemma 2 , we obtain the following expression for the j-th component of the Lie bracket $[\mathbf{F}, \mathbf{D}_0]$:

$$[\mathbf{F}, \mathbf{D}_0] \cdot \mathbf{e}_j = \nabla \mathbf{F} \cdot \mathbf{e}_j \, \mathbf{D}_0 - \nabla \mathbf{D}_0 \cdot \mathbf{e}_j \, \mathbf{F} = (k + t_j)\mathbf{F} \cdot \mathbf{e}_j - t_j \mathbf{F} \cdot \mathbf{e}_j = k\mathbf{F}_j.$$

3.2 Quasi-homogeneous Normal Form for Vector Fields

To adapt the procedure for determining normal forms using quasi-homogeneous expansions, we first explain how obtain such expansion for system (1).

Let us include a parameter ε by the scaling $\mathbf{x} = E\tilde{x}$, with $\tilde{x} \in \mathbb{R}^n$. Thus, we get the system $\dot{\tilde{x}} = E^{-1}\mathbf{F}(E\tilde{x})$. Developing in powers of ε, we can write this system in the form

$$\dot{\tilde{x}} = \mathbf{F}_r(\tilde{x})\varepsilon^r + \mathbf{F}_{r+1}(\tilde{x})\varepsilon^{r+1} + \cdots,$$

where it is easy to prove that $\mathbf{F}_j \in \mathcal{Q}_j^{\mathbf{t}}$. Taking $\varepsilon = 1$, we write system (1) as a sum of quasi-homogeneous polynomials of type $\mathbf{t}$:

$$\dot{\mathbf{x}} = \mathbf{F}_r(\mathbf{x}) + \mathbf{F}_{r+1}(\mathbf{x}) + \cdots. \tag{15}$$

Definition 4 The lowest-degree nonzero term $\mathbf{F}_r$ in the quasi-homogeneous expansion (15) is called the *principal part* of the vector field $\mathbf{F}$ with respect to the type $\mathbf{t}$.

The quasi-homogeneous degree r of the principal part $\mathbf{F}_r$ is a integer number (i.e., in general $r \notin \mathbb{N}$). For example, let us consider a two-dimensional vector field with a nilpotent singularity:

$$\dot{x} = y + a_{20}x^2 + a_{11}xy + a_{02}y^2 + \cdots ,$$
$$\dot{y} = b_{20}x^2 + b_{11}xy + b_{02}y^2 + \cdots ,$$

and take the type $\mathbf{t} = (1, 3)$. It can be verified that the principal part of the above vector field $\mathbf{F}$ with respect to this type is $\mathbf{F}_r(x, y) = (0, b_{20}x^2)^T$, where $r = -1$.

It is a simple matter to show that the classical Taylor expansion is a particular case of quasi-homogeneous expansion that corresponds to the unity type $\mathbf{t} = (1, 1, \ldots, 1)$. In this case, there is a shift in the degree: linear terms have quasi-homogeneous degree 0, quadratic terms have quasi-homogeneous degree 1, and so on.

Next, we describe the process to obtain a normal form for system (15). Let us consider $k \geq 1$ and perform a near-identity transformation $\mathbf{x} = \mathbf{y} + \mathbf{P}_k(\mathbf{y})$, where $\mathbf{P}_k \in \mathcal{Q}_k^{\mathbf{t}}$. The transformed system is

$$\dot{\mathbf{y}} = \mathbf{G}(\mathbf{y}) = (I + D\mathbf{P}_k(\mathbf{y}))^{-1} \sum_{j \geq r} \mathbf{F}_j\,(\mathbf{y} + \mathbf{P}_k(\mathbf{y}))\,. \tag{16}$$

Let us consider the quasi-homogeneous expansion in terms of type $\mathbf{t}$ for the above system:

$$\dot{\mathbf{y}} = \mathbf{G}_r(\mathbf{y}) + \mathbf{G}_{r+1}(\mathbf{y}) + \cdots , \tag{17}$$

with $\mathbf{G}_k(\mathbf{y}) \in \mathcal{Q}_k^{\mathbf{t}}$, for all $k \geq r$. The following result holds:

Proposition 3 *With the above notation,*

- $\mathbf{G}_j = \mathbf{F}_j$, *for* $j = r, r+1, \ldots, r+k-1$,
- $\mathbf{G}_{r+k} = \mathbf{F}_{r+k} - (D\mathbf{P}_k\mathbf{F}_r - D\mathbf{F}_r\mathbf{P}_k) = \mathbf{F}_{r+k} - [\mathbf{P}_k, \mathbf{F}_r]$.

Proof Performing the scaling $\mathbf{y} = E\tilde{y}$, system (16) becomes $\dot{\tilde{y}} = E^{-1}\mathbf{G}(E\tilde{y})$. We observe that

$$\begin{aligned}
\mathbf{G}(E\mathbf{y}) &= (I + D\mathbf{P}_k(E\mathbf{y}))^{-1} \sum_{j \geq r} \mathbf{F}_j\,(E\mathbf{y} + \mathbf{P}_k(E\mathbf{y}))\\
&= \left(E\left(I + \varepsilon^k D\mathbf{P}_k(\mathbf{y})\right)E^{-1}\right)^{-1} \sum_{j \geq r} E\varepsilon^j \mathbf{F}_j\left(\mathbf{y} + \varepsilon^k \mathbf{P}_k(\mathbf{y})\right)\\
&= E\left(I - \varepsilon^k D\mathbf{P}_k(\mathbf{y}) + \varepsilon^{2k}\,(D\mathbf{P}_k(\mathbf{y}))^2 - \cdots\right) \sum_{j \geq r} \varepsilon^j \mathbf{F}_j\left(\mathbf{y} + \varepsilon^k \mathbf{P}_k(\mathbf{y})\right).
\end{aligned}$$

Moreover,

$$\mathbf{F}_j\left(\mathbf{y}+\varepsilon^k\mathbf{P}_k(\mathbf{y})\right)=\mathbf{F}_j(\mathbf{y})+D\mathbf{F}_j(\mathbf{y})\mathbf{P}_k(\mathbf{y})\varepsilon^k+\mathcal{O}\left(\varepsilon^{k+1}\right).$$

Therefore,

$$\begin{aligned}
E^{-1}\mathbf{G}(E\mathbf{y}) &= \left(I-\varepsilon^k D\mathbf{P}_k(\mathbf{y})+\varepsilon^{2k}\left(D\mathbf{P}_k(\mathbf{y})\right)^2-\cdots\right)\sum_{j\geq r}\varepsilon^j\mathbf{F}_j\left(\mathbf{y}+\varepsilon^k\mathbf{P}_k(\mathbf{y})\right)\\
&= \left(I-\varepsilon^k D\mathbf{P}_k(\mathbf{y})\right)\sum_{j\geq r}\varepsilon^j\left(\mathbf{F}_j(\mathbf{y})+D\mathbf{F}_j(\mathbf{y})\mathbf{P}_k(\mathbf{y})\varepsilon^k\right)+\mathcal{O}\left(\varepsilon^{r+k+1}\right)\\
&= \sum_{j\geq r}\varepsilon^j\left(\mathbf{F}_j(\mathbf{y})+D\mathbf{F}_j(\mathbf{y})\mathbf{P}_k(\mathbf{y})\varepsilon^k\right)-\varepsilon^{r+k}D\mathbf{P}_k(\mathbf{y})\mathbf{F}_r(\mathbf{y})+\mathcal{O}\left(\varepsilon^{r+k+1}\right)\\
&= \mathbf{F}_r(\mathbf{y})\varepsilon^r+\mathbf{F}_{r+1}(\mathbf{y})\varepsilon^{r+1}+\cdots+\mathbf{F}_{r+k-1}(\mathbf{y})\varepsilon^{r+k-1}\\
&\quad+\left(\mathbf{F}_{r+k}(\mathbf{y})+D\mathbf{F}_r(\mathbf{y})\mathbf{P}_k(\mathbf{y})-D\mathbf{P}_k(\mathbf{y})\mathbf{F}_r(\mathbf{y})\right)\varepsilon^{r+k}+\mathcal{O}\left(\varepsilon^{r+k+1}\right).
\end{aligned}$$

This result suggests to introduce the *homological operator*

$$\begin{aligned}
\mathbf{L}_{r+k}:&\mathcal{Q}_k^{\mathbf{t}}\longrightarrow\mathcal{Q}_{r+k}^{\mathbf{t}}\\
&\mathbf{P}_k\rightarrow\mathbf{L}_{r+k}(\mathbf{P}_k)=D\mathbf{P}_k\mathbf{F}_r-D\mathbf{F}_r\mathbf{P}_k=[\mathbf{P}_k,\mathbf{F}_r].
\end{aligned}\tag{18}$$

This is a linear operator that only depends on the principal part $\mathbf{F}_r$.

Proposition 3 states that the quasi-homogeneous terms up to order $r+k-1$ do not change, and the quasi-homogeneous term of degree $r+k$ in the transformed vector field is

$$\mathbf{G}_{r+k}=\mathbf{F}_{r+k}-[\mathbf{P}_k,\mathbf{F}_r]=\mathbf{F}_{r+k}-\mathbf{L}_{r+k}(\mathbf{P}_k).$$

Following the same ideas of the classical Normal Form Theory, we can achieve that $\mathbf{G}_{r+k}$ belongs to a complementary subspace to the range of the linear operator $\mathbf{L}_{r+k}$, simply by annihilating the part of $\mathbf{F}_{r+k}$ that belongs to the range of the linear operator $\mathbf{L}_{r+k}$, through an appropriate choice of $\mathbf{P}_k$.

Performing the near-identity transformations described above for $k=1,2,\ldots,$ we obtain the following result.

Theorem 2 *There exists a* $\mathcal{C}^\infty$*-diffeomorphism* Φ *verifying* $\Phi(\mathbf{0})=\mathbf{0}$ *and* $D\Phi(\mathbf{0})=I$ *such that the change of variables* $\mathbf{x}=\Phi(\mathbf{y})$ *transforms system* (15) *into* (17) *where* $\mathbf{G}_r=\mathbf{F}_r$ *and* $\mathbf{G}_{r+k}\in\mathrm{Cor}(\mathbf{L}_{r+k})$, *a complementary subspace to the range of the linear operator* $\mathbf{L}_{r+k}$, *for all* $k\geq 1$. *In this case, we say that* (6) *is a quasi-homogeneous normal form for system* (3) *corresponding to type* $\mathbf{t}$.

4 Decompositions of Quasi-homogeneous Vector Fields

Let us introduce some definitions.

Definition 5 We denote the Hamiltonian vector field corresponding to a Hamilton function h by

$$\mathbf{X}_h := \left(-\frac{\partial h}{\partial y}, \frac{\partial h}{\partial x}\right)^T.$$

It is easy to show that $\mathbf{X}_h \in \mathcal{Q}_k^t$ if, and only if $h \in \mathcal{P}^t_{k+|\hat{\mathbf{t}}|}$.

We define the wedge product of two vector fields $\mathbf{F} = (P, Q)^T$ and $\mathbf{G} = (\tilde{P}, \tilde{Q})^T$ by $\mathbf{F} \wedge \mathbf{G} := P\tilde{Q} - Q\tilde{P}$. We notice that if $\mathbf{F} \in \mathcal{Q}_k^{\mathbf{t}}$, $\mathbf{G} \in \mathcal{Q}_l^{\mathbf{t}}$, then $\mathbf{F} \wedge \mathbf{G} \in \mathcal{P}^{\mathbf{t}}_{k+l+|\mathbf{t}|}$.

Recall that we are interested in obtaining normal forms for the tridimensional system (2), whose principal part is

$$\mathbf{F}_r(x, y) = \begin{pmatrix} -\frac{\partial h}{\partial y}(x, y) \\ \frac{\partial h}{\partial x}(x, y) \\ f(x, y) \end{pmatrix} = \begin{pmatrix} \mathbf{X}_h(x, y) \\ f(x, y) \end{pmatrix} \in \mathcal{Q}_r^{\mathbf{t}}. \tag{19}$$

This means that

$$\mathbf{X}_h \in \mathcal{Q}_r^{\hat{\mathbf{t}}} \left(\text{i.e., } h \in \mathcal{P}^{\hat{\mathbf{t}}}_{r+|\hat{\mathbf{t}}|}\right), \text{ and } f \in \mathcal{P}^{\hat{\mathbf{t}}}_{r+t_3}.$$

In this section we present two decompositions for planar quasi-homogeneous vector fields that allow us generate a new decomposition for three-dimensional vector fields.

Looking at the principal part (19), we observe that we need to work with two and three-dimensional quasi-homogeneous vector fields. As we are using at once functions of two and three variables, we need to distinguish its quasi-homogeneity type. For this reason, given the type $\mathbf{t} = (t_1, t_2, t_3)$, we will denote $\hat{\mathbf{t}} = (t_1, t_2)$, that will appear when we deal with functions depending on two variables. In the same way, we denote the planar gradient by $\hat{\nabla} := \left(\frac{\partial}{\partial x}, \frac{\partial}{\partial y}\right)^T$ in order to distinguish it from the tridimensional gradient $\nabla := \left(\frac{\partial}{\partial x}, \frac{\partial}{\partial y}, \frac{\partial}{\partial z}\right)^T$.

4.1 Decompositions of Planar Quasi-homogeneous Vector Fields

Given a type $\hat{\mathbf{t}}$, any planar quasi-homogenous vector field can be decomposed uniquely as the sum of two quasi-homogeneous vector field: one of them having zero divergence (*conservative part*) and the other one with divergence equal to the original vector field (*dissipative part*). This decomposition, called *conservative-dissipative splitting*, can be seen in more details in Algaba et al. [12].

Proposition 4 *Let us assume that* $\mathbf{P}_k \in \mathcal{Q}_k^{\hat{\mathbf{t}}}$. *Then, there exist unique polynomials* $\mu_k \in \mathcal{P}_k^{\hat{\mathbf{t}}}$ *and* $h_{k+|\hat{\mathbf{t}}|} \in \mathcal{P}_{k+|\hat{\mathbf{t}}|}^{\hat{\mathbf{t}}}$ *such that:*

$$\mathbf{P}_k = \mathbf{X}_{h_{k+|\hat{\mathbf{t}}|}} + \mu_k \mathbf{D}_0, \tag{20}$$

where $h_{k+|\hat{\mathbf{t}}|} = \frac{1}{k+|\hat{\mathbf{t}}|}(\mathbf{D}_0 \wedge \mathbf{P}_k)$ *and* $\mu_k = \frac{1}{k+|\hat{\mathbf{t}}|}\mathrm{div}(\mathbf{P}_k)$.

Proof First, we prove the unicity. Let us suppose that there are $\mu \in \mathcal{P}_k^t$ and $h \in \mathcal{P}_{k+|t|}^t$ verifying (20). Then:

$$\begin{aligned}
\mathrm{div}(\mathbf{P}_k) &= \mathrm{div}(\mathbf{X}_h) + \mathrm{div}(\mu\mathbf{D}_0) = 0 + \frac{\partial}{\partial x}(\mu t_1 x) + \frac{\partial}{\partial y}(\mu t_2 y) \\
&= \frac{\partial \mu}{\partial x} t_1 x + + \frac{\partial \mu}{\partial y} t_2 y + \mu|t| = \hat{\nabla}\mu \cdot \mathbf{D}_0 + \mu|\hat{\mathbf{t}}| = \mu \cdot k + \mu|\hat{\mathbf{t}}| = (k + |\hat{\mathbf{t}}|)\mu.
\end{aligned}$$

$$\mathbf{D}_0 \wedge \mathbf{P}_k = \mathbf{D}_0 \wedge \mathbf{X}_h = \hat{\nabla} h \cdot \mathbf{D}_0 = (k + |\hat{\mathbf{t}}|)h,$$

(above, we have used Lemma 2).

Next, we prove the existence. Using again Lemma 2, we obtain the following expression for the first component of the vector field $\mathbf{P}_k$:

$$\begin{aligned}
(\mathbf{X}_h + \mu\mathbf{D}_0) \cdot \mathbf{e}_1 &= -\frac{\partial h}{\partial y} + \mu t_1 x = \frac{t_2 \mathbf{P}_k \cdot \mathbf{e}_1 + (k + t_1)\mathbf{P}_k \cdot \mathbf{e}_1}{k + |\hat{\mathbf{t}}|} \\
&= \frac{k + t_1 + t_2}{k + |\hat{\mathbf{t}}|}\mathbf{P}_k \cdot \mathbf{e}_1 = \mathbf{P}_k \cdot \mathbf{e}_1.
\end{aligned}$$

The result for the second component follows analogously.

Next lemmas analyze the conservative-dissipative splitting for some kind of vector fields, namely the Lie product of two quasi-homogeneous Hamiltonian vector fields and the product of a quasi-homogeneous scalar function and a quasi-homogeneous Hamiltonian vector field.

Lemma 5 *Let us consider* $p \in \mathcal{P}_k^{\hat{\mathbf{t}}}$, $p \in \mathcal{P}_e^{\hat{\mathbf{t}}}$. *Then:*

(a) $[\mathbf{X}_p, \mathbf{X}_q] = \mathbf{X}_f$, *with* $f = \hat{\nabla} p \cdot \mathbf{X}_q \in \mathcal{P}_{k+l}^{\hat{\mathbf{t}}}$.
(b) $p\mathbf{X}_q = \mathbf{X}_h + \mu\mathbf{D}_0$, *with* $h = \frac{l+|\hat{\mathbf{t}}|}{k+l+|\hat{\mathbf{t}}|}pq$ *and* $\mu = \frac{1}{k+l+|\hat{\mathbf{t}}|}\hat{\nabla} p\mathbf{X}_q$.

Proof (a) We have:

$$\begin{aligned}
[\mathbf{X}_p, \mathbf{X}_q] &= \begin{pmatrix} p_{yx}q_y - p_{yy}q_x - q_{yx}p_y + q_{yy}p_x \\ -p_{xx}q_y + p_{xy}q_x + q_{xx}p_y - q_{xy}p_x \end{pmatrix} = \begin{pmatrix} \frac{\partial}{\partial y}(p_x q_y - p_y q_x) \\ -\frac{\partial}{\partial x}(p_x q_y - p_y q_x) \end{pmatrix} \\
&= -\mathbf{X}_{p_x q_y - p_y q_x} = \mathbf{X}_{\hat{\nabla} p \cdot \mathbf{X}_q}.
\end{aligned}$$

(b) From Proposition 4, the conservative part of $p\mathbf{X}_q$ is

$$h = \frac{1}{k+l+|\hat{\mathbf{t}}|}[\mathbf{D}_0 \wedge (\mathbf{X}_h + \mu \mathbf{D}_0)] = \frac{1}{k+l+|\hat{\mathbf{t}}|}[\mathbf{D}_0 \wedge (p\mathbf{X}_q)] = \frac{l+|\hat{\mathbf{t}}|}{k+l+|\hat{\mathbf{t}}|} pq,$$

and the dissipative part is

$$\mu = \frac{1}{k+l+|\hat{\mathbf{t}}|}\operatorname{div}(\mathbf{X}_h + \mu \mathbf{D}_0) = \frac{1}{k+l+|\hat{\mathbf{t}}|}\operatorname{div}(p\mathbf{X}_q) = \frac{1}{k+l+|\hat{\mathbf{t}}|}\hat{\nabla} p\mathbf{X}_q.$$

Next, we present a new decomposition of planar quasi-homogeneous vector fields (for more details, see Algaba et al. [8]). To this end, we show that the space $\mathcal{Q}_k^{\hat{\mathbf{t}}}$ can be decomposed as a direct sum of three subspaces. Let us define

$$h\mathcal{P}_{k-r}^{\hat{\mathbf{t}}} = \{h(x, y)\gamma(x, y) \in \mathcal{P}_{k+|\hat{\mathbf{t}}|}^{\hat{\mathbf{t}}} : \gamma \in \mathcal{P}_{k-r}^{\hat{\mathbf{t}}}\}, \tag{21}$$

and let denote by $\Delta_{k+|\hat{\mathbf{t}}|}$ a complementary subspace of $h\mathcal{P}_{k-r}^{\hat{\mathbf{t}}}$, i.e.:

$$\mathcal{P}_{k+|\hat{\mathbf{t}}|}^{\hat{\mathbf{t}}} = \Delta_{k+|\hat{\mathbf{t}}|} \oplus h\mathcal{P}_{k-r}^{\hat{\mathbf{t}}}.$$

Let us also define the subspaces

$$\mathscr{F}_k^{\hat{\mathbf{t}}} := \left\{\lambda\, \mathbf{F}_r \in \mathcal{Q}_k^{\hat{\mathbf{t}}} : \lambda \in \mathcal{P}_{k-r}^{\hat{\mathbf{t}}}\right\}.$$
$$\mathscr{D}_k^{\hat{\mathbf{t}}} := \left\{\eta\, \mathbf{D}_0 \in \mathcal{Q}_k^{\hat{\mathbf{t}}} : \eta \in \mathcal{P}_k^{\hat{\mathbf{t}}}\right\}.$$
$$\mathscr{C}_k^{\hat{\mathbf{t}}} := \left\{\mathbf{X}_g \in \mathcal{Q}_k^{\hat{\mathbf{t}}} : g \in \Delta_{k+|\hat{\mathbf{t}}|}\right\}.$$

The following propositions provides the quoted decomposition, and their proof can be found in Algaba et al. [8].

Proposition 5 *Let us assume that* $\mathbf{F}_r = \mathbf{X}_h + \mu \mathbf{D}_0$ *and* $h \in \mathcal{P}_{r+|\hat{\mathbf{t}}|}^{\hat{\mathbf{t}}} \setminus \{0\}$. *Then*

$$\mathcal{Q}_k^{\hat{\mathbf{t}}} = \mathscr{C}_k^{\hat{\mathbf{t}}} \oplus \mathscr{D}_k^{\hat{\mathbf{t}}} \oplus \mathscr{F}_k^{\hat{\mathbf{t}}}.$$

Moreover, given $\mathbf{P}_k \in \mathcal{Q}_k^{\hat{\mathbf{t}}}$, *there exist unique polynomials* $g \in \Delta_{k+|\hat{\mathbf{t}}|}$, $\eta \in \mathcal{P}_k^{\hat{\mathbf{t}}}$ *and* $\lambda \in \mathcal{P}_{k-r}^{\hat{\mathbf{t}}}$, *such that*

$$\mathbf{P}_k = \mathbf{X}_g + \eta \mathbf{D}_0 + \lambda \mathbf{F}_r, \tag{22}$$

where

$$g = \frac{Proy_{\Delta_{k+|\hat{\mathbf{t}}|}}(\mathbf{D}_0 \wedge \mathbf{P}_k)}{k+|\hat{\mathbf{t}}|},$$
$$\lambda = \frac{Proy_{h\mathcal{P}^{\hat{\mathbf{t}}}_{k-r}}(\mathbf{D}_0 \wedge \mathbf{P}_k)}{(r+|\hat{\mathbf{t}}|)h},$$
$$\eta = \frac{\operatorname{div}(\mathbf{P}_k) - \hat{\nabla}\lambda\mathbf{F}_r - \lambda\operatorname{div}(\mathbf{F}_r)}{k+|\hat{\mathbf{t}}|}.$$

4.2 Decompositions of Three Dimensional Quasi-homogeneous Vector Fields

Next, we present a decomposition similar to (22), that applies to three-dimensional quasi-homogeneous vector fields. For this purpose, we define the following set:

$$h\mathcal{P}^{\mathbf{t}}_{k-r} = \{h(x,y)\gamma(x,y,z) \in \mathcal{P}^{\mathbf{t}}_{k+|\hat{\mathbf{t}}|} : \gamma \in \mathcal{P}^{\mathbf{t}}_{k-r}\},$$

(compare with to (21)), and denote by $\Delta_{k+|\hat{\mathbf{t}}|}$ a complementary subspace to $h\mathcal{P}^{\mathbf{t}}_{k-r}$ in $\mathcal{P}^{\mathbf{t}}_{k+|\hat{\mathbf{t}}|}$.

Let us introduce the following subspaces:

- $\mathscr{C}^{\mathbf{t}}_k = \left\{ \begin{pmatrix} \mathbf{X}_g \\ \hline 0 \end{pmatrix} \in \mathcal{Q}^{\mathbf{t}}_k \ : \ g \in \Delta_{k+|\hat{\mathbf{t}}|},\ g(0,0,z) = 0 \right\}$, where $\mathbf{X}_g = \begin{pmatrix} -\frac{\partial g(x,y,z)}{\partial y} \\ \frac{\partial g(x,y,z)}{\partial x} \end{pmatrix}$.
- $\mathscr{D}^{\mathbf{t}}_k = \left\{ \begin{pmatrix} \mu\mathbf{D}_0 \\ \hline 0 \end{pmatrix} \in \mathcal{Q}^{\mathbf{t}}_k \ : \ \mu \in \mathcal{P}^{\mathbf{t}}_k \right\}$.
- $\mathscr{F}^{\mathbf{t}}_k = \left\{ \begin{pmatrix} \lambda\mathbf{X}_h \\ \hline 0 \end{pmatrix} \in \mathcal{Q}^{\mathbf{t}}_k \ : \ \lambda \in \mathcal{P}^{\mathbf{t}}_{k-r} \right\}$.
- $\mathscr{G}^{\mathbf{t}}_k = \left\{ \begin{pmatrix} 0 \\ \hline \varsigma \end{pmatrix} \in \mathcal{Q}^{\mathbf{t}}_k \ : \ \varsigma \in \mathcal{P}^{\mathbf{t}}_{k+t_3} \right\}$.

Next proposition generalizes the decomposition given in Proposition 5 for three-dimensional vector fields. The proof can be can be found in Algaba et al. [8].

Proposition 6 *Let us consider* $\mathbf{F}_r \in \mathcal{Q}^{\mathbf{t}}_r$ *given in* (19)*, where* $h \not\equiv 0$*. Then*

$$\mathcal{Q}^{\mathbf{t}}_k = \mathscr{C}^{\mathbf{t}}_k \oplus \mathscr{D}^{\mathbf{t}}_k \oplus \mathscr{F}^{\mathbf{t}}_k \oplus \mathscr{G}^{\mathbf{t}}_k.$$

Moreover, given $\mathbf{P}_k \in \mathcal{Q}^{\mathbf{t}}_k$*, there exist unique polynomials* $g_{k+|\hat{\mathbf{t}}|} \in \Delta_{k+|\hat{\mathbf{t}}|} \subset \mathcal{P}^{\mathbf{t}}_{k+|\hat{\mathbf{t}}|}$*,* $\mu_k \in \mathcal{P}^{\mathbf{t}}_k$*,* $\lambda_{k-r} \in \mathcal{P}^{\mathbf{t}}_{k-r}$*,* $\varsigma_{k+t_3} \in \mathcal{P}^{\mathbf{t}}_{k+t_3}$*, with* $g_{k+|\hat{\mathbf{t}}|}(0,0,z) = 0$*, such that:*

$$\mathbf{P}_k = \begin{pmatrix} \mathbf{X}_{g_{k+|\hat{\mathbf{t}}|}} \\ \hline 0 \end{pmatrix} + \begin{pmatrix} \mu_k\mathbf{D}_0 \\ \hline 0 \end{pmatrix} + \begin{pmatrix} \lambda_{k-r}\mathbf{X}_h \\ \hline 0 \end{pmatrix} + \begin{pmatrix} \mathbf{0} \\ \hline \varsigma_{k+t_3} \end{pmatrix}.$$

Next lemma, that we state without proof (it can be found in Algaba et al. [8]) will be used later in order to obtain a matrix representation for the homological operator.

Lemma 6 *The following properties hold:*

1. *If* $g_{k+|\hat{\mathbf{t}}|} \in \Delta_{k+|\hat{\mathbf{t}}|}$, *with* $g_{k+|\hat{\mathbf{t}}|}(0,0,z) = 0$, *then*

$$\left[\begin{pmatrix}\mathbf{X}_{g_{k+|\hat{\mathbf{t}}|}}\\ \hline 0\end{pmatrix}, \mathbf{F}_r\right] = \begin{pmatrix}\mathbf{X}_{\tilde{g}_{r+k+|\hat{\mathbf{t}}|}}\\ \hline 0\end{pmatrix} + \begin{pmatrix}\tilde{\mu}_{r+k}\mathbf{D}_0\\ \hline 0\end{pmatrix} + \begin{pmatrix}\tilde{\lambda}_k\mathbf{X}_h\\ \hline 0\end{pmatrix} - \begin{pmatrix}\mathbf{0}\\ \hline \hat{\nabla} f \cdot \mathbf{X}_{g_{k+|\hat{\mathbf{t}}|}}\end{pmatrix},$$

where

$$\begin{aligned}
\tilde{g}_{r+k+|\hat{\mathbf{t}}|} &= \mathrm{Proy}_{\Delta_{r+k+|\hat{\mathbf{t}}|}}\left(\hat{\nabla} g_{k+|\hat{\mathbf{t}}|} \cdot \mathbf{X}_h + \tfrac{k+|\hat{\mathbf{t}}|}{r+k+|\mathbf{t}|} f \frac{\partial g_{k+|\hat{\mathbf{t}}|}}{\partial z}\right) \in \Delta_{r+k+|\hat{\mathbf{t}}|}, \\
\tilde{\mu}_{r+k} &= \tfrac{1}{r+k+|\hat{\mathbf{t}}|}\left(\hat{\nabla} f \cdot \mathbf{X}_{\frac{\partial g_{k+|\hat{\mathbf{t}}|}}{\partial z}} - \hat{\nabla}\tilde{\lambda}_k \cdot \mathbf{X}_h\right) \in \mathcal{P}^{\mathbf{t}}_{r+k}, \qquad (23) \\
\tilde{\lambda}_k &= \tfrac{r+k+|\hat{\mathbf{t}}|}{(r+|\hat{\mathbf{t}}|)h} \mathrm{Proy}_{h\,\mathcal{P}^{\hat{\mathbf{t}}}_k}\left(\hat{\nabla} g_{k+|\hat{\mathbf{t}}|} \cdot \mathbf{X}_h + \tfrac{k+|\hat{\mathbf{t}}|}{r+k+|\mathbf{t}|} f \frac{\partial g_{k+|\hat{\mathbf{t}}|}}{\partial z}\right) \in \mathcal{P}^{\mathbf{t}}_k.
\end{aligned}$$

2. *If* $\mu_k \in \mathcal{P}^{\mathbf{t}}_k$, *then* $\left[\begin{pmatrix}\mu_k\mathbf{D}_0\\ \hline 0\end{pmatrix}, \mathbf{F}_r\right] = \begin{pmatrix}(\nabla\mu_k \cdot \mathbf{F}_r)\mathbf{D}_0\\ \hline 0\end{pmatrix} - \begin{pmatrix}r\mu_k\mathbf{X}_h\\ \hline 0\end{pmatrix} - \begin{pmatrix}\mathbf{0}\\ \hline (r+t_3)\mu_k f\end{pmatrix}.$
3. *If* $\lambda_{k-r} \in \mathcal{P}^{\mathbf{t}}_{k-r}$, *then* $\left[\begin{pmatrix}\lambda_{k-r}\mathbf{X}_h\\ \hline 0\end{pmatrix}, \mathbf{F}_r\right] = \begin{pmatrix}(\nabla\lambda_{k-r} \cdot \mathbf{F}_r)\mathbf{X}_h\\ \hline 0\end{pmatrix} - \begin{pmatrix}\mathbf{0}\\ \hline \lambda_{k-r}\hat{\nabla} f \cdot \mathbf{X}_h\end{pmatrix}.$
4. *If* $\varsigma_{k+t_3} \in \mathcal{P}^{\mathbf{t}}_{k+t_3}$, *then* $\left[\begin{pmatrix}\mathbf{0}\\ \hline \varsigma_{k+t_3}\end{pmatrix}, \mathbf{F}_r\right] = \begin{pmatrix}\mathbf{0}\\ \hline \nabla\varsigma_{k+t_3} \cdot \mathbf{F}_r\end{pmatrix}.$

5 Three-Dimensional Normal Forms

In this section, we plain to obtain the normal form for the tridimensional system (2), whose principal part is given in (19), by using the decomposition obtained in the previous section.

We start analyzing the homological operator (18)

$$\mathbf{L}_{r+k} : \mathcal{Q}^{\mathbf{t}}_k \longrightarrow \mathcal{Q}^{\mathbf{t}}_{r+k},$$

defined by $\mathbf{L}_{r+k}(\mathbf{P}_k) = [\mathbf{P}_k, \mathbf{F}_r]$.

From Proposition 6, this operator can be expresed as

$$\mathbf{L}_{r+k} : \mathscr{C}^{\mathbf{t}}_k \oplus \mathscr{D}^{\mathbf{t}}_k \oplus \mathscr{F}^{\mathbf{t}}_k \oplus \mathscr{G}^{\mathbf{t}}_k \longrightarrow \mathscr{C}^{\mathbf{t}}_{r+k} \oplus \mathscr{D}^{\mathbf{t}}_{r+k} \oplus \mathscr{F}^{\mathbf{t}}_{r+k} \oplus \mathscr{G}^{\mathbf{t}}_{r+k},$$

where, if we write $\mathbf{P}_k = \begin{pmatrix}\mathbf{X}_{g_{k+|\mathbf{t}|}}\\ \hline 0\end{pmatrix} + \begin{pmatrix}\mu_k\mathbf{D}_0\\ \hline 0\end{pmatrix} + \begin{pmatrix}\lambda_{k-r}\mathbf{X}_h\\ \hline 0\end{pmatrix} + \begin{pmatrix}\mathbf{0}\\ \hline \varsigma_{k+t_3}\end{pmatrix}$, then

$$\mathbf{L}_{r+k}\left(\mathbf{P}_k\right)=\left[\left(\frac{\mathbf{X}_{g_{k+|\mathbf{t}|}}}{0}\right),\mathbf{F}_r\right]+\left[\left(\frac{\mu_k\mathbf{D}_0}{0}\right),\mathbf{F}_r\right]+\left[\left(\frac{\lambda_{k-r}\mathbf{X}_h}{0}\right),\mathbf{F}_r\right]+\left[\left(\frac{\mathbf{0}}{\varsigma_{k+t_3}}\right),\mathbf{F}_r\right].$$

From Lemma 6 we obtain the following matricial structure for this operator:

$\left(\frac{\mathbf{X}_{\tilde{g}_{r+k+\lvert\hat{\mathbf{t}}\rvert}}}{0}\right)$	0	0	0	$\mathscr{C}^{\mathbf{t}}_{r+k}$
$\left(\frac{\tilde{\mu}_{r+k}\mathbf{D}_0}{0}\right)$	$\left(\frac{(\nabla\mu_k\cdot\mathbf{F}_r)\mathbf{D}_0}{0}\right)$	0	0	$\mathscr{D}^{\mathbf{t}}_{r+k}$
$\left(\frac{\tilde{\lambda}_k\mathbf{X}_h}{0}\right)$	$\left(\frac{r\,\mu_k\mathbf{X}_h}{0}\right)$	$\left(\frac{(\nabla\lambda_{k-r}\cdot\mathbf{F}_r)\mathbf{X}_h}{0}\right)$	0	$\mathscr{F}^{\mathbf{t}}_{r+k}$
$\left(\frac{\mathbf{0}}{-\hat{\nabla}f\cdot\mathbf{X}_{g_{k+\lvert\mathbf{t}\rvert}}}\right)$	$\left(\frac{\mathbf{0}}{-(r+t_3)\mu_k f}\right)$	$\left(\frac{\mathbf{0}}{-\lambda_{k-r}\,\hat{\nabla}f\cdot\mathbf{X}_h}\right)$	$\left(\frac{\mathbf{0}}{\nabla\varsigma_{k+t_3}\cdot\mathbf{F}_r}\right)$	$\mathscr{G}^{\mathbf{t}}_{r+k}$
$\left(\frac{\mathbf{X}_{g_{k+\lvert\mathbf{t}\rvert}}}{0}\right)\in\mathscr{C}^{\mathbf{t}}_k$	$\left(\frac{\mu_k\mathbf{D}_0}{0}\right)\in\mathscr{D}^{\mathbf{t}}_k$	$\left(\frac{\lambda_{k-r}\mathbf{X}_h}{0}\right)\in\mathscr{F}^{\mathbf{t}}_k$	$\left(\frac{\mathbf{0}}{\varsigma_{k+t_3}}\right)\in\mathscr{G}^{\mathbf{t}}_k$	

where $\tilde{g}_{r+k+|\hat{\mathbf{t}}|}$, $\tilde{\mu}_{r+k}$, $\tilde{\lambda}_k$ are given in (23).

Let us introduce the Lie derivative operator associated to the principal part $\mathbf{F}_r$ given in (19) of the vector field (2), by:

$$\begin{aligned}\ell_k :&\mathcal{P}^{\mathbf{t}}_{k-r}\longrightarrow\mathcal{P}^{\mathbf{t}}_k\\ &\mu_{k-r}\rightarrow\nabla\mu_{k-r}\cdot\mathbf{F}_r.\end{aligned}\tag{24}$$

Moreover, let us define the following modified Lie derivative operator:

$$\begin{aligned}\ell_{k,A} :&\mathcal{P}^{\mathbf{t}}_{k-r}\longrightarrow\mathcal{P}^{\mathbf{t}}_k\\ &\mu_{k-r}\longrightarrow\hat{\nabla}\mu_{k-r}\cdot\mathbf{X}_h+A\,f\,\frac{\partial\mu_{k-r}}{\partial z}.\end{aligned}$$

Observe that $\ell_{k,A}$ is the Lie derivative operator corresponding to the vector field $\mathbf{F}_r+\left(\frac{\mathbf{0}}{(A-1)f}\right)$. Therefore, taking $A=1$, we get the operator ℓ_k defined in (24).

Using this new operator we can write the above matrix as

$\left(\frac{\mathbf{X}_{\tilde{g}_{r+k+\lvert\hat{\mathbf{t}}\rvert}}}{0}\right)$	0	0	0	$\mathscr{C}^{\mathbf{t}}_{r+k}$
$\left(\frac{\tilde{\mu}_{r+k}\,\mathbf{D}_0}{0}\right)$	$\left(\frac{\ell_{r+k}(\mu_k)\,\mathbf{D}_0}{0}\right)$	0	0	$\mathscr{D}^{\mathbf{t}}_{r+k}$
$\left(\frac{\tilde{\lambda}_k\mathbf{X}_h}{0}\right)$	$\left(\frac{r\,\mu_k\mathbf{X}_h}{0}\right)$	$\left(\frac{\ell_{r+k}(\lambda_{k-r})\,\mathbf{X}_h}{0}\right)$	0	$\mathscr{F}^{\mathbf{t}}_{r+k}$
$\left(\frac{\mathbf{0}}{-\hat{\nabla}f\cdot\mathbf{X}_{g_{k+\lvert\hat{\mathbf{t}}\rvert}}}\right)$	$\left(\frac{\mathbf{0}}{-(r+t_3)\mu_k f}\right)$	$\left(\frac{\mathbf{0}}{-\lambda_{k-r}\,\hat{\nabla}f\cdot\mathbf{X}_h}\right)$	$\left(\frac{\mathbf{0}}{\ell_{r+k+t_3}(\varsigma_{k+t_3})}\right)$	$\mathscr{G}^{\mathbf{t}}_{r+k}$
$\left(\frac{\mathbf{X}_{g_{k+\lvert\hat{\mathbf{t}}\rvert}}}{0}\right)\in\mathscr{C}^{\mathbf{t}}_k$	$\left(\frac{\mu_k\mathbf{D}_0}{0}\right)\in\mathscr{D}^{\mathbf{t}}_k$	$\left(\frac{\lambda_{k-r}\mathbf{X}_h}{0}\right)\in\mathscr{F}^{\mathbf{t}}_k$	$\left(\frac{\mathbf{0}}{\varsigma_{k+t_3}}\right)\in\mathscr{G}^{\mathbf{t}}_k$	

where

$$\tilde{g}_{r+k+|\hat{\mathbf{t}}|} = \text{Proy}_{\Delta_{r+k+|\hat{\mathbf{t}}|}}\left(\ell_{r+k+|\hat{\mathbf{t}}|,A_0}\left(g_{k+|\hat{\mathbf{t}}|}\right)\right),$$
$$\tilde{\mu}_{r+k} = \frac{1}{r+k+|\hat{\mathbf{t}}|}\left(\hat{\nabla} f \cdot \mathbf{X}_{\frac{\partial g_{k+|\mathbf{t}|}}{\partial z}} - \hat{\nabla}\tilde{\lambda}_k \cdot \mathbf{X}_h\right),$$
$$\tilde{\lambda}_k = \frac{r+k+|\hat{\mathbf{t}}|}{(r+|\hat{\mathbf{t}}|)h}\text{Proy}_{h\,\mathcal{P}_k^{\mathbf{t}}}\left(\ell_{r+k+|\hat{\mathbf{t}}|,A_0}\left(g_{k+|\hat{\mathbf{t}}|}\right)\right),$$

being $A_0 = \frac{k+|\hat{\mathbf{t}}|}{r+k+|\mathbf{t}|}$.

From the structure of the above matrix is deduced the following proposition.

Proposition 7 *Let us consider* $\mathbf{F}_r$ *given in* (19) *with* $h \not\equiv 0$. *Then, a complementary space of the range of the homological operator* $\mathbf{L}_{r+k}$ *is given by*

$$\left(\frac{\mathbf{X}_{\text{Cor}(\ell_{r+k,A_0})\cap\Delta_{r+k+|\hat{\mathbf{t}}|}}}{0}\right) \oplus \left(\frac{\text{Cor}(\ell_{r+k})\,\mathbf{D}_0}{0}\right) \oplus \left(\frac{\text{Cor}(\ell_k)\,\mathbf{X}_h}{0}\right) \oplus \left(\frac{\mathbf{0}}{\text{Cor}(\ell_{r+k+t_3})}\right),$$

where $A_0 = \frac{k+|\hat{\mathbf{t}}|}{r+k+|\mathbf{t}|}$ *and* $\text{Cor}(\ell_{r+k,A_0})$ *is a complementary subspace to* $\text{Range}(\ell_{r+k,A_0})$ *in* $\mathcal{P}^{\mathbf{t}}_{r+k}$, *such that* $\text{Cor}(\ell_{r+k,A_0}) \cap \Delta_{r+k+|\hat{\mathbf{t}}|}$ *has maximal dimension.*

Next theorem provides a formal normal form for system (2).

Theorem 3 *A formal normal form for system* (2) *is* $\dot{\mathbf{x}} = \mathbf{G}(\mathbf{x})$, *where*

$$\mathbf{G} = \mathbf{F}_r + \sum_{k\geq 1}\left(\left(\frac{\mathbf{X}_{g_{r+k+|\hat{\mathbf{t}}|}}}{0}\right) + \left(\frac{\mu_{r+k}\,\mathbf{D}_0}{0}\right) + \left(\frac{\lambda_k\,\mathbf{X}_h}{0}\right) + \left(\frac{\mathbf{0}}{\varsigma_{r+k+t_3}}\right)\right),$$

with $g_{r+k+|\hat{\mathbf{t}}|} \in \text{Cor}(\ell_{r+k+|\hat{\mathbf{t}}|,A_0}) \cap \Delta_{r+k+|\hat{\mathbf{t}}|}$, $\mu_{r+k} \in \text{Cor}(\ell_{r+k})$, $\lambda_k \in \text{Cor}(\ell_k)$, $\varsigma_{r+k+t_3} \in \text{Cor}(\ell_{r+k+t_3})$; *being* $A_0 = \frac{k+|\hat{\mathbf{t}}|}{r+k+|\mathbf{t}|}$ *and* $\text{Cor}(\ell_{r+k,A_0})$ *a complementary subspace to* $\text{Range}(\ell_{r+k,A_0})$ *in* $\mathcal{P}^{\mathbf{t}}_{r+k}$, *such that* $\text{Cor}(\ell_{r+k,A_0}) \cap \Delta_{r+k+|\hat{\mathbf{t}}|}$ *has maximal dimension.*

5.1 The Operator $\ell_{r+k,A}$

In this section, we will study the linear operator $\ell_{r+k,A}$ for $A \neq 0$, since the co-range of the homological operator $\mathbf{L}_{r+k}$ depends on the co-range of this operator. We recall that $\ell_{r+k,1} = \ell_{r+k}$ is the Lie derivative operator ℓ_k defined in (24), which is associated to the principal part (19) of the tridimensional vector field (2).

Let us introduce the Lie derivative operator associated to the planar Hamiltonian vector field $\mathbf{X}_h$:

$$\begin{aligned} \hat{\ell}_k : &\mathcal{P}^{\hat{\mathbf{t}}}_{k-r} \longrightarrow \mathcal{P}^{\hat{\mathbf{t}}}_k \\ &\mu_{k-r} \longrightarrow \hat{\ell}_k\left(\mu_{k-r}\right) = \hat{\nabla}\mu_{k-r} \cdot \mathbf{X}_h. \end{aligned} \tag{25}$$

Let us consider the subspace $\text{Ker}(\hat{\ell}_{r+k})$. We observe that the elements of this subspace are the $(r+k)$-degree quasi-homogeneous polynomial first integrals of $\mathbf{X}_h$. The first integrals of $\mathbf{X}_h$ are of the form $C\, h^{\alpha}$, where $C, \alpha \in \mathbb{R}$. Then, impossing that they are polynomials, we obtain that $\alpha = \frac{k}{r+|\hat{\mathbf{t}}|}$ and

$$\text{Ker}(\hat{\ell}_{r+k}) = \text{Span}\left\{h^{\frac{k}{r+|\hat{\mathbf{t}}|}}\right\}, \text{ if } k \text{ is a multiple of } r+|\hat{\mathbf{t}}|.$$

Otherwise, we have $\text{Ker}(\hat{\ell}_{r+k}) = \{\mathbf{0}\}$.

Let us denote by $\mathscr{E}_k$ a complementary subspace of $\text{Ker}(\hat{\ell}_{r+k})$ in $\mathcal{P}^{\hat{\mathbf{t}}}_k$, i.e.,

$$\mathcal{P}^{\hat{\mathbf{t}}}_k = \mathscr{E}_k \oplus \text{Ker}(\hat{\ell}_{r+k})$$

Let also define the subspace $\widetilde{\mathscr{E}_l} = z^l\, \mathscr{E}_{k-lt_3}$ and $\widetilde{\mathscr{K}_l} = z^l\, \text{Ker}(\hat{\ell}_{r+k-lt_3})$. Then:

$$\mathcal{P}^{\mathbf{t}}_k = \bigoplus_{l=0}^{k_1} z^l\, \mathcal{P}^{\hat{\mathbf{t}}}_{k-lt_3} = \bigoplus_{l=0}^{k_1} z^l\left(\mathscr{E}_{k-lt_3} \oplus \text{Ker}(\ell_{r+k-lt_3})\right) = \bigoplus_{l=0}^{k_1} \widetilde{\mathscr{E}_l} \oplus \widetilde{\mathscr{K}_l},$$

where $k_1 = \lfloor k/t_3 \rfloor$ ($\lfloor \cdot \rfloor$ is the *floor* function).

Hence, given $p_k \in \mathcal{P}^{\hat{\mathbf{t}}}_k$, we can express it in the form

$$p_k(x,y) = p_k^{(1)}(x,y) + p_k^{(2)}(x,y), \tag{26}$$

where $p_k^{(1)} \in \mathscr{E}_k$ and $p_k^{(2)} \in \text{Ker}(\hat{\ell}_{r+k})$.

Given $\mu_k \in \mathcal{P}^{\mathbf{t}}_k$, it is easy to show that it can be expressed as

$$\mu_k(x,y,z) = \sum_{l=0}^{\lfloor k/t_3 \rfloor} z^l\, p_{k-lt_3}(x,y),$$

with $p_{k-lt_3} \in \mathcal{P}^{\hat{\mathbf{t}}}_{k-lt_3}$. Let us denote by k_1, k_2 the quotient and the remainder of the division $k \div t_3$, that is:

$$k = k_1 t_3 + k_2, \text{ with } 0 \le k_2 < t_3. \tag{27}$$

Then, using (26), we can write

$$\mu_k(x,y,z) = \sum_{l=0}^{k_1} z^l\, p_{k-lt_3}(x,y) = \sum_{l=0}^{k_1} z^l\, p_{(k_1-l)t_3+k_2}(x,y)$$
$$= \sum_{l=0}^{k_1} z^l \left(p^{(1)}_{(k_1-l)t_3+k_2}(x,y) + p^{(2)}_{(k_1-l)t_3+k_2}(x,y)\right),$$

where $p^{(1)}_{(k_1-l)t_3+k_2} \in \mathscr{E}_{(k_1-l)t_3+k_2}$ and $p^{(2)}_{(k_1-l)t_3+k_2} \in \mathrm{Ker}(\hat{\ell}_{r+(k_1-l)t_3+k_2})$.

After some computations, we obtain

$$\ell_{r+k,A}(\mu_k) = \hat{\nabla}\mu_k \cdot \mathbf{X}_h + A\,f\,\frac{\partial \mu_k}{\partial z}$$
$$= z^{k_1}\hat{\nabla} p^{(1)}_{k_2} \cdot \mathbf{X}_h + \sum_{l=0}^{k_1-1} z^l \left(\hat{\nabla} p^{(1)}_{(k_1-l)t_3+k_2} \cdot \mathbf{X}_h\right.$$
$$\left. +(l+1)z^l A f p^{(2)}_{(k_1-l-1)t_3+k_2}\right) + \sum_{l=0}^{k_1} l z^{l-1} A f p^{(1)}_{(k_1-l)t_3+k_2}.$$

Consequently, we have the following matrix expression for $\ell_{r+k,A}$:

0	0	0	0	0	0	$z^{k_1+1}\,\mathcal{P}^{\mathbf{t}}_{r+k-(k_1+1)t_3}$
d_{k_1}	0	0	0	0	0	$z^{k_1}\,\mathcal{P}^{\hat{\mathbf{t}}}_{r+k-k_1t_3}$
0	d_{k_1-1}	0	0	0	0	$z^{k_1-1}\,\mathcal{P}^{\hat{\mathbf{t}}}_{r+k-(k_1-1)t_3}$
$\vdots$	$\vdots$	$\ddots$	$\ddots$	$\vdots$	$\vdots$	$\vdots$
0	$\vdots$	$\vdots$	d_1	0	0	$z^1\,\mathcal{P}^{\hat{\mathbf{t}}}_{r+k-t_3}$
0	0	$\cdots$	0	d_0	0	$z^0\,\mathcal{P}^{\hat{\mathbf{t}}}_{r+k}$
$\widetilde{\mathscr{E}_{k_1}}$	$\widetilde{\mathscr{E}_{k_1-1} \oplus \mathscr{K}_{k_1}}$	$\cdots$	$\widetilde{\mathscr{E}_1 \oplus \mathscr{K}_2}$	$\widetilde{\mathscr{E}_0 \oplus \mathscr{K}_1}$	$\widetilde{\mathscr{K}_0}$	

Here, we have denoted

$$d_{k_1} = z^{k_1}\,\hat{\nabla} p^{(1)}_{k_2} \cdot \mathbf{X}_h = z^{k_1}\,\hat{\ell}_{r+k_2}\left(p^{(1)}_{k_2}\right), \text{ and}$$
$$d_{k_1-l} = z^{k_1-l}\left(\hat{\nabla} p^{(1)}_{lt_3+k_2} \cdot \mathbf{X}_h + (k_1-l+1)\,A\,f\,p^{(2)}_{(l-1)t_3+k_2}\right)$$
$$= z^{k_1-l}\left(\hat{\ell}_{r+lt_3+k_2}\left(p^{(1)}_{lt_3+k_2}\right) + (k_1-l+1)\,A\,f\,p^{(2)}_{(l-1)t_3+k_2}\right),$$

for $l = 1, \ldots, k_1$.

Moreover, using that $p^{(2)}_{(l-1)t_3+k_2} \in \mathrm{Ker}(\hat{\ell}_{r+(l-1)t_3+k_2})$, we have $p^{(2)}_{(k_1-l-1)t_3+k_2} \in \mathrm{Span}\left\{h^{\frac{(l-1)t_3+k_2}{r+|\hat{\mathbf{t}}|}}\right\}$ if $(l-1)t_3+k_2$ is a multiple of $r+|\hat{\mathbf{t}}|$, or zero otherwise.

Remark 1 The elements of $\widetilde{\mathscr{K}_0} = \text{Ker}(\hat{\ell}_{r+k})$ are not used in the matrix. Moreover, if $d_{k_1-l} = 0$ for some l, then

$$\hat{\nabla} p^{(1)}_{lt_3+k_2} \cdot \mathbf{X}_h = -(k_1 - l + 1)\, A\, f\, p^{(2)}_{(l-1)t_3+k_2} \in \text{Range}(\hat{\ell}_{r+lt_3+k_2}) \cap \text{Cor}(\hat{\ell}_{r+lt_3+k_2}).$$

Therefore, $p^{(1)}_{lt_3+k_2} = p^{(1)}_{(l-1)t_3+k_2} = 0$ and

$$\text{Ker}\left(\ell_{r+k,A}\right) = \text{Ker}\left(\hat{\ell}_{r+k}\right) = \begin{cases} \text{Span}\{h^n\}, & \text{if } k = n(r + |\hat{\mathbf{t}}|), \\ \{0\}, & \text{otherwise} \end{cases}$$

From the structure of the above matrix, we obtain the following result:

Theorem 4 *Let us consider the subspace $\mathscr{V}_{r+k}$ defined by*

$$\mathscr{V}_{r+k} = \text{Range}(\hat{\ell}_{r+k}) \oplus \text{Span}\left\{ f\, h^{\frac{k-t_3}{r+|\hat{\mathbf{t}}|}} \right\}, \text{ if } k - t_3 \text{ is a multiple of } r + |\hat{\mathbf{t}}|, \text{ or}$$

$$\mathscr{V}_{r+k} = \text{Range}(\hat{\ell}_{r+k}), \text{ if } k - t_3 \text{ is not a multiple of } r + |\hat{\mathbf{t}}|.$$

Let us denote by $\widetilde{\mathscr{V}}_{r+k}$ a complementary subspace of $\mathscr{V}_{r+k}$. Then, a complementary subspace to the range of the operator $\ell_{r+k,A}$ is

$$\text{Cor}(\ell_{r+k,A}) = z^{k_1+1} \mathcal{P}^{\mathbf{t}}_{r+k-(k_1+1)t_3} \oplus z^{k_1} \text{Cor}(\hat{\ell}_{r+k_2}) \oplus \sum_{l=0}^{k_1-1} z^l\, \widetilde{\mathscr{V}}_{r+k-lt_3},$$

where $k_1 = \lfloor k/t_3 \rfloor$, $k_2 = k - k_1 t_3$.

Remark 2 We claim that, without loss of generality, we can assume that $f \in \text{Cor}\left(\hat{\ell}_{r+t_3}\right)$: a complementary of the range of the Lie derivative operator $\hat{\ell}_{r+t_3}$. Namely, if we transform the system (2) by the change of variables $\tilde{x} = x$, $\tilde{y} = y$, $\tilde{z} = z - p_{t_3}(x, y)$, with $p_{t_3} \in \mathcal{P}^{\hat{\mathbf{t}}}_{t_3}$, then the principal part of the transformed system becomes

$$\widetilde{\mathbf{F}}_r = \begin{pmatrix} \mathbf{X}_h \\ f - \hat{\nabla} p_{t_3} \cdot \mathbf{X}_h \end{pmatrix} = \begin{pmatrix} \mathbf{X}_h \\ f - \hat{\ell}_{r+t_3}\left(p_{t_3}\right) \end{pmatrix},$$

and we can annihilate the part of f that belongs to the range of $\hat{\ell}_{r+t_3}$ by selecting p_{t_3} adequately.

Moreover, through this work we assume that the factorization of h in $\mathbb{C}[x, y]$ only has simple factors. In Proposition 3.18 of Algaba et al. [9] has been shown that, under this generic assumption, we have $\text{Cor}\left(\hat{\ell}_{r+k+|\hat{\mathbf{t}}|}\right) = h\text{Cor}\left(\hat{\ell}_k\right)$. This property is important because, in this case, the study of the co-range of this operator can be accomplished in a finite number of steps.

Remark 3 If $k - t_3$ is a multiple of $r + |\hat{\mathbf{t}}|$, then $\text{Range}(\hat{\ell}_{r+k}) \cap \text{Span}\left\{f\, h^{\frac{k-t_3}{r+|\hat{\mathbf{t}}|}}\right\} = \{0\}$. Indeed, assuming that $f \in \text{Cor}\left(\hat{\ell}_{r+t_3}\right)$, from Proposition 3.18 of Algaba et al. [9] we have $f\, h^{\frac{k-t_3}{r+|\hat{\mathbf{t}}|}} \in \text{Cor}(\hat{\ell}_{r+k})$.

Remark 4 Although the linear operator $\ell_{r+k,A}$ depends on A, the above complementary subspace $\text{Cor}(\ell_{r+k,A})$ does not depend on A, whenever $A \neq 0$. Hence, in the applications we can select $A = 1$. In others words, we can substitute $\text{Cor}(\ell_{r+k,A})$ by $\text{Cor}(\ell_{r+k})$.

Remark 5 Observe that, if $f \equiv 0$, then $\tilde{\mathcal{V}}_{r+k} = \text{Cor}(\hat{\ell}_{r+k})$.

6 Normal Form for a Triple-Zero Singularity

Finally, we study the normal form for a triple-zero singularity with geometric multiplicity two (i.e., they correspond to the coupling of a Takens-Bogdanov and a saddle-node singularities). We consider the following system, expanded in quasi-homogeneous terms of type $\mathbf{t} = (2, 3, 5)$:

$$\dot{\mathbf{x}} = \mathbf{F}_0 + \mathbf{F}_1 + \cdots, \tag{28}$$

where the principal part is

$$\mathbf{F}_1 = \begin{pmatrix} y \\ x^2 \\ \hline x^3/3 - y^2/2 \end{pmatrix} = \begin{pmatrix} \mathbf{X}_h(x, y) \\ \hline h \end{pmatrix} \in \mathcal{Q}_1^{\mathbf{t}}.$$

We notice that $\hat{\mathbf{t}} = (2, 3)$, $r = 1$ and $h = x^3/3 - y^2/2 \in \mathcal{P}_6^{\hat{\mathbf{t}}}$.

Firstly, following the ideas presented in Sect. 5.1, let us obtain bases for the complementary subspaces $\text{Cor}(\ell_{r+k})$.

The Lie derivative operator $\hat{\ell}_k$ is studied in Algaba et al. [6]. It is obtained that a complementary subspace to the range of this operator is

$\text{Cor}(\hat{\ell}_k) = \text{Span}\left\{h^{\frac{k}{6}}\right\}$ if k is a multiple of 6,
$\text{Cor}(\hat{\ell}_k) = \text{Span}\left\{xh^{\frac{k-2}{6}}\right\}$ if $k - 2$ is a multiple of 6, and
$\text{Cor}(\hat{\ell}_k) = \{0\}$ otherwise.

According to Theorem 4, we must distinguish if $k - t_3 = k - 5$ is a multiple of $r + |\hat{\mathbf{t}}| = 6$. This suggest us to write

$$k = 5(6k_1 + k_2) + k_3,$$

with $0 \le k_2 < 6$ and $0 \le k_3 < 5$. We notice that $\frac{k-t_3}{r+|\hat{\mathbf{t}}|} = \frac{5(6k_1+k_2-1)+k_3}{6}$.

Let us consider the subspace

$$\mathscr{V}_k = \mathrm{Cor}(\hat{\ell}_k) \oplus \mathrm{Span}\left\{h^{\frac{k-5}{6}}\right\} \text{ if } k-5 \text{ is a multiple of 6, or}$$
$$\mathscr{V}_k = \mathrm{Cor}(\hat{\ell}_k) \text{ otherwise,}$$

and denote by $\tilde{\mathscr{V}}_k$ a complementary subspace of $\mathscr{V}_k$ in $\mathcal{P}_k^{\mathbf{t}}$.

Then, we have

$$\mathrm{Cor}(\ell_{r+k}) = z^{6k_1+k_2+1}\mathcal{P}^{\mathbf{t}}_{k_3-4} \oplus z^{6k_1+k_2}\mathrm{Cor}(\hat{\ell}_{1+k_3}) \oplus \sum_{i=0}^{6k_1+k_2-1} z^i \tilde{\mathscr{V}}_{k+1-\varsigma i}. \qquad (29)$$

Next result presents bases for the subspaces $\mathrm{Cor}(\ell_{1+k})$ for the different values of k_2 and k_3.

Proposition 8 *Let us write $k = 5(6k_1 + k_2) + k_3$, with $0 \le k_2 < 6$ and $0 \le k_3 < 5$. Then, the following are bases for the subspaces* $\mathrm{Cor}(\ell_{1+k})$*:*

k_2	k_3	$\mathrm{Cor}(\ell_{1+k})$
$0 \le k_2 \le 4$	$k_3 = 0$	$\mathrm{Span}\left\{xz^{6(k_1-l)+k_2+1}h^{5l-1} : l = 1, \ldots, k_1\right\}$.
$k_2 = 5$	$k_3 = 0$	$\mathrm{Span}\left\{xz^{6(k_1-l)+k_2+1}h^{5l-1} : l = 1, \ldots, k_1+1\right\}$
$0 \le k_2 \le 5$	$k_3 = 1$	$\mathrm{Span}\left\{xz^{6(k_1-l)+k_2+6}h^{5l-5} : l = 1, \ldots, k_1+1\right\}$
$k_2 = 0$	$k_3 = 2$	$\mathrm{Span}\left\{xz^{6(k_1-l)+k_2+5}h^{5l-4} : l = 1, \ldots, k_1\right\}$
$1 \le k_2 \le 5$	$k_3 = 2$	$\mathrm{Span}\left\{xz^{6(k_1-l)+k_2+5}h^{5l-4} : l = 1, \ldots, k_1+1\right\}$
$0 \le k_2 \le 1$	$k_3 = 3$	$\mathrm{Span}\left\{xz^{6(k_1-l)+k_2+4}h^{5l-3} : l = 1, \ldots, k_1\right\}$
$2 \le k_2 \le 5$	$k_3 = 3$	$\mathrm{Span}\left\{xz^{6(k_1-l)+k_2+4}h^{5l-3} : l = 1, \ldots, k_1+1\right\}$
$0 \le k_2 \le 2$	$k_3 = 4$	$\mathrm{Span}\left\{z^{6k_1+k_2+1};\ xz^{6(k_1-l)+k_2+3}h^{5l-2} : l = 1, \ldots, k_1\right\}$
$3 \le k_2 \le 5$	$k_3 = 4$	$\mathrm{Span}\left\{z^{6k_1+k_2+1};\ xz^{6(k_1-l)+k_2+3}h^{5l-2} : l = 1, \ldots, k_1+1\right\}$

Proof We consider the case $k_3 = 0$, or equivalently $k = 5(6k_1 + k_2)$. For the remaining values ($k_3 = 1, 2, 3, 4, 5, 6$), we omit the proof since it is similar.

For $k_3 = 0$, the expression (29) becomes

$$\mathrm{Cor}(\ell_{1+5(6k_1+k_2)}) = \sum_{i=0}^{6k_1+k_2-1} z^i \tilde{\mathscr{V}}_{5(6k_1+k_2-i)+1}.$$

By writing $i = 6m + n$ with $m \in \mathbb{N}_0$ and $0 \le n < 6$, we have

$$\begin{aligned}\mathrm{Cor}(\ell_{1+5(6k_1+k_2)}) &= \sum_{m=0}^{k_1-1}\sum_{n=0}^{5} z^{6m+n}\tilde{\mathscr{V}}_{5(6k_1+k_2-6m-n)+1} + \sum_{n=0}^{k_2-1} z^{6k_1+n}\tilde{\mathscr{V}}_{5(6k_1+k_2-6k_1-n)+1}\\ &= \sum_{m=0}^{k_1-1}\sum_{n=0}^{5} z^{6m+n}\tilde{\mathscr{V}}_{5(6(k_1-m)+k_2-n)+1} + \sum_{n=0}^{k_2-1} z^{6k_1+n}\tilde{\mathscr{V}}_{5(k_2-n)+1}.\end{aligned}$$

(a) If $k_2 = 0$, then

$$\mathrm{Cor}(\ell_{1+5(6k_1)}) = \sum_{m=0}^{k_1-1}\sum_{n=0}^{5} z^{6m+n}\tilde{\mathscr{V}}_{5(6(k_1-m)-n)+1}.$$

We distinguish two cases:

(a.1) If $n = 5$, then $\tilde{\mathscr{V}}_{5(6(k_1-m)-5)+1}$ is a complementary subspace to

$$\mathscr{V}_{5(6(k_1-m)-5)+1} = \mathrm{Range}(\hat{\ell}_{6(5(k_1-m)-4)}) \oplus \mathrm{Span}\left\{h^{5(k_1-m)-4}\right\}.$$

As $\mathrm{Cor}(\hat{\ell}_{6(5(k_1-m)-4)}) = \mathrm{Span}\left\{h^{5(k_1-m)-4}\right\}$, we get $\tilde{\mathscr{V}}_{5(6(k_1-m)-5)+1} = \{0\}$.

(a.2) If $n \neq 5$, then $\tilde{\mathscr{V}}_{5(6(k_1-m)-n)+1}$ is a complementary subspace to

$$\mathscr{V}_{5(6(k_1-m)-n)+1} = \mathrm{Range}(\hat{\ell}_{5(6(k_1-m)-n)+1}).$$

Hence $\tilde{\mathscr{V}}_{5(6(k_1-m)-n)+1} = \mathrm{Cor}(\hat{\ell}_{5(6(k_1-m)-n)+1})$ and the co-ranges are the trivial space, except for $n = 1$ where $\mathrm{Cor}(\hat{\ell}_{6(5(k_1-m)-1)+2}) = \mathrm{Span}\left\{xh^{5(k_1-m)-1}\right\}$. In summary, for $k_2 = 0$ we obtain

$$\mathrm{Cor}(\ell_{1+30k_1}) = \mathrm{Span}\left\{z^{6m+1}xh^{5(k_1-m)-1} : m = 0, \ldots, k_1 - 1\right\}.$$

(b) If $k_2 = 1$, then

$$\mathrm{Cor}(\ell_{1+5(6k_1+1)}) = \sum_{m=0}^{k_1-1}\sum_{n=0}^{5} z^{6m+n}\tilde{\mathscr{V}}_{5(6(k_1-m)+1-n)+1}.$$

We distinguish two cases:

(b.1) If $n = 0$, then $\tilde{\mathscr{V}}_{5(6(k_1-m)+1)+1}$ is a complementary subspace to

$$\mathscr{V}_{5(6(k_1-m)+1)+1} = \mathrm{Range}(\hat{\ell}_{6(5(k_1-m)+1)}) \oplus \mathrm{Span}\left\{h^{5(k_1-m)+1}\right\}.$$

As $\mathrm{Cor}(\hat{\ell}_{6(5(k_1-m)+1)}) = \mathrm{Span}\left\{h^{5(k_1-m)+1}\right\}$, we get $\tilde{\mathscr{V}}_{5(6(k_1-m)+1)+1} = \{0\}$.

(b.2) If $n \neq 0$, then $\tilde{\mathscr{V}}_{5(6(k_1-m)+1-n)+1}$ is a complementary subspace to

$$\mathscr{V}_{5(6(k_1-m)+1-n)+1} = \mathrm{Range}(\hat{\ell}_{5(6(k_1-m)+1-n)+1}).$$

Hence $\tilde{\mathscr{V}}_{5(6(k_1-m)+1-n)+1} = \mathrm{Cor}(\hat{\ell}_{5(6(k_1-m)+1-n)+1})$ which is the trivial space, except for $n = 2$ where $\mathrm{Cor}(\hat{\ell}_{6(5(k_1-m)-1)+2}) = \mathrm{Span}\left\{xh^{5(k_1-m)-1}\right\}$. In summary, for $k_2 = 1$ we obtain

$$\mathrm{Cor}(\ell_{6+30k_1}) = \mathrm{Span}\left\{z^{6m+2}xh^{5(k_1-m)-1} : m = 0, \ldots, k_1 - 1\right\}.$$

(c) If $k_2 = 2$, then

$$\mathrm{Cor}(\ell_{1+5(6k_1+2)}) = \sum_{m=0}^{k_1-1}\sum_{n=0}^{5} z^{6m+n}\tilde{\mathscr{V}}_{5(6(k_1-m)+2-n)+1} + \sum_{n=0}^{1} z^{6k_1+n}\tilde{\mathscr{V}}_{5(2-n)+1}.$$

We notice that $\tilde{\mathscr{V}}_{5(2-n)+1} = \{0\}$ for $n = 0, 1$. To determine $\tilde{\mathscr{V}}_{5(6(k_1-m)+2-n)+1}$, we distinguish two cases:

(c.1) If $n = 1$, then $\tilde{\mathscr{V}}_{5(6(k_1-m)+1)+1}$ is a complementary subspace to

$$\mathscr{V}_{5(6(k_1-m)+1)+1} = \mathrm{Range}(\hat{\ell}_{6(5(k_1-m)+1)}) \oplus \mathrm{Span}\left\{h^{5(k_1-m)+1}\right\}.$$

As $\mathrm{Cor}(\hat{\ell}_{6(5(k_1-m)+1)}) = \mathrm{Span}\left\{h^{5(k_1-m)+1}\right\}$, we get $\tilde{\mathscr{V}}_{5(6(k_1-m)+1)+1} = \{0\}$.

(c.2) If $n \neq 1$ then $\tilde{\mathscr{V}}_{5(6(k_1-m)+2-n)+1}$ is a complementary subspace to

$$\mathscr{V}_{5(6(k_1-m)+2-n)+1} = \mathrm{Range}(\hat{\ell}_{5(6(k_1-m)+2-n)+1}).$$

Thus $\tilde{\mathscr{V}}_{5(6(k_1-m)+2-n)+1} = \mathrm{Cor}(\hat{\ell}_{5(6(k_1-m)+2-n)+1})$ which is the trivial space, except for $n = 3$ where $\mathrm{Cor}(\hat{\ell}_{6(5(k_1-m)-1)+2}) = \mathrm{Span}\left\{xh^{5(k_1-m)-1}\right\}$.
In summary, for $k_2 = 2$ we have

$$\mathrm{Cor}(\ell_{11+30k_1}) = \mathrm{Span}\left\{z^{6m+3}xh^{5(k_1-m)-1} : m = 0, \ldots, k_1 - 1\right\}.$$

(d) If $k_2 = 3$, then

$$\mathrm{Cor}(\ell_{1+5(6k_1+3)}) = \sum_{m=0}^{k_1-1}\sum_{n=0}^{5} z^{6m+n}\tilde{\mathscr{V}}_{5(6(k_1-m)+3-n)+1} + \sum_{n=0}^{2} z^{6k_1+n}\tilde{\mathscr{V}}_{5(3-n)+1}.$$

Observe that $\tilde{\mathscr{V}}_{5(3-n)+1} = \{0\}$ for $n = 0, 1, 2$. To characterize $\tilde{\mathscr{V}}_{5(6(k_1-m)+3-n)+1}$, we distinguish two cases:

(d.1) If $n = 2$ then $\tilde{\mathscr{V}}_{5(6(k_1-m)+1)+1}$ is a complementary subspace to

$$\mathscr{V}_{5(6(k_1-m)+1)+1} = \mathrm{Range}(\hat{\ell}_{6(5(k_1-m)+1)}) \oplus \mathrm{Span}\left\{h^{5(k_1-m)+1}\right\}.$$

As $\mathrm{Cor}(\hat{\ell}_{6(5(k_1-m)+1)}) = \mathrm{Span}\left\{h^{5(k_1-m)+1}\right\}$, we get $\tilde{\mathscr{V}}_{5(6(k_1-m)+1)+1} = \{0\}$.

(d.2) If $n \neq 2$ then $\tilde{\mathscr{V}}_{5(6(k_1-m)+3-n)+1}$ is a complementary subspace to

$$\mathscr{V}_{5(6(k_1-m)+3-n)+1} = \mathrm{Range}(\hat{\ell}_{5(6(k_1-m)+3-n)+1}).$$

Hence $\tilde{\mathscr{V}}_{5(6(k_1-m)+3-n)+1} = \mathrm{Cor}(\hat{\ell}_{5(6(k_1-m)+3-n)+1})$ which is the trivial space, except for $n = 4$, where $\mathrm{Cor}(\hat{\ell}_{6(5(k_1-m)-1)+2}) = \mathrm{Span}\left\{xh^{5(k_1-m)-1}\right\}$. In summary, for $k_2 = 3$ we obtain

$$\mathrm{Cor}(\ell_{16+30k_1}) = \mathrm{Span}\left\{z^{6m+4} x h^{5(k_1-m)-1} : m = 0, \ldots, k_1 - 1\right\}.$$

(e) If $k_2 = 4$, then

$$\mathrm{Cor}(\ell_{1+5(6k_1+4)}) = \sum_{m=0}^{k_1-1} \sum_{n=0}^{5} z^{6m+n} \tilde{\mathscr{V}}_{5(6(k_1-m)+4-n)+1} + \sum_{n=0}^{3} z^{6k_1+n} \tilde{\mathscr{V}}_{5(4-n)+1}.$$

We notice that $\tilde{\mathscr{V}}_{5(4-n)+1} = \{0\}$ for $n = 0, 1, 2, 3$. To determine $\tilde{\mathscr{V}}_{5(6(k_1-m)+4-n)+1}$, we distinguish two cases:

(e.1) If $n = 3$ then $\tilde{\mathscr{V}}_{5(6(k_1-m)+1)+1}$ is a complementary subspace to

$$\mathscr{V}_{5(6(k_1-m)+1)+1} = \mathrm{Range}(\hat{\ell}_{6(5(k_1-m)+1)}) \oplus \mathrm{Span}\left\{h^{5(k_1-m)+1}\right\}.$$

As $\mathrm{Cor}(\hat{\ell}_{6(5(k_1-m)+1)}) = \mathrm{Span}\left\{h^{5(k_1-m)+1}\right\}$, we get $\tilde{\mathscr{V}}_{5(6(k_1-m)+1)+1} = \{0\}$.

(e.2) If $n \neq 3$ then $\tilde{\mathscr{V}}_{5(6(k_1-m)+4-n)+1}$ is a complementary subspace to

$$\mathscr{V}_{5(6(k_1-m)+4-n)+1} = \mathrm{Range}(\hat{\ell}_{5(6(k_1-m)+4-n)+1}).$$

Hence, $\tilde{\mathscr{V}}_{5(6(k_1-m)+4-n)+1} = \mathrm{Cor}(\hat{\ell}_{5(6(k_1-m)+4-n)+1})$ which is the trivial space, except for $n = 5$ where $\mathrm{Cor}(\hat{\ell}_{6(5(k_1-m)-1)+2}) = \mathrm{Span}\left\{xh^{5(k_1-m)-1}\right\}$. In summary, for $k_2 = 4$ we have

$$\mathrm{Cor}(\ell_{21+30k_1}) = \mathrm{Span}\left\{z^{6m+5} x h^{5(k_1-m)-1} : m = 0, \ldots, k_1 - 1\right\}.$$

(f) If $k_2 = 5$, then

$$\mathrm{Cor}(\ell_{1+5(6k_1+5)}) = \sum_{m=0}^{k_1-1} \sum_{n=0}^{5} z^{6m+n} \tilde{\mathscr{V}}_{5(6(k_1-m)+5-n)+1} + \sum_{n=0}^{4} z^{6k_1+n} \tilde{\mathscr{V}}_{5(5-n)+1}.$$

We have $\tilde{\mathscr{V}}_{5(5-n)+1} = \{0\}$ if $n = 1, 2, 3, 4$. For $n = 0$, we obtain $\tilde{\mathscr{V}}_{5(5-n)+1} = \tilde{\mathscr{V}}_{5\cdot 5+1} = \mathrm{Cor}(\hat{\ell}_{6\cdot 4+2}) = \mathrm{Span}\left\{xh^4\right\}$. To characterize $\tilde{\mathscr{V}}_{5(6(k_1-m)+5-n)+1}$, we distinguish two cases:

(f.1) If $n = 4$ then $\tilde{\mathscr{V}}_{5(6(k_1-m)+1)+1}$ is a complementary subspace to

$$\mathscr{V}_{5(6(k_1-m)+1)+1} = \mathrm{Range}(\hat{\ell}_{6(5(k_1-m)+1)}) \oplus \mathrm{Span}\left\{h^{5(k_1-m)+1}\right\}.$$

As $\mathrm{Cor}(\hat{\ell}_{6(5(k_1-m)+1)}) = \mathrm{Span}\left\{h^{5(k_1-m)+1}\right\}$, we get $\tilde{\mathscr{V}}_{5(6(k_1-m)+1)+1} = \{0\}$.

(f.2) If $n \neq 4$ then $\tilde{\mathscr{V}}_{5(6(k_1-m)+5-n)+1}$ is a complementary subspace to

$$\mathscr{V}_{5(6(k_1-m)+5-n)+1} = \mathrm{Range}(\hat{\ell}_{5(6(k_1-m)+5-n)+1}).$$

Thus, $\tilde{\mathscr{V}}_{5(6(k_1-m)+5-n)+1} = \mathrm{Cor}(\hat{\ell}_{5(6(k_1-m)+5-n)+1})$ which is the trivial space, except for $n = 0$ where $\mathrm{Cor}(\hat{\ell}_{6(5(k_1-m)+4)+2}) = \mathrm{Span}\left\{xh^{5(k_1-m)+4}\right\}$.

In summary, for $k_2 = 5$, we obtain

$$\mathrm{Cor}(\ell_{26+30k_1}) = \mathrm{Span}\left\{z^{6m}xh^{5(k_1-m)+4} : m = 0, \dots, k_1\right\}.$$

From Theorem 3 and Proposition 8, we can state the following result:

Theorem 5 *A formal normal form for system* (28) *is:*

$$\dot{\mathbf{x}} = \begin{pmatrix} y \\ x^2 \\ \hline x^3/3 - y^2/2 \end{pmatrix} + \begin{pmatrix} \mathbf{X}_g \\ \hline 0 \end{pmatrix} + \begin{pmatrix} \mu\mathbf{D}_0 \\ \hline 0 \end{pmatrix} + \begin{pmatrix} \lambda\mathbf{X}_h \\ \hline 0 \end{pmatrix} + \begin{pmatrix} 0 \\ \hline \varsigma \end{pmatrix},$$

where

$$g = \sum_{k=1}^{\infty}\sum_{l=0}^{5} \alpha_{k,l} x z^{6k+l},$$

$$\begin{aligned}
\mu = {} & z\Phi(z) + xz^5 h\Phi_{0,2}(z^6, h^5) + xh^4\Phi_{5,0}(z^6, h^5) + \sum_{k=0}^{5} xz^{k+1}h^4\Phi_{k,0}(z^6, h^5) \\
& + \sum_{k=0}^{5} xz^k\Phi_{k,1}(z^6, h^5) + \sum_{k=1}^{5} xz^{k-1}h\Phi_{k,2}(z^6, h^5) + \sum_{k=0}^{1} xz^{k+4}h^2\Phi_{k,3}(z^6, h^5) \\
& + \sum_{k=2}^{5} xz^{k-2}h^2\Phi_{k,3}(z^6, h^5) + \sum_{k=0}^{2} xz^{k+3}h^3\Phi_{k,4}(z^6, h^5) + \sum_{k=3}^{5} xz^{k-3}h^3\Phi_{k,4}(z^6, h^5),
\end{aligned}$$

$$\begin{aligned}
\varsigma = {} & z^2\Psi(z) + xh^4\Psi_{4,0}(z^6, h^5) + xzh^4\Psi_{5,0}(z^6, h^5) + x\Psi_{5,1}(z^6, h^5) + xz^5h^2\Psi_{0,3}(z^6, h^5) \\
& + \sum_{k=0}^{3} xz^{k+2}h^4\Psi_{k,0}(z^6, h^5) + \sum_{k=0}^{4} xz^{k+1}\Psi_{k,1}(z^6, h^5) + \sum_{k=0}^{5} xz^k h\Psi_{k,2}(z^6, h^5) \\
& + \sum_{k=1}^{5} xz^{k-1}h^2\Psi_{k,3}(z^6, h^5) + \sum_{k=0}^{1} xz^{k+4}h^3\Psi_{k,4}(z^6, h^5) + \sum_{k=2}^{5} xz^{k-2}h^3\Psi_{k,4}(z^6, h^5),
\end{aligned}$$

with $\alpha_{k,l} \in \mathbb{R}$, *and* $\Phi(\cdot)$, $\Psi(\cdot)$, $\Phi_{k,l}(\cdot,\cdot)$, $\Psi_{k,l}(\cdot,\cdot)$, *are power series that vanish at the origin and* λ *has the same structure that* μ.

Acknowledgements This work has been partially supported by *Ministerio de Ciencia y Tecnología, Plan Nacional I+D+I* co-financed with FEDER funds, in the frame of the project MTM2014-56272-C2-02, and by *Consejería de Educación y Ciencia de la Junta de Andalucía* (FQM-276 and P12-FQM-1658).

References

1. Algaba, A., Freire, E., Gamero, E.: Hypernormal form for the Hopf-zero bifurcation. Int. J. Bifurc. Chaos **8**, 1855–1887 (1998)
2. Algaba, A., Freire, E., Gamero, E.: Hypernormal forms for equilibria of vector fields. Codimension one linear degeneracies. Rocky Mt. J. Math. **29**, 13–45 (1999)
3. Algaba, A., Freire, E., Gamero, E.: Characterizing and computing normal forms using lie transforms: a survey. Computation of normal forms and applications. Dyn. Contin. Discret. Impulsive Syst. **8**(4), 449–475 (2001)
4. Algaba, A., Freire, E., Gamero, E.: Computing simplest normal forms for the Takens-Bogdanov singularity. Qual. Theory Dyn. Syst. **3**(2), 377–435 (2002)
5. Algaba, A., Freire, E., Gamero, E., García, C.: Quasi-homogeneous normal forms. J. Comput. Appl. Math. **150**, 193–216 (2003)
6. Algaba, A., Freire, E., Gamero, E., García, C.: Quasi-homogeneous normal forms for null linear part. Dyn. Contin. Discret. Impulsive Syst. **10**, 247–261 (2003)
7. Algaba, A., Freire, E., Gamero, E., García, C.: An algorithm for computing quasi-homogeneous formal normal forms under equivalence. Acta Apl. Math. **80**, 335–359 (2004)
8. Algaba, A., Fuentes, N., Gamero, E., García, C.: Normal forms for a class of three-dimensional suspended hamiltonian planar systems. Preprint. (2016)
9. Algaba, A., Gamero, E., García, C.: The integrability problem for a class of planar systems. Nonlinearity **22**, 395–420 (2009)
10. Algaba, A., García, C., Giné, J.: Analytic integrability for some degenerate planar systems. Commun. Pure Appl. Anal. **12**, 2797–2809 (2013)
11. Algaba, A., García, C., Giné, J.: Analytic integrability for some degenerate planar vector fields. J. Differ. Equ. **257**, 549–565 (2014)
12. Algaba, A., García, C., Reyes, M.: Integrability for two dimensional quasi-homogeneous polynomial differential systems. Rocky Mt. J. Math. **41**(1), 1–22 (2011)
13. Baider, A.: Unique normal forms for vector fields and Hamiltonians. J. Differ. Equ. **78**(1), 33–52 (1989)
14. Baider, A., Sanders, J.A.: Further reduction of the Takens-Bogdanov normal form. J. Differ. Equ. **99**(2), 205–244 (1992)
15. Basov, V., Slutskaya, V.: Generalized normal forms of two-dimensional real systems of ordinary differential equations with a quasi-homogeneous polynomial in the unperturbatid part. Differ. Uravn Protsessy Upr. **4**, 108–133 (2010)
16. Chen, G., Wang, D., Yang, J.: Unique normal forms for Hopf-zero vector fields. C. R. Acad. Sci. Paris (Ser. I), 345–348 (2003)
17. Chen, G., Wang, D., Yang, J.: Unique orbital normal forms for vector fields of Hopf-Zero singularity. J. Dyn. Diff. Equ. **17**, 3–20 (2005)
18. Chow, S., Hale, J.K.: Methods of Bifurcation Theory. Springer, New York (1982)
19. Chow, S., Li, C., Wang, D.: Normal Forms and Bifurcations of Planar Vector Fields. Cambridge University Press (1994)
20. Chua, L.O., Kokubu, H.: Normal forms of nonlinear vector fields—part I: theory and algorithm. IEEE Trans. Circuits Syst. CAS **35**, 863–880 (1988)
21. Elphick, C., Tirapegui, E., Brachet, M.E., Coullet, P., Iooss, G.: A simple global characterization for normal forms of singular vector fields. Phys. D **29**, 95–127 (1987)
22. Gazor, M., Mokhtari, F.: Volume-preserving normal forms for Hopf-zero singularity. Nonlinearity **26**, 2809–2832 (2013)
23. Golubitsky, M., Shaeffer, D.G.: Singularities and Groups in Bifurcation Theory, vol. I. Springer, New York (1985)
24. Guckenheimer, J., Holmes, P.: Nonlinear Oscilations, Dynamical System and Bifurcations of Vector Fields. Springer, New York (1983)
25. Iooss, G., Adelmeyer, M.: Topics in Bifurcation Theory and Applications. World Scientific, Singapore (1992)

26. Kokubu, H., Oka, H., Wang, D.: Linear grading function and further reduction of normal forms. J. Differ. Equ. **132**, 293–318 (1996)
27. Lombardi, E., Stolovich, L.: Normal forms of analytic perturbations of quasi-homogeneous vector fields: rigidity, invariant analytic set and exponentially small approximation. Ann. Sci. Ec. Norm. Sup. pp. 659–718 (2010)
28. Strozyna, E.: Normal forms for germs of vector fields with quadratic leading part. The polynomial first integral case. J. Differ. Equ. **259**, 6718–6748 (2015)
29. Strozyna, E., Zoladek, H.: The complete formal normal form for the Bogdanov-takens singularity. Mosc. Math. J. **15**, 141–178 (2015)
30. Takens, F.: Singularities of vector fields. Publ. Math. Inst. Hautes Éstudes Sci **43**, 47–100 (1974)
31. Ushiki, S.: Normal forms for singularities of vector fields. Jpn. J. Appl. Math. **1**, 1–37 (1984)
32. Vanderbauwhede, A.: Centre manifolds, normal forms and elementay bifurcations. Dyn. Rep. **2**, 89–169 (1989)
33. Wang, D., Li, J., Huang, M., Jiang, Y.: Unique normal form of Bogdanov-Takens singularities. J. Differ. Equ. **163**, 223–238 (2000)

Piecewise-Linear (PWL) Canard Dynamics

Simplifying Singular Perturbation Theory in the Canard Regime Using Piecewise-linear Systems

Mathieu Desroches, Soledad Fernández-García, Martin Krupa, Rafel Prohens and Antonio E. Teruel

Abstract In this chapter we gather recent results on piecewise-linear (PWL) slow-fast dynamical systems in the canard regime. By focusing on minimal systems in $\mathbb{R}^2$ (one slow and one fast variables) and $\mathbb{R}^3$ (two slow and one fast variables), we prove the existence of (maximal) canard solutions and show that the main salient features from smooth systems is preserved. We also highlight how the PWL setup carries a level of simplification of singular perturbation theory in the canard regime, which makes it more amenable to present it to various audiences at an introductory level. Finally, we present a PWL version of Fenichel theorems about slow manifolds, which are valid in the normally hyperbolic regime and in any dimension, which also offers a simplified framework for such persistence results.

M. Desroches (✉) · M. Krupa
MathNeuro Team, Inria Sophia Antipolis Research Centre, 2004 route des Lucioles BP93, 06902 Sophia Antipolis Cedex, France
e-mail: mathieu.desroches@inria.fr

S. Fernández-García
Departamento EDAN, Facultad de Matemáticas, University of Sevilla, C/ Tarfia s/n., 41012 Sevilla, Spain
e-mail: soledad@us.es

M. Krupa
Université Côte d'Azur (UCA), Nice, France
e-mail: maciej.krupa@inria.fr

M. Krupa
Laboratoire J. A. Dieudonné, Université de Nice Sophia Antipolis, 06108 Nice Cedex 02, France

R. Prohens · A. E. Teruel
Departament de Matemàtiques i Informàtica, Universitat de les Illes Balears, Carretera de Valldemossa km. 7.5., 07122 Palma de Mallorca, Spain
e-mail: rafel.prohens@uib.es

A. E. Teruel
e-mail: antonioe.teruel@uib.es

V. Carmona et al. (eds.), *Nonlinear Systems, Vol. 1*,
Understanding Complex Systems, https://doi.org/10.1007/978-3-319-66766-9_3

Keywords Piecewise-linear systems · Singularly perturbed systems
Canard solution · Slow manifolds

1 Introduction

Singularly perturbed systems of ordinary differential equations are written in standard form as

$$\varepsilon\dot{\mathbf{x}} = \varepsilon\frac{d\mathbf{x}}{dt} = \mathbf{f}(\mathbf{x}, \mathbf{y}, \varepsilon), \quad \dot{\mathbf{y}} = \frac{d\mathbf{y}}{dt} = \mathbf{g}(\mathbf{x}, \mathbf{y}, \varepsilon), \tag{1}$$

where $(\mathbf{x}, \mathbf{y}) \in \mathbb{R}^q \times \mathbb{R}^s$ are the state variables, $\mathbf{f}$ and $\mathbf{g}$ are sufficiently smooth functions and $0 < \varepsilon \ll 1$ is a small parameter. From the expression above, the coordinates of $\mathbf{x}$ and $\mathbf{y}$ evolve with a different speed, provided that ε is small enough. Thus, the coordinates of $\mathbf{x}$ are called fast variables, while the coordinates of $\mathbf{y}$ are called slow variables. The time variable t is referred to as the slow time.

Changing the time t to the fast time $\tau = t/\varepsilon$, system (1) is written as

$$\mathbf{x}' = \frac{d\mathbf{x}}{d\tau} = \mathbf{f}(\mathbf{x}, \mathbf{y}, \varepsilon), \quad \mathbf{y}' = \frac{d\mathbf{y}}{d\tau} = \varepsilon\mathbf{g}(\mathbf{x}, \mathbf{y}, \varepsilon). \tag{2}$$

Systems (1) and (2) are differentiably equivalent and their phase portraits are the same. Both dynamics exhibit an slow-fast explicit splitting. In this setting, systems (1) and (2) are called slow-fast systems. Often, system (1) is referred to as the slow system whereas system (2) is called the fast system.

Fenichel's geometric theory [11] allows to analyse the dynamics of the perturbed system (1) by combining the behaviour of the singular orbits, corresponding to the limiting cases given by $\varepsilon = 0$. In particular, by setting $\varepsilon = 0$ in Eqs. (1) and (2), we get respectively the differential algebraic equation (DAE)

$$\mathbf{0} = \mathbf{f}(\mathbf{x}, \mathbf{y}, 0), \quad \dot{\mathbf{y}} = \mathbf{g}(\mathbf{x}, \mathbf{y}, 0), \tag{3}$$

typically referred to as the slow subsystem or reduced problem, and the fast subsystem or layer problem

$$\mathbf{x}' = \mathbf{f}(\mathbf{x}, \mathbf{y}, 0), \quad \mathbf{y}' = \mathbf{0}. \tag{4}$$

The reduced problem consists of an s-dimensional vector field defined on the critical manifold $\mathscr{S} = \{(\mathbf{x}, \mathbf{y}) \in \mathbb{R}^{q+s} \mid \mathbf{f}(\mathbf{x}, \mathbf{y}, 0) = \mathbf{0}\}$, which is assumed to be an s-dimensional manifold. Regarding the layer problem, its dynamical behaviour takes place along q-dimensional fibers which are formed by considering $\mathbf{y}$ constant. Hence, both limiting problems have dimension lower than that of the perturbed system. Moreover, the critical manifold $\mathscr{S}$ plays a key role in both limiting problems: it is the phase space of the reduced systems and it corresponds to singular points of the layer problem. A singular point $(\mathbf{x}_0, \mathbf{y}_0) \in \mathcal{S}$ is said to be normally hyperbolic if the eigenvalues of the Jacobian matrix $\mathbf{f_x}(\mathbf{x}_0, \mathbf{y}_0, 0)$ have nonzero real part.

The flow of the reduced problem can be analysed by differentiating the equation of the critical manifold $\mathcal{S}$ with respect to the slow time t, which yields

$$\dot{\mathbf{x}} = \pm \mathbf{f_x}^{-1} \mathbf{f_y} \mathbf{g}(\mathbf{x}, \mathbf{y}, 0), \quad \dot{\mathbf{y}} = \mathbf{g}(\mathbf{x}, \mathbf{y}, 0), \tag{5}$$

where $\mathbf{f_x}$ (resp. $\mathbf{f_y}$) denotes the differential of $\mathbf{f}$ with respect to the fast (resp. slow) variables. Then, system (5) is clearly singular at non-hyperbolic points (in particular in the fold set $\mathcal{F}$), which can be remedied by rescaling time by a factor $\pm\det(\mathbf{f_x})$ (owing to Kramer's rule). This brings the so-called *desingularised reduced system* (DRS)

$$\dot{\mathbf{x}} = \mathbf{f_y} \mathbf{g}(\mathbf{x}, \mathbf{y}, 0), \quad \dot{\mathbf{y}} = \pm \det(\mathbf{f_x}) \mathbf{g}(\mathbf{x}, \mathbf{y}, 0), \tag{6}$$

which by construction is regular everywhere including in the fold set, and has the same orbits as the reduced system with simply an opposite direction along the repelling sheet $\mathcal{S}^r$ of the critical manifold.

Consider $\mathcal{S}_0 \subset \mathcal{S}$ a compact set such that every point in $\mathcal{S}_0$ is a normally hyperbolic singular point. From Fenichel's Theorems [11], the submanifold $\mathcal{S}_0$ persists as a locally invariant slow manifold $\mathcal{S}_\varepsilon$, of the perturbed system (1) for every small enough ε. Moreover, the restriction of the flow of the perturbed system (1) to the slow manifold $\mathcal{S}_\varepsilon$ is a small smooth perturbation of the flow of the reduced problem (3). Fenichel also proved that there exists an invariant foliation with basis $\mathcal{S}_\varepsilon$ with the dynamics along each fiber being a small smooth perturbation of the flow of the layer problem. See also Jones [18], for a survey on geometric singular perturbation theory (GSPT).

In Sect. 2, we show that these results apply when the assumption of smoothness of the vector field is relaxed. In particular, we state a variant of Fenichel theorem in the context of piecewise-linear slow-fast systems, with slow dynamics given by a linear differential equation and a critical manifold given by the graph of a piecewise linear (PWL) function. A key aspect of this result is that, due to the PWL setting, explicit formulation are obtained for canonical linear slow manifolds.

Following Fenichel results, under normal hyperbolicity conditions, orbits of the perturbed system (1) are composed by slow and fast segments. The former ones evolving close to the flow defined over the slow manifold, while the latter ones are following the flow defined over the fast fibers. A general question is: *what does remain of this dynamical behaviour when normal hyperbolicity is lost?* in particular, at points $(\mathbf{x}_0, \mathbf{y}_0) \in \mathcal{S}$ where the critical manifold is folded, that is, at which the determinant of the Jacobian matrix $\mathbf{f_x}(\mathbf{x}_0, \mathbf{y}_0, 0)$ vanishes. Typically, when the critical manifold $\mathcal{S}$ folds, then the fold set (a point or a curve in the most examples treated here) separates branches with different stability properties. Consequently, attracting (resp. repelling or saddle-type) branches of $\mathcal{S}$ perturb to attracting (resp. repelling or saddle-type) slow manifolds $\mathcal{S}^a_\varepsilon$ (resp. $\mathcal{S}^r_\varepsilon$ or $\mathcal{S}^s_\varepsilon$). Then, in the vicinity of the fold set of $\mathcal{S}$, conditions can be obtained for slow manifolds with different stability to connect, hence allowing for the existence of orbits which closely follow an attracting slow manifold $\mathcal{S}_\varepsilon^a$, pass close to fold set of $\mathcal{S}$, and subsequently follow closely a repelling slow manifold, $\mathcal{S}_\varepsilon^r$. These orbits are called *canards* and they play a crucial role in explaining complicated slow-fast dynamics organising the transition between stationary and relaxation regimes in planar systems (see Sect. 3) or the transition between different oscillatory regimes (see Sect. 4).

The aforementioned conditions for slow manifolds with different stability to connect, are obtained by the linear analysis of certain equilibria of the DRS (6), namely those lying on the fold set $\mathcal{F}$ and hence satisfying $\det(\mathbf{f}_\mathbf{x}) = 0$. Such equilibria are called *folded singularities* and they appear due to the (singular) time rescaling which transforms (5) into (6). Note that folded singularities are not equilibria of the slow flow. Therefore, depending on the local behaviour in a neighbourhood of the folded singularity, trajectories starting on $\mathcal{S}^a$ may cross them in finite time and continue flowing along $\mathcal{S}^r$, which is a singular canard behaviour. These singular canards allow for the existence of canard solutions in the original system for small enough $\varepsilon > 0$; see Sect. 4 for details.

Seminal and classical papers on canards in planar systems are those of Benoît et al. [3], Dumortier and Roussarie [10], and Krupa and Szmolyan [21, 22]. Regarding canards in higher-dimensional systems with (at least) two slow and one fast variables, see [2, 4, 28]; a recent survey can be found in [8].

Singularly perturbed PWL systems exhibiting canard dynamics are considered in Sects. 3 and 4 in the two and three dimensional cases with two slow variables, respectively; a brief summary of initial results on an example of three-dimensional case with two fast variables (in the context of *bursting*) is briefly mentioned in the conclusion section. Through these examples one can conclude that the PWL framework is able to reproduce all salient dynamical features present in the smooth case, both qualitatively and quantitatively, while allowing for a substantial level of simplification. What is more, these examples also suggest elements that naturally appear in the PWL setting and help revisiting unsettled questions from the smooth case. To paraphrase Diener in [9], the natural biotope of canards is that of PWL vector field. At least, it seemingly appears as the simplest environment in which one can understand the essence of canard dynamics while dropping all unnecessary refinements.

2 Canonical Fenichel Slow Manifolds

Under suitable conditions (most importantly, normal hyperbolicity of the unperturbed manifold), Fenichel theory guarantees the existence of slow manifolds perturbing from the critical manifold, when $\varepsilon > 0$ is sufficiently small. These slow manifolds are locally invariant by the flow of the smooth system (1); however, just like centre manifolds, Fenichel slow manifolds are not necessarily unique. A proof of these results can be found in [18, 19].

In this section a version of Fenichel Theorem is stated in the context of slow-fast PWL systems. In particular, we consider slow-fast PWL systems with s slow and 1

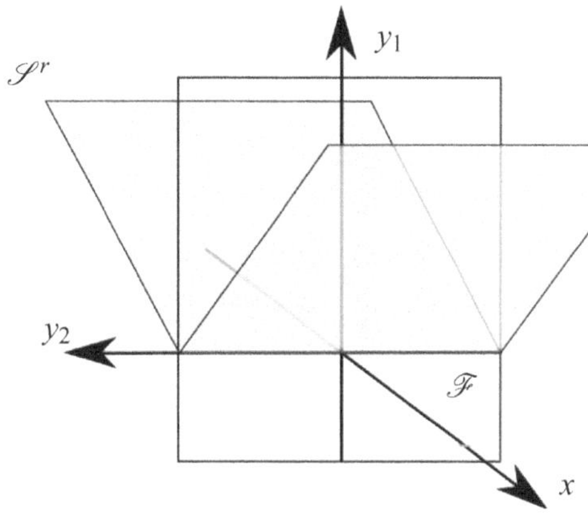

Fig. 1 3D representation of the critical manifold $\mathscr{S}$ of system (7). From the fast subsystem, it can be noticed that the attracting branch of the critical manifold $\mathscr{S}^a$ corresponds to the half–plane contained in the half–space $\{x > 0\}$, the repelling branch of the critical manifold $\mathscr{S}^r$ corresponds to the half–plane contained in the half–space $\{x < 0\}$, and the fold manifold $\mathscr{F}$ corresponds to the segment contained in the switching manifold $\{x = 0\}$

fast variables, of the form

$$\begin{cases} x' = -|x| + \mathbf{e}_1^T \mathbf{y}, \\ \mathbf{y}' = \varepsilon(\mathbf{a}x + A\mathbf{y} + \mathbf{b}), \end{cases} \tag{7}$$

where $A = (a_{ij})_{1\le i,j\le s}$ is an $s \times s$ real matrix, $\mathbf{e}_1$ is the first element of the canonical basis in $\mathbb{R}^s$, $\mathbf{a} = (a_i)^T_{1\le i,j\le s}$ and $\mathbf{b} = (b_i)^T_{1\le i,j\le s}$ are vectors in $\mathbb{R}^s$, and the superscript T stands for the transposed vector. Note that system (7) is linear on each side of the switching manifold $\{x = 0\}$.

The critical manifold associated with system (7) is $\mathscr{S} = \{(x, \mathbf{y}) : |x| = \mathbf{e}_1^T \mathbf{y}\}$. It is formed by the union of the two s-dimensional manifolds

$$\begin{aligned} \mathscr{S}^a &= \{(x, \mathbf{y}) : x > 0,\ x = \mathbf{e}_1^T \mathbf{y}\}, \\ \mathscr{S}^r &= \{(x, \mathbf{y}) : x < 0,\ x = -\mathbf{e}_1^T \mathbf{y}\}, \end{aligned} \tag{8}$$

connected by the $(s-1)$-dimensional fold set $\mathscr{F} = \{(0, \mathbf{y}) : \mathbf{e}_1^T \mathbf{y} = 0\}$, see Fig. 1 for a three-dimensional representation.

The critical manifold $\mathscr{S}$ is normally hyperbolic, except in the fold set $\mathscr{F}$; the branch $\mathscr{S}^a$ is attracting and the branch $\mathscr{S}^r$ is repelling. Since the vector field defined by (7) is smooth in each of the half–spaces $\{x > 0\}$ and $\{x < 0\}$ (it is linear), Fenichel theory applies locally to each of these systems. Therefore compact submanifolds of the two branches $\mathscr{S}^a$ and $\mathscr{S}^r$ persist under the flow of system (7) as locally invariant slow manifolds for small enough $\varepsilon > 0$.

A strong gain of using the PWL setting is that one can prove the persistence of the entire manifolds $\mathscr{S}^a$ and $\mathscr{S}^r$ as locally invariant slow manifolds, and not just compact submanifolds; we denote these slow manifolds by $\mathscr{S}^a_\varepsilon$ and $\mathscr{S}^r_\varepsilon$, respectively. Since these manifolds are contained in the half–spaces where the system is linear, their dynamical behaviour can be deduced from the spectra of the corresponding matrices, that is,

$$B_\varepsilon^+ = \begin{pmatrix} -1 & \mathbf{e}_1^T \\ \varepsilon\mathbf{a} & \varepsilon A \end{pmatrix} \quad \text{and} \quad B_\varepsilon^- = \begin{pmatrix} 1 & \mathbf{e}_1^T \\ \varepsilon\mathbf{a} & \varepsilon A \end{pmatrix}, \tag{9}$$

respectively.

Following [27], we can obtain explicit equations for these slow manifolds by proceeding as follows. The spectrum of $B_\varepsilon^\pm$ decomposes into two parts: one composed by a real eigenvalue of $O(1)$ and the other one formed by s eigenvalues (counted with multiplicity) of $O(\varepsilon)$. We consider the spectra of both B_ε^+ and B_ε^- simultaneously (see [27, Lemma 3]) and write the eigenvalues as

$$\lambda_1^\pm = \mp 1 + O(\varepsilon) \quad \text{and} \quad \lambda_k^\pm = \beta_k^\pm \varepsilon + O(\varepsilon^2),\ k = 2, \ldots, s+1.$$

The eigenvalue $\lambda_1^\pm$ is responsible for the fast dynamics whereas the s eigenvalues $\lambda_k^\pm$ are responsible for the slow dynamics. Consequently, for small enough $\varepsilon > 0$ the slow dynamics in the half–space $\{x > 0\}$ is restricted to the half–hyperplane defined by the generalized eigenvectors associated with the eigenvalues $\{\lambda_k^+\}_{k=2}^{s+1}$. From [27, Lemma 5], we conclude that the slow manifold in $\{x \geq 0\}$ is given by the half–hyperplane

$$\mathscr{S}_\varepsilon^a = \left\{(x, \mathbf{y}) \in \mathbb{R}^n : x \geq 0,\ x = \mathbf{e}_1^T(\varepsilon A - \lambda_1^+ I)^{-1}\left(\frac{\varepsilon}{\lambda_1^+}\mathbf{b} + \mathbf{y}\right)\right\}, \tag{10}$$

see [27] for more details. As mentioned above, the fast dynamics in $\{x > 0\}$ is organized by the fast negative eigenvalue λ_1^+. Therefore, $\mathscr{S}_\varepsilon^a$ is an attracting slow manifold.

Similarly, the slow dynamics in the half–space $\{x < 0\}$ is restricted to the half–hyperplane defined by the generalized eigenvectors associated with the eigenvalues $\{\lambda_k^-\}_{k=2}^{s+1}$ and given by

$$\mathscr{S}_\varepsilon^r = \left\{(x, \mathbf{y}) \in \mathbb{R}^n : x \leq 0,\ x = \mathbf{e}_1^T(\varepsilon A - \lambda_1^- I)^{-1}\left(\frac{\varepsilon}{\lambda_1^-}\mathbf{b} + \mathbf{y}\right)\right\}. \tag{11}$$

In this case, the fast dynamics is organized by the positive eigenvalue λ_1^-, and hence, $\mathscr{S}_\varepsilon^r$ is a repelling slow manifold.

A 3-dimensional representation of the slow manifolds is shown in Fig. 2. The slow manifolds in the 3-dimensional PWL system have been explicitly computed in [26].

One can prove that $\mathscr{S}_\varepsilon^{a,r}$ are Fenichel slow manifolds and that they possess similar properties as Fenichel slow manifolds in smooth systems. Therefore, one can extend Fenichel's theorem to the case of PWL system, as stated below; see [26, 27] for a proof of this result.

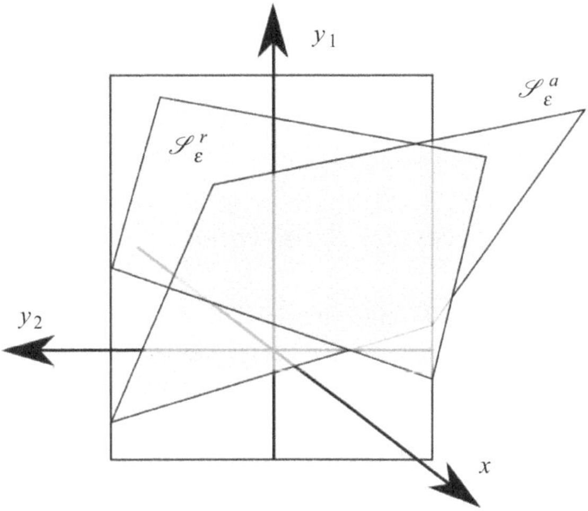

Fig. 2 3D representation of the attracting slow manifold $\mathscr{S}^a_\varepsilon$ and the repelling slow manifold $\mathscr{S}^r_\varepsilon$ of system (7)

Theorem 1 (Fenichel theorem for PWL systems). *For $\varepsilon > 0$ and sufficiently small, the manifolds $\mathscr{S}^{a,r}_\varepsilon$ satisfy the following:*

(a) $\mathscr{S}^{a,r}_\varepsilon$ *is locally invariant under the flow of system* (7).
(b) The restriction of the flow of system (7) *to* $\mathscr{S}^{a,r}_\varepsilon$ *is a regular perturbation of the flow of the reduced problem defined on the critical manifold* $\mathscr{S}$.
(c) $\mathscr{S}^r_\varepsilon$ *is a repelling slow manifold and* $\mathscr{S}^a_\varepsilon$ *is an attracting slow manifold.*
(d) Given a compact subset $\hat{\mathscr{S}}^a$ *(resp.* $\hat{\mathscr{S}}^r$*) of* $\mathscr{S}^a$ *(resp.* $\mathscr{S}^r$*), there exists a compact subset* $\hat{\mathscr{S}}^a_\varepsilon$ *(resp.* $\hat{\mathscr{S}}^r_\varepsilon$*) of the slow manifold* $\mathscr{S}^a_\varepsilon$ *(resp.* $\mathscr{S}^r_\varepsilon$*) which is diffeomorphic to* $\hat{\mathscr{S}}^a$ *(resp.* $\hat{\mathscr{S}}^r$*) such that* $d_H(\hat{\mathscr{S}}^a_\varepsilon, \hat{\mathscr{S}}^a) = O(\varepsilon)$ *(resp.* $d_H(\hat{\mathscr{S}}^r_\varepsilon, \hat{\mathscr{S}}^r) = O(\varepsilon)$*), where* d_H *denotes the Hausdorff distance.*

2.1 Simplification in the PWL Setting

Contrary to the original Fenichel's Theorem, Theorem 1 offers an explicit expression for $\mathscr{S}^{a,r}_\varepsilon$. These slow manifolds are *canonical* in the sense that they are uniquely defined, they are the only linear slow manifolds as well as the only ones on which the dynamics has no influence from the fast eigenvalues. In other words, solutions on any other invariant manifold contain a component of the form $e^{t\lambda_1^\pm}$. Hence, as soon as this component becomes dominant, the orbit is not part of a slow manifold any more. That is why all other (nonlinear) slow manifolds are only locally invariant. Hence, $\mathscr{S}^{a,r}_\varepsilon$ are canonically slow and, to a certain extent, they are the "best" Fenichel manifolds that one can hope for in any singularly perturbed system. This is a major difference with the smooth case and the existence of such unique linear slow manifolds offers a key advantage. Indeed, their explicit equations (10) and (11) are very useful to locate maximal canard solutions, which are specific orbits passing from the attracting slow manifold $\mathscr{S}_\varepsilon^a$ to the repelling slow manifold $\mathscr{S}_\varepsilon^r$; see Sects. 3 and 4 below and [6, 26, 27] for details.

2.2 A Necessary Perturbation to Obtain Canard Dynamics

Within the PWL setup presented in the previous section, one can entirely reproduce relaxation oscillations that are typical in van der Pol (VDP) type systems. One only needs to consider in place of the cubic critical manifold of the VDP system, the graph of a piecewise-linear function with three pieces; in other words, one approximates the quadratic fold points of the VDP critical manifold by corners. Then, relaxation oscillations can be generated and their properties are perfectly similar to those generated by the VDP system. Moreover, when the slow nullcline is non-vertical, then the resulting PWL caricature of the VDP system is typically called the McKean model and it has been thoroughly studied in the relaxation regime since the early 1970s [23]. In fact, the McKean model is a caricature of the so-called FitzHugh-Nagumo model, which amounts to the VDP system where the slow nullcline is not vertical and gives a simple phenomenological model of action potential generation in neurons [15, 25]. However, when attempting to reproduce canard dynamics from the VDP system, approximating the quadratic fold points of the associated critical manifold by corners is not sufficient and a refinement is required in order to recover the slow passage from the attracting side to the repelling side of the critical manifold, as we explain below. Indeed, since the late 1990s with the work of Arima et al. [1], it is known that three linearity zones are needed to approximate locally the van der Pol system in order to get canard dynamics. In order words, the PWL critical manifold must locally have three segments in order to correctly approximate the quadratic critical manifold of the van der Pol system and open the possibility for canard cycles to appear.

We now consider the following slow-fast PWL systems,

$$\begin{cases} x' = -|x|_\delta + \mathbf{e}_1^T \mathbf{y}, \\ \mathbf{y}' = \varepsilon(\mathbf{a}x + A\mathbf{y} + \mathbf{b}), \end{cases} \tag{12}$$

where the generalized absolute value function $|x|_{\boldsymbol{\delta}}$ is defined as follows

$$|x|_{\boldsymbol{\delta}} = \begin{cases} -x - (m+1)\delta^- & x \leq -\delta^-, \\ mx & -\delta^- \leq x \leq \delta^+, \\ x + (m-1)\delta^+ & \delta^+ \leq x, \end{cases}$$

s and $\boldsymbol{\delta} = (\delta^-, m, \delta^+)$ is a continuous function of ε such that $\boldsymbol{\delta}(0) = \mathbf{0}$.

Since the function $|x|_{\mathbf{0}}$ coincides with the absolute value function, the layer and the reduced problems associated with system (12) are identical to those associated with system (7). Thus, the critical manifold $\mathscr{S}$ also coincides with that defined for system (7).

For $\varepsilon > 0$ small enough, system (12) is a slow-fast PWL system, locally linear in the three closed regions $\{x \leq -\delta^-\}$, $\{-\delta^- \leq x \leq \delta^+\}$, and $\{x \geq \delta^+\}$, which we will refer to from now on as the left, central and right zones, respectively. Therefore, the dynamics of system (12) in the lateral regions can be deduced from the spectra of the matrices (9), whereas in the central region it is deduced from the spectrum of the matrix

$$B_\varepsilon^0 = \begin{pmatrix} m & \mathbf{e}_1^T \\ \varepsilon \mathbf{a} & \varepsilon A \end{pmatrix}.$$

As previously shown, the slow behaviour in the lateral regions is reduced to linear manifolds, which are defined by the eigenvectors associated with the slow eigenvalues. Therefore, the canonical slow manifolds in the lateral regions are parallel to those defined in (10) and (11), namely we have

$$\mathscr{S}_\varepsilon^a = \left\{ (x, \mathbf{y}) \in \mathbb{R}^n : x \geq \delta^+,\ x = \mathbf{e}_1^T (\varepsilon A - \lambda_1^+ I)^{-1} \left(\frac{\varepsilon}{\lambda_1^+} \mathbf{b} + \mathbf{y} \right) - \frac{(1-m)\delta^+}{\lambda_1^+} \right\},$$
$$\mathscr{S}_\varepsilon^r = \left\{ (x, \mathbf{y}) \in \mathbb{R}^n : x \leq -\delta^-,\ x = \mathbf{e}_1^T (\varepsilon A - \lambda_1^- I)^{-1} \left(\frac{\varepsilon}{\lambda_1^-} \mathbf{b} + \mathbf{y} \right) - \frac{(1+m)\delta^-}{\lambda_1^-} \right\}. \tag{13}$$

Note that the eigenvalue λ_1^+ is the fast one in the right zone $\{x \geq \delta^+\}$ and the eigenvalue λ_1^- is the fast one in the left zone $\{x \leq -\delta^-\}$.

Regarding the dynamics in the central region, every non zero eigenvalue of B_ε^0 is $O(\varepsilon^\alpha)$, $\alpha \in \mathbb{R}$. Hence, in order for the flight time in the central region not to diverge to infinity as $\varepsilon \to 0$, we assume that δ^+ and δ^- have greater order in ε than the smaller non-zero eigenvalue. Recent work has allowed to refine the results from [1] and find that the optimal of the central zone is $O(\sqrt{\varepsilon})$ [6].

Arima and co-authors have computed numerically small (so-called *headless*) canard cycles in a PWL approximation of the van der Pol system similar to (12) as well as large canard cycles (so-called *canards with head*) in a four-zone system. They gave arguments to justify the need for the three-piece critical manifold in order to make sense of a repelling slow manifold and, hence, find canards. However, they did not develop GSPT arguments proving the existence of canards in this planar context and they did not investigate three-dimensional canard systems either. This has been done more recently in [6, 13]. We summarise the results obtained in these two papers in the next two sections, which will then be entirely focused on canards in slow-fast PWL systems.

3 Canard Explosion

Canard dynamics can be loosely described as a complicated mix of local passage (near non-normally hyperbolic regions of the critical manifold) and global return (or reinjection) mechanism, which allows for recurrent dynamics. Note that need not be part of recurrent dynamics. Their main feature is this local passage and it is well approximated by linear dynamics, as we shall see below. In the smooth case, the dynamics during this local passage is organised by the Weber equation, obtained (after a coupled of changes of variables) by linearising the system along the so-called "weak canard" (axis of rotation for trajectories during this local passage). It is interesting to notice that solutions to the Weber equation can be expressed in terms of parabolic cylinder functions, while in the three-dimensional PWL slow-fast system that we propose in Sect. 4, solutions in the central zone (approximating the local passage) are organised by invariant cylinders. In this context, canard- induced mixed-mode oscillations (MMOs, see Sect. 4) are a combination of small-amplitude oscillations (SAOs) near a fold curve, a passage near a repelling slow manifold and a global return that reinjects trajectory on an attracting slow manifold. In this context, the SAO part can be purely explained by linear dynamics. As already mentioned, generally speaking canards arise due to connections between an attracting slow manifold and a repelling slow manifold. These are rare events and the associated connecting orbits are called *maximal canards*; other canard solutions exist in an exponentially small neighbourhood of maximal canards. It turns out that connections between two such slow manifolds can be perfectly reproduced with PWL systems, however with an additional linearity zone in between them so as to make the passage and, hence, the connection possible. Therefore, by decomposing the dynamics of canard systems in phase space into several linear zones, one cannot only reproduce all canard phenomena—planar, three-dimensional, canard-induced MMOs, canard-induced bursting trajectories, etc.—but also properly zoom into these intricate dynamics and extract their essential features. Overall, one can say that the PWL setup offers a simpler alternative to method of "geometric desingularisation" (blow-up) studied in smooth canard systems [10, 21, 22, 28]. To this extent, the small zone in between the attracting region and the repelling region can be seen (in a loose sense) as a blow-up of the corner that one would naturally use to approximate, in the relaxation regime, the quadratic fold of the van der Pol oscillator with a PWL system.

In this section, we summarise the results published in [13] about canard cycles in planar slow-fast PWL systems. We consider the following planar version of system (12)

$$\begin{cases} x' = -|x|_\delta + y, \\ y' = \varepsilon(a - x), \end{cases} \tag{14}$$

for the parameter vector

$$\delta = \left(\frac{\varepsilon}{1-b}, -b, \frac{\varepsilon}{1+b}\right),$$

where $|b| < 2\sqrt{\varepsilon}$ and $a \in (-\varepsilon/(1-b), \varepsilon/(1+b))$.

System (14) possesses exactly one equilibrium point which is in central zone namely, $\mathbf{p}^C = (a, -ab)^T$. The topological type of this equilibrium depends on parameter b: if $b = 0$, the equilibrium point $\mathbf{p}^C$ is a center; If $b < 0$ (resp. $b > 0$) it is a stable (resp. unstable) focus.

Using the method described in Sect. 2, we can compute the canonical slow manifolds $\mathscr{S}_\varepsilon^{a,r}$ associated with system (14) and obtain formulas equivalent to (13), namely,

$$\mathscr{S}_\varepsilon^a = \left\{(x, y) \in \mathbb{R}^2 : x \geq \frac{\varepsilon}{1+b},\ y = -\frac{\varepsilon}{\lambda_2^+}(x-a) + a - \varepsilon\right\},$$
$$\mathscr{S}_\varepsilon^r = \left\{(x, y) \in \mathbb{R}^2 : x \leq -\frac{\varepsilon}{1-b},\ y = -\frac{\varepsilon}{\lambda_2^-}(x-a) - a - \varepsilon\right\}. \tag{15}$$

Here $\lambda_2^+ < 0$ and $\lambda_2^- > 0$ are the slow eigenvalues of the matrices B_ε^+ and B_ε^- defining the lateral linear systems, equivalent to (9).

When parameters a and b are zero, system (14) is reversible with respect to the involution $x \mapsto -x$ and the time change $t \mapsto -t$. In such a case, the slow manifolds $\mathscr{S}_\varepsilon^{a,r}$ are also images of one another under this involution and this time change. Therefore, they can connect by forming a maximal canard. The maximal canard splits the phase plane into two regions: one region contains the equilibrium point and is foliated by periodic orbits (see Fig. 3), the other one is fully foliated by unbounded orbits ([13, Theorem 4.1]).

By perturbing this non-generic situation, it is possible to find a curve $a = \tilde{a}(b, \varepsilon)$ in parameter space, with Taylor series expansion at $b = 0$ given by

$$a = \tilde{a}(b, \varepsilon) = \left(\frac{\pi}{4}\sqrt{\varepsilon} + O(\varepsilon)\right) b + O(b^3), \tag{16}$$

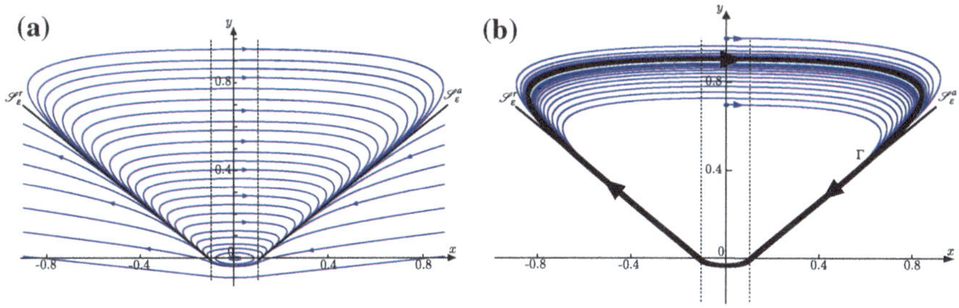

Fig. 3 **a** Continuum of canard periodic orbits bounded by the canonical slow manifolds $\mathscr{S}_\varepsilon^{a,r}$ (adapted from [13, Fig. 4]). **b** Stable (headless) canard limit cycle together with the canonical slow manifolds and two trajectories approaching the canard cycle in forward time (adapted from [13, Fig. 5])

such that, the maximal canard orbit persists ([13, Theorem 4.2 and Prop. 4.4]).

By breaking the connection that corresponds to the maximal canard, one can prove the existence of a family of canard cycles in system (14), as stated below (see [13, Theorems 4.3 and 4.5] for a proof).

Theorem 2 *For each point* $(0, y_0)$ *with* $y_0 > 0$, *there exists* $U \subset \mathbb{R}^2$ *containing* $(b, \varepsilon) = (0, 0)$, *such that, for* $(b, \varepsilon) \in U \cap \{\varepsilon > 0\}$, *there exists a curve* $a(b, \varepsilon)$ *in parameter space, with the same first terms in its Taylor series expansion as* $\tilde{a}(b, \varepsilon)$, *such that system* (14) *possesses a canard cycle passing through* $(0, y_0)$. *Moreover, the family of canard limit cycles is asymptotically stable if* $b > 0$ *and unstable if* $b < 0$.

The main conclusion one can draw from the results stated so far is that canard phenomena in the PWL framework and in the classical (smooth) context have very similar features. In particular in the way they are born: a Hopf bifurcation in the smooth case, and a two-zonal Hopf-like [16, 17] bifurcation in the PWL case, which occurs when the real equilibrium point enters the central zone from either of the lateral zones, by suitably moving parameter a. If we then consider $b = 2\tilde{b}\sqrt{\varepsilon}$, then we obtain the following:

- if $b > 0$, a two-zonal supercritical Hopf-like bifurcation takes place when the equilibrium enters the central zone from the right, at

$$a_H^R(b, \varepsilon) = \frac{\varepsilon}{1 + b} = \varepsilon + O(\varepsilon^{3/2});$$

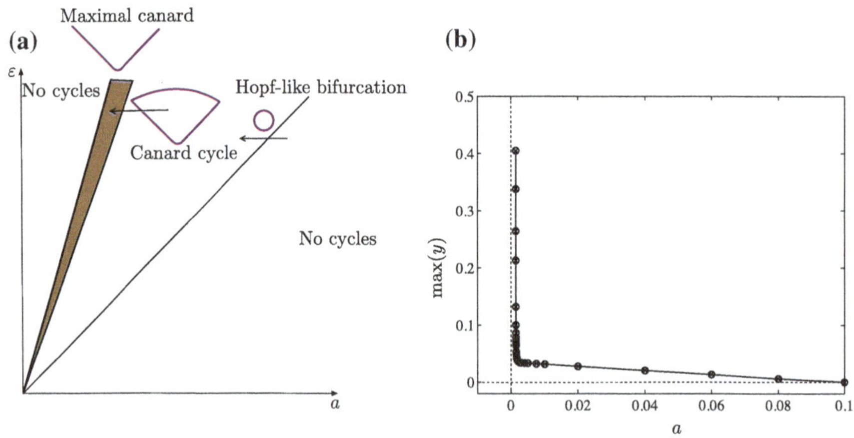

Fig. 4 **a** Bifurcation diagram for $b > 0$. Consider $\varepsilon > 0$ fixed and $a > 0$ in the rightmost sector. By decreasing a, a Hopf-like bifurcation takes place, giving rise to a small stable limit cycle. The limit cycle is growing as a decreases. When a reaches the grey-shaded region, the limit cycle becomes a canard cycle. Along the leftmost line the family of canard cycles ends at a maximal canard connection. **b** Explosive branch of limit cycles obtained by direct simulation when varying parameter a for fixed $\varepsilon = 0.1$ and $b = 0.009944$ (adapted from [13, Fig. 6])

- if $b < 0$, a two-zonal subcritical Hopf-like bifurcation takes place when the equilibrium enters the central zone from the left, at

$$a_H^L(b,\varepsilon) = -\frac{\varepsilon}{1-b} = -\varepsilon + O(\varepsilon^{3/2}).$$

In both cases, subsequently to the Hopf-like bifurcation, the amplitude of the two-zonal limit cycle grows linearly in the two regions until it becomes a three-zonal limit cycle. Then, the third linear system affects dramatically the dynamics and the limit cycle starts to grow very rapidly, *explosively* (in terms of parameter variation), as it becomes a canard. While a increases within an exponentially small range, the amplitude of the canard cycle increases by an $O(1)$ amount until the a-value where the maximal canard occurs, $\tilde{a}(b,\varepsilon)$; see (16). Past the maximal canard, the limit cycle disappears. Figure 4 presents the bifurcation diagram corresponding to the case $b > 0$.

4 Folded Singularities and Their Canards

Canards have been also extensively studied in systems with more than one slow and one fast variables. In particular, much is known about maximal canards in the context of three-dimensional systems with two slow variables where they appear through the presence of folded singularities, which are the equivalent to the 3D setting of canard points in VDP type systems; see [2, 4, 8, 28]. This section presents a summary of results recently published in [6] about folded singularities in 3D PWL slow-fast systems.

We consider the following PWL slow-fast system

$$\begin{cases} x' = |x|_\delta - y, \\ y' = \varepsilon(p_1 x + p_2 z), \\ z' = \varepsilon p_3 \end{cases} \tag{17}$$

which then corresponds to system (12) after the change of variable $x \mapsto -x$ and with

$$\mathbf{a} = \begin{pmatrix} -p_1 \\ 0 \end{pmatrix}, \quad A = \begin{pmatrix} 0 & p_2 \\ 0 & 0 \end{pmatrix}, \quad \mathbf{b} = \begin{pmatrix} 0 \\ p_3 \end{pmatrix},$$

and $\boldsymbol{\delta} = (\delta, 0, \delta)$ for a given $\delta > 0$.

Given that system (17) can be recasted as system (12) for suitable values of matrix coefficients, we can readily apply the Fenichel analysis performed on this latter system and conclude that system (17) possesses canonical Fenichel slow manifolds $\mathscr{S}_\varepsilon^{a,r}$ given by Eq. (13). Slow-fast systems of this form are minimal three-dimensional systems with two slow variables displaying canard dynamics. Minimal here means that the z-dynamics is a simple slow drift. Therefore system (17) can be seen as a two-dimensional canard (VDP type) system where the parameter controlling the slow nullcline moves slowly. Such systems were first studied in [2] where conditions for connections between attracting and repelling slow manifold, that is, for maximal canards, were obtained in link with the presence of special points located along the fold curve and called folded singularities. These special points arise due to the existence of a non-normally hyperbolic set (the fold set) on the critical manifold, and they are defined in the singular limit, from the flow of the reduced system (3) (also referred to as the *slow flow*). Indeed, typical points on the fold set are jump points in the sense that the slow flow is directed towards the fold set on both sides (attracting and repelling) of such points without being defined at these points; this is because the slow flow is typically singular on the fold set. When perturbing in $\varepsilon > 0$, these points give rise to relaxation dynamics. However, one can find algebraic conditions for which points on the fold set at which the slow flow is not singular and has the same direction on both sides, hence allowing for a passage from one side of the critical manifold to the other. These special points are folded singularities (or folded equilibria) and they can be obtained as true equilibria of the DRS (6), which has the same trajectories as the slow flow but with opposite orientation on the repelling side; therefore, an equilibrium of the DRS corresponds to a point of the reduced system at which the slow flow crosses from $\mathscr{S}^a$ to $\mathscr{S}^r$. When switching on ε with a small-enough positive value, these points give rise to canard dynamics: "true" canards if the trajectory goes from attracting to repelling and "false" (or *faux*) canards when the trajectory goes in the opposite direction. Depending on the topological type of the equilibrium point of the DRS on the fold set, one has folded equilibria of folded, saddle, focus, etc., type; see [8] for details on folded singularities and associated maximal canards.

In order to find maximal canards due to folded singularities (essentially of node and saddle type) in PWL slow-fast systems like (17), we first have to be more precise on the size of δ relative to ε so as to match an important result from the smooth case, namely that the number of maximal canards that exist for small enough ε near a folded singularity, does not depend on the specific value of ε. In particular in the folded-node scenario where multiple maximal canard can appear with both segments along $\mathscr{S}^a_\varepsilon$ and $\mathscr{S}^r_\varepsilon$ and small-amplitude oscillations (SAOs) in the fold region. The maximal number of SAO determines the type of canard solution and does not depend upon the specific value of ε. By construction, the system that we consider in the central zone is linear and the angular velocity is constant, hence the number of oscillations that trajectories make in this zone is obtained by considering the time it takes to go from one boundary of this zone to the other. This give a formula involving both δ and ε and forces a particular relationship between the two in order for this value to be independent of ε. Namely, δ has to scale like $\sqrt{\varepsilon}$; more precisely, we find that δ has to be equal to $\pi\sqrt{\varepsilon}$ for which the maximal rotation number μ is then $\dfrac{p_1\sqrt{p_1}}{|p_2 p_3|}$. This is a key result as it fixes the optimal size of the central zone in order to match results from the smooth case, that is, the optimal distance between $\mathscr{S}^a_\varepsilon$ and $\mathscr{S}^r_\varepsilon$ in order to find connections (maximal canards) entirely similar to those found in smooth slow-fast systems. We find that this optimal distance is $O(\sqrt{\varepsilon})$, which interestingly agree with the well-known result from the smooth case allowing to extend the Fenichel slow manifold up to a similar distance to the fold set before establishing conditions for them to intersect along maximal canards (using blow-up); see [4, 21, 28].

Once we have the correct scaling for the size of the central zone, we can obtain conditions for maximal canards to exist in the folded-node and in the folded-saddle cases, and verify that we have a complete similarity with these cases in smooth systems. This result is gathered in the following proposition, whose proofs are detailed in [6].

Proposition 1 *Consider system* (17) *with* $p_3 > 0$, $\delta = \pi\sqrt{\varepsilon}$ *and* ε *small enough. Assuming that different maximal canards have different flight times, the following statements hold.*

(a) Maximal canards γ *are reversible orbits.*

(b) If $p_1 > 0$ *and* $p_2 < 0$, *for every integer* k *with* $0 \leq k \leq [\mu]$, *where* μ *is the maximal rotation number, there exists a maximal canard* γ_k *intersecting the switching plane* $\{x = -\delta\}$ *at* $\mathbf{p}_k = (-\delta, y_k, z_k)$ *where*

$$\begin{aligned} y_k &= -\left(\left(k + \frac{1}{2}\right)\frac{p_2 p_3}{\sqrt{p_1}} + p_1\right)\pi\varepsilon^{\frac{3}{2}} - p_2 p_3 \varepsilon^2 + O(\varepsilon^{\frac{5}{2}}), \\ z_k &= -\left(k + \frac{1}{2}\right)\frac{p_3}{\sqrt{p_1}}\pi\sqrt{\varepsilon} + O(\varepsilon). \end{aligned} \tag{18}$$

Moreover, γ_k *turns k times around the weak canard* γ_w.

(c) If $p_1 > 0$ *and* $p_2 > 0$, *there exists a unique maximal canard* γ_0 *intersecting the switching plane at* $\mathbf{p}_0 = (-\delta, y_0, z_0)$ *where the coordinates* y_0 *and* z_0 *satisfy Eq.* (18) *with* $k = 0$.

(d) If $p_1 < 0$, *there are no maximal canards.*

This result establishes the existence of maximal canards near folded singularities of system (17) that are qualitatively and quantitatively similar to those found in the smooth case [8]. In the case of a folded saddle, only one maximal canard persists for small enough $\varepsilon > 0$; near a folded node, many more do and, except for the simplest one called the *strong canard*, all other maximal canards have SAOs in the central zone and they are called *secondary canards*. In order to characterise these maximal canards, one can write a series expansion in ε. However, we can exploit the PWL structure a bit further and exhibit "selected" maximal canards by taking special values of δ (within the correct scaling), each special δ-value giving rise to one selected maximal canard for which the series expansion is exact and all terms but the very first few vanish. We show four examples of such exact maximal canard solutions in Fig. 5, for four specific values of δ selecting the strong canard and the first three secondary canards, respectively. This is one more aspect of simplification brought about by the PWL framework and that could potentially be exploited in applications.

The above results enable a full comparison between three-dimensional smooth and PWL slow-fast systems in terms of number and geometry of maximal canard solutions near both folded-node and folded-saddle singularities. However, it does

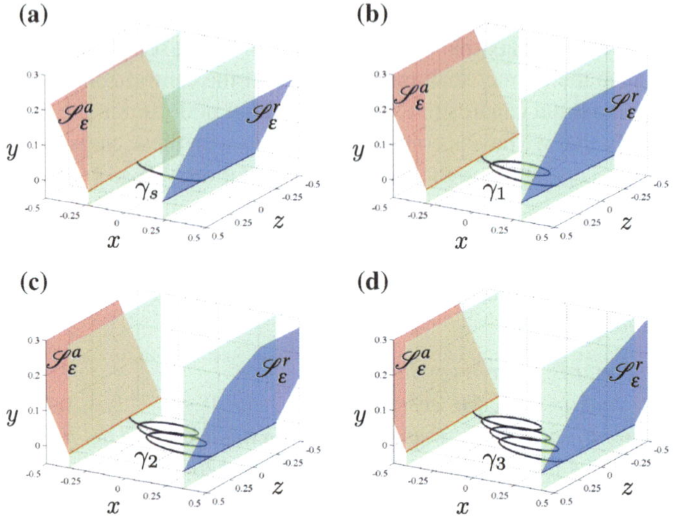

Fig. 5 Canonical slow manifolds and selected maximal canards (only central segment shown) with 0, 1, 2 and 3 SAOs in panels (**a**) to (**d**), respectively; γ_s is then the strong canard, and γ_i the ith secondary canard ($i = 1, 2, 3$). Also shown are the switching planes at $\{x = \pm\delta\}$ (modified from [6, Fig. 4.2])

not address the question of what are folded singularities in the PWL context, that is, how to define them in systems (17). This was the other main result from [6], where we introduce a strategy in order to identify the equivalent of folded singularities for three-dimensional PWL slow-fast systems with two slow variables. In brief, the main issue is to properly define the slow flow—for the $\varepsilon = 0$ limit of the fast-time system obtained from (17) by a time rescaling as described in the introduction—and exhibit conditions for singular trajectories passing from $\mathcal{S}^a$ to $\mathcal{S}^r$, that is, singular canards which all intersect at the folded singularity. The linearity zone in question is of course the central zone, which we take to be of size $\delta = O(\sqrt{\varepsilon})$ in order to keep similar properties as in the smooth case. However, this implies that $\delta \to 0$ as $\varepsilon \to 0$ and therefore the central zone shrinks to the switching planes in the singular limit, hiding information about the slow flow and, hence, about folded singularities. In order to remedy this, we artificially keep the central zone open in the singular limit and consider inside the $\varepsilon \to 0$ limit of the flow defined in there for $\varepsilon > 0$. This limiting flow allows us to find conditions, depending on p_1, p_2 and p_3, to obtain singular phase portraits entirely compatible to those of smooth singular systems near folded-node, folded-saddle and folded-saddle-node singularities, hence allowing to define the equivalent of these points in the PWL context; see [6, Sect. 4.4] for details.

Finally, we can also highlight the level of simplification brought by the PWL setting in this three-dimensional context. As explained above, we can "select" exact maximal canard solutions by appropriately choosing the value of δ within the correct scaling in ε. This offers almost for free maximal canards, of any type (strong

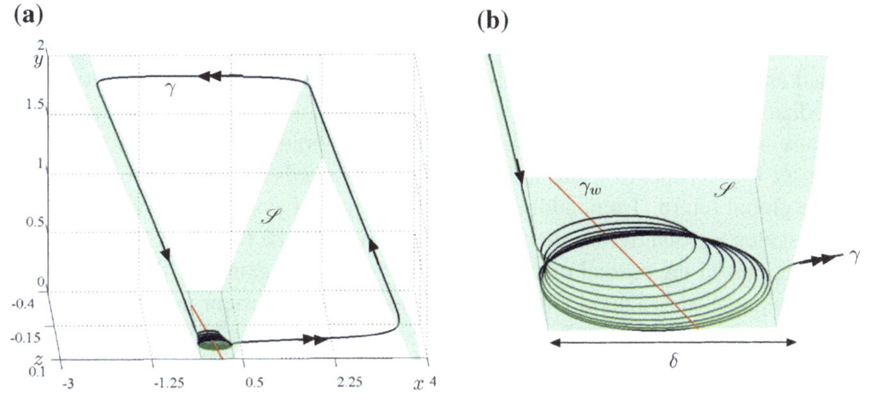

Fig. 6 Robust MMO solution γ of the three-dimensional slow-fast PWL system (17) with added linear terms in the z-equation and a fourth piece on the critical manifold, in order to ensure a global return mechanism on top of the local passage through a folded node. Also shown are the critical manifold $\mathscr{S}$ as well as the so-called weak canard (axis of rotation) γ_w (modified from [6, Fig. 5.1])

or secondary), with a simple and explicit time parametrisation only depending on system parameters and ε. This is substantial gain from the smooth case, where no such explicit canards are available. We anticipate that this could be of potential in applications as maximal canards form boundaries between different activity regimes and, hence, a direct analytical access to them could help to understand and control such a dynamical system. Another gain obtained from the PWL setting in this three-dimensional case is to be able to revisit unresolved questions from the smooth case. As explained in [6], we could prove with very simple arguments that the so-called "weak canard" associated with a folded node—a solution that plays the role of rotation axis for secondary canards and whose existence as a maximal canard was not proven in smooth system—is in fact not a maximal canard. Indeed, we can compute explicitly the perturbation for small $\varepsilon > 0$ of the rotation axis defined in the central zone (in which the dynamics in restricted to the cylindrical leaves of a foliated structure) and verify that this trajectory enters the left zone at an $O(\varepsilon)$ distance to the canonical slow manifold, which makes it impossible for this trajectory to be a maximal canard since for that to happen it would need to be exponentially close to the canonical slow manifold. Given the total parallel of the canard structure in system (17) compared to smooth minimal systems for folded nodes, we conjecture that this result of the non-existence of this special trajectory as a maximal canard is also valid in the smooth context. The second question that we could revisit from the smooth concerns the possibility for SAOs near a folded-saddle singularity, which was not reported in previous studies and has been developed in the smooth case in an independent paper soon to appear in [24]. This result came naturally and easily from the PWL setting.

As an opening towards future work, we close this section by mentioning the possibility for constructing robust MMO systems by using the local dynamics previous defined and analysed, near folded nodes. It suffices to add a fourth linearity zone, immediately to the left zone of system (17), that is, adding a fourth piece to the previous critical manifold so that the new one has a corner line in the new switching plane; see Fig. 6 panel (a). This is because one only needs a relaxation segment during the global return. Then, adding to the z-equation of system (17) suitable linear terms creates a global return mechanism that re-injects the trajectory near the right attracting part of the critical manifold so that it can pass again near the folded node while making SAOs. Direct simulation of this extended system indicated that one indeed obtains an MMO limit cycle whose SAOs are organised by the folded node. This is only a numerical example and we plan to prove the existence of such MMOs as well as investigate their bifurcation structure in future work.

5 Summary and Perspective

In this chapter, we have presented a compendium of recent results on PWL slow-fast systems displaying canard solutions, both in the planar and the three-dimensional cases, with more general results on Fenichel slow manifolds, valid in any dimensions. Our work [6, 13, 26, 27] summarised here demonstrates that one can recover all essential results from the smooth case while gaining a substantial level of simplification in the way objects are defined and in their essential properties. This ground work has allowed us to construct minimal slow-fast systems in the PWL setup possessing maximal canard solutions by using a mix of local passage near the equivalent of the fold set and globally defined slow manifolds. These results can be used to construct complex oscillations using PWL vector fields adequately designed. We have shown the case of an MMO system with canard-induced behaviour in Fig. 6. Another gain of the PWL setting is to be able to better control a given system through a more quantitative knowledge of its dynamics, which we have applied to a four-dimensional model of secreting neuron demonstrating the use of this approach [12, 14]. We have also obtained preliminary results on canard-induced bursting oscillations, that is, in the context of slow-fast systems with one slow and two fast variables. In [5], we constructed minimal PWL slow-fast systems in order to reproduce a *spike-adding canard explosion* and all salient features of square-wave bursting organised by canard solutions [7]. Surprisingly, we proved that this scenario could not be obtained in such a minimal setup with the assumption of continuity of the vector field across all linearity zones. This is a first step into investigating canard-induced complex oscillations of bursting type; ongoing and future work will involve looking at other forms of bursting with PWL systems, in particular elliptic bursting, whose understanding is a key stage towards studying canard phenomena within fast oscillations, namely, *torus canards* [20]. This case is interesting given that numerous questions are still open in the smooth case and, hence, where we believe that the PWL framework could once more prove useful with its simplification power without altering the essential dynamics.

Acknowledgements SFG is supported by the University of Seville VPPI-US and partially supported by Proyectos de Excelencia de la Junta de Andalucía under Grant No. P12-FQM-1658 and Ministerio de Economía y Competitividad under Grant No. MTM2015-65608-P. RP and AET are supported by the Spanish Ministerio de Economía y Competitividad through project MTM2014-54275-P.

References

1. Arima, N., Okazaki, H., Nakano, H.: A generation mechanism of canards in a piecewise linear system. IEICE T. Fundam. Electr. **80**, 447–453 (1997)
2. Benoît, E.: Canards et enlacements. Publications Mathématiques de l'IHÉS **72**, 63–91 (1990)
3. Benoît, E., Callot, J.L., Diener, F., Diener, M.: Chasse au canard. Collect. Math. **32**, 37–119 (1981)
4. Brøns, M., Krupa, M., Wechselberger, M.: Mixed mode oscillations due to the generalized canard phenomenon. Fields Inst. Commun. **49**, 39–63 (2006)
5. Desroches, M., Fernández-García, S., Krupa, M.: Canards in a minimal piecewise-linear square-wave burster. Chaos **26**(7), 073,111 (2016)
6. Desroches, M., Guillamon, A., Ponce, E., Prohens, R., Rodrigues, S., Teruel, A.E.: Canards, folded nodes, and mixed-mode oscillations in piecewise-linear slow-fast systems. SIAM Rev. **58**(4), 653–691 (2016)
7. Desroches, M., Kaper, T.J., Krupa, M.: Mixed-mode bursting oscillations: dynamics created by a slow passage through spike-adding canard explosion in a square-wave burster. Chaos **23**(4), 046,106 (2013)
8. Desroches, M., Guckenheimer, J.M., Krauskopf, B., Kuehn, C., Osinga, H.M., Wechselberger, M.: Mixed-mode oscillations with multiple time scales. SIAM Rev. **54**, 211–288 (2012)
9. Diener, M.: The canard unchained or how fast/slow dynamical systems bifurcate. The Math. Intell. **6**(3), 38–49 (1984)
10. Dumortier, F., Roussarie, R.: Canards cycles and center manifolds. Mem. Am. Math. Soc. **557**, 1131–1162 (1996)
11. Fenichel, N.: Geometric singular perturbation theory for ordinary differential equations. J. Differ. Equ. **31**, 53–98 (1979)
12. Fernández-García, S., Desroches, M., Krupa, M., Clément, F.: A multiple time scale coupling of piecewise linear oscillators. application to a neuroendocrine system. SIAM J. Appl. Dyn. Syst. **14**(2), 643–673 (2015)
13. Fernández-García, S., Desroches, M., Krupa, M., Teruel, A.E.: Canard solutions in planar piecewise linear systems with three zones. Dyn. Syst. A.I.J. **31**, 173–197 (2016)
14. Fernández-García, S., Krupa, M., Clément, F.: Mixed-mode oscillations in a piecewise linear system with multiple time scale coupling. Phys. D **332**, 9–22 (2016)
15. FitzHugh, R.: Impulses and physiological states in theoretical models of nerve membrane. Biophys. J. **1**(6), 445–466 (1961)
16. Freire, E., Ponce E., Rodrigo. F., Torres, F.: Bifurcation sets of continuous piecewise linear systems with two zones. J. Bifur. Chaos Appl. Sci. Eng. **8**, 2073–2097 (1998)
17. Freire, E., Ponce, E., Torres, F.: Hopf-like bifurcations in planar piecewise linear systems. Publ. Mat. **41**, 135–148 (1997)
18. Jones, C.K.R.T.: Geometric Singular Perturbation Theory. Springer, Berlin, Heidelberg (1995)

19. Kaper, T.: Systems theory for singular perturbation problems. In: O'Malley, R.E. Jr., Cronin, J. (eds.) Analyzing Multiscale Phenomena Using Singular Perturbation Methods; Proceedings of Symposia in Applied Mathematics, vol. 56, pp. 8–132; Am. Math. Soc. (1999)
20. Kramer, M.A., Traub, R.D., Kopell, N.J.: New dynamics in cerebellar purkinje cells: torus canards. Phys. Rev. Lett. **101**(6), 068,103 (2008)
21. Krupa, M., Szmolyan, P.: Extending geometric singular perturbation theory to nonhyperbolic points-fold and canard points in two dimensions. SIAM J. Math. Anal. **33**, 286–314 (2001)
22. Krupa, M., Szmolyan, P.: Relaxation oscillations and canard explosion. J. Differ. Equ. **174**, 312–368 (2001)
23. McKean, H.P.: Nagumo's equation. Adv. Math. **4**(3), 209–223 (1970)
24. Mitry, J., Wechselberger, M.: Folded saddles and faux canards. SIAM J. Appl. Dyn. Syst. **16**, 546–596 (2017)
25. Nagumo, J., Arimoto, S., Yoshizawa, S.: An active pulse transmission line simulating nerve axon. Proc. IRE **50**(10), 2061–2070 (1962)
26. Prohens, R., Teruel, A.E.: Canard trajectories in 3d piecewise linear systems. Discret. Contin. Dyn. Syst. **33**(3), 4595–4611 (2013)
27. Prohens, R., Teruel, A.E., Vich, C.: Slow-fast n-dimensional piecewise-linear differential systems. J. Differ. Equ. **260**, 1865–1892 (2016)
28. Wechselberger, M.: Existence and bifurcation of canards in $\mathbb{R}^3$ in the case of a folded node. SIAM J. Appl. Dyn. Syst. **4**, 101–139 (2005)

Part II
Wave Equations

Solitary Waves in the Nonlinear Dirac Equation

Jesús Cuevas-Maraver, Nabile Boussaïd, Andrew Comech, Ruomeng Lan, Panayotis G. Kevrekidis and Avadh Saxena

Abstract In the present work, we consider the existence, stability, and dynamics of solitary waves in the nonlinear Dirac equation. We start by introducing the Soler model of self-interacting spinors, and discuss its localized waveforms in one, two, and three spatial dimensions and the equations they satisfy. We present the associated explicit solutions in one dimension and numerically obtain their analogues in higher dimensions. The stability is subsequently discussed from a theoretical perspective

J. Cuevas-Maraver (✉)
Grupo de Física No Lineal, Departamento de Física Aplicada I, Escuela Politécnica Superior, Universidad de Sevilla, C/ Virgen de África, 7, 41011 Sevilla, Spain
e-mail: jcuevas@us.es

J. Cuevas-Maraver
Instituto de Matemáticas de la Universidad de Sevilla (IMUS), Edificio Celestino Mutis. Avda. Reina Mercedes s/n, 41012 Sevilla, Spain

N. Boussaïd
Université Bourgogne Franche-Comté, 25030 Besançon CEDEX, France
e-mail: nabile.boussaid@univ-fcomte.fr

A. Comech · R. Lan
Department of Mathematics, Texas A&M University, College Station, TX 77843-3368, USA
e-mail: comech@math.tamu.edu

A. Comech
IITP, Moscow 127994, Russia

P. G. Kevrekidis
Department of Mathematics and Statistics, University of Massachusetts, Amherst, MA 01003-4515, USA
e-mail: kevrekid@math.umass.edu

A. Saxena
Los Alamos National Laboratory, Center for Nonlinear Studies and Theoretical Division, Los Alamos, NM 87545, USA
e-mail: avadh@lanl.gov

V. Carmona et al. (eds.), *Nonlinear Systems, Vol. 1*,
Understanding Complex Systems, https://doi.org/10.1007/978-3-319-66766-9_4

and then complemented with numerical computations. Finally, the dynamics of the solutions is explored and compared to its non-relativistic analogue, which is the nonlinear Schrödinger equation.

Keywords Solitons · Solitary waves · Vortices · Nonlinear Dirac equation Stability · Soler model

1 Introduction

In the last three decades, there has been an enormous interest in the study of waves in nonlinear dispersive media. Arguably, two of the most paradigmatic equations that describe such waves are the nonlinear Schrödinger equation (NLS) and the sine–Gordon equation. The first among these equations covers a broad range of settings including atomic physics [130, 131], nonlinear optics [99, 101], condensed matter physics, and mathematical physics [2, 156]. The sine–Gordon equation also covers settings in condensed matter physics and mathematical physics apart from high-energy physics models [31, 50]. A principal focus of the relevant properties of these equations has been the study of the existence, stability, and dynamics of solitary waves (i.e., spatially localized waves supported by the nonlinearity and dispersion), both in lower-dimensional settings (such as one-dimensional solitons and multi-solitons) and in higher dimensional settings (vortices, vortex rings, and related structures) [53, 99].

By comparison, far less attention has been paid to the nonlinear Dirac equation (NLD), despite its presence for 90 years in the realm of high energy physics. The nonlinear Dirac equation with scalar-type self-interaction was initially introduced by Ivanenko [90]. Following the ideas of Finkelstein et al. [73], Heisenberg [87] used this NLD model in an attempt to formulate a unified theory of elementary particles. In 1958, a completely integrable one-dimensional model known as the Massive Thirring Model (MTM) [158], based on vector-type self-interaction of spinor field, was introduced. This model possesses solitary wave solutions. Curiously, the fundamental solutions of the MTM can be transformed into solitons of the sine–Gordon equation by means of a bosonization process [38]. In 1970, Soler re-introduced Ivanenko's model with scalar-type self-interaction in the context of extended nucleons [152] and also provided the numerical analysis of solitary wave solutions. The one-dimensional version of the Soler model, known as the Gross–Neveu model [78], was introduced in 1974 as a toy model of quark confinement in quantum chromodynamics, and explicit solitary wave solutions in the corresponding massive model were found by Lee et al. [104]. We can not complete this quick review of NLD models in high-energy physics without mentioning the recent work of [121] (see also [114]), where a variant of the NLD is applied to the study of neutrino oscillations. Related systems are the Dirac–Maxwell system [23, 41, 60, 79, 166], the Einstein–Dirac system [140, 155], and Einstein–Dirac–Maxwell system [141]. In quantum chemistry, the Dirac–Hartree–Fock model [63, 64, 105] takes into account the fermionic properties of electrons (describing the exchange interaction, which is a fundamental effect of purely quantum nature) and is used for accurate computation of the electronic energy [134, 164]; this model has also started to receive mathematical attention [63–65].

In recent years, the study of nonlinear Dirac type models has received a renewed thrust for a variety of reasons. Both one-dimensional [27, 29, 40, 128] and two-dimensional variants of the model have been examined from the perspective of solitary wave solutions and their stability [17, 27, 127]. Furthermore, the computational examination of solitary wave solutions and their dynamics has provided numerous insights on the NLD [44, 47, 149, 168] also on its variant in the presence of external fields [120]. Thirdly, the NLD is emerging as the model of relevance in a variety of settings including the dynamics of Bose–Einstein condensates in honeycomb lattices [81, 82], as well as in atomically thin 2D Dirac materials [71, 167]. Relevant examples include, but are not limited to graphene, silicene, germanene, borophene, and transition metal dichalcogenides [111]. One of the most intriguing applications of the model has arisen in recent years in so-called photonic graphene [1, 3], i.e., the examination of light propagation in honeycomb photorefractive lattices. This direction has led to the investigation of conical diffraction [124] and the exploration of nonlinear phenomena [1, 3]. It is worth highlighting, however, that the nonlinearity in this context *breaks* the Lorentz symmetry (that is, such models are not invariant under Lorentz transformations; for the explicit form of the Lorentz transformations of the spinor fields see e.g. [20, 157]).

Another recent application involving the interplay of the Dirac and Schrödinger operators is that of spin-orbit coupled Bose–Einstein condensates [51]. Their experimental realization [103, 107, 133] and interplay with nonlinearity (due to interatomic interactions) has led to the exploration of a diverse host of nonlinear states including bright, dark and gap solitons [4, 5, 72, 95, 170], self-trapped states [117], vortices [136, 137, 169], Skyrmions [96], as well as Dirac monopoles [42]. The complex interplay of these different effects may even stabilize vortex solitons against collapse in free space, under attractive interactions [144].

From a mathematical perspective, Dirac models are described by systems (rather than by scalar equations) that correspond to the Hamiltonian functionals unbounded from below. This unboundedness makes all the aspects of the analysis of these models (well-posedness, existence of localized solutions, stability, numerical simulations) much more challenging. This has fueled an increasing interest in the nonlinear Dirac equation and more general models of self-interacting spinor fields, with many results on the existence of solitary waves [36, 62, 118] and well-posedness in $(3+1)$D [59, 108] and in $(1+1)$D [33, 89, 109, 125, 147].[1] The stability of solitary wave solutions of the nonlinear Dirac equation was approached via numerical simulations [8, 9, 11, 29, 37, 120, 135, 168] and via heuristic arguments [21, 22, 44, 115, 154], but it is still not settled. Recently, the first stability results in the context of self-interacting spinor fields started appearing [24, 25, 27–29, 40, 127].

---

[1]With the notation $(N+1)D$ we want to denote that the system possesses $N+1$ dimensions, with N spatial ones plus time.

The NLD can also be viewed as a relativistic generalization (or extension) of the NLS, or, alternatively, the NLS can be seen as a special case limit of the NLD at the low-energy limit. Nevertheless, it has turned out that the Dirac equation as a result of its matrix nature and the fact that it is only first order in spatial derivatives (as opposed to second order in the NLS) has proven far more computationally (and theoretically) challenging, on a number of grounds, than its NLS counterpart. This difficulty has hindered the progress in the study of solitary waves, particularly in two-dimensional and three-dimensional settings. However, recent developments are gradually enabling the study of the stability and dynamical properties of solitary waves in two-dimensional and even three-dimensional Soler models; see [49] for a relevant example. Clearly, however, this process requires numerous additional steps that will present several challenges over the coming years.

The aim of this chapter is to give a review of recent results developed by the authors and their collaborators in the last few years, as well as to present a basic framework of the NLD theory, mainly focused on the Soler model and its variants; this is our principal workhorse model. The content of the chapter covers a wide spectrum of results ranging from existence and stability of solitary waves to numerical methods and dynamics of unstable solutions.

This chapter is organized as follows: in Sect. 2 we start with an introduction to the main nonlinear Dirac equation, namely the Soler model, and tractable expressions for the determination of solitary waves and linearizations at solitary waves in one, two, and three spatial dimensions. Section 3 is devoted to the existence properties of solitary waves and numerical methods for their calculation. Stability analysis from a theoretical and numerical point of view is the topic of Sects. 4 and 5, respectively. The dynamics of solitary waves is analyzed in Sect. 6. We finalize the chapter with a summary of the considered results and an outlook on future directions on solitary waves in the NLD equations.

2 The Soler Model of Self-interacting Spinors

In this section we start with the linear Dirac equation and move on to the Soler model as a principal, Lorentz-invariant variant of the model with scalar self-interactions. We give explicit expressions of linearization at solitary waves in one-, two-, and three-dimensional cases.

2.1 The Dirac Equation

In December 1927, Paul Dirac arrived at the idea of the first-order relativistically invariant equation [57] that describes massive spin-1/2 relativistic fermions in $(3 + 1)$ space-time dimensions:

$$i\hbar\frac{\partial}{\partial t}\psi(t,x) = \left(-i\hbar c\alpha\cdot\nabla + mc^2\beta\right)\psi(t,x), \qquad \psi(t,x)\in\mathbb{C}^4, \qquad x\in\mathbb{R}^3,$$

with ψ being the spinor-valued wavefunction, $\alpha\cdot\nabla = \sum_{j=1}^{3}\alpha^j\frac{\partial}{\partial x^j}$, and $m\ge 0$ the mass of the particle. As usual, we choose in what follows the physical units so that Planck's constant $\hbar$ and the speed of light c are both equal to one. The self-adjoint 4×4 matrices α^j, $1\le j\le 3$, and β satisfy

$$\{\alpha^j,\alpha^k\} = 2\delta_{jk}I_4, \qquad \{\alpha^j,\beta\} = 0, \qquad \beta^2 = I_4,$$

with I_N being the $N\times N$ identity matrix and $\{A,B\} = AB+BA$ the anticommutator. According to the Dirac–Pauli theorem (see [57, 122, 165], [157, Lemma 2.25], and also [97, Theorem 7] for a general version in odd spatial dimensions), different choices of the matrices α^j and β are equivalent. The most common choice, known as the Dirac–Pauli representation, is

$$\alpha^j = \begin{pmatrix} 0 & \sigma_j \\ \sigma_j & 0 \end{pmatrix}, \qquad \beta = \begin{pmatrix} I_2 & 0 \\ 0 & -I_2 \end{pmatrix},$$

with the Pauli matrices given by

$$\sigma_1 = \begin{pmatrix} 0 & 1 \\ 1 & 0 \end{pmatrix}, \qquad \sigma_2 = \begin{pmatrix} 0 & -i \\ i & 0 \end{pmatrix}, \qquad \sigma_3 = \begin{pmatrix} 1 & 0 \\ 0 & -1 \end{pmatrix}. \tag{1}$$

In the covariant form, the Dirac equation is written as

$$i\gamma^\mu\partial_\mu\psi = m\psi,$$

where $\gamma^\mu\partial_\mu = \sum_{\mu=0}^{3}\gamma^\mu\partial_\mu$, $\partial_0\equiv\partial_t$, with γ^μ being the Dirac γ-matrices

$$\gamma^0 = \beta, \qquad \gamma^j = \beta\alpha^j = \begin{pmatrix} 0 & \sigma_j \\ -\sigma_j & 0 \end{pmatrix}, \quad j = 1,2,3.$$

Matrices γ^μ fulfill the anticommutation relation $\{\gamma^\mu,\gamma^\nu\} = 2\eta^{\mu\nu}I_4$, with $\eta^{\mu\nu}$ being the Minkowski tensor [54]. In other words, $(\gamma^0)^2 = I_4$ and $(\gamma^1)^2 = (\gamma^2)^2 = (\gamma^3)^2 = -I_4$. There exists another matrix which anticommutes with γ^0 and γ^j, $1\le j\le 3$, which plays an important role in the parity transformation. It is the γ^5 matrix, defined by

$$\gamma^5 = i\gamma^0\gamma^1\gamma^2\gamma^3 = \begin{pmatrix} 0 & I_2 \\ I_2 & 0 \end{pmatrix}.$$

This matrix is self-adjoint and satisfies $(\gamma^5)^2 = I_4$.

One can immediately generalize the ideas of Dirac to an arbitrary spatial dimension $n \geq 1$, writing the Dirac equation

$$i\partial_t\psi = D_m\psi \equiv -i\sum_{j=1}^{n}\alpha^j\partial_j\psi + \beta m\psi, \qquad \psi(t,x) \in \mathbb{C}^N, \quad x \in \mathbb{R}^n,$$

with α^j, $1 \leq j \leq n$, and β being self-adjoint matrices satisfying the relations

$$\{\alpha^j, \alpha^k\} = 2\delta_{jk}I_N, \quad \{\alpha^j, \beta\} = 0, \quad (\alpha^j)^2 = \beta^2 = I_N; \qquad 1 \leq j,k \leq n.$$

The smallest number of spinor components N for the spatial dimension $n \geq 1$ is obtained in the Clifford algebra theory (see e.g. [70, Chap. 1, Sect. 5.3]) and is given by

$$N = 2^{\lfloor(n+1)/2\rfloor}. \tag{2}$$

Notice that this relation implies that in the three-dimensional case ($n = 3$), the number of spinor components N must be at least four.

The Dirac equation is derived from the following Lagrangian density:

$$\mathscr{L}_{\text{Dirac}} = \bar{\psi}(i\gamma^\mu\partial_\mu - m)\psi,$$

where the so-called *Dirac conjugate* $\bar{\psi}$ is defined by

$$\bar{\psi} \equiv \psi^*\gamma^0,$$

with ψ^* the Hermitian conjugate of ψ.

2.2 The Soler Model

In 1938, Russian physicist Dmitri Ivanenko proposed a nonlinear model of self-interacting electrons, introducing the nonlinear term $(\bar{\psi}\psi)\psi$ into the Dirac equation [90]. This self-interaction term is based on the quantity $\bar{\psi}\psi = \psi^*\beta\psi$ which transforms as a scalar under Lorentz transformations. In 1970, Spanish physicist Mario Soler re-introduced this model in order to study, from a classical point of view, extended nucleons interacting with their own electromagnetic field [152, 153]. Now this equation (or, rather, its version with an arbitrary function of $\bar{\psi}\psi$) is known as the Soler model [36, 62, 118]:

$$i\partial_t\psi = D_m\psi - f(\bar{\psi}\psi)\beta\psi, \qquad \psi(t,x) \in \mathbb{C}^N, \quad x \in \mathbb{R}^n, \tag{3}$$

or, in the covariant form,

$$i\gamma^\mu \partial_\mu \psi = (m - f(\bar{\psi}\psi))\psi,$$

where $f \in C(\mathbb{R}) \cap C^1(\mathbb{R} \setminus \{0\})$, $f(0) = 0$. Equation (3) admits solitary wave solutions of the form $\psi(t, x) = \phi_\omega(x)e^{-i\omega t}$, with $\phi_\omega(x)$ exponentially localized in space [26, 36, 62, 118, 152, 162]. In addition, the equation is a $\mathbf{U}(1)$-invariant, relativistically invariant hamiltonian system, with the Hamiltonian represented by the density

$$\mathscr{H}_{\text{Soler}}(\psi) = \psi^* D_m \psi - F(\psi^* \beta \psi), \tag{4}$$

with

$$F(s) = \int_0^s f(t)\,dt$$

the antiderivative of f. Because of the $\psi^* D_m \psi$-term, this Hamiltonian functional is unbounded from below. The Soler model (3) is also characterized by the Lagrangian density

$$\mathscr{L}_{\text{Soler}} = \bar{\psi}(i\gamma^\mu \partial_\mu - m)\psi + F(\bar{\psi}\psi).$$

The $\mathbf{U}(1)$-symmetry of the Soler equation leads to the conservation of the value of the charge functional, given by

$$Q(\psi(t)) = \int_{\mathbb{R}^n} \psi(t, x)^* \psi(t, x)\,dx.$$

which is conserved in time. If $\psi(t, x)$ is a solution to (3), then both the charge $Q(\psi(t))$ and the energy $E(\psi(t)) = \int \mathscr{H}_{\text{Soler}}(\psi(t, x))\,dx$ are conserved in time (formally; that is, as long as ψ is sufficiently smooth, allowing one the integration by parts).

A common choice of the nonlinearity is $f(s) = |s|^k$, $k > 0$; this leads to $F(s) = s|s|^k/(k + 1)$. We note that the absolute value is needed when k is not an integer since the quantity $s = \bar{\psi}\psi$ could be negative. Let us mention that for $k \in (0, 1)$, the function $f(s) = |s|^k$ is not differentiable at $s = 0$, which leads to certain difficulties in the construction of the solitary waves; see [26].

We want to remark that the cubic Soler model

$$i\partial_t \psi = D_m \psi - \bar{\psi}\psi\beta\psi, \tag{5}$$

which appeared in [90, 152], differs from (3) with $f(s) = |s|^k$, $k = 1$:

$$i\partial_t \psi = D_m \psi - |\bar{\psi}\psi|\beta\psi. \tag{6}$$

Both Eqs. (5) and (6) are relativistically invariant Hamiltonian systems. In particular, they are invariant under the time reversal and parity transformation, which are elements of the full Lorentz group, given respectively by (see e.g. [20])

$$\psi_T(t,x) = i\gamma^1\gamma^3 K\psi(-t,x),$$

with $K : \mathbb{C}^4 \to \mathbb{C}^4$ the complex conjugation, and

$$\psi_P(t,x) = \gamma^0\psi(t,-x).$$

At the same time, since $\bar{\psi}_C\psi_C = -\bar{\psi}\psi$, where the charge conjugation is given by [20]

$$\psi_C(t,x) = -i\gamma^2 K\psi(t,x).$$

Equation (6) is invariant under the charge conjugation, while Eq. (5) is not. Let us mention that the choice of unitary factors in all these three transformations is not important.

We also point out that the stationary waves $\phi_\omega e^{-i\omega t}$ constructed in [36] in the three-dimensional case satisfy $\bar{\phi}_\omega\phi_\omega > 0$ for all $x \in \mathbb{R}^3$, thus being solutions to both (5) and (6).

2.3 One-Dimensional Soler Model

The Soler model in one spatial dimension, Eq. (3) with $n = 1$, is also known as the Gross–Neveu model [78]. According to relation (2), one can take $N = 2$, so that the wavefunction is represented by a bi-spinor (i.e., a spinor with only two complex components). We will choose $\alpha^1 = -\sigma_2$, $\beta = \sigma_3$. In this case, the nonlinear Dirac equation (3) can be written as a system of coupled partial differential equations of the form

$$\begin{aligned} i\partial_t\psi_1 &= \partial_x\psi_2 + (m - f(|\psi_1|^2 - |\psi_2|^2))\psi_1, \\ i\partial_t\psi_2 &= -\partial_x\psi_1 - (m - f(|\psi_1|^2 - |\psi_2|^2))\psi_2, \end{aligned} \tag{7}$$

where $\psi_1, \psi_2 \in \mathbb{C}$ denote the two components of $\psi(t,x) \in \mathbb{C}^2$.

The focus of the present chapter is on solitary wave solutions. To this aim, we will search for standing waves of the form

$$\psi(t,x) = \phi_\omega(x)e^{-i\omega t}, \qquad \phi_\omega(x) = \begin{bmatrix} v(x,\omega) \\ u(x,\omega) \end{bmatrix} \in \mathbb{R}^2,$$

with $v(x,\omega)$ and $u(x,\omega)$ satisfying

$$\begin{aligned} \omega v &= \partial_x u + [m - f(v^2 - u^2)]v, \\ \omega u &= -\partial_x v - [m - f(v^2 - u^2)]u. \end{aligned} \tag{8}$$

Once such standing wave solutions are calculated using the methods explained in Sect. 3.2, their linear stability is considered by means of a Bogoliubov–de Gennes (BdG) linearized stability analysis. That is, given a solitary wave solution $\phi_\omega(x)e^{-i\omega t}$ with $\phi_\omega(x) \in \mathbb{R}^2$, we consider its perturbation in the form $\psi(t,x) = (\phi_\omega(x) + \rho(t,x))e^{-i\omega t}$, with $\rho(t,x) \in \mathbb{C}^2$. Then, the linearized equations on $R(t,x) = [\mathrm{Re}(\rho), \mathrm{Im}(\rho)]^T \in \mathbb{R}^4$ can be written as (see e.g. [29])

$$\partial_t R = \mathscr{A}_\omega R, \tag{9}$$

with

$$\mathscr{A}_\omega = \begin{bmatrix} 0 & L_-(\omega) \\ -L_+(\omega) & 0 \end{bmatrix}, \tag{10}$$

where $L_+(\omega)$ and $L_-(\omega)$ are the following self-adjoint operators:

$$\begin{aligned} L_-(\omega) &= \begin{pmatrix} m - f(\tau) - \omega & \partial_x \\ -\partial_x & -m + f(\tau) - \omega \end{pmatrix}, \\ L_+(\omega) &= L_-(\omega) - 2f'(\tau)\begin{pmatrix} v^2 & -vu \\ -vu & u^2 \end{pmatrix}, \end{aligned}$$

with $f(\tau)$ and $f'(\tau)$ evaluated at $\tau \equiv v^2 - u^2$.

The potential presence of an eigenvalue with non-zero real part in the spectrum of $\mathscr{A}_\omega$ suggests the dynamical instability; the corresponding solitary wave is called linearly unstable. If all the eigenvalues are purely imaginary, then the solitary wave is called spectrally (neutrally) stable.

2.4 Two-Dimensional Soler Model

Taking into account the relation given by expression (2), in two spatial dimensions one can again consider two-component spinors. Following [39], a convenient choice for α and β matrices is $\alpha^1 = \sigma_1$, $\alpha^2 = \sigma_2$, $\beta = \sigma_3$. With this in mind, Eq. (3) is expressed as

$$\begin{aligned} i\partial_t\psi_1 &= -(i\partial_x + \partial_y)\psi_2 + [m - f(|\psi_1|^2 - |\psi_2|^2)]\psi_1, \\ i\partial_t\psi_2 &= -(i\partial_x - \partial_y)\psi_1 - [m - f(|\psi_1|^2 - |\psi_2|^2)]\psi_2. \end{aligned} \tag{11}$$

In order to simplify further analysis, we use the polar coordinates $r = |\mathbf{r}|$ and θ; then Eq. (11) takes the form

$$\begin{aligned} i\partial_t\psi_1 &= -e^{-i\theta}\left(i\partial_r + \frac{\partial_\theta}{r}\right)\psi_2 + [m - f(|\psi_1|^2 - |\psi_2|^2)]\psi_1, \\ i\partial_t\psi_2 &= -e^{i\theta}\left(i\partial_r - \frac{\partial_\theta}{r}\right)\psi_1 - [m - f(|\psi_1|^2 - |\psi_2|^2)]\psi_2, \end{aligned} \tag{12}$$

with $r \in (0, \infty)$ and $\theta \in [0, 2\pi)$. The form of this equation suggests searching for standing wave solutions in the form

$$\psi(t, \mathbf{r}) = \phi_\omega(\mathbf{r})e^{-i\omega t}, \qquad \phi_\omega(\mathbf{r}) = \begin{bmatrix} v(r, \omega)e^{iS\theta} \\ i\,u(r, \omega)e^{i(S+1)\theta} \end{bmatrix}, \tag{13}$$

with $v(r, \omega)$ and $u(r, \omega)$ real-valued. The value $S \in \mathbb{Z}$ is associated with the vorticity of the first spinor component. Thus, according to Eqs. (12) and (13), the equations for the stationary solutions read as follows:

$$\begin{aligned} \omega v &= \left(\partial_r + \frac{S+1}{r}\right)u + [m - f(v^2 - u^2)]v\,, \\ \omega u &= -\left(\partial_r - \frac{S}{r}\right)v - [m - f(v^2 - u^2)]u. \end{aligned} \tag{14}$$

This set of equations only depends on the radial coordinate r. The absence of angular coordinates turns the determination of stationary solutions into an effectively one-dimensional problem, substantially simplifying the numerics.

To examine the spectral stability of a solitary wave, we consider a solution ψ in the form of a perturbed solitary wave:

$$\psi(t, \mathbf{r}) = \begin{bmatrix} \big(v(r, \omega) + \xi_1(t, r, \theta) + i\eta_1(t, r, \theta)\big)e^{iS\theta} \\ i\big(u(r, \omega) + \xi_2(t, r, \theta) + i\eta_2(t, r, \theta)\big)e^{i(S+1)\theta} \end{bmatrix} e^{-i\omega t},$$

with small perturbations $\xi(t, r, \theta) = [\xi_1, \xi_2]^T \in \mathbb{R}^2$, $\eta(t, r, \theta) = [\eta_1, \eta_2]^T \in \mathbb{R}^2$.

The linearized equation on $R(t, r, \theta) = [\xi_1, \xi_2, \eta_1, \eta_2]^T \in \mathbb{R}^4$ has the form

$$\partial_t R = \mathscr{A}_\omega R, \tag{15}$$

with $\mathscr{A}_\omega(r, \theta, \partial_r, \partial_\theta)$ a matrix-valued first order differential operator

$$\mathscr{A}_\omega(r, \theta, \partial_r, \partial_\theta) = \begin{bmatrix} -\sigma_1 \frac{\partial_\theta}{r} & L_-(\omega) \\ -L_+(\omega) & -\sigma_1 \frac{\partial_\theta}{r} \end{bmatrix}, \tag{16}$$

where

$$L_-(\omega) = \begin{pmatrix} m - f(\tau) - \omega & \partial_r + \frac{S+1}{r} \\ -(\partial_r - \frac{S}{r}) & -m + f(\tau) - \omega \end{pmatrix},$$

$$L_+(\omega) = L_-(\omega) - 2f'(\tau) \begin{pmatrix} v^2 & -vu \\ -vu & u^2 \end{pmatrix},$$

with $f(\tau)$ and $f'(\tau)$ evaluated at $\tau \equiv v^2 - u^2$.

To find the spectrum of the operator $\mathscr{A}_\omega$, we consider it in the space of $\mathbb{C}^4$-valued functions. The key observation which facilitates a computation of the spectrum is that the explicit form (16) of $\mathscr{A}_\omega$ contains r, ∂_r, ∂_θ, but not θ. As a consequence, $\mathscr{A}_\omega$ is invariant in the spaces which correspond to the Fourier decomposition with respect to θ,

$$\mathscr{X}_q = \left\{ [a_1(r); a_2(r); b_1(r); b_2(r)] \, e^{iq\theta} \right\}, \qquad q \in \mathbb{Z}.$$

The restriction of $\mathscr{A}_\omega$ to each such subspace is given by

$$\mathscr{A}_{\omega,q}(r, \partial_r) = \mathscr{A}_\omega|_{\mathscr{X}_q} = \begin{bmatrix} -\sigma_1 \frac{iq}{r} & L_-(\omega) \\ -L_+(\omega) & -\sigma_1 \frac{iq}{r} \end{bmatrix}, \qquad q \in \mathbb{Z}, \tag{17}$$

and this allows to compute the spectrum of $\mathscr{A}_\omega$ as the union of spectra of the one-dimensional spectral problems,

$$\sigma\left(\mathscr{A}_\omega\right) = \overline{\bigcup_{q \in \mathbb{Z}} \sigma\left(\mathscr{A}_{\omega,q}\right)},$$

where the operators $\mathscr{A}_{\omega,q}$ do not contain the angular variable.

2.5 Three-Dimensional Soler Model

In three spatial dimensions, it is convenient to consider Eq. (3) in spherical coordinates. We consider the 4-spinor solitary waves in the form of the Wakano Ansatz [166]:

$$\psi(t,\mathbf{r}) = \phi_\omega(\mathbf{r})e^{-i\omega t}, \qquad \phi_\omega(\mathbf{r}) = \begin{bmatrix} v(r,\omega)\begin{pmatrix}1\\0\end{pmatrix} \\ iu(r,\omega)\begin{pmatrix}\cos\theta\\ e^{i\varphi}\sin\theta\end{pmatrix}\end{bmatrix},$$

with real-valued $v(r,\omega)$, $u(r,\omega)$ satisfying

$$\begin{aligned} \omega v &= \left(\partial_r + \frac{2}{r}\right)u + [m - f(v^2-u^2)^k]v\,, \\ \omega u &= -\,\partial_r v - [m - f(v^2-u^2)^k]u. \end{aligned} \tag{18}$$

To study the linearization operator in the invariant space which has the same angular structure as the solitary waves, we consider the perturbed solutions in the form

$$\psi(t,\mathbf{r}) = \begin{bmatrix} (v(r,\omega)+\xi_1(t,r)+i\eta_1(t,r))\begin{pmatrix}1\\0\end{pmatrix} \\ i(u(r,\omega)+\xi_2(t,r)+i\eta_2(t,r))\begin{pmatrix}\cos\theta\\ e^{i\varphi}\sin\theta\end{pmatrix}\end{bmatrix} e^{-i\omega t},$$

with real-valued $\xi = [\xi_1, \xi_2]^T \in \mathbb{R}^2$, $\eta = [\eta_1, \eta_2]^T \in \mathbb{R}^2$ (note that the considered perturbation only depends on r but not on the angular variables). The linearized equation on $R(t,r) = (\xi_1, \xi_2, \eta_1, \eta_2)^T$ is similar to Eqs. (15) and (16):

$$\partial_t R = \mathscr{A}_\omega R \quad \text{with} \quad \mathscr{A}_\omega = \begin{bmatrix} 0 & L_-(\omega) \\ -L_+(\omega) & 0 \end{bmatrix},$$

and

$$\begin{aligned} L_-(\omega) &= \begin{pmatrix} m - f(\tau) - \omega & \partial_r + \frac{2}{r} \\ -\partial_r & -m + f(\tau) - \omega \end{pmatrix}, \\ L_+(\omega) &= L_-(\omega) - 2f'(\tau)\begin{pmatrix} v^2 & -vu \\ -vu & u^2 \end{pmatrix}, \end{aligned}$$

with $f(\tau)$ and $f'(\tau)$ evaluated at $\tau \equiv v^2 - u^2$. We point out that above we only considered the setup for finding the spectrum of the restriction of the linearization operator onto a particular ("radial") invariant subspace. The resulting spectrum is presented on Fig. 19 below.

2.6 One-Dimensional $\mathscr{PT}$-symmetric Soler Model

To finish this section, we present a connection between the budding area of research of open $\mathscr{PT}$-symmetric systems and the nonlinear Dirac equation. Open systems that feature a combination of gain and loss have become a focal point of numerous recent studies [12, 14, 74, 113]. A principal reason for this was the introduction of the notion of $\mathscr{P}\mathscr{T}$-symmetry and its proposition by Bender (and collaborators) as an alternative to the postulate of hermiticity in quantum mechanics. Its principal realization, though, came at the level of optical systems [102, 112, 142], where they were experimentally implemented in the context of waveguide arrays [80, 139, 143]. Since this first set of efforts, numerous additional experiments emerged in the context of electronic circuits [145, 146], in whispering-gallery microcavities [129], as well as in mechanical settings [13].

In Ref. [49], we introduced a generalized $\mathscr{P}\mathscr{T}$-symmetric 1D Soler model, which in covariant form reads as follows:

$$\left(i\gamma^{\mu}\partial_{\mu} - m + g\left(\bar{\psi}\psi\right)^{k} + \gamma\gamma^{5}\right)\psi = 0. \tag{19}$$

Alternatively, written in the standard form as a function of the bispinor components $\psi = [\psi_1(x), \psi_2(x)]^T$, the equations assume the following form:

$$\begin{aligned} i\partial_t\psi_1 &= \partial_x\psi_2 - g(|\psi_1|^2 - |\psi_2|^2)^k\psi_1 + m\psi_1 + i\gamma\psi_2, \\ i\partial_t\psi_2 &= -\partial_x\psi_1 + g(|\psi_1|^2 - |\psi_2|^2)^k\psi_2 - m\psi_2 + i\gamma\psi_1. \end{aligned} \tag{20}$$

Equations (20) are $\mathscr{P}\mathscr{T}$-symmetric due to their invariance under the transformation

$$\mathscr{P} : x \to -x, \qquad \psi_1 \to \psi_1, \qquad \psi_2 \to -\psi_2$$

and

$$\mathscr{T} : t \to -t, \qquad \psi_1 \to \psi_1^*, \qquad \psi_2 \to \psi_2^*.$$

Comparing the present model to the standard Soler setting, we note the inclusion of the gain-loss term proportional to γ through the Dirac matrix γ^5 (cf. [12]) multiplying the spinor ψ in (19). For our two-component spinors, γ^5 is represented by the Pauli matrix
σ_1 (1).

It is straightforward to see that in the linear case (of $g = 0$), plane waves $\psi_1(t, x) = Ae^{i(\kappa x-\omega t)}$ and $\psi_2(t, x) = iBe^{i(\kappa x-\omega t)}$ are solutions provided the dispersion relation $\omega = \pm\sqrt{m^2 + \kappa^2 - \gamma^2}$ is satisfied. Not only does this formula have the characteristic Dirac form, but it also is consistent with the equivalence of the linear $\mathscr{P}\mathscr{T}$-Dirac equation with effective mass $\tilde{m} = \sqrt{m^2 - \gamma^2}$, as per the above discussion.

To determine the stability of stationary solitary wave solutions, we consider infinitesimal perturbations to potentially calculated numerically solutions $u(x), v(x)$ of the form:

$$\begin{aligned} \psi_1(t, x) &= e^{-i\omega t}\left[v(x) + \delta(a_1(x)e^{\lambda t} + b_1^*(x)e^{\lambda^* t})\right], \\ \psi_2(t, x) &= e^{-i\omega t}\left[u(x) + \delta(a_2(x)e^{\lambda t} + b_2^*(x)e^{\lambda^* t})\right], \end{aligned}$$

where δ denotes a formal small parameter. The relevant linearization equations are derived to order $O(\delta)$ [by substitution of the above Ansatz into Eq. (20)] and are subsequently solved as a matrix eigenvalue problem

$$\lambda[a_1(x), a_2(x), b_1(x), b_2(x)]^T = \mathcal{M}[a_1(x), a_2(x), b_1(x), b_2(x)]^T,$$

with $\mathcal{M}$ being

$$\mathcal{M} = \begin{pmatrix} L_1 & L_2 \\ -L_2^* & -L_1^* \end{pmatrix} - i\gamma \begin{pmatrix} \gamma^5 & 0 \\ 0 & \gamma^5 \end{pmatrix}$$

and

$$L_1 = \begin{pmatrix} f(|v|^2 - |u|^2) - m + \Lambda & -\partial_x \\ \partial_x & m - f(|v|^2 - |u|^2) + \Lambda \end{pmatrix} + f'(|v|^2 - |u|^2) \begin{pmatrix} |v|^2 & -v^* u \\ -v^* u & |u|^2 \end{pmatrix},$$

$$L_2 = f'(|v|^2 - |u|^2) \begin{pmatrix} v^2 & -vu \\ -vu & u^2 \end{pmatrix}.$$

From the dynamical equations (20) it is straightforward to show that the charge is not preserved. Instead, the following "moment equation" is satisfied:

$$\frac{dQ}{dt} = 4\gamma \int \mathrm{Re}(V^* U)\, dx\,. \tag{21}$$

Note that in the case of a stationary state, $dQ/dt = 0$ and charge is conserved.

Although the charge is not generally conserved, remarkably there is a conserved quantity in the form of the energy:

$$E = \frac{1}{2} \int \left[\psi_1^* \partial_x \psi_2 - \psi_2^* \partial_x \psi_1 + m(|\psi_1|^2 - |\psi_2|^2) - \frac{g}{k+1} [|\psi_1|^2 - |\psi_2|^2]^{k+1} \right] dx\,.$$

Notice the absence of a term proportional to γ in this expression. Nevertheless, this energy is conserved not only for $\gamma = 0$, but also for $\gamma \neq 0$, a feature that appears to be due to the use of the matrix γ^5 to introduce the $\mathcal{PT}$-symmetry. As an additional observation associated with the unconventional nature of this system, we note that considering only the time-dependence and the terms proportional to γ, the resulting form reads: $i\partial_t \psi_1 = i\gamma\psi_2$ and $i\partial_t \psi_2 = i\gamma\psi_1$. This makes the two components appear as if they have both gain and loss.

Numerical results for this model can be found in [49].

3 Solitary Waves: Exact Solutions and Numerical Methods

Solitary wave solutions of the form $\phi_\omega(x)e^{-i\omega t}$, $\omega \in (0, m)$, are known to exist in (3) and in other important systems based on the Dirac equation (see e.g. the review [61]). In the one-dimensional case, for pure power nonlinearity, the solutions are available in a closed form; see Sect. 3.1. However, for higher-dimensional cases, solitary wave and vortex solutions must be obtained by means of numerical methods. These methods can also be applied to 1D models with a general nonlinearity f in (7) when the solutions are not available in a closed form.

3.1 *One-Dimensional Soler Model: Exact Solutions*

In [104] it was shown for the cubic nonlinearity, i.e., $k = 1$ in (7), and later in [37, 44, 120] for generic value $k > 0$, that the solitary wave solutions can be found in a closed form for any $\omega \in (0, m)$:

$$v(x) = \cosh(k\beta x)\sqrt{\frac{(m+\omega)}{m+\omega\cosh(2k\beta x)}}\left[\frac{(k+1)\beta^2}{m+\omega\cosh(2k\beta x)}\right]^{1/2k},$$

$$u(x) = \sinh(k\beta x)\sqrt{\frac{(m-\omega)}{m+\omega\cosh(2k\beta x)}}\left[\frac{(k+1)\beta^2}{m+\omega\cosh(2k\beta x)}\right]^{1/2k}, \qquad (22)$$

where $\beta = \sqrt{m^2 - \omega^2}$. In the special case of $k = 1$, waveforms in Eq. (22) reduce to

$$v(x) = \frac{\sqrt{2(m-\omega)}}{[1-\mu\tanh^2(\beta x)]\cosh(\beta x)}, \quad u(x) = \frac{\sqrt{2\mu(m-\omega)}\tanh(\beta x)}{[1-\mu\tanh^2(\beta x)]\cosh(\beta x)},$$

with $\mu = (m-\omega)/(m+\omega)$. Figure 1 shows the profiles of solitary waves given by the expression (22) for $k = 1$ and $k = 3$. Notice that the first component of the spinor, $v(x)$, is spatially even, whereas the second component $u(x)$ is spatially odd. Moreover, $v^2(x) - u^2(x) > 0$ for all $x \in \mathbb{R}$, so that the solitary waves satisfy the nonlinear Dirac equation (3) (with $n = 1$) with both $f(s) = s$ and $f(s) = |s|$. Evaluating ρ'' at $x = 0$, one can check that the charge density profiles $\rho(x) = \phi_\omega(x)^*\phi_\omega(x)$ (cf. (3)) become double-humped for $\omega \le \omega_h(k)$, with $\omega_h(k) = mk/(k+1)$. The dependence of the charge and energy with respect to ω for different values of $k \in \mathbb{N}$ are shown in Fig. 2.

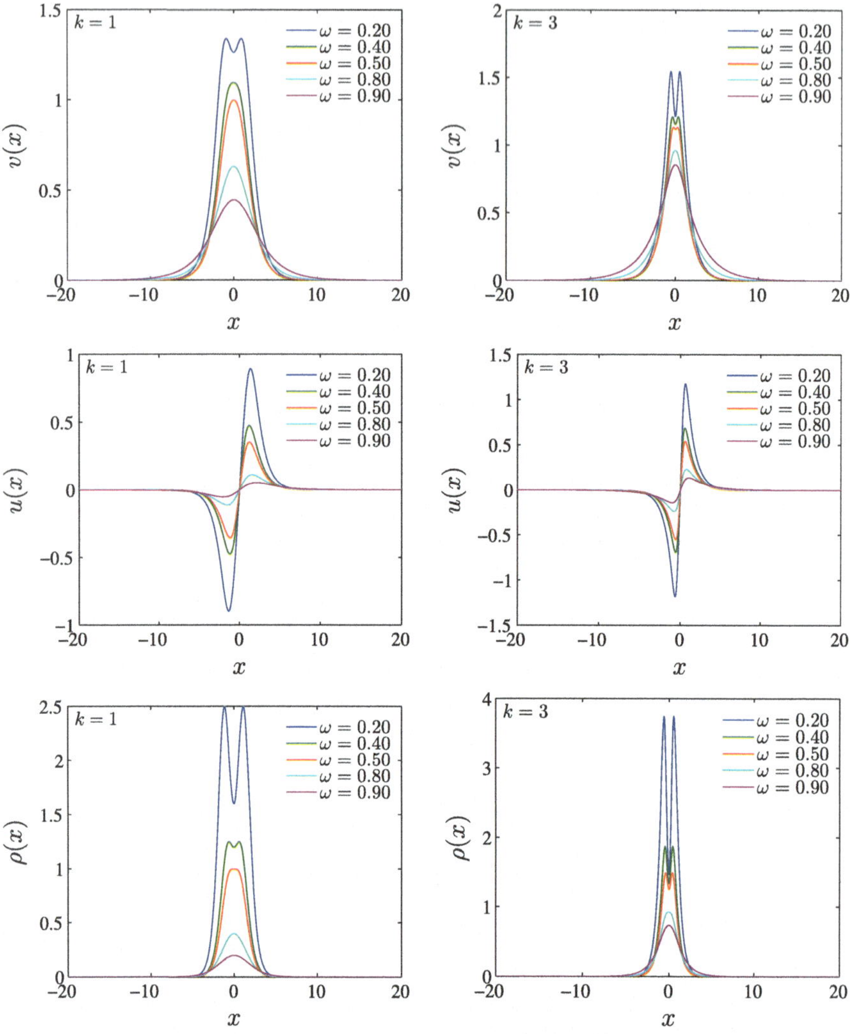

Fig. 1 Profile of solitary waves in the 1D Soler model. Figures depict the first and second spinor components together with the solitary wave density. Left (right) panels correspond to $k = 1$ ($k = 3$)

3.2 *Two-Dimensional Soler Model: Numerical Solutions*

No explicit solitary wave solutions are known for the Soler model in 2D (12). For this reason, one must rely on numerical results. We show in Sect. 3.2.1 the numerical methods used for the numerical determination of stationary solutions in (14). These methods can easily be adapted for numerically solving the Soler 3D model (18) (in the particular case of zero vorticity) and for finding solitary wave solutions in 1D

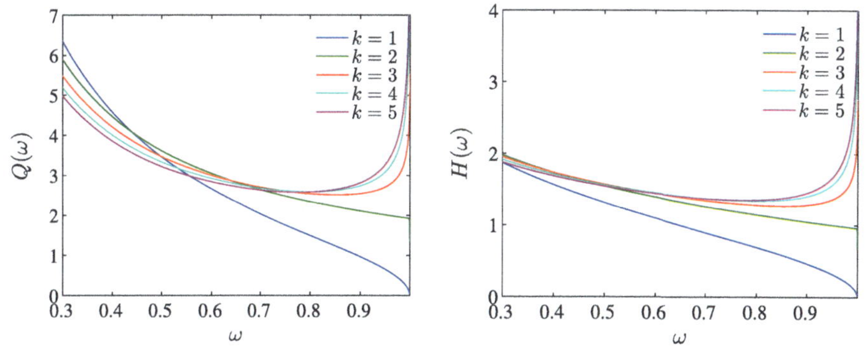

Fig. 2 Charge and energy of solitary waves (left and right panel, respectively) as functions of the frequency ω for $1 \leq k \leq 5$. Notice the existence of a minimum in the curve $Q(\omega)$ for $k > 2$, which is related to the change in stability properties (see Sect. 5)

models where additional terms to the equation (8) have been added, such as external fields [120] or in the $\mathscr{P}\mathscr{T}$-symmetric Soler model [48].

3.2.1 Brief Summary of Spectral Methods

Prior to explaining the numerical methods used for calculating stationary solutions, we will proceed to present a summary of spectral methods needed for dealing with derivatives in continuum settings. For a detailed discussion on these methods, the reader is directed to [30] and references therein.

Spectral methods arise due to the necessity of calculating spatial derivatives with higher accuracy than that given by finite difference methods. As shown in [46], finite difference methods cannot be used for the stability and dynamics analysis of solitary waves in the NLD equation.

In order to implement spectral derivatives, a differentiation matrix $\mathbf{D} \equiv \{D_{n,m}\}$ must be given together with N collocation[2] (i.e., grid) points $\mathbf{x} \equiv \{x_n\}$, $n = 1, 2, \ldots, N$, which are not necessarily equi-spaced. Thus, if the *spectral* derivative of a function $\mathbf{f}(\mathbf{x}) \equiv \{f_n(x_n)\}$ needs to be calculated, it can be cast as:

$$f'(x) = \partial_x f(x) \leftrightarrow f'_n = \sum_{m=1}^{N} D_{n,m} f_m,$$

where $f_m \equiv f(x_m)$ and $f'_n \equiv f'(x_n)$. If $x \in [-L, L]$ and the boundary conditions are periodic, the Fourier collocation can be used. In this case,

[2]Notice that this value of N is not related to the dimension of the NLD, although the same symbol is used in both cases.

$$x_n = \frac{2L}{N}\left(n - \frac{N}{2}\right), \quad n = 1, 2, \ldots, N \tag{23}$$

with N even. The differentiation matrix is

$$D_{n,m} = \begin{cases} 0 & \text{if } n = m, \\ \dfrac{\pi}{2L} \dfrac{(-1)^{n+m}}{\tan[(x_n - x_m)/2]} & \text{if } n \neq m. \end{cases}$$

Notice that doing the multiplication $\mathbf{Df}$ is equivalent to performing the following pair of Discrete Fourier Transform applications:

$$\mathbf{Df} = \mathscr{F}^{-1}\left(i\mathbf{k}\mathscr{F}(\mathbf{f})\right), \tag{24}$$

with $\mathscr{F}$ and $\mathscr{F}^{-1}$ denoting, respectively, the direct and inverse discrete Fourier transform [160]. The vector wavenumber $\mathbf{k} = \{k_n\}$ is defined as:

$$k_n = \begin{cases} \dfrac{n\pi}{L} & \text{if } n < N/2, \\ 0 & \text{if } n = N/2. \end{cases}$$

The computation of the direct and inverse discrete Fourier transforms, which is useful in simulations, can be accomplished by the Fast Fourier Transform. On the contrary, the differentiation matrix is used for finding the Jacobian and stability matrices. Notice that the grid for a finite difference discretization is the same as in the Fourier collocation; and, in addition, there is a differentiation matrix for the finite difference method, i.e.,

$$D_{n,m} = \frac{1}{2h}\left(\delta_{m,n+1} - \delta_{m,n-1} + \delta_{n,1}\delta_{m,N} - \delta_{n,N}\delta_{m,1}\right), \qquad h = \frac{2L}{N}, \tag{25}$$

with δ being Kronecker's delta. It can be observed from the above discussion that in the Fourier spectral method, the banded differentiation matrix of the finite difference method is substituted by a dense matrix, or, in other words, a nearest-neighbor interaction is exchanged with a long-range one. The lack of sparsity of differentiation matrices is one of the drawbacks of spectral methods, especially when having to diagonalize large systems. However, they have the advantage of needing (a considerably) smaller number of grid points N for getting the same accuracy as with finite difference methods.

For fixed (Dirichlet) boundary conditions, the Chebyshev spectral methods are the most suitable ones. There are several collocation schemes, the Gauss–Lobatto being the most extensively used:

$$x_n = L\cos\left(\frac{n\pi}{N+1}\right), \quad n = 1, 2, \ldots, N,$$

with N being even or odd. The differentiation matrix is

$$D_{n,m} = \begin{cases} \dfrac{x_n}{2L(1-x_n^2)} & \text{if } n = m, \\ \dfrac{(-1)^{n+m}}{L\cos(x_n - x_m)} & \text{if } n \neq m. \end{cases}$$

The significant drawback of Chebyshev collocation is that the discretization matrix possesses a great number of spurious eigenvalues [30]. They are approximately equal to $N/2$. These spurious eigenvalues also have a significant non-zero real part, which increases when N grows. This fact naturally reduces the efficiency of the method when performing numerical time-integration. However, it gives a higher accuracy than the Fourier collocation method when determining the spectrum of the stability matrix (see e.g. [46]).

Several modifications must be introduced when applying spectral methods to polar coordinates. They basically rely on overcoming the difficulty of not having Dirichlet boundary conditions at $r = 0$ and the singularity of the equations at that point. In addition, in the case of the Dirac equation, the spinor components can be either symmetric or anti-symmetric in their radial dependence, so the method described in [88, 160] must be modified accordingly. As shown in the previously mentioned references, the radial derivative of a general function $f(r, \theta)$ can be expressed as:

$$\partial_r f(r_n, \theta) = \sum_{m=1}^{N} D_{n,m} f(r_m, \theta) + D_{n,2N-m} f(r_m, \theta + \pi). \tag{26}$$

Notice that in this case, the collocation points must be taken as

$$r_n = L\cos\left(\frac{n\pi}{2N+1}\right), \quad n = 1, 2, \dots, 2N,$$

but only the first N points are taken so that the domain of the radial coordinate does not include $r = 0$. Analogously the differentiation matrix would possess now $2N \times 2N$ components, but only the upper half of the matrix, of size $N \times 2N$ is used.

If the function that must be derived is symmetric or anti-symmetric, i.e., $f(r, \theta + \pi) = \pm f(r, \theta)$, with the upper (lower) sign corresponding to the (anti-)symmetric function, Eq. (26) can be written as follows:

$$\partial_r f(r_n, \theta) = \sum_{m=1}^{N} \left[\left(D_{n,m} \pm D_{n,2N-m} \right) f(r_m, \theta) \right]. \tag{27}$$

Thus, the differentiation matrix has a different form depending on whether $f(r, \theta)$ is symmetric or anti-symmetric:

$$\partial_r \mathbf{f}(\mathbf{r}, \theta) = \mathbf{D}^{(\pm)}\mathbf{f} \quad \text{if} \quad f(r, \theta) = \pm f(r, \theta + \pi),$$

with $\mathbf{r} \equiv \{r_n\}$, $\mathbf{f}(\mathbf{r}) \equiv \{f(r_n)\}$ and $\mathbf{D}^{(\pm)}\mathbf{f}$ defined as in (27).

3.2.2 Fixed Point Methods

Among the numerical methods available for solving nonlinear systems of equations we have chosen to use fixed point methods, such as the Newton–Raphson one [132], which requires the transformation of the set of two coupled ordinary differential equations (14) into a set of $2N$ algebraic equations; this is performed by defining the set of collocation points $\mathbf{r} \equiv \{r_n\}$, and transforming the derivatives into multiplication of the differentiation matrices $\mathbf{D}^{(1)}$ and $\mathbf{D}^{(2)}$ (to be defined below) times the vectors $\mathbf{u} \equiv \{u_n\}$ and $\mathbf{v} \equiv \{v_n\}$, respectively, being $u_n \equiv u(r_n)$ and $v_n \equiv v(r_n)$ as explained in the previous subsection. Thus, the discrete version of (14) reads:

$$F_n^{(1)} \equiv (m-\omega)v_n - g\tau_n^k v_n + \sum_m D_{nm}^{(2)} u_m + \frac{S+1}{r_n} u_n = 0,$$

$$F_n^{(2)} \equiv (m+\omega)u_n - g\tau_n^k u_n + \sum_m D_{nm}^{(1)} v_m + \frac{S}{r_n} v_n = 0,$$

with $\tau_n \equiv v_n^2 - u_n^2$. It is important to notice that matrices $\mathbf{D}^{(1)}$ and $\mathbf{D}^{(2)}$ correspond to either $\mathbf{D}^{(+)}$ or $\mathbf{D}^{(-)}$, depending on the symmetry of $\mathbf{v}$ and $\mathbf{u}$, which, at the same time, depend on the value of the vorticity S. If S is even, then $\mathbf{v}$ and $\mathbf{u}$ are symmetric and antisymmetric, respectively, being $\mathbf{D}^{(1)} = \mathbf{D}^{(+)}$ and $\mathbf{D}^{(2)} = \mathbf{D}^{(-)}$. On the contrary, if S is odd, then $\mathbf{u}$ is symmetric and $\mathbf{v}$ is antisymmetric, being $\mathbf{D}^{(1)} = \mathbf{D}^{(-)}$ and $\mathbf{D}^{(2)} = \mathbf{D}^{(+)}$.

In order to find the roots of the vector function $\mathbf{F} = (\{F_n^{(1)}\}, \{F_n^{(2)}\})^T$, an analytical expression of the Jacobian matrix

$$\mathbf{J} = \begin{pmatrix} \dfrac{\partial \mathbf{F}^{(1)}}{\partial \mathbf{u}} & \dfrac{\partial \mathbf{F}^{(1)}}{\partial \mathbf{v}} \\ \dfrac{\partial \mathbf{F}^{(2)}}{\partial \mathbf{u}} & \dfrac{\partial \mathbf{F}^{(2)}}{\partial \mathbf{v}} \end{pmatrix} = \begin{pmatrix} (m-\omega) - g\tau^{k-1}[2kv^2 + \tau] & 2kg\mathbf{vu}\tau^{k-1} + D^{(2)} + \dfrac{S+1}{r} \\ -2kg\mathbf{vu}\tau^{k-1} + D^{(1)} - \dfrac{S}{r} & (m+\omega) - g\tau^{k-1}[\tau - 2kv^2] \end{pmatrix}$$

must be introduced, with the derivatives expressed by means of spectral methods and the matrix is evaluated at the corresponding grid points. The roots of $\mathbf{F}$, $\Phi = (\mathbf{v}, \mathbf{u})^T$, are found by successive application of the iteration $\Phi \to \Phi - \mathbf{J}^{-1}\mathbf{F}$ until convergence is attained. In our case, we have chosen as convergence condition that $\|\mathbf{F}\|_\infty < 10^{-10}$.

Spectral stability is analyzed by evaluating the functions appearing in matrix $\mathscr{A}_\omega$ of Eq. (16) at the collocation points and substituting the partial derivatives by the corresponding differentiation matrices. At this point, one must be very cautious because, as also occurred with the Jacobian, there will be two different differentiation matrices in our problem. Now $L_-(\omega)$ will be represented by the following matrix:

$$L_-(\omega) = \begin{pmatrix} f(\tau) - \omega & D^{(2)} + \frac{S+1}{r} \\ -(D^{(1)} - \frac{S}{r}) & -f(\tau) - \omega \end{pmatrix}.$$

3.2.3 Solitary Waves and Vortices

This section deals with the numerically found profiles for solitary waves ($S = 0$) and vortices ($S = 1$) in the two-dimensional Soler model. Figure 3 shows, in radial coordinates, the profiles of each component of $S = 0$ solitary waves with $k = 1$ and $k = 2$; the left panels of Fig. 4 depict those components for $S = 1$ vortices. As explained in Sect. 3.2.1, the first spinor component is spatially symmetric whereas the second component is anti-symmetric as long as the vorticity of the first component,

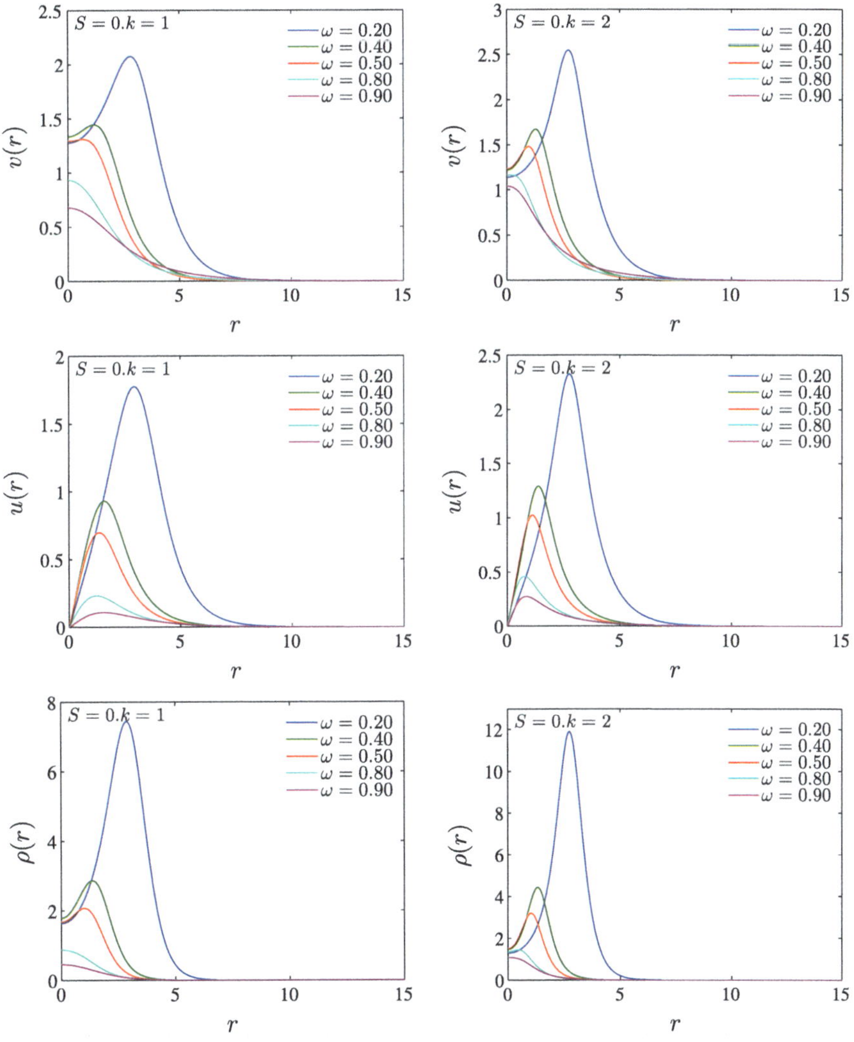

Fig. 3 Radial profile of $S = 0$ solitary waves in the 2D Soler model. Figures depict the first and second spinor components together with the solution density. Left (right) panels correspond to $k = 1$ ($k = 2$)

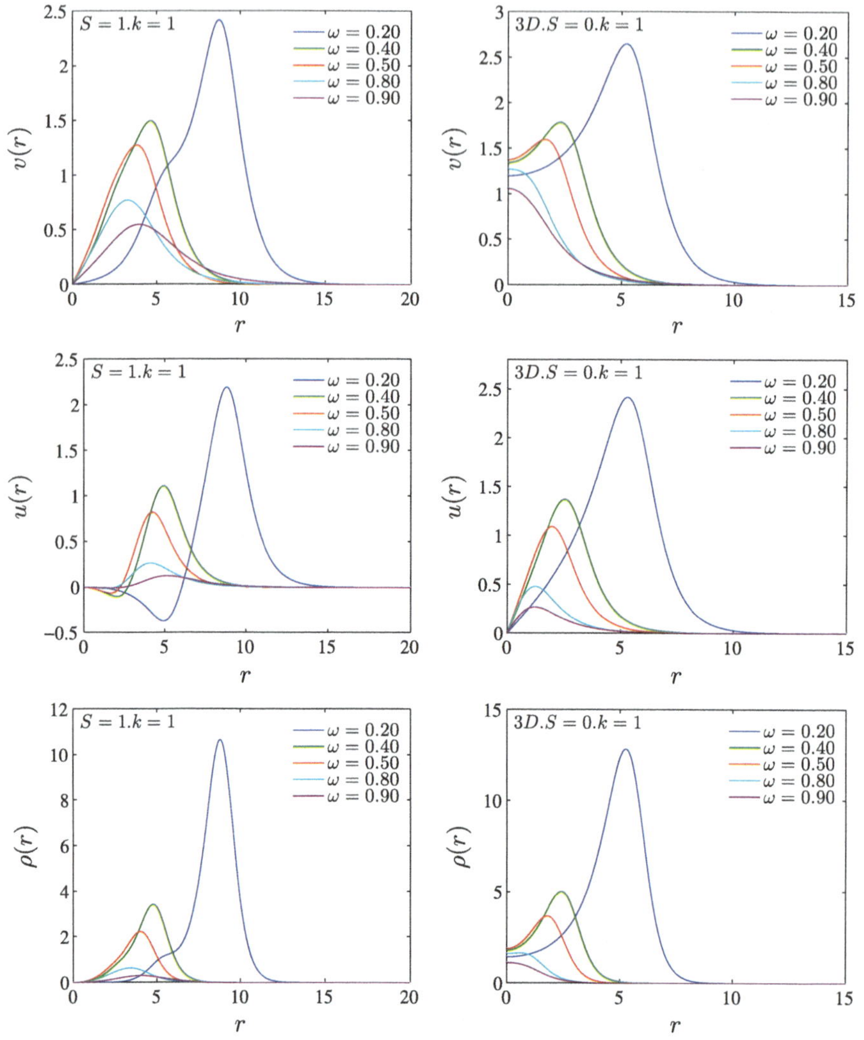

Fig. 4 (Left panels) Radial profile of $S = 1$ vortices in the cubic 2D Soler model. (Right panels) Radial profile of $S = 0$ solitary waves in the cubic 3D Soler model. Figures depict the first and second spinor components together with the solution density

$S \in \mathbb{Z}$, is even. The spatial symmetry is inverted if S is odd. Notice also that in the $S = 0$ case, the solution profile has a hump for $r > 0$ whenever ω is below a critical value. It manifests as the transformation of the solitary wave density from a circle to a ring. The ring radius increases when ω decreases, becoming infinite when $\omega \to 0$. For this reason, computations are progressively more demanding for smaller values of ω.

The right panels of Fig. 4 show the radial profile of $S = 0$ solitary waves in 3D. We have not included solitary waves with higher vorticity because, as explained in

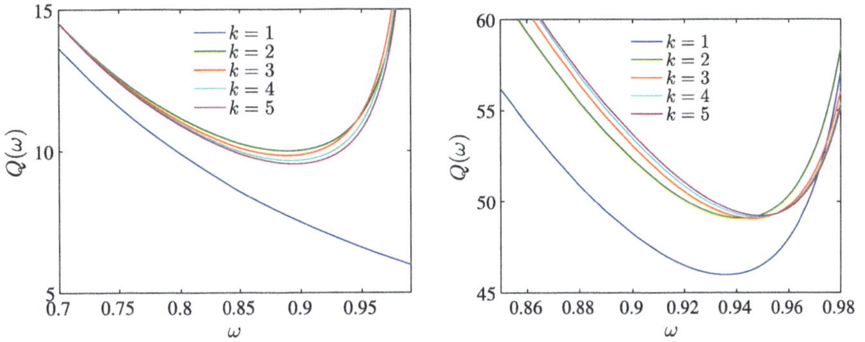

Fig. 5 Dependence of the charge of $S = 0$ solitary waves in the 2D and 3D Soler models (left and right panels, respectively) with respect to the frequency for $1 \leq k \leq 5$. Notice the existence of a minimum in the 2D (3D) curve for $k > 1$ ($k \geq 1$), which will be related to stability changes (see Sect. 5)

Sect. 2.5, the Soler equation in radial coordinates can only be expressed in the $S = 0$ case. Figure 5 shows the charge for the $S = 0$ solitary waves in the 2D and 3D Soler models for different values of k.

It is worth mentioning that, despite the absence of an explicit analytical form of 3D solitary waves, their existence has been rigorously proven in [36, 62, 118].

4 Stability of Solitary Waves: Theoretical Results

In Sect. 2, we presented the equation governing the linear stability analysis of stationary solutions. In the present section, we will show the theoretical background related to spectral and orbital stability. Many of the results proposed herein will be numerically checked in Sect. 5.

4.1 Spectral Stability of Solitary Waves

Prior to proceeding to the spectral stability analysis, we introduce some definitions.

The linearization of (3) at a solitary wave solution $\psi(t, x) = \phi_\omega(x)e^{-i\omega t}$ is represented by non-self-adjoint operators of the form

$$J(D_m - \omega + V(x, \omega)), \qquad \text{with } J \text{ skew-adjoint}, \; J^2 = -1, \tag{28}$$

where the matrix J commutes with D_m but not necessarily with the potential $V(x, \omega)$.

We say that the solitary wave is spectrally stable if the spectrum of its linearization operator has no points with positive real part. The spectral stability is the weakest type of stability; it does not necessarily lead to actual, dynamical one. The essential spectrum is easy to analyze: the application of Weyl's theorem (see e.g. [138, Theorem XIII.14, Corollary 2]) shows that the essential spectrum of the operator corresponding to the linearization at a solitary wave starts at $\pm(m - |\omega|)i$ and extends to $\pm\infty i$. Thus, the spectral stability of the corresponding solitary wave would be a corollary of the absence of eigenvalues with positive real part in the spectrum of $J(D_m - \omega + V(\omega))$ in (28). The major difficulties in identifying the point spectrum $\sigma_p\big(J(D_m - \omega + V(\omega))\big)$ are due to the spectrum of D_m extending to both $\pm\infty$; this prevents us from using standard tools developed in the NLS context.

In the absence of linear stability (that is when the linearized system is not dynamically stable), one expects to be able to prove *orbital instability*, in the sense of [77]; in [76], such instability is proved in the context of the nonlinear Schrödinger equation; such results are still absent for the nonlinear Dirac equation.

Since the isolated eigenvalues depend continuously on the perturbation, it is convenient to trace the location of "unstable" eigenvalues (eigenvalues with positive real part) considering ω as a parameter. One wants to know how and when the "unstable" eigenvalues may emerge from the imaginary axis, particularly from the essential spectrum; that is, at which critical values of ω the solitary waves start developing an instability. Below, we describe the possible scenarios.

4.1.1 Instability Scenario 1: Collision of Eigenvalues

The well-known Vakhitov–Kolokolov stability criterion [161] keeps track of the collision of purely imaginary eigenvalues at the origin and a subsequent birth of a positive and a negative eigenvalue. This criterion was discovered in the context of nonlinear Schrödinger equations, in relation to ground state solitary waves $\phi_\omega(x)e^{-i\omega t}$ ("ground state" in the sense that $\phi_\omega(x)$ is strictly positive; for more details, see [16]). When $\partial_\omega Q(\omega) < 0$, with $Q(\omega) = \|\phi_\omega\|^2_{L^2}$ being the charge of the solitary wave (3), then the linearization at a solitary wave has purely imaginary spectrum; when $\partial_\omega Q(\omega) > 0$, there are two real (one positive, one negative) eigenvalues of the linearization operator. The vanishing of the quantity $\partial_\omega Q(\omega)$ at some value of ω indicates the parameter value for the collision of eigenvalues, when the Jordan block corresponding to the zero eigenvalue has a jump of two in its size. A nice feature of the linearization at a ground state solitary wave in the nonlinear Schrödinger equation is that its spectrum belongs to the imaginary axis, with some eigenvalues possibly located on the real axis; thus, the collision of eigenvalues at $\lambda = 0$ is the only way the spectral instability could develop. In the NLD context, such a collision does not necessarily occur at $\lambda = 0$; both situations as in Fig. 6 are possible.

In [18], it was shown that in NLD (and similar fermionic systems) the collision of eigenvalues at the origin and a subsequent transition to instability is characterized not only by the Vakhitov–Kolokolov condition $dQ/d\omega = 0$, but also by the condition $E(\omega) = 0$, where E is the value of the energy functional on the corresponding solitary wave.

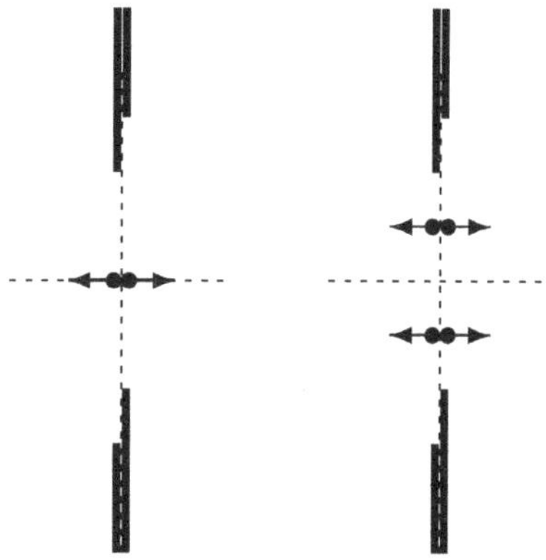

Fig. 6 Birth of "unstable" eigenvalues out of collisions of imaginary eigenvalues. When the frequency ω of the solitary wave $\phi_\omega e^{-i\omega t}$ changes, the "unstable", positive-real-part eigenvalues in the linearized equation could be born from the collisions of discrete imaginary eigenvalues

Theorem 1 *The algebraic multiplicity of the eigenvalue* $\lambda = 0$ *of the linearization* $\mathscr{A}_\omega$ *at the solitary wave* $\phi_\omega(x)e^{-i\omega t}$ *has a jump of (at least) 2 when at a particular value of* ω *either* $\partial_\omega Q(\phi_\omega) = 0$ *or* $E(\phi_\omega) = 0$, *with* $Q(\phi_\omega)$ *and* $E(\phi_\omega)$ *being the charge and the energy of the solitary wave* $\phi_\omega(x)e^{-i\omega t}$.

The eigenvalues with positive real part could also be born from the collision of purely imaginary eigenvalues at some point in the spectral gap but away from the origin. This is expected to lead to an oscillatory instability due to the resulting complex eigenvalues; we have recently observed this scenario in the cubic Soler model in two spatial dimensions [49]. Presently we do not have a criterion for such a collision of eigenvalues.

4.1.2 Instability Scenario 2: Bifurcations from the Essential Spectrum

The most peculiar feature of the linearization at a solitary wave in the NLD context is the possibility of bifurcations of eigenvalues with nonzero real part off the imaginary axis, out of the bulk of the essential spectrum.

The article [27] gives a thorough analytical study of eigenvalues of the Dirac operators, focusing on whether and how such eigenvalues can bifurcate from the essential spectrum. Generalizing the Jensen–Kato approach [91] to the context of the Dirac operators, it was shown in [27, Theorem 2.15] that for $|\omega| < m$ the bifurcations from the essential spectrum are only possible from embedded eigenvalues (Fig. 7, center), with the following exceptions: the bifurcation could start at the embedded thresholds located at $\pm i(m + |\omega|)$ (Fig. 7, left), or they could start at $\lambda = \pm im$ when $\omega = 0$ (Fig. 7, right; this situation corresponds to the collision of thresholds). Indeed, bifurcations from the embedded thresholds have been observed in a one-dimensional NLD-type model of coupled-mode equations [11, 37]. The bifurcations from the collision of thresholds at $\pm im$ (when $\omega = 0$) were demonstrated in [94] in the context of the perturbed massive Thirring model.

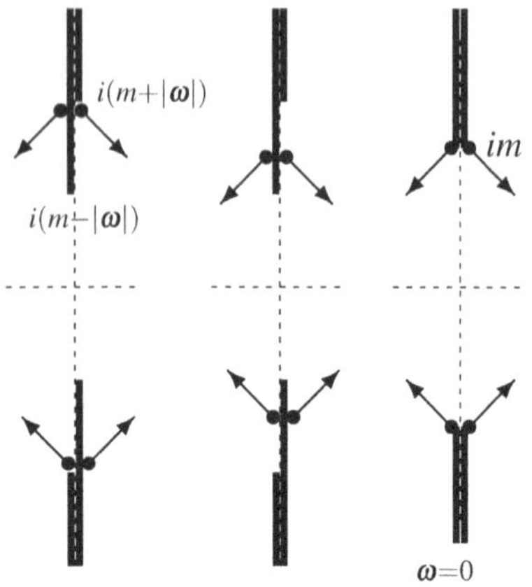

Fig. 7 Possible bifurcations from the essential spectrum. Theoretically, when $|\omega| < m$, the nonzero-real-part eigenvalues could be born from the embedded thresholds at $\pm i(m + |\omega|)$, from the embedded eigenvalue in the bulk of the essential spectrum between the threshold and the embedded threshold, and from the collision of the thresholds at $\pm im$ when $\omega = 0$

One can use the Carleman–Berthier–Georgescu estimates [19] to prove that there are no embedded eigenvalues (hence no bifurcations) in the portion of the essential spectrum outside of the embedded thresholds [27].

As to the bifurcations from the embedded eigenvalues before the embedded thresholds, as in Fig. 7 (center), we do not have any such examples in the NLD context, although such examples could be produced for Dirac operators of the form (28) (with V kept self-adjoint).

4.1.3 Instability Scenario 3: Bifurcations from the Nonrelativistic Limit

The nonzero-real-part eigenvalues could be present in the spectrum of the linearization operators at small amplitude solitary waves for all $\omega \lesssim m$, being born "from the nonrelativistic limit". It was proved in [27, Theorem 2.19], under very mild assumptions, that the bifurcations of eigenvalues for ω departing from $\pm m$ are only possible from the thresholds $\lambda = 0$ and $\lambda = \pm 2mi$; see Fig. 8.

We now undertake a detailed study of these bifurcations; let us concentrate on the case $\lambda = 0$. It is of no surprise that the behaviour of eigenvalues of the linearized operator near $\lambda = 0$, in the nonrelativistic limit $\omega \lesssim m$, follows closely the pattern which one observes in the nonlinear Schrödinger equation with the same nonlinearity. In other words, if the linearizations of the nonlinear Dirac equation at solitary waves with $\omega \lesssim m$ admit a family of eigenvalues Λ_ω which continuously depends on ω, such that $\Lambda_\omega \to 0$ as $\omega \to m$, then this family is merely a deformation of an eigenvalue family $\Lambda_\omega^{\mathrm{NLS}}$ of the linearization of the nonlinear Schrödinger equation with the same nonlinearity (linearized at corresponding solitary waves). To make this rigorous, one considers the spectral problem for the linearization at a solitary wave with $\omega \lesssim m$, applies the rescaling with respect to $m - \omega \ll 1$, and uses the reduction based on the Schur complement method, recovering in the nonrelativistic limit $\omega \to m$ the linearization of the nonlinear Schrödinger equation, and then applying

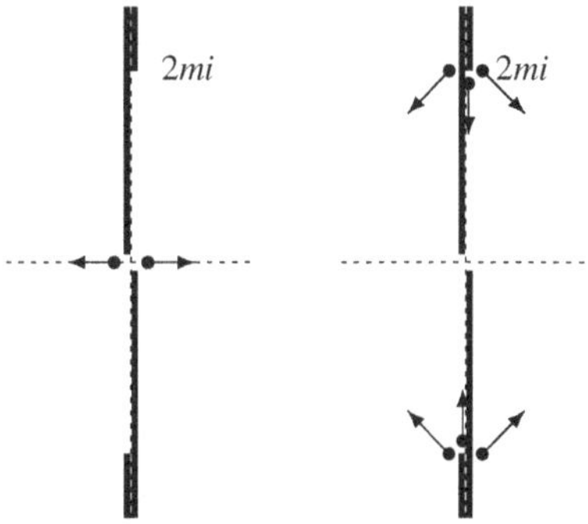

Fig. 8 Bifurcations from $\lambda = 0$ and hypothetical bifurcations from $\lambda = \pm 2mi$ in the nonrelativistic limit, $\omega \lesssim m$. The nonzero-real-part eigenvalues could be present in the spectrum of the linearization at a solitary wave $\phi_\omega e^{-i\omega t}$ for ω arbitrarily close to m; these eigenvalues would have to be located near $\lambda = 0$ or near the embedded threshold at $\lambda = \pm 2mi$

the Rayleigh–Schrödinger perturbation theory; in [39], this approach was developed to prove the linear instability of small amplitude solitary waves $\phi_\omega(x)e^{-i\omega t}$ in the "charge-supercritical" NLD, in the nonrelativistic limit $\omega \lesssim m$.

Theorem 2 *Assume that $f(s) = |s|^k$, where $k \in \mathbb{N}$ satisfies $k > 2/n$ (and $k < 2$ for $n = 3$). Then there is $\omega_1 < m$ such that the solitary wave solutions $\phi_\omega(x)e^{-i\omega t}$ (in the form of the Wakano Ansatz from Sect. 2.5) to NLD are linearly unstable for $\omega \in (\omega_1, m)$. More precisely, let $\mathscr{A}_\omega$ be the linearization of the nonlinear Dirac equation at a solitary wave $\phi_\omega(x)e^{-i\omega t}$. Then for $\omega \in (\omega_1, m)$ there are eigenvalues*

$$\pm\lambda_\omega \in \sigma_p(\mathscr{A}_\omega), \qquad \lambda_\omega > 0, \qquad \lambda_\omega = O(m - \omega).$$

Let us remark here that the restriction in the above theorem that k is a natural number was needed to make sure that the solitary wave family of the form of the Wakano Ansatz indeed exists. Theorem 2 extends to $f(s) = a|s|^k + O(|s|^K), a > 0$, with $k \in (2/n, 2/(n-2))$ ($k > 2/n$ when $n \le 3$) and $K > k$. The existence of the corresponding families of solitary waves was proved in [26]. In that article, a general construction was given for small amplitude solitary waves in the nonlinear Dirac equation, deriving the asymptotics which we will need in the forthcoming stability analysis of such solitary waves. This is a general result proved for nonlinearities which are not necessarily smooth, thus applicable to e.g. critical and subcritical nonlinearities.

We point out that the instability stated in Theorem 2 is in a formal agreement with the Vakhitov–Kolokolov stability criterion [161]; one has $dQ(\omega)/d\omega > 0$ for $\omega \lesssim m$. Conversely, we expect that the presence of eigenvalues with nonzero real part in the vicinity of $\lambda = 0$ for $\omega \lesssim m$, is prohibited by the Vakhitov–Kolokolov stability criterion $\frac{dQ(\omega)}{d\omega} < 0$, $\omega \lesssim m$.

Similarly to how the NLS corresponds to the nonrelativistic limit of NLD, in the nonrelativistic limit of the Dirac–Maxwell system one arrives at the Choquard equation [106]; see [41] and the references therein. The Choquard equation is known to be spectrally (in fact, even orbitally) stable [35]; we expect that this implies absence of unstable eigenvalues bifurcating from the origin in the Dirac–Maxwell system.

As we pointed out above, in the nonrelativistic limit $\omega \lesssim m$, there could be eigenvalue families of the linearization of the nonlinear Dirac operator bifurcating not only from the origin, but also from the embedded threshold (that is, such that $\lim_{\omega\to m} \Lambda_i(\omega) = \pm 2mi$). Rescaling and using the Schur complement approach shows that there could be at most $N/2$ such families bifurcating from each of $\pm 2mi$, with N the number of components of a spinor field (in 3D Dirac, one takes $N = 4$). Could these eigenvalues go off the imaginary axis into the complex plane? While for the nonlinear Dirac equations with a general nonlinearity the answer to this question is unknown, in the Soler model we can exclude this scenario. One can show that there are exact eigenvalues $\lambda_{\pm}(\omega) = \pm 2\omega i$, each being of multiplicity $N/2$; thus, we know exactly what happens to the eigenvalues which bifurcate from $\pm 2mi$, and expect no bifurcations of eigenvalues off the imaginary axis. The details are given in [28].

Let us finish with a very important result: the existence of eigenvalues $\pm 2\omega i$ of the linearization at a solitary wave in the Soler model (3) is a consequence of having bi-frequency solitary wave solutions in the Soler model, in any dimension and for any choice of f in (3). For more details, see [28].

4.2 *Orbital and Asymptotic Stability of Solitary Waves*

The spectral analysis is one aspect of global analysis of the dynamical stability. In principle any spectral instability around a stationary solution should lead to a dynamical instability, namely the stationary solution is orbitally unstable. The contrapuntal statement that a stable stationary state has a spectrally stable linearized operator needs to be analyzed carefully.

If the Dirac operator D_m is perturbed by some zero-order external potential, the perturbation theory provides tools which allow one to analyze the linear stability of linearized operators of the form (28). Still some important restrictions on the potential appear (decay, regularity, and absence of resonances). Even if the perturbation analysis needs some work, it is much less involved compared to the complete spectral characterization of the linearized operator. This opens the gates to the analysis of the nonlinear stability.

Prior to a bibliographical review of the available works in this direction, we make a remark. While in many models the orbital stability is obtained by using the energy as some kind of a Lyapunov functional, this is no longer possible for models of Dirac type since the energy is sign-indefinite. Even if there are some conserved quantities which allow one to control certain negative directions of the Hessian of the energy, the latter are in infinite number ("infinite Morse index") and in most cases the conservation laws are not enough. The route "use linear stability to prove the asymptotic stability" seems to be the only one available for the sign-indefinite systems such as nonlinear Dirac, Dirac–Hartree–Fock, and others. As a result, due to the strong indefiniteness of the Dirac operator (the energy conservation does not lead to any bounds on the $H^{1/2}$-norm), we do not know how to prove the *orbital stability* [77] but via proving the asymptotic stability first. The only exceptional case in nonlinear Dirac-type systems seems to be the completely integrable massive Thirring model in one spatial dimension [158], where additional conserved quantities arising from the complete integrability allow one to prove orbital stability of solitary waves [43, 127]. Note that these conserved quantities are used not to control the negative directions but rather to construct a new Lyapunov functional. More precisely, by [127], there is a functional R defined on $H^1(\mathbb{R}, \mathbb{C}^2)$ (which contains terms dependent on powers of components of $\psi \in \mathbb{C}^2$ of order up to six) which is (formally) conserved for solutions to the massive Thirring model, and it is further shown that there is $\omega_0 \in (0, m]$ such that for $\omega \in (-\omega_0, \omega_0)$ the solitary wave amplitude is a local minimizer of R in H^1 under the charge and momentum conservation, and hence the corresponding solitary wave is orbitally stable in $H^1(\mathbb{R}, \mathbb{C}^2)$. Moreover, in [43], using the global existence of L^2-solutions for the (cubic) massive Thirring model [33], the orbital stability of solitary waves in $L^2(\mathbb{R})$ has been shown, with the proof based on the auto-Bäcklund transformation. Now we turn to the asymptotic stability. In [40], the asymptotic stability was proved for the small energy perturbations to solitary waves in the Gross–Neveu model. The model is taken with particular pure-power nonlinearities when all the assumptions on the spectral and linear stability of solitary waves have been verified directly. This is, referring to the previous discussion, also the "proof of concept": it is shown that there are translation-invariant systems based on the Dirac operator which are asymptotically stable; this is in spite of the energy functional being unbounded from below.

First results on asymptotic stability were obtained in [24, 25] in the case $n = 3$, in the external potential. There, the spectrum of the linear part of the equation $D_m + V$ is supposed to be, beside the essential spectrum $\mathbb{R} \setminus (-m, m)$, formed by two simple eigenvalues; let us denote them by λ_0 and λ_1, with $\lambda_0 < \lambda_1$. From the associated eigenspaces, there is a bifurcation of small solitary waves for the nonlinear equation. The corresponding linearized operators are exponentially localized small perturbations of $D_m + V$, so that the perturbation theory allows a precise knowledge of the resulting spectral stability. Depending on the distance from λ_0 to λ_1 compared to the distance from λ_0 to the essential spectrum, the resulting point spectrum for the linearized operator may be discrete and purely imaginary and hence spectrally stable, or instead it may have nonzero-real-part eigenvalues if a "nonlinear Fermi Golden Rule" assumption is satisfied (similarly to the Schrödinger case, see [32, 150, 151]); in the latter case, linear and dynamical instabilities can occur. In the former case, the linear stability follows from the spectral one via the perturbation theory. In any case, using the dispersive properties for perturbations of D_m, there is a stable manifold of real codimension 2. Due to the presence of nonzero discrete modes, even in the linearly stable case, the dynamical stability is not guaranteed. Before considering the results on the dynamics outside this manifold, for perturbations along the remaining two real directions, one could ask what might happen if $D_m + V$ had only one eigenvalue. The answer follows quite immediately with the ideas from [24, 25]. In this case, there is only one family of solitary waves and it is asymptotically stable. Notice that the asymptotic profile is possibly another solitary wave but close to the perturbed one. In the one-dimensional case, this was studied properly in [127]. Note that the one-dimensional framework suffers from relatively weak dispersion which makes the analysis of the stabilization process more delicate. As for the dynamics outside the above-mentioned stable manifold, the techniques rely on the analysis of nonlinear resonances between discrete isolated modes and the essential spectrum where the dispersion takes place. This requires the normal form analysis in order to isolate the leading resonant interactions. The former is possible only if the "nonlinear Fermi Golden Rule" is imposed. Such an analysis was done in [17] but in a slightly different framework: instead of considering the perturbative case the authors chose the translation-invariant case, imposing a series of assumptions that lead to the spectral stability of solitary waves. These assumptions are verified in some perturbative context with $V \neq 0$. This case is analyzed in [45]. The asymptotic stability approach from [17, 40, 127] is developed under important restrictions on the types of admissible perturbations. These restrictions are needed to avoid the translation invariance and, most importantly, to prohibit the perturbations in the direction of exceptional eigenvalues $\pm 2\omega i$ of the linearization operator at a solitary wave $\phi_\omega(x)e^{-i\omega t}$. These eigenvalues are a feature of the Soler model (see [52, 75]); they are present in the spectrum for any nonlinearity f in the Soler model (3), see [29, 52, 75]. These eigenvalues are embedded into the essential spectrum when $|\omega| > m/3$ and violate the "nonlinear Fermi Golden Rule": they do not "interact" (that is, do not resonate) with the essential spectrum; the energy from the corresponding modes does not disperse to infinity. This does not allow the standard approach to proving the asymptotic stability.

5 Stability of Solitary Waves: Numerical Results

Once the theoretical background on linear stability has been presented, we review in this section some very recent numerical results on this topic. To this aim, we first include a brief introduction to the Evans function formalism [66], and then, detailed results based on numerical analysis of BdG-like spectral stability are shown for both 1D and 2D Soler models.

Let us recall some notation regarding the spectral stability, as we will make an extensive use of them in what follows. The *essential spectrum* corresponds to $\lambda \in i(-\infty, |\omega| - m] \cup i[-|\omega| + m, \infty)$. Embedded eigenvalues can be in the region $\lambda \in \pm i[-|\omega| + m, |\omega| + m]$ of the essential spectrum; for abbreviation, we denote this region as the *embedded spectrum* and the remaining part of the essential spectrum as *non-embedded spectrum*.

In what follows, without lack of generality we will take $g = m = 1$ unless stated otherwise.

5.1 Evans Function Approach to the Analysis of Spectral Stability

The study of the spectral stability of the cubic 1D Soler model was performed in [29], with the aid of the Evans function technique. This was the first definitive linear stability result (as well as the first *definite* stability result) in the context of the nonlinear Dirac equation.

Let us give more details. In order to compute $\sigma\ (\mathscr{A}_\omega)$ we can employ the Evans function which provides an efficient tool to locate the point spectrum. The Evans function was first introduced by Evans [66–69] in his study of the stability of nerve impulses. In his work, Evans defined $D(\lambda)$ to represent the determinant of eigenvalue problems associated with traveling waves of a class of nerve impulse models. $D(\lambda)$ was constructed to detect the intersections of the subspace of solutions decaying exponentially to the right and the subspace of solutions decaying exponentially to the left. Jones [93] used Evans' idea to study the stability of a singularly perturbed FitzHugh–Nagumo system. Jones called it the Evans function, and the notation $E(\lambda)$ is now common. The first general definition of the Evans function was given by Alexander et al. [6] in their study of the stability for traveling waves of a semi-linear parabolic system. Pego and Weinstein [123] expanded on Jones' construction of Evans function to study the linear instability of solitary waves in the Korteweg–de Vries equation (KdV), the Benjamin–Bona–Mahoney equation (BBM), and the Boussinesq equation. Generally, the Evans function for a differential operator $\mathscr{D}$ is an analytic function such that $E(\lambda) = 0$ if and only if λ is an eigenvalue of $\mathscr{D}$, and the order of zero is equal to the algebraic multiplicity of the eigenvalue.

Let us give a simple example which illustrates the nature of the Evans function. Consider the stationary Schrödinger equation

$$-\lambda^2 u(x) = Hu(x), \tag{29}$$

where $H = -\partial_x^2 + V$ with $V \in C(\mathbb{R}), \operatorname{supp}(V) \subset (-1, 1)$. For $\lambda \in \mathbb{C} \setminus \{0\}, \operatorname{Re}(\lambda) > 0$, it has the solutions $J_+(\lambda, x)$ and $J_-(\lambda, x)$, $x \in \mathbb{R}$, defined by their behaviour at $\pm\infty$:

$$J_+(\lambda, x) = e^{-\lambda x}, \quad x \geq 1; \qquad J_-(\lambda, x) = e^{+\lambda x}, \quad x \leq -1.$$

We should note that J_+ and J_- decay exponentially as $x \to \pm\infty$, respectively, and they have the same asymptotics at $\pm\infty$ as the solutions to the equation

$$-\lambda^2 u(x) = H_0 u(x),$$

where $H_0 = -\partial_x^2$, which agrees with H on $\mathbb{R} \setminus [-1, 1]$. We call J_+ and J_- the Jost solution to (29) and define the Evans function to be the Wronskian of J_+ and J_-:

$$E(\lambda) = W(J_+, J_-)(x, \lambda) = J_+(x, \lambda)\partial_x J_-(x, \lambda) - J_-(x, \lambda)\partial_x J_+(x, \lambda),$$

where the right-hand side depends only on λ. Vanishing of E at some particular $\lambda \in \mathbb{C}$, $\mathrm{Re}(\lambda) < 0$ shows that the Jost solutions J_+ and J_- are linearly dependent, and there is $c \in \mathbb{C} \setminus \{0\}$ such that

$$\phi(x) = \begin{cases} J_+(x, \lambda), & x \geq 0 \\ cJ_-(x, \lambda), & x < 0 \end{cases}$$

is C^1 and thus is an eigenfunction corresponding to an eigenvalue λ^2 of H.

The construction for the one-dimensional Soler model is done by decomposing $L^2(\mathbb{R}, \mathbb{C}^4)$ into two invariant subspaces for the operator $\mathscr{A}_\omega$ introduced in (9): the "even" subspace, with even first and third components and with odd second and fourth components, and the "odd" subspace, with odd first and third components and with even second and fourth components; the direct sum of the "even" and "odd" subspaces coincides with $L^2(\mathbb{R}, \mathbb{C}^4)$. The Evans function corresponding to the "even" subspace is defined by

$$E_{even}(\lambda) = \det\left(R_1, R_3, J_1, J_2\right), \tag{30}$$

where $R_j(x)$, $1 \leq j \leq 4$, are the solutions to the equation $\lambda R = \mathscr{A}_\omega R$ with the initial data

$$R_j|_{x=0} = \mathbf{e}_j, \qquad 1 \leq j \leq 4,$$

where $\mathbf{e}_j$, $1 \leq j \leq 4$, is the standard basis in $\mathbb{C}^4$. J_1 and J_2 are the Jost solutions of $\mathscr{A}_\omega$, which are defined as the solutions to $\lambda\Psi = \mathscr{A}_\omega\Psi$ with the same asymptotics at $+\infty$ as the solutions to $\lambda\Psi = (\mathbf{D}_m - \omega)\Psi$ which decay as $x \to +\infty$, where

$$\mathbf{D}_m = \begin{bmatrix} D_m & 0 \\ 0 & D_m \end{bmatrix}, \quad D_m = \begin{bmatrix} m & \partial_x \\ -\partial_x & -m \end{bmatrix} = -i(-\sigma_2)\partial_x + m\sigma_3.$$

The Evans function corresponding to the "odd" subspace is constructed by using in (30) functions R_2 and R_4 instead of R_1 and R_3. We note that, by Liouville's formula, the right-hand side in (30) does not depend on x.

Figure 9 shows the zeros of the Evans function which are plotted alongside with the essential spectrum for the linearization at the solitary waves in the 1D Soler model.

Later, in [49], it was observed that the linearized operator admits invariant subspaces which correspond to spinorial spherical harmonics. This allows one to factorize the operator, essentially reducing the consideration to a one-dimensional setting, and to perform a complete numerical analysis of the linearized stability in the nonlinear Dirac equation in two spatial dimensions and give partial results in three dimensions, basing our approach on both the Evans function technique and the linear stability analysis using spectral methods.

For the two-dimensional Soler model, we can use the same process as the one-dimensional case to construct the Evans function. Recall (see (17)) that $\mathscr{A}_\omega$ acts

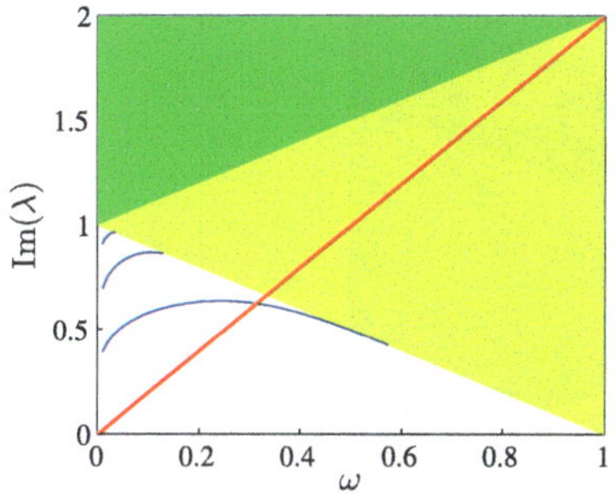

Fig. 9 Eigenvalues corresponding to zeros of the Evans function in the upper half of the spectral gap as a function of ω. Yellow area represents the part of the continuous spectrum that corresponds to $iL_{\pm}(\omega)$, while the green area represents the (doubly-covered) part of the continuous spectrum corresponding to both $iL_{\pm}(\omega)$ and $-iL_{\pm}(\omega)$. The eigenvalues $\lambda = 2\omega i$ (red straight line) are embedded into the essential spectrum for $\omega > m/3$

invariantly on $\mathscr{X}_q$ for each $q \in \mathbb{Z}$ and $\mathscr{A}_{\omega,q} = \mathscr{A}_\omega|_{\mathscr{X}_q}$. We consider the case $S = 0$. The Evans function for each $\mathscr{A}_{\omega,q}$ is defined by

$$E_q(\lambda) = \det(R_q^+, R_q^-, Y_1, Y_2).$$

Here R_q^+ and R_q^- are linearly independent solutions to the equation $\lambda R = \mathscr{A}_{\omega,q} R$ with the following linearly independent initial data at $r = 0$

$$\begin{bmatrix} -\lambda - (\omega + f_0)\frac{iq}{|q|} \\ 0 \\ i|q| \\ q \end{bmatrix} \quad \text{and} \quad \begin{bmatrix} 0 \\ -i\lambda\frac{q}{|q|} + \omega + f_0 \\ |q| \\ iq \end{bmatrix},$$

where $f_0 = m - g\left(u^2(0) - v^2(0)\right)$. The Jost solutions Y_1 and Y_2 of $\mathscr{A}_{\omega,q}$ are defined as the solution to $\lambda Y = \mathscr{A}_{\omega,q} Y$ with the same asymptotics at $+\infty$ as the solutions to $\lambda Y = \mathbf{D}_q Y$ where

$$\mathbf{D}_q = \begin{bmatrix} -\sigma_1\frac{iq}{r} & D_m - \omega I_2 \\ -D_m + \omega I_2 & -\sigma_1\frac{iq}{r} \end{bmatrix}, \qquad q \in \mathbb{Z}.$$

5.2 *Bogoliubov–de Gennes Analysis: The One-Dimensional Case*

Let us recall from the analysis shown in Sect. 4 that near the non-relativistic limit ($\omega \lesssim m$), the stability of solitary waves formally agrees with the Vakhitov–Kolokolov stability criterion $\partial_\omega Q(\phi_\omega) < 0$ [161]. In particular, there is no positive eigenvalue

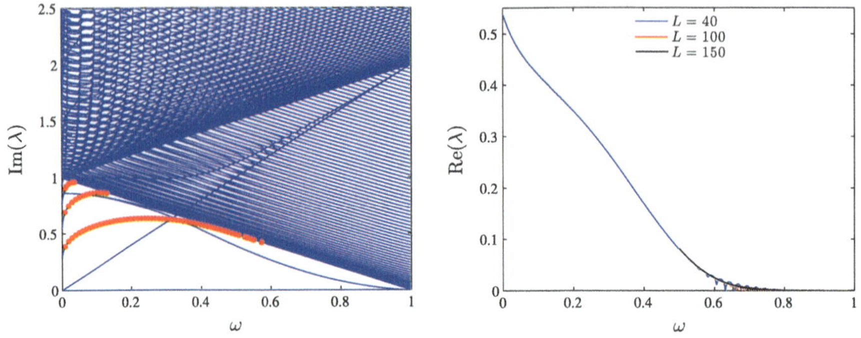

Fig. 10 Spectrum of the stability matrix (10) for solitary waves in domain $[-L, L]$ with $L = 40$ obtained using finite differences with $N = 800$ grid points in the cubic ($k = 1$) case. Dots correspond to Evans function predictions. Right panel displays only the maximum values of $\mathrm{Re}(\lambda)$ (i.e., the growth rates) and includes the values for $L = 40$, $L = 100$, and $L = 150$

emerging from $\lambda = 0$ for $\omega \lesssim m$ as long as $k \leq 2$ (and, consequently, the solitary waves are spectrally stable), while in the case $k > 2$ there is a pair of (a positive and a negative) eigenvalues which result in linear instability. As it turns out, in the one-dimensional case, the Vakhitov–Kolokolov stability criterion agrees with the observed stability of solitary waves not only in the nonrelativistic limit, but for all frequencies $\omega \in (0, m)$, as our numerical calculations show below. Evans function analysis presented above also shows that solitary waves do not present oscillatory instabilities (i.e., there are no complex λ's with nonzero real part) in the 1D case; the instability could only develop when eigenvalues collide and bifurcate from the origin. Additionally, for any k, the existence of an eigenvalue $\lambda = \pm 2\omega i$ is a consequence of the $\mathbf{SU}(1, 1)$-invariance of the Soler model [52, 75]. This mode, which does not give rise to any instability, is embedded into the essential spectrum for $\omega \in (m/3, m)$ (see Fig. 10).

Let us mention that it was shown in [120, 149] that attempts to apply Derrick's argument [56] to stability of solitary waves in the context of the nonlinear Dirac equation [22, 154]—in particular, the so-called Bogolubsky criterion—do not seem to work. This is not particularly surprising, given that Derrick's empirical argument, based on singling out one family of perturbations of a solitary wave and checking whether the solitary wave corresponds to the energy minimum on this curve, was introduced in the context of the second order systems, appealing to our Newtonian-world intuition. Apparently, this approach does not necessarily work in the context of the first order systems, such as the Dirac equation.

Examining the finite difference discretization of the 1D Soler model, we replace the spatial derivatives $\partial_x f(x)$ in (7) by the central difference $(f_{n+1} - f_{n-1})/(2h)$. This method is tantamount to using the collocation points of (23) and (25) with N collocation points, a domain $x \in [-L, L]$ and $h = 2L/N$. Figure 10 shows the stability eigenvalues for $k = 1$ and $L = 40$ while the spacing $h = 0.1$. We do not consider here instabilities that disappear in the infinite domain, continuum limit.

In the case of small ω, the solitary waves are identified as unstable. The "size" of the instability decreases as the frequency is increased. This is because of a mode colliding with the essential spectrum around $\omega \approx 0.037$ and leading to the formation of eigenfrequency quartets. For $\omega \approx 0.632$, the stability of the solitary wave is briefly restored, only to be lost again at $\omega \approx 0.634$. Subsequently, "bubbles" of instability arise (with decreasing amplitude as ω is increased). To identify the relevant trend, we have examined in the right panel of Fig. 10 the cases of different length for $L = 40$, 100 and 150. As L increases, so does the number of bubbles, while the width decreases, with their envelope tending to zero when ω approaches 1, in a way reminiscent of the corresponding scenario for dark solitons in the discrete nonlinear Schrödinger equation (DNLS) case [92]. From this trend, it is not straightforward to infer whether the unstable solution becomes stabilized at a critical ω or whether it is asymptotically approaching the stable NLS limit of $\omega \to 1$.

In order to find out a strategy which assures a spectral accuracy of BdG stability analysis which is also correlated to the Evans' function analysis, we used spectral collocation methods in [46]. We utilized two case examples of such methods therein: the Fourier Spectral Collocation Method, which implicitly enforces periodic boundary conditions, and the Chebyshev Spectral Collocation Method, which enforces (homogeneous) Dirichlet boundary conditions (see Sect. 3.2). The advantage of the Finite Difference Method with respect to the other ones concerns the fact that the resulting stability matrix is sparse. In the computations performed in that work and that will be presented below, $N = 800$ collocation points were taken in a domain $[-L, L]$, with a discretization parameter $h = 1/(2L)$; this value coincides with the distance between grid points in the Fourier collocation and finite difference methods, but not in the Chebyshev collocation as the grid points are not equidistant. Increasing the node numbers to $N = 1200$ does not seem to qualitatively improve the findings.

In Fig. 11 we examine the dependence of the imaginary part of the eigenvalues λ with respect to the frequency ω of the solution for both spectral methods in the cubic case of $k = 1$. In addition to the $\lambda = \pm 2\omega i$ mode, the different methods have

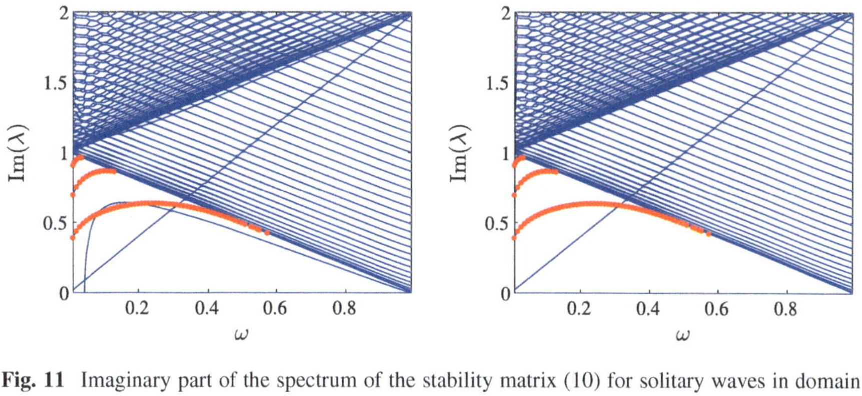

Fig. 11 Imaginary part of the spectrum of the stability matrix (10) for solitary waves in domain $[-L, L]$ with $L = 40$ obtained using Fourier (left) and Chebyshev (right) spectral collocation method with $N = 800$ grid points in the cubic ($k = 1$) case. Dots correspond to Evans function predictions

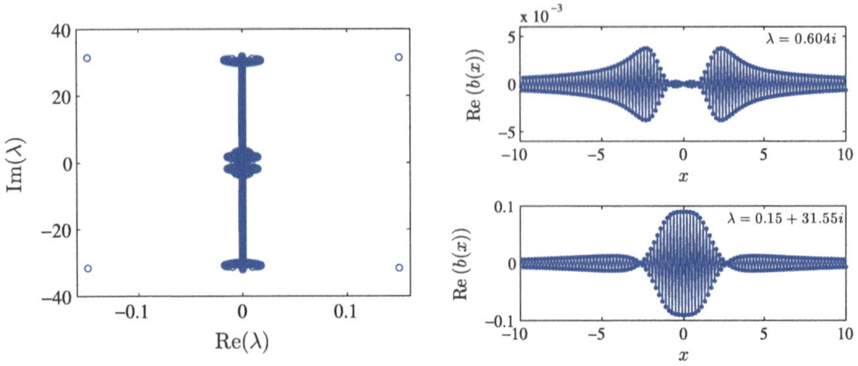

Fig. 12 Spectral plane of a solitary wave with $\omega = 0.1$ (cubic case, $L = 40$, and $N = 800$) obtained using the Fourier spectral collocation method (left panel). The typical profile of two modes corresponding to spurious eigenvalues is depicted in the right panel. In particular, we have included the mode with $\text{Re}(\lambda) = 0$ which does not arise in the Evans' function analysis of Sect. 5.1 together with the largest real part eigenvalue, which is also spurious

additional modes which can be compared also with the Evans function analysis outcome of Fig. 9. We thus find that the comparison of the Fourier spectral collocation method with the Evans function analysis (Fig. 9) seems qualitatively (and even quantitatively) to yield very good agreement with the exception of a mode that seems to initially grow steeply (for small ω) and subsequently to slowly asymptote to the band edge (as ω increases). This mode is shown in the right panel of Fig. 12, while the left panel of the figure illustrates a prototypical example of the Fourier spectral collocation method spectrum for $\omega = 0.1$. From the above panel, we can immediately infer that this mode is, in fact, spurious and an outcome of the discretization as it carries a staggered profile that cannot be supported in the continuum limit. In the left panel of the same figure, we can see the existence of additional spurious modes forming bubbles of complex eigenvalues. However, the fact that these bubbles are occurring at the eigenvalues of the continuous spectrum assures us that these are spurious instabilities due to the finite size of the domain and ones which disappear in the $L \to \infty$, $h \to 0$ limit. This is confirmed by Fig. 13 which shows that as we decrease h (and increase the number of lattice sites, approaching the continuum limit for a given domain size) the growth rate of such spuriously unstable eigenmodes accordingly decreases.

Remarkably, the finite difference spectrum of Fig. 10 is the one that seems most "distant" from the findings of the Evans function method. While all four of the internal modes of the latter spectrum seem to be captured by the finite difference method, three additional modes create a nontrivial disparity. Two of them are in fact "benign" and maintain an eigenvalue below the band edge of the continuous spectrum for all values of $\omega \in (0, m)$. However, as explained in [47], we also observe the existence of an eigenmode embedded in the essential spectrum. Unfortunately, this mode is accompanied by a real part in the corresponding eigenvalue and hence gives rise to a spurious instability. Figure 14 presents a graph analogous to Fig. 12 but for the finite

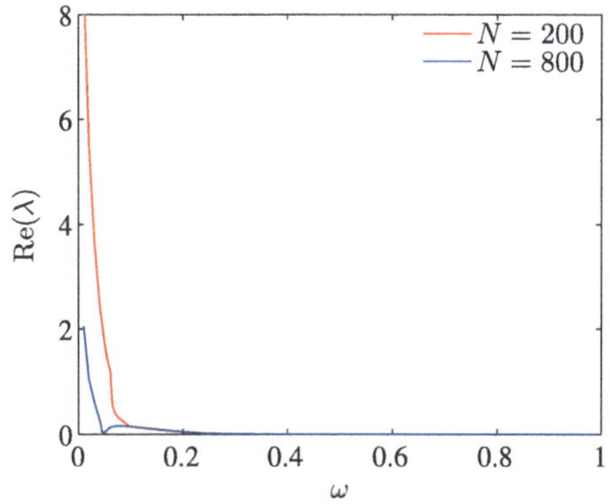

Fig. 13 Growth rates (i.e., maximum of the real part of the eigenvalues) for a solitary wave with $L = 40$ in the cubic case using the Fourier spectral collocation method. The number of grid points is either $N = 800$ or $N = 200$

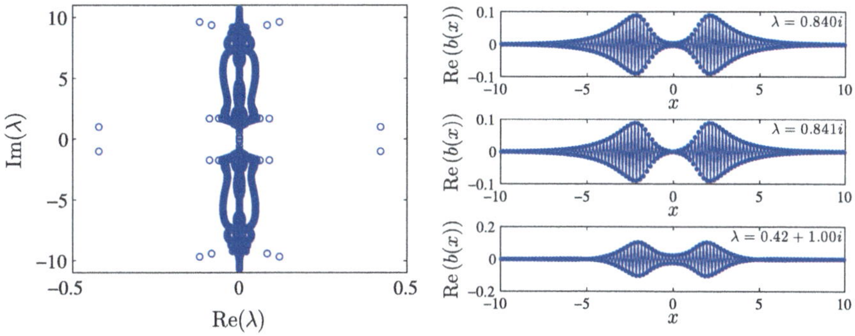

Fig. 14 Spectral plane of a solitary wave with $\omega = 0.1$, $L = 40$ and $N = 800$ in the cubic case, using finite difference discretization (left panel). The typical profile of three modes corresponding to spurious eigenvalues is depicted in the right panel. In particular, we have included the two modes with $\text{Re}(\lambda) = 0$ which do not arise in the Evans function analysis together with the embedded spurious mode

difference method. The undesirable unstable mode, as well as additional spurious modes are explicitly indicated through the eigenvector profiles of the right panel.

The scenario of the Chebyshev spectral collocation method bears advantages and disadvantages in its own right. Although it gives an accurate result for the imaginary part of the eigenvalues, their real part grows for large $\text{Im}(\lambda)$, as is also shown in Fig. 15. Additionally, as indicated in [30], approximately half of the values of the spectrum are spurious within the Chebyshev collocation method, so they should be excluded from consideration. Furthermore, one can observe that in this case as well, spurious instability bubbles arise (see the right panel of Fig. 15), yet we have checked that these disappear in the continuum limit of $h \to 0$.

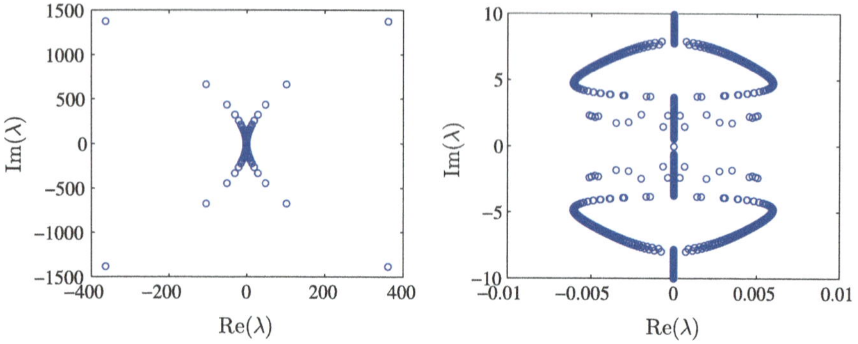

Fig. 15 Spectral plane of a solitary wave with $\omega = 0.4$, $L = 40$, and $N = 800$ in the cubic case, using the Chebyshev spectral collocation method. The right panel is a zoom of that on the left, illustrating the weak, spurious instabilities (which disappear as the continuum limit is approached)

As a final aspect of the spectral considerations that we provide herein, we have examined the instability that arises e.g. from the Chebyshev spectral collocation method for larger values of k. Recall that the Chebyshev spectral collocation method predicts (at least as regards the point spectrum out of the non-embedded spectrum) that there is no instability for any ω in the case of $k = 1$, in agreement with the Evans function analysis and [126]. The method identifies an instability for such point spectrum eigenvalues *only* for $k > 2$. The relevant instability predicted numerically in the k-ω plane is illustrated in Fig. 16. We note that this instability is precisely captured by the Vakhitov–Kolokolov criterion, i.e., it precisely corresponds to the condition $\partial_\omega Q(\omega) = 0$, in agreement with [18]. Hence, by analogy with the nonrelativistic limit $\omega \to m = 1$, we expect this to be an instability associated with the collapse of the latter model (however, we will observe a key dynamical difference, in comparison to the NLS, in Sect. 6). Nevertheless, it is relevant to point out here that the NLD, contrary to the NLS, does not exhibit an instability for all ω when $k > 2$. The instability is instead limited to $\omega > \omega_c(k)$, as characterized by the curve of Fig. 16. Hence, it can be inferred that the instability is mitigated by the relativistic limit of the NLD and only occurs in an interval of frequency values including the non-relativistic limit $\omega \to m = 1$, yet not encompassing the full range of available frequencies in the relativistic case.

5.3 *Bogoliubov–de Gennes Analysis: The Two- and Three-Dimensional Cases*

From the experience acquired with the study of the stability of solitary waves in one spatial dimension, it is clear that a Chebyshev spectral collocation method must be followed in order to analyze the stability in higher-dimensional solitary waves. This is the approach followed in the present section, which summarizes the results of [49].

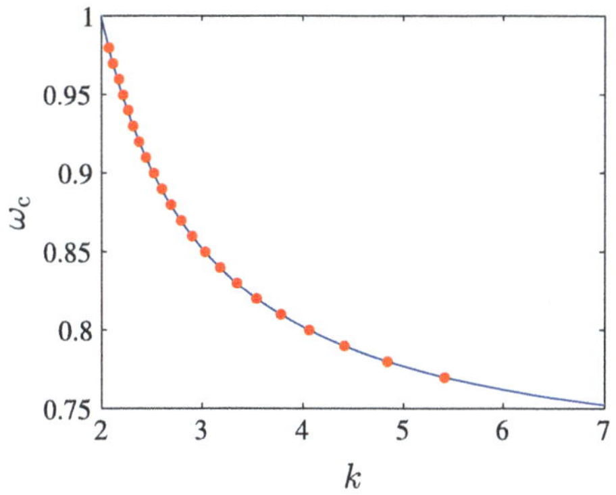

Fig. 16 Exponential bifurcation loci in the ω_c-k plane for the 1D Soler model. The solitary waves under the curve are linearly (spectrally) stable, while the ones above the curve are linearly unstable. Full line corresponds to the application of the Vakhitov–Kolokolov criterion (i.e., points for which $\partial_\omega Q(\omega) = 0$), whereas the dots correspond to the stability calculations

Let us remember that the spectrum of $\mathscr{A}_\omega$ is the union of spectra of the one-dimensional spectral problems (17): $\sigma(\mathscr{A}_\omega) = \overline{\cup_{q\in\mathbb{Z}}\sigma\left(\mathscr{A}_{\omega,q}\right)}$. In our numerics we have analyzed values of $q \in [-6, 6]$, although the main phenomenology is captured by $|q| \le 4$ and those are the values shown in the next figures for the sake of better visualization.

We start by considering the stability of $S = 0$ structures in the case of cubic nonlinearity. In Fig. 17, we can see in the top panels how the real and imaginary parts of the eigenvalues depend on ω. Based on this we can make some observations. Contrary to the case of the 2D NLS where the zero eigenvalues are degenerate [156], in the present NLD setting this degeneracy is lifted. As the frequency decreases, the pair of $q = 0$ eigenvalues depart from the origin (where they are at $\omega = 1$). Since these eigenvalues become marginally stable, the mechanism of charge-critical NLS self-similar blowup [119] is no longer "available". Moreover, the translation and gauge symmetry lead to two pairs of eigenvalues at the origin, for both fundamental and excited solutions. An additional ($\mathbf{SU}(1,1)$) symmetry is responsible for the presence of eigenvalues $\lambda = \pm 2\omega i$. It is relevant to note that as ω is decreased for $\omega < 0.121$, Hamiltonian-Hopf bifurcations for $|q| = 2$ lead to a complex eigenvalue quartet, with an additional one arising for $|q| = 3$ at $\omega = 0.0885$ and so on.

On the other hand, vortex solutions with $S > 0$ were found to be generically unstable due to quartets of complex eigenvalues. This was true not only for $S = 1$ solutions, but also for cases with $S \ge 2$ that we do not analyze further. Notice that the eigenvalues $\lambda = \pm 2\omega i$ generally correspond to the particular mode with $q = \mp(2S + 1)$.

Importantly, the quintic ($k = 2$) NLD model was also found to possess stable intervals in two dimensions. Here, the NLS limit is itself unstable and in fact the relevant instability emerges (for the $q = 0$ perturbations) already for $\omega > 0.890$. In

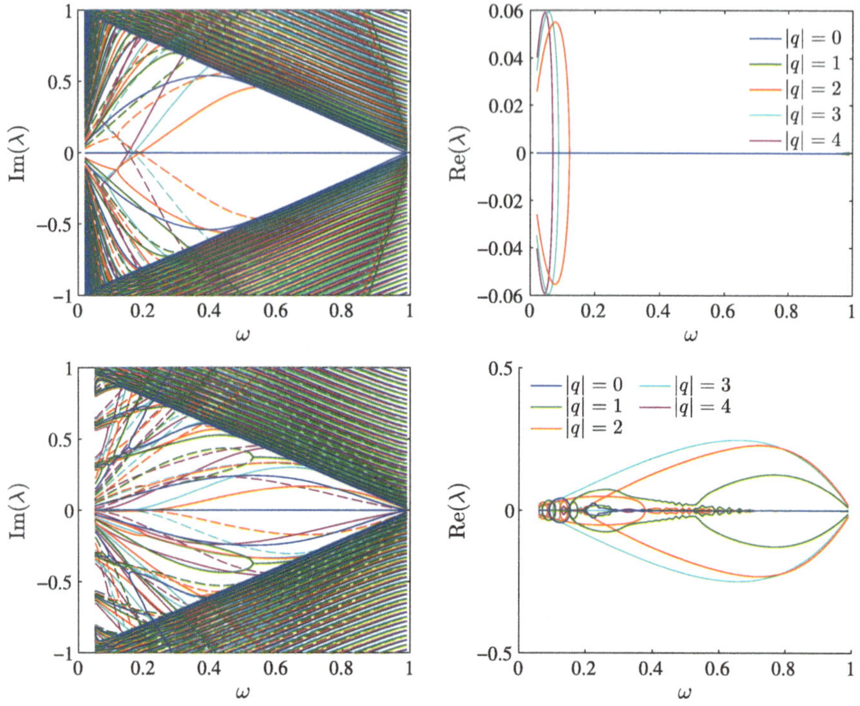

Fig. 17 Dependence of the (left) imaginary and (right) real part of the eigenvalues with respect to the frequency ω of solitary waves in the 2D Soler model with cubic ($k = 1$) nonlinearity. Top (respectively, bottom) panels correspond to $S = 0$ solitary waves ($S = 1$ vortices). For the sake of clarity, we only included the values $|q| \leq 4$. Full (dashed) lines in left panels represent the eigenvalues for $q \geq 0$ ($q < 0$). The correspondence between colors and $|q|$ is indicated in the legend of right panels

the small ω realm, the instability sets in (via Hamiltonian Hopf bifurcations again) for $\omega < 0.312$ (Fig. 18).

Perhaps even more remarkably, Fig. 19 shows that the radial perturbations do not destabilize NLD solitary waves *even* in the case of 3D Soler models for suitable frequency intervals (entirely contrary to what is the case for the non-relativistic NLS limit). More specifically, the stability to radial perturbations arises below some dimension-dependent critical value $\omega_c = \omega_c(n, k)$, with n being the number of spatial dimensions.

Figure 20 shows those critical frequencies as a function of the nonlinearity parameter k for $n = 2$ and $n = 3$. For $\omega \in (\omega_c, 1)$, the NLD solitary waves are linearly unstable. Below ω_c the linear instability disappears. In the particular case of cubic ($k = 1$) 3D Soler model, we have that $\omega_c \approx 0.936$. This value was identified by Soler in his original paper [152] as the parametric value of the occurrence of the energy and charge minimum.

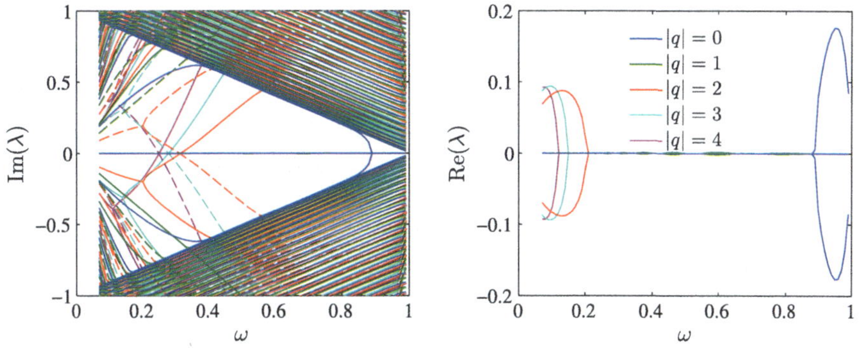

Fig. 18 Dependence of the (left) imaginary and (right) real part of the eigenvalues with respect to the frequency ω of $S = 0$ solitary waves in the 2D Soler model with quintic ($k = 2$) nonlinearity. For the sake of clarity, we only included the values $|q| \leq 4$. Full (dashed) lines in the left panel represent the eigenvalues for $q \geq 0$ ($q < 0$). The correspondence between colors and $|q|$ is indicated in the legend of the right panel

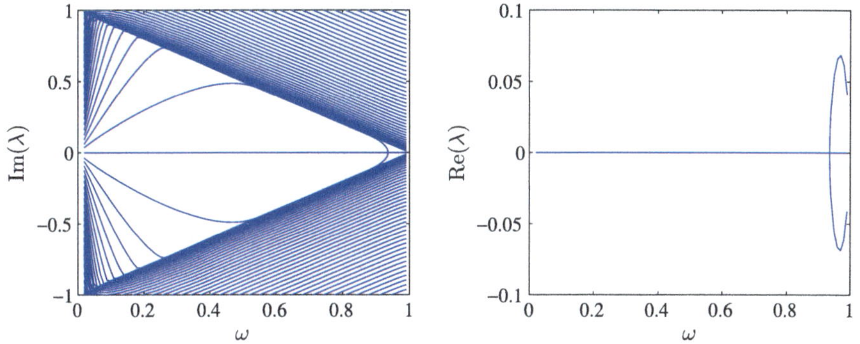

Fig. 19 Dependence with respect to ω of the (left) imaginary and (right) real part of the eigenvalues of the one-dimensional invariant radial subspace of solitary waves in the 3D Soler model with cubic ($k = 1$) nonlinearity

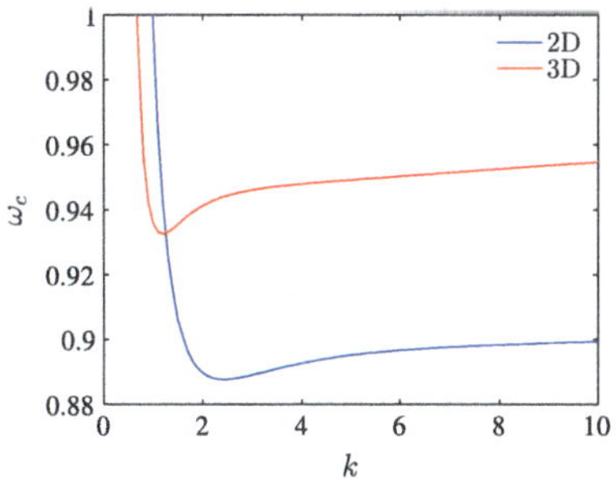

Fig. 20 Critical frequency for radially-symmetric exponential bifurcations in the 2D and 3D Soler model, as a function of the exponent k. For $(\omega_c, 1)$, the solitary waves are linearly unstable. For $k \leq 2/n$, with n being the system dimension, there is no linear instability for $\omega \lesssim 1$, according to the Vakhitov–Kolokolov criterion (see Sect. 4)

6 Dynamics

Once the stability properties of solitary waves and vortices of the Soler model have been elucidated, it is now natural to turn our attention towards the observation of their dynamical properties. In the one-dimensional case, we will analyze some integration schemes in order to observe their suitability for simulation of solitary waves in non-linear Dirac equations. In addition, the dynamics of unstable solutions in equations with high-order instabilities (i.e., $k > 1$) will be shown. Finally, the dynamics of unstable solitary waves and vortices for the 2D Soler model will be considered.

6.1 One-Dimensional Solutions

This subsection is divided into two parts. In the first one, we will show the evolution of stable solitary waves within several numerical integrators in the cubic ($k = 1$) Soler model. The second part deals with the evolution of unstable solitary waves with $k > 1$. Most of the results presented herein are taken from [46].

6.1.1 Stable Solutions

We turn here our attention to the implications of spectral collocation methods to the nonlinear dynamical evolution problem. We focus on the case of $k = 1$. Given the large (yet spurious) growth rate of the modes emerging from the Chebyshev spectral collocation method and the spurious point spectrum instability of the finite difference method, for our dynamical considerations, we will focus our attention to the Fourier spectral collocation method results. As discussed in Sect. 5.2, in that method too, there exist spurious modes which, as expected, are found to affect the corresponding dynamics. As a dynamical outcome of these modes, the solitary waves are found to be destroyed after a suitably long evolution time, although the time for this feature is controllably longer in comparison to the one observed in [149]. This, in turn, suggests the expected stability of the solitary wave solutions, in accordance with what was proposed in Sect. 5.

As a prototypical diagnostic of the dynamical stability of solitary waves in a finite domain $[-L, L]$, we have monitored the L^2-error in a similar fashion as in [149]:

$$\varepsilon_2(t) = \left(\int |\rho(t,x) - \rho(0,x)|^2 \, dx \right)^{1/2},$$

with $\rho = \psi^* \psi$ being the charge density.

A first approach to the dynamics problem is accomplished by choosing a fixed-step 4th order Runge–Kutta method. We observe that the lifetime is longer when the frequency ω is fixed and the domain length L is increased. This is associated with

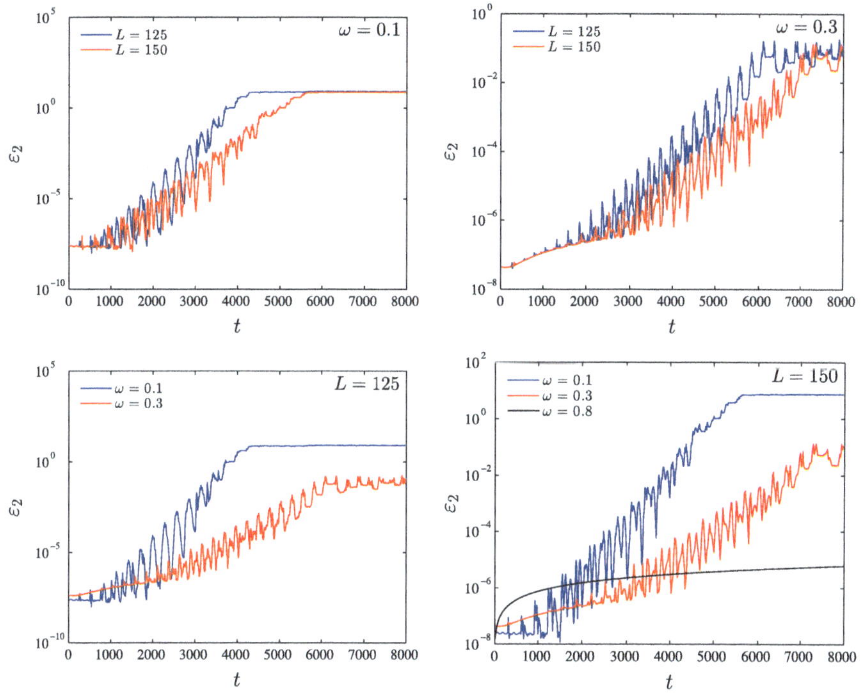

Fig. 21 Stable solitary waves simulations in cubic 1D Soler model using a 4th-order Runge–Kutta integrator with a Fourier spectral collocation method. The norm error is compared for different domain sizes and frequencies. In every case, the time step of the integrator is $\Delta t = 0.05$

the decrease of the size of spurious instability bubbles, as we approach the infinite domain limit. A similar decrease of the growth rate is observed for a given L, when the discretization spacing h is decreased (i.e., as the continuum limit is approached), in accordance with the spectral picture of Fig. 13. In addition, if L is fixed, the lifetime is longer when ω is increased. These facts are summarized in Fig. 21. Obviously, this is in consonance with earlier observations such as those of [149], however, our ability to expand upon the lifetimes as the domain and discretization parameters are suitably tuned suggests that in the infinite domain, continuum limit such instabilities could be made to disappear upon suitable selection of the numerical scheme. As a final comment, we note that the growth rates observed in Fig. 21 are consonant with the maximal (yet spurious) instability growth identified in Fig. 13. This is yet another indication that this growth featured in the time dynamics is a spurious by-product of the discretization scheme, rather than a true feature of the corresponding continuum problem.

In Table 1 we compare the critical time for which $\varepsilon_2 > 10^{-3}$ within the Fourier spectral collocation method and the corresponding time for the 4th order operator splitting algorithm used in [149] for frequencies $\omega = 0.1$ and $\omega = 0.5$ and different domain lengths L. As can be seen from the comparison, although in some cases

Table 1 Comparison between the critical times for which $\varepsilon_2 > 10^{-3}$ using the Fourier spectral collocation method with a 4th-order Runge–Kutta integrator (t_1) and the operator splitting method of [149] (t_2)

L	$\omega = 0.1$		$\omega = 0.5$	
	t_1	t_2	t_1	t_2
50	1220	121	5620	6614
75	1320	122	8480	8724
100	1990	122	14,660	9937
125	2540	120	14,660	11,670
150	3120	122	14,660	13,560

(e.g. for $\omega = 0.5$ and $L = 50$) the observed destabilization may happen later for the scheme of [149], generally the Fourier spectral collocation method code explored herein allows to enhance the wave lifetime, in some cases by an order of magnitude. This can be further improved by tweaking parameters such as h and the time spacing of the integrator Δt, as discussed above. Hence, our conclusion is that despite the artificial instabilities existing in the spectral picture and their dynamical manifestation, it is anticipated that the continuum, real line variant of the problem is spectrally stable for all $\omega \in (0, m)$ in the case of $k = 1$.

A tweak to the problem could be, on the one hand, to use adaptive step-size integrators [85]. The case of 4th–5th order Dormand–Prince integrator [58] does not improve significantly the solitary wave lifetime. On the other hand, when using a 2nd–3rd order Runge–Kutta integrator supplemented by a TR-BDF2 scheme (i.e., a trapezoidal rule step as a first stage and a backward differentiation formula as a second stage) [148], many of the spurious eigenvalues can be damped out and the lifetimes are strongly enhanced.

6.1.2 Unstable Solutions for High-Order Nonlinearity

Having observed that the solitary wave solutions of the problem with $k = 1$ are dynamically stable, we now turn our attention to the dynamics associated with the instability in the case $k > 2$, for $\omega > \omega_c(k)$, as per Fig. 16. Figure 22 shows the evolution of an exponentially unstable solitary wave with $k = 3$ and $\omega = 0.9$. We can observe the existence of oscillations around a stable fixed point. This fixed point approximately corresponds to the solitary wave with frequency $\omega \approx 0.82$, for which the solution is spectrally stable. This is in stark contrast with the supercritical dynamics of the nonlinear Schrödinger equation. There, the instability directly leads to collapse and an indefinite growth of the amplitude of the solution. On the contrary, in the case of the Soler model, for any value of k for which the solution may become unstable, there exists (for the same k) an interval of spectrally stable states of the same type. Hence, the solution to the Soler model does not escape towards collapse

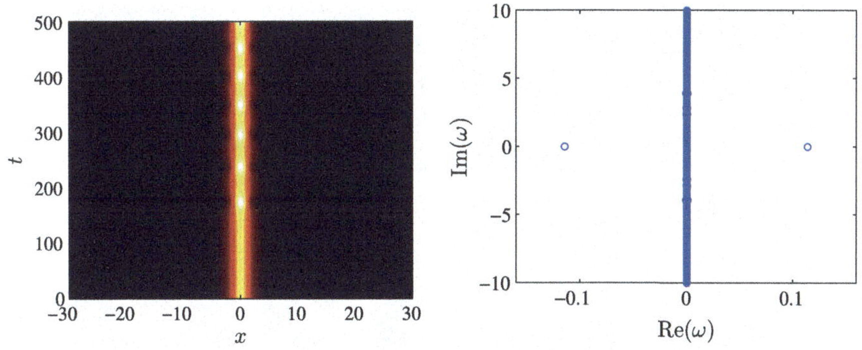

Fig. 22 (Left) Time evolution of a 1D solitary wave with nonlinearity exponent $k = 3$ and frequency $\omega = 3$. (Right) Spectral plane of the solitary wave whose evolution is traced in the left panel

but rather departs from the vicinity of the unstable fixed point solution and finds itself orbiting around a center, i.e., a stable solitary wave structure.

6.2 *Two-Dimensional Solutions*

This subsection reviews the results on the dynamics of 2D solitary waves and vortices shown in [49]. In order to simulate their dynamics, Chebyshev spectral methods and finite difference methods are not the most suitable ones, because of the presence of many spurious eigenvalues, and the dimensionality of the problem makes the TR-BDF2 schemes difficult to implement because of the high memory requirements. Thus, it seems that the optimal way to proceed is to use a Fourier spectral collocation method, which, as shown for the 1D problem, works fairly well as long as the frequency ω is not close to zero.

Consequently, periodic boundary conditions must be supplied to our problem. This is less straightforward when working in polar coordinates in the domain $(0, L) \times [0, 2\pi)$. For this reason, we opt to work with a purely 2D problem in rectangular coordinates in the domain $(-L, L] \times (-L, L]$. The simulations we show below have been performed with a Dormand–Prince numerical integrator using such a spectral collocation scheme with the aid of Fast Fourier Transforms (24).

A prototypical example of the evolution of the instability of the fundamental solitary waves for $k = 1$ is shown in Fig. 23. As can be observed, the radial symmetry in the density of $S = 0$ solitary waves is spontaneously broken and, as a result, the coherent structures acquire an elliptical shape, rotating around the density center of the original coherent structure, as may be expected by the quadrupolar ($q = 2$) nature of the unstable mode (see spectrum at the left panel of Fig. 24). The dynamical outcome of $S = 1$ vortices for $k = 1$ is shown in Fig. 25, whose instability (see spectral plane at right panel of Fig. 24) leads to the splitting into three smaller ones;

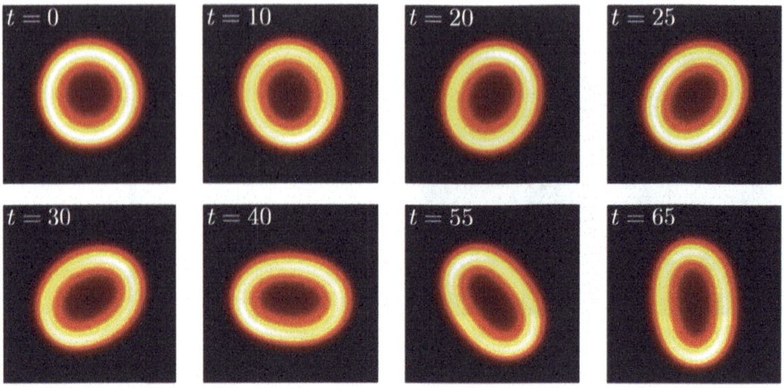

Fig. 23 Snapshots showing the evolution of the density of an unstable $S = 0$ solitary wave with $\omega = 0.12$ in the cubic 2D Soler model. The solitary wave which initially had a circular shape becomes elliptical and rotates around the center of the original solitary wave

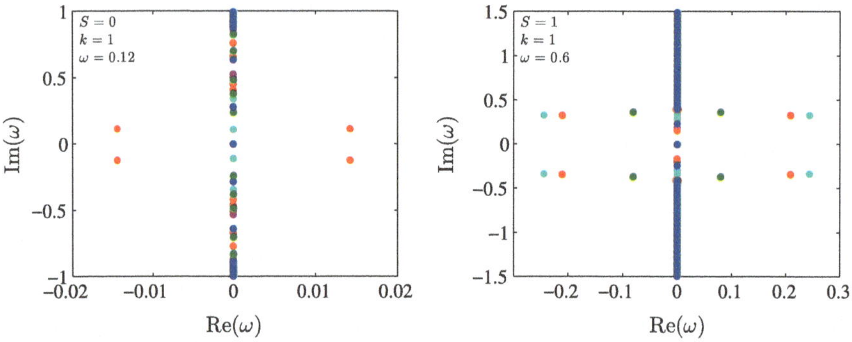

Fig. 24 Spectral planes of the unstable solitary waves whose dynamics are depicted in Figs. 23 and 25 (left and right panels, respectively). Each color represents a different value of q as in Fig. 17

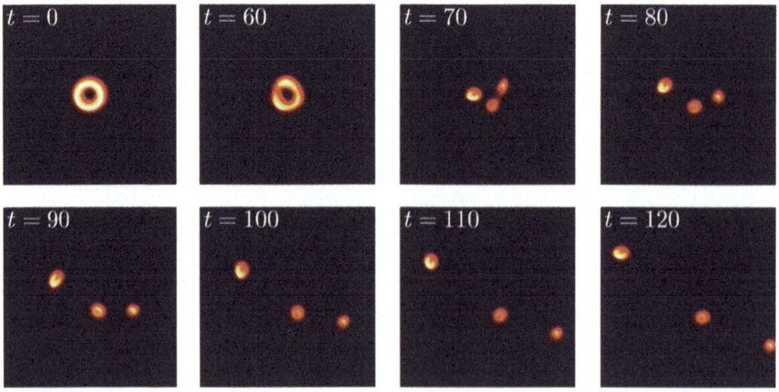

Fig. 25 Snapshots showing the evolution of the density of an unstable $S = 1$ vortex with $\omega = 0.6$ in the cubic 2D Soler model

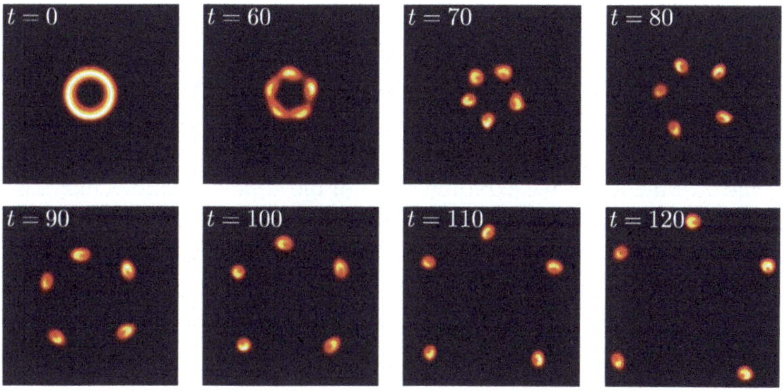

Fig. 26 Snapshots showing the evolution of the density of an unstable $S = 2$ vortex with $\omega = 0.6$ in the cubic 2D Soler model

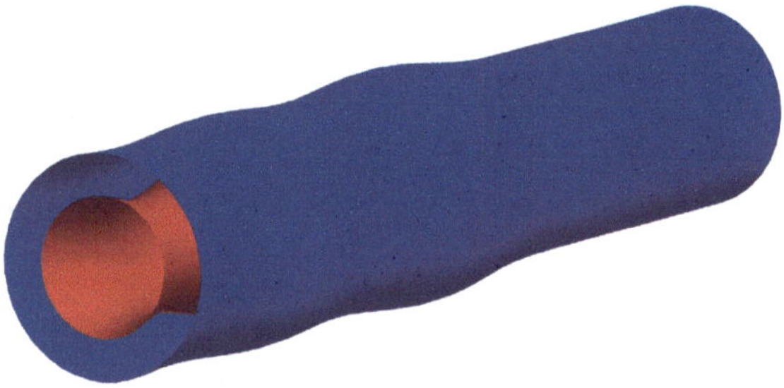

Fig. 27 Isosurface for the density of the $S = 0$ solitary wave in the quintic 2D Soler model with $\omega = 0.94$

in particular, the first spinor splits into fundamental structures, while the second one leads to the formation of vortices in a way preserving the total vorticity across the components. In the case of an $S = 2$ vortex, five similar structures arise, again preserving the total vorticity (see Fig. 26).

As a final example, we showcase the effect of radial $k = 2$ perturbations in the unstable case of $\omega > \omega_c$ (see Fig. 27). We can see that the relevant dynamics amounts to a breathing pattern, without featuring collapse (similarly e.g. to Fig. 22).

7 Summary and Outlook

In the present work, we have reviewed some of the principal properties of the nonlinear Dirac equation and its similarities, as well as differences, in comparison to its extensively studied cousin, namely the nonlinear Schrödinger equation. We have discussed nonlinear models that possess solitary wave solutions and vortices, and have placed particular emphasis on their spectral stability, also mentioning the orbital and asymptotic stability thereof and the corresponding issues that arise. We have seen that especially in higher dimensions the stability properties of solitary waves in the nonlinear Dirac equation can be fundamentally different from the NLS case, and may not feature collapse scenarios. Moreover, solitary waves may be spectrally stable for suitable parametric (i.e., frequency) regimes. For the three-dimensional case, the stability properties are just starting to be explored (in suitable subspaces), yet this problem is extremely interesting, also due to its connections with the dynamics. In the context of the latter, we explored some of the delicate features that arise from different types of discretizations (finite-difference, Fourier and Chebyshev spectral schemes) and the implications for the evolutionary dynamics. Generally, we hope to have exposed some of the significant complications arising in dynamically propagating such a system, especially when trying to do so for long time scales.

From every perspective that we can think of, nonlinear Dirac systems pose significant challenges ahead of us. From the point of view of the mathematical analysis, understanding the spectral properties observed herein and their dynamical implications is already a formidable problem. Computing efficiently and systematically both the solutions and their linearization eigenvalues emerges as a significant and upcoming challenge. This is especially true in three spatial dimensions. Devising numerical schemes—possibly based on integrable (semi-discrete or genuinely discrete) variants of the model—could prove to be of paramount importance towards future robust computations of the dynamics. Finally, combining some of the cutting edge themes in nonlinear waves (such as for instance rogue waves [100]) with relevant scenarios involving Dirac-type nonlinear models opens another highly promising vein of research for future studies [55].

Acknowledgements The research of Andrew Comech was carried out at the Institute for Information Transmission Problems of the Russian Academy of Sciences at the expense of the Russian Foundation for Sciences (project 14-50-00150). J.C.-M. thanks financial support from MAT2016-79866-R project (AEI/FEDER, UE). P.G.K. gratefully acknowledges the support of NSF-PHY-1602994, the Alexander von Humboldt Foundation, the Stavros Niarchos Foundation via the Greek Diaspora Fellowship Program, and the ERC under FP7, Marie Curie Actions, People, International Research Staff Exchange Scheme (IRSES-605096). This work was supported in part by the U.S. Department of Energy.

References

1. Ablowitz, M.J., Nixon, S.D., Zhu, Y.: Conical diffraction in honeycomb lattices. Phys. Rev. A **79**, 053830 (2009)
2. Ablowitz, M.J., Prinari, B., Trubatch, A.D.: Discrete and Continuous Nonlinear Schrödinger Systems. Cambridge University Press, Cambridge (2004)
3. Ablowitz, M.J., Zhu, Y.: Evolution of Bloch-mode envelopes in two-dimensional generalized honeycomb lattices. Phys. Rev. A **82**, 013840 (2010)
4. Achilleos, V., Frantzeskakis, D.J., Kevrekidis, P.G., Pelinovsky, D.E.: Matter-wave bright solitons in spin-orbit coupled Bose–Einstein condensates. Phys. Rev. Lett. **110**, 264101 (2013)
5. Achilleos, V., Stockhofe, J., Kevrekidis, P.G., Frantzeskakis, D.J., Schmelcher, P.: Matter-wave dark solitons and their excitation spectra in spin-orbit coupled Bose–Einstein condensates. Europhys. Lett. **103**(2), 20002 (2013)

6. Alexander, J., Gardner, R., Jones, C.: A topological invariant arising in the stability analysis of traveling waves. J. Reine Angew. Math. **410**, 167–212 (1990)
7. Alfimov, G.L., Kevrekidis, P.G., Konotop, V.V., Salerno, M.: Wannier functions analysis of the nonlinear Schrödinger equation with a periodic potential. Phys. Rev. E **66**, 046608 (2002)
8. Alvarez, A., Soler, M.: Energetic stability criterion for a nonlinear spinorial model. Phys. Rev. Lett. **50**, 1230–1233 (1983)
9. Alvarez, A., Soler, M.: Stability of the minimum solitary wave of a nonlinear spinorial model. Phys. Rev. D **34**, 644–645 (1986)
10. Aubry, S.: Discrete breathers: localization and transfer of energy in discrete Hamiltonian nonlinear systems. Physica D **216**, 1–30 (2006)
11. Barashenkov, I.V., Pelinovsky, D.E., Zemlyanaya, E.V.: Vibrations and oscillatory instabilities of gap solitons. Phys. Rev. Lett. **80**, 5117–5120 (1998)
12. Bender, C.M.: Making sense of non-Hermitian Hamiltonians. Rep. Prog. Phys. **70**, 947–1018 (2007)
13. Bender, C.M., Berntson, B.K., Parker, D., Samuel, E.: Observation of PT phase transition in a simple mechanical system. Am. J. Phys. **81**, 173–179 (2013)
14. Bender, C.M., Fring, A., Günther, U., Jones, H.: Special issue: quantum physics with non-hermitian operators. J. Phys. A Math. Theory **45**(44), 020201 (2012)
15. Bender, C.M., Jones, H.F., Rivers, R.J.: Dual PT-symmetric quantum field theories. Phys. Lett. B **625**, 333–340 (2005)
16. Berestycki, H., Lions, P.L.: Nonlinear scalar field equations. I. Existence of a ground state. Arch. Ration. Mech. Anal. **82**(4), 313–345 (1983)
17. Berkolaiko, G., Comech, A.: On spectral stability of solitary waves of nonlinear dirac equation in 1D. Math. Model. Nat. Phenom. **7**, 13–31 (2012)
18. Berkolaiko, G., Comech, A., Sukhtayev, A.: Vakhitov–Kolokolov and energy vanishing conditions for linear instability of solitary waves in models of classical self-interacting spinor fields. Nonlinearity **28**(3), 577–592 (2015)
19. Berthier, A., Georgescu, V.: On the point spectrum of Dirac operators. J. Funct. Anal. **71**(2), 309–338 (1987)
20. Bjorken, J., Drell, S.: Relativistic Quantum Mechanics. McGraw-Hill, New York (1964)
21. Blanchard, P., Stubbe, J., Vázquez, L.: Stability of nonlinear spinor fields with application to the Gross–Neveu model. Phys. Rev. D **36**, 2422–2428 (1987)
22. Bogolubsky, I.L.: On spinor soliton stability. Phys. Lett. A **73**, 87–90 (1979)
23. Bournaveas, N.: Local existence for the Maxwell–Dirac equations in three space dimensions. Commun. Partial Differ. Equ. **21**(5–6), 693–720 (1996)
24. Boussaïd, N.: Stable directions for small nonlinear Dirac standing waves. Commun. Math. Phys. **268**(3), 757–817 (2006)
25. Boussaïd, N.: On the asymptotic stability of small nonlinear Dirac standing waves in a resonant case. SIAM J. Math. Anal. **40**(4), 1621–1670 (2008)
26. Boussaïd, N., Comech, A.: On spectral stability of the nonlinear Dirac equation. J. Funct. Anal. **271**(6), 1462–1524 (2016)
27. Boussaïd, N., Comech, A.: Spectral stability of weakly relativistic solitary waves of the Dirac equation with the Soler-type nonlinearity (2016)
28. Boussaïd, N., Comech, A.: Spectral stability of weakly relativistic solitary waves of the Dirac equation with the Soler-type nonlinearity (2017). To appear
29. Boussaïd, N., Cuccagna, S.: On stability of standing waves of nonlinear Dirac equations. Commun. Part. Diff. Equ. **37**, 1001–1056 (2012)
30. Boyd, J.P.: Chebyshev and Fourier Spectral Methods, 2nd edn. Dover, New York (2001)
31. Braun, O.M., Kivshar, Y.S.: The Frenkel–Kontorova Model. Springer Nature (2004)
32. Buslaev, V.S., Perel′man, G.S.: On the stability of solitary waves for nonlinear Schrödinger equations. In: Nonlinear Evolution Equations. Am. Math. Soc. Trans. Ser. (Am. Math. Soc., Providence, RI.) **164**(2), 75–98 (1995)
33. Candy, T.: Global existence for an L^2 critical nonlinear Dirac equation in one dimension. Adv. Differ. Equ. **16**(7–8), 643–666 (2011)

34. Carretero-González, R., Talley, J.D., Chong, C., Malomed, B.A.: Multistable solitons in the cubic-quintic discrete nonlinear Schrödinger equation. Physica D **216**, 77–89 (2006)
35. Cazenave, T., Lions, P.L.: Orbital stability of standing waves for some nonlinear Schrödinger equations. Commun. Math. Phys. **85**(4), 549–561 (1982)
36. Cazenave, T., Vazquez, L.: Existence of localized solutions for a classical nonlinear Dirac field. Commun. Math. Phys. **105**, 35–47 (1986)
37. Chugunova, M., Pelinovsky, D.: Block-diagonalization of the symmetric first-order coupled-mode system. SIAM J. Appl. Dyn. Syst. **5**(1), 66–83 (2006)
38. Coleman, S.: Quantum sine-Gordon equation as the massive Thirring model. Phys. Rev. D **11**, 2088–2097 (1975)
39. Comech, A., Guan, M., Gustafson, S.: On linear instability of solitary waves for the nonlinear Dirac equation. Ann. Inst. H. Poincaré - AN **31**, 639–654 (2014)
40. Comech, A., Phan, T.V., Stefanov, A.: Asymptotic stability of solitary waves in generalized Gross-Neveu model. Ann. Inst. H. Poincaré - AN **34**, 157–196 (2017)
41. Comech, A., Stuart, D.: Small solitary waves in the Dirac–Maxwell system (2012). ArXiv:1210.7261
42. Conduit, G.J.: Line of Dirac monopoles embedded in a Bose–Einstein condensate. Phys. Rev. A **86**, 021605(R) (2012)
43. Contreras, A., Pelinovsky, D.E., Shimabukuro, Y.: L^2 orbital stability of Dirac solitons in the massive Thirring model. Commun. Partial Differ. Equ. **41**, 227–255 (2016)
44. Cooper, F., Khare, A., Mihaila, B., Saxena, A.: Solitary waves in the nonlinear Dirac equation with arbitrary nonlinearity. Phys. Rev. E **82**, 036604 (2010)
45. Cuccagna, S., Tarulli, M.: On stabilization of small solutions in the nonlinear Dirac equation with a trapping potential. J. Math. Anal. Appl. **436**(2), 1332–1368 (2016)
46. Cuevas-Maraver, J., Kevrekidis, P., Saxena, A., Cooper, F., Mertens, F.: Solitary waves in the nonlinear Dirac equation at the continuum limit: stability and dynamics. In: Ordinary and Partial Differential Equations. Nova Science Publishers, New York (2015)
47. Cuevas-Maraver, J., Kevrekidis, P.G., Saxena, A.: Solitary waves in a discrete nonlinear Dirac equation. J. Phys. A: Math. Theory **48**, 055204 (2015)
48. Cuevas-Maraver, J., Kevrekidis, P.G., Saxena, A., Comech, A., Lan, R.: Stability of solitary waves and vortices in a 2D nonlinear Dirac model. Phys. Rev. Lett. **116**, 214101 (2016)
49. Cuevas-Maraver, J., Kevrekidis, P.G., Saxena, A., Cooper, F., Khare, A., Comech, A., Bender, C.M.: Solitary waves of a PT-symmetric nonlinear Dirac equation. IEEE J. Sel. Top. Quantum Electron. **22**, 5000109 (2016)
50. Cuevas-Maraver, J., Kevrekidis, P.G., Williams, F. (eds.): The Sine-Gordon Model and its Applications. Springer International Publishing (2014)
51. Dalibard, J., Gerbier, F., Juzeliunas, G., Öhberg, P.: Colloquium: artificial gauge potentials for neutral atoms. Rev. Mod. Phys. **83**, 1523–1543 (2011)
52. Darby, D., Ruijgrok, T.W.: A noncompact gauge group for the Dirac equation. Acta Phys. Polon. B **10**, 959–973 (1979)
53. Dauxois, T., Peyrard, M.: Physics of Solitons. Cambridge University Press, Cambridge (2006)
54. De Wit, B., Smith, J.: Field Theory in Particle Physics. North Holland Physics Publishing, New York (1986)
55. Degasperis, A., Wabnitz, S., Aceves, A.: Bragg grating rogue wave. Phys. Lett. A **379**, 1067–1070 (2015)
56. Derrick, G.H.: Comments on nonlinear wave equations as models for elementary particles. J. Math. Phys. **5**, 1252–1254 (1964)
57. Dirac, P.: The quantum theory of the electron. I. Proc. R. Soc. Lond. A **117**, 610–624 (1928)
58. Dormand, J.R., Prince, P.J.: A family of embedded Runge–Kutta formulae. J. Comput. Appl. Math. **6**, 19–26 (1980)
59. Escobedo, M., Vega, L.: A semilinear Dirac equation in $H^s(R^3)$ for $s>1$. SIAM J. Math. Anal. **28**(2), 338–362 (1997)
60. Esteban, M.J., Georgiev, V., Séré, É.: Stationary solutions of the Maxwell–Dirac and the Klein–Gordon–Dirac equations. Calc. Var. Partial Differ. Equ. **4**(3), 265–281 (1996)

61. Esteban, M.J., Lewin, M., Séré, É.: Variational methods in relativistic quantum mechanics. Bull. Amer. Math. Soc. (N.S.) **45**(4), 535–593 (2008)
62. Esteban, M.J., Séré, E.: Stationary states of the nonlinear Dirac equation: a variational approach. Commun. Math. Phys. **171**, 323–350 (1995)
63. Esteban, M.J., Séré, É.: Solutions of the Dirac–Fock equations for atoms and molecules. Commun. Math. Phys. **203**(3), 499–530 (1999)
64. Esteban, M.J., Séré, E.: Nonrelativistic limit of the Dirac–Fock equations. Ann. Henri Poincaré **2**(5), 941–961 (2001)
65. Esteban, M.J., Séré, É.: Dirac–Fock models for atoms and molecules and related topics. In: XIVth International Congress on Mathematical Physics, pp. 21–28. World Scientific Publishing, Hackensack, NJ (2005)
66. Evans, J.: Nerve axon equations, I: Linear approximations. Indiana U. Math. J. **21**, 877–955 (1972)
67. Evans, J.: Nerve axon equations, II: Stability at rest. Indiana U. Math. J. **22**, 75–90 (1972)
68. Evans, J.: Nerve axon equations, III: Stability of the nerve impulse. Indiana U. Math. J. **22**, 577–594 (1972)
69. Evans, J.: Nerve axon equations, IV: The stable and unstable impulse. Indiana U. Math. J. **24**, 1169–1190 (1975)
70. Fedosov, B.V.: Index theorems. In: Partial Differential Equations, VIII Encyclopaedia Mathematical Sciences, vol. 65. Springer-Verlag, Berlin (1996)
71. Feng, B., Sugino, O., Liu, R.Y., Zhang, J., Yukawa, R., Kawamura, M., Iimori, T., Kim, H., Hasegawa, Y., Li, H., Chen, L., Wu, K., Kumigashira, H., Komori, F., Chiang, T.C., Meng, S., Matsuda, I.: Dirac fermions in borophene. Phys. Rev. Lett. **118**, 096401 (2017)
72. Fialko, O., Brand, J., Zülicke, U.: Hidden long-range order in a two-dimensional spin-orbit coupled bose gas. Phys. Rev. A **85**, 051605(R) (2012)
73. Finkelstein, R., Lelevier, R., Ruderman, M.: Nonlinear spinor fields. Phys. Rev. **83**, 326–332 (1951)
74. Fring, A., Jones, H., Znojil, M.: Papers dedicated to the subject of the 6th international workshop on pseudo-Hermitian Hamiltonians in quantum physics (PHHQPVI). J. Phys. A: Math. Theory **41**(44) (2008)
75. Galindo, A.: A remarkable invariance of classical Dirac Lagrangians. Lett. Nuovo Cimento **20**, 210–212 (1977)
76. Georgiev, V., Ohta, M.: Nonlinear instability of linearly unstable standing waves for nonlinear Schrödinger equations. J. Math. Soc. Jpn. **64**(2), 533–548 (2012)
77. Grillakis, M., Shatah, J., Strauss, W.: Stability theory of solitary waves in the presence of symmetry. I. J. Funct. Anal. **74**(1), 160–197 (1987)
78. Gross, D.J., Neveu, A.: Dynamical symmetry breaking in asymptotically free field theories. Phys. Rev. D **10**, 3235–3253 (1974)
79. Gross, L.: The Cauchy problem for the coupled Maxwell and Dirac equations. Commun. Pure Appl. Math. **19**, 1–15 (1966)
80. Guo, A., Salamo, G.J., Duchesne, D., Morandotti, R., Volatier-Ravat, M., Aimez, V., Siviloglou, G.A., Christodoulides, D.N.: Observation of PT-symmetry breaking in complex optical potentials. Phys. Rev. Lett. **103**, 093902 (2009)
81. Haddad, L.H., Carr, L.D.: The nonlinear Dirac equation in Bose–Einstein condensates: vortex solutions and spectra in a weak harmonic trap. New J. Phys. **17**, 113011 (2015)
82. Haddad, L.H., O'Hara, K.M., Carr, L.D.: Nonlinear Dirac equation in Bose–Einstein condensates: preparation and stability of relativistic vortices. Phys. Rev. A **91**, 043609 (2015)
83. Haddad, L.H., Weaver, C.M., Carr, L.D.: The nonlinear Dirac equation in Bose–Einstein condensates: I. Relativistic solitons in armchair nanoribbon optical lattices. New J. Phys. **17**, 063044 (2015)
84. Hadzievski, L., Maluckov, A., Stepić, M., Kip, D.: Power controlled soliton stability and steering in lattices with saturable nonlinearity. Phys. Rev. Lett. **93**, 033901 (2004)
85. Hairer, E., Nørsett, S.P., Wanner, G.: Solving Ordinary Differential Equations I. Springer Series in Computational Mathematics, vol. 8. Springer, Berlin (1993)

86. Hamner, C., Zhang, Y., Khamehchi, M.A., Davis, M.J., Engels, P.: In a one-dimensional optical lattice, spin-orbit-coupled Bose–Einstein condensates. Phys. Rev. Lett. **114**, 070401 (2014)
87. Heisenberg, W.: Quantum theory of fields and elementary particles. Rev. Mod. Phys. **29**, 269–278 (1957)
88. Herring, G., Carr, L.D., Carretero-González, R., Kevrekidis, P.G., Frantzeskakis, D.J.: Radially symmetric nonlinear states of harmonically trapped Bose–Einstein condensates. Phys. Rev. A **77**, 023625 (2008)
89. Huh, H.: Global solutions to Gross–Neveu equation. Lett. Math. Phys. **103**(8), 927–931 (2013)
90. Ivanenko, D.D.: Notes to the theory of interaction via particles. Sov. Phys. JETP **13**, 141 (1938)
91. Jensen, A., Kato, T.: Spectral properties of Schrödinger operators and time-decay of the wave functions. Duke Math. J. **46**(3), 583–611 (1979)
92. Johansson, M., Kivshar, Y.S.: Discreteness-induced oscillatory instabilities of dark solitons. Phys. Rev. Lett. **82**, 85–88 (1999)
93. Jones, C.: Stability of the travelling wave solutions of the Fitzhugh–Nagumo system. Trans. AMS **286**(2), 431–469 (1984)
94. Kapitula, T., Sandstede, B.: Edge bifurcations for near integrable systems via Evans function techniques. SIAM J. Math. Anal. **33**(5), 1117–1143 (2002)
95. Kartashov, Y.V., Konotop, V.V., Abdullaev, F.K.: Gap solitons in a spin-orbit-coupled Bose–Einstein condensate. Phys. Rev. Lett. **111**, 060402 (2013)
96. Kawakami, T., Mizushima, T., Nitta, M., Machida, K.: Stable skyrmions in SU(2) gauged Bose–Einstein condensates. Phys. Rev. Lett. **109**, 015301 (2012)
97. Kestelman, H.: Anticommuting linear transformations. Canad. J. Math. **13**, 614–624 (1961)
98. Kevrekidis, P.G.: The Discrete Nonlinear Schrödinger Equation: Mathematical Analysis, Numerical Computations and Physical Perspectives, vol. 232. Springer-Verlag, Heidelberg (2009)
99. Kevrekidis, P.G., Frantzeskakis, D.J., Carretero-González, R.: The defocusing nonlinear Schrödinger equation: from dark solitons, to vortices and vortex rings. SIAM, Philadelphia (2015)
100. Kharif, C., Pelinovsky, E., Slunyaev, A.: Rogue Waves in the Ocean. Springer-Verlag, Berlin (2009)
101. Kivshar, Y.S., Agrawal, G.P.: Optical solitons: from fibers to photonic crystals. Academic Press, San Diego (2003)
102. Klaiman, S., Günther, U., Moiseyev, N.: Visualization of branch points in PT-symmetric waveguides. Phys. Rev. Lett. **101**, 080402 (2008)
103. LeBlanc, L.J., Beeler, M.C., Jiménez-García, K., Perry, A.R., Sugawa, S., Williams, R.A., Spielman, I.B.: Direct observation of zitterbewegung in a Bose–Einstein condensate. New J. Phys. **15**, 073011 (2013)
104. Lee, S.Y., Kuo, T.K., Gavrielides, A.: Exact localized solutions of two-dimensional field theories of massive fermions with Fermi interactions. Phys. Rev. D **12**, 2249–2253 (1975)
105. Lee, Y.S., McLean, A.D.: Relativistic effects on R_e and D_e in AgH and AuH from all-electron Dirac–Hartree–Fock calculations. J. Chem. Phys. **76**(1), 735–736 (1982)
106. Lieb, E.H.: Existence and uniqueness of the minimizing solution of Choquard's nonlinear equation. Stud. Appl. Math. **57**(2), 93–105 (1977)
107. Lin, Y.J., Jiménez-García, K., Spielman, I.B.: Spin-orbit-coupled Bose–Einstein condensates. Nature **471**, 83–86 (2011)
108. Machihara, S., Nakamura, M., Nakanishi, K., Ozawa, T.: Endpoint Strichartz estimates and global solutions for the nonlinear Dirac equation. J. Funct. Anal. **219**(1), 1–20 (2005)
109. Machihara, S., Nakanishi, K., Tsugawa, K.: Well-posedness for nonlinear Dirac equations in one dimension. Kyoto J. Math. **50**(2), 403–451 (2010)
110. MacKay, R.S., Aubry, S.: Proof of existence of breathers for time-reversible or Hamiltonian networks of weakly coupled oscillators. Nonlinearity **7**, 1623–1643 (1994)

111. Mak, K.F., Lee, C., Hone, J., Shan, J., Heinz, T.F.: Atomically thin MoS_2: a new direct-gap semiconductor. Phys. Rev. Lett. **105**, 136805 (2010)
112. Makris, K.G., El-Ganainy, R., Christodoulides, D.N., Musslimani, Z.H.: Beam dynamics in PT-symmetric optical lattices. Phys. Rev. Lett. **100**, 103904 (2008)
113. Makris, K.G., El-Ganainy, R., Christodoulides, D.N., Musslimani, Z.H.: PT-symmetric periodic optical potentials. Int. J. Theor. Phys. **50**, 1019–1041 (2011)
114. Marini, A., Longhi, S., Biancalana, F.: Optical simulation of neutrino oscillations in binary waveguide arrays. Phys. Rev. Lett. **113**, 150401 (2014)
115. Mathieu, P., Morris, T.F.: Charged spinor solitons. Can. J. Phys. **64**(3), 232–238 (1986)
116. Melvin, T.R.O., Champneys, A.R., Kevrekidis, P.G., Cuevas, J.: Radiationless traveling vaves in saturable nonlinear Schrödinger lattices. Phys. Rev. Lett. **97**, 124101 (2006)
117. Merkl, M., Jacob, A., Zimmer, F.E., Öhberg, P., Santos, L.: Chiral confinement in quasirelativistic Bose–Einstein condensates. Phys. Rev. Lett. **104**, 073603 (2010)
118. Merle, F.: Existence of stationary states for nonlinear Dirac equations. J. Differ. Equ. **74**, 50–68 (1988)
119. Merle, F.: Construction of solutions with exactly k blow-up points for the Schrödinger equation with critical nonlinearity. Commun. Math. Phys. **129**, 223–240 (1990)
120. Mertens, F.G., Quintero, N.R., Cooper, F., Khare, A., Saxena, A.: Nonlinear dirac equation solitary waves in external fields. Phys. Rev. E **86**, 046602 (2012)
121. Ng, W., Parwani, R.: Nonlinear Dirac equations. SIGMA **3**, 023 (2009)
122. Pauli, W.: Contributions mathématiques à la théorie des matrices de Dirac. Ann. Inst. H. Poincaré **6**, 109–136 (1936)
123. Pego, R.L., Weinstein, M.I.: Eigenvalues, and instabilities of solitary waves. Philos. Trans. Roy. Soc. Lond. Ser. A **340**(1656), 47–94 (1992)
124. Peleg, O., Bartal, G., Freedman, B., Manela, O., Segev, M., Christodoulides, D.N.: Conical diffraction and gap solitons in honeycomb photonic lattices. Phys. Rev. Lett. **98**, 103901 (2007)
125. Pelinovsky, D.: Survey on global existence in the nonlinear Dirac equations in one spatial dimension. In: Harmonic Analysis and Nonlinear Partial Differential Equations, RIMS Kôkyûroku Bessatsu, B26, pp. 37–50. Res. Inst. Math. Sci. (RIMS), Kyoto (2011)
126. Pelinovsky, D., Shimabukuro, Y.: Transverse instability of line solitons in massive Dirac equations. J. Nonlinear Sci. **26**, 365–403 (2016)
127. Pelinovsky, D.E., Shimabukuro, Y.: Orbital stability of Dirac solitons. Lett. Math. Phys. **104**, 21–41 (2014)
128. Pelinovsky, D.E., Stefanov, A.: Asymptotic stability of small gap solitons in nonlinear Dirac equations. J. Math. Phys. **53**, 073705 (2012)
129. Peng, B., Özdemir, S.K., Lei, F., Monifi, F., Gianfreda, M., Long, G.L., Fan, S., Nori, F., Bender, C.M., Yang, L.: Parity-time-symmetric whispering-gallery microcavities. Nat. Phys. **10**, 394–398 (2014)
130. Pethick, C.J., Smith, H.: Bose–Einstein Condensation in Dilute Gases. Cambridge University Press, Cambridge (2002)
131. Pitaevskii, L.P., Stringari, S.: Bose–Einstein condensation. Oxford University Press, Oxford (2003)
132. Press, W.H., Flannery, B.P., Teukolsky, S.A., Vetterling, W.T.: Numerical Recipes: The Art of Scientific Computing, 3rd edn. Cambridge University Press, Cambridge (1986)
133. Qu, C., Hamner, C., Gong, M., Zhang, C., Engels, P.: Observation of zitterbewegung in a spin-orbit-coupled Bose–Einstein condensate. Phys. Rev. A **88**, 021064(R) (2013)
134. Quiney, H.M., Glushkov, V.N., Wilson, S.: The Dirac equation in the algebraic approximation. IX. Matrix Dirac–Hartree–Fock calculations for the HeH and BeH ground states using distributed gaussian basis sets. Int. J. Quantum Chem. **99**(6), 950–962 (2004)
135. Rañada, A.F., Rañada, M.F., Soler, M., Vázquez, L.: Classical electrodynamics of a nonlinear Dirac field with anomalous magnetic moment. Phys. Rev. D **10**(2), 517–525 (1974)
136. Radić, J., Sedrakyan, T.A., Spielman, I.B., Galitski, V.: Vortices in spin-orbit-coupled Bose–Einstein condensates. Phys. Rev. A **84**, 063604 (2011)

137. Ramachandhran, B., Opanchuk, B., Liu, X.J., Pu, H., Drummond, P.D., Hu, H.: Half-quantum vortex state in a spin-orbit-coupled Bose–Einstein condensate. Phys. Rev. A **85**, 023606 (2012)
138. Reed, M., Simon, B.: Methods of Modern Mathematical Physics. IV. Analysis of Operators. Academic Press [Harcourt Brace Jovanovich Publishers], New York (1978)
139. Regensburger, A., Bersch, C., Miri, M.A., Onishchukov, G., Christodoulides, D.N., Peschel, U.: Parity-time synthetic photonic lattices. Nature **488**, 167–171 (2012)
140. Rota Nodari, S.: Perturbation method for particle-like solutions of the Einstein–Dirac equations. Ann. Henri Poincaré **10**(7), 1377–1393 (2010)
141. Rota Nodari, S.: Perturbation method for particle-like solutions of the Einstein–Dirac–Maxwell equations. C. R. Math. Acad. Sci. Paris **348**(13–14), 791–794 (2010)
142. Ruschhaupt, A., Delgado, F., Muga, J.G.: Physical realization of PT-symmetric potential scattering in a planar slab waveguide. J. Phys. A: Math. Gen. **38**, L171–L176 (2005)
143. Rüter, C.E., Makris, K.G., El-Ganainy, R., Christodoulides, D.N., Segev, M., Kip, D.: Observation of parity-time symmetry in optics. Nat. Phys. **6**, 192–195 (2010)
144. Sakaguchi, H., Li, B., Malomed, B.A.: Creation of two-dimensional composite solitons in spin-orbit-coupled self-attractive Bose–Einstein condensates in free space. Phys. Rev. E **89**, 032920 (2014)
145. Schindler, J., Li, A., Zheng, M.C., Ellis, F.M., Kottos, T.: Experimental study of active LRC circuits with PT-symmetries. Phys. Rev. A **84**, 040101 (2011)
146. Schindler, J., Lin, Z., Lee, J.M., Ramezani, H., Ellis, F.M., Kottos, T.: PT-symmetric electronics. J. Phys. A: Math. Theory **45**, 444029 (2012)
147. Selberg, S., Tesfahun, A.: Low regularity well-posedness for some nonlinear Dirac equations in one space dimension. Differ. Integr. Equ. **23**(3–4), 265–278 (2010)
148. Shampine, L.F., Hosea, M.E.: Analysis and implementation of TR-BDF2. Appl. Num. Math. **20**, 21–37 (1996)
149. Shao, S., Quintero, N.R., Mertens, F.G., Cooper, F., Khare, A., Saxena, A.: Stability of solitary waves in the nonlinear Dirac equation with arbitrary nonlinearity. Phys. Rev. E **90**, 032915 (2014)
150. Sigal, I.M.: Nonlinear wave and Schrödinger equations. I. Instability of periodic and quasiperiodic solutions. Commun. Math. Phys. **153**(2), 297–320 (1993)
151. Soffer, A., Weinstein, M.I.: Resonances, radiation damping and instability in Hamiltonian nonlinear wave equations. Invent. Math. **136**(1), 9–74 (1999)
152. Soler, M.: Classical, stable, nonlinear spinor field with positive rest energy. Phys. Rev. D **1**, 2766–2769 (1970)
153. Soler, M.: Classical electrodynamics for a nonlinear spinor field: perturbative and exact approaches. Phys. Rev. D **8**, 3424–3429 (1973)
154. Strauss, W.A., Vázquez, L.: Stability under dilations of nonlinear spinor fields. Phys. Rev. D **34**(2), 641–643 (1986)
155. Stuart, D.: Existence and Newtonian limit of nonlinear bound states in the Einstein–Dirac system. J. Math. Phys. **51**(3), 032501 (2010)
156. Sulem, C., Sulem, P.L.: The Nonlinear Schrödinger Equation. Springer-Verlag, New York (1999)
157. Thaller, B.: The Dirac Equation. Texts and Monographs in Physics. Springer-Verlag, Berlin (1992)
158. Thirring, W.E.: A soluble relativistic field theory. Ann. Phys. **3**, 91–112 (1958)
159. Tran, T.X., Longhi, S., Biancalana, F.: Optical analogue of relativistic Dirac solitons in binary waveguide arrays. Ann. Phys. **340**, 179–187 (2014)
160. Trefethen, L.N.: Spectral Methods in Matlab. SIAM, Philadelphia (2000)
161. Vakhitov, N.G., Kolokolov, A.A.: Stationary solutions of the wave equation in a medium with nonlinearity saturation. Radiophys. Quantum Electron. **16**, 783–789 (1973)
162. Vázquez, L.: Localised solutions of a non-linear spinor field. J. Phys. A: Math. Gen. **10**, 1361–1368 (1977)
163. Vicencio, R.A., Johansson, M.: Discrete soliton mobility in two-dimensional waveguide arrays with saturable nonlinearity. Phys. Rev. E **73**, 046602 (2006)

164. Visscher, L., Dyall, K.: Dirac–Fock atomic enectronic structore calculations using different nuclear charge distributions. At. Data Nucl. Data Tables **67**(2), 207–224 (1997)
165. van der Waerden, B.: Group Theory and Quantum Mechanics. Springer-Verlag, New York (1974)
166. Wakano, M.: Intensely localized solutions of the classical Dirac–Maxwell field equations. Prog. Theory Phys. **35**, 1117–1141 (1966)
167. Wehling, T.O., Black-Schaffer, A.M., Balatsky, A.V.: Dirac materials. Adv. Phys. **63**, 1–76 (2014)
168. Xu, J., Shao, S., Tang, H.: Numerical methods for nonlinear Dirac equation. J. Comput. Phys. **245**, 131–149 (2013)
169. Xu, X.Q., Han, J.H.: Spin-orbit coupled Bose–Einstein condensate under rotation. Phys. Rev. Lett. **107**, 200401 (2011)
170. Xu, Y., Zhang, Y., Wu, B.: Bright solitons in spin-orbit-coupled Bose–Einstein condensates. Phys. Rev. A **87**, 013614 (2013)

On Nonlinear Schrödinger Equation as a Model for Dark Matter

Comments on Galactic Collisions, Supermassive Black Holes and Analogue Laboratory Implementations

Angel Paredes and Humberto Michinel

Abstract In this chapter, we present an overview of the problem of dark matter and the scalar field dark matter model, which assumes the existence of a cosmological matter wave describing a condensate of ultralight axions. The mathematical description is in terms of a nonlinear Schrödinger-Poisson system of equations. We introduce the framework in a pedagogical way, for readers interested in nonlinear science assuming no prior knowledge of cosmology. We describe a split-step pseudospectral numerical method which is useful to compute the evolution in time of dark matter distributions. We then discuss two aspects of the model: an explanation of the so-called offsets between dark matter and stars in galactic clusters and the laws relating supermassive black holes and dark matter distributions. Finally, we emphasize the formal connections to particular situations of other physical systems, including cold atom Bose-Einstein condensates and laser beam propagation in thermo-optical media, which may lead to tabletop laboratory analogues of cosmological phenomena.

Keywords Dark matter · Axion-like particle · Solitons · Nonlinear Schrödinger equation · Schrödinger-Poisson equation · Nonlocal nonlinearities · Scalar field dark matter

1 Introduction

These first decades of the twenty-first century are an exciting time for cosmology. Understanding the forces that have shaped our Universe and control its evolution from galactic to cosmological scales is one of the most important open problems of

A. Paredes (✉)
Área de óptica, Departamento de Física Aplicada, Campus de As Lagoas, 32004 Ourense, Spain
e-mail: angel.paredes@uvigo.es

H. Michinel
Área de óptica, Escola de Enxeñaría Aeroespacial, Campus de As Lagoas, 32004 Ourense, Spain
e-mail: hmichinel@uvigo.es

V. Carmona et al. (eds.), *Nonlinear Systems, Vol. 1*,
Understanding Complex Systems, https://doi.org/10.1007/978-3-319-66766-9_5

fundamental physics. A definitive solution will need increasingly precise observations and experiments. The good news is that many such experiments are underway or are projected for the near future [9]. Thus, there is a clear need of research on models that take into account and explain new cosmological data.

In this chapter, we intend to summarize the current status of cosmology for readers who, despite not being experts in dark matter theories, have a background in the nonlinear Schrödinger equation (NLSE) and its variants (like the authors ourselves). In particular, we will discuss the role of the Schrödinger-Poisson equation (SPE) in different cosmological scenarios and provide some concrete examples of its usage for particular dark matter problems. The goal is to introduce this framework to physicists with different expertise, with the hope that mathematical methods or physical intuitions developed in other contexts can provide insights on certain aspects of the quickly developing field of astrophysics and cosmology, and vice versa. We believe that numerical computations of Schrödinger-Poisson in the cosmological context can be of importance in the coming years. We emphasize that different versions of nonlinear Schrödinger equations have been thoroughly discussed in many physical and mathematical frameworks as, just to mention a few, optics, ocean waves, cold atoms or even finance. Understanding formal coincidences between the equations utilized in different fields can spark collaborations and new ideas.

In Sect. 2, we present an overview of the present understanding of cosmology and some of its open problems. In Sect. 3, we introduce the scalar field dark matter model and the role of Schrödinger-Poisson equation. In Sect. 4, we introduce some numerical methods that can be used to solve it. In this framework, we then discuss some aspects in the scalar field dark matter context: collisions of galaxies in relation to galactic offsets [81] (Sect. 5) and supermassive black holes (Sect. 6). Section 7 is devoted to the exploration of formal analogies with other disciplines of nonlinear physics where similar equations appear. All the sections are (mostly) independent from each other and can be read separately. Finally, in Sect. 8, we summarize the presentation of this chapter.

2 A Glimpse of Our Current Understanding of Cosmology

Cosmology is the branch of science that studies the Universe as a whole, and it has a long history from Ptolemy and Copernicus to Einstein and the Hubble telescope. Since the second half of the twentieth century and especially in the last two or three decades, it has experienced a revolution, relying on increasingly richer and more precise observations. We start here by providing a qualitative overview of what is, at present, the standard model of cosmology and some of the facts that still require a better explanation. We certainly do not intend to be rigorous nor exhaustive in the presentation or the referencing. There are excellent books that provide comprehensive introductions to this topic, including those honored by time as, e.g., [83, 108] or those including the developments associated to the most recent observations [63, 93].

The mathematical framework to study the evolution of the Universe is that of Einstein's theory of general relativity. An extremely simplifying assumption is that of the *cosmological principle*. It states that, when considering large scales, no observer sits at a special place; that is, the Universe is homogeneous and isotropic. This principle agrees rather well with observations when for large scales we mean hundreds of Megaparsecs (Mpc). (One parsec $\approx$ 3.26 light years $\approx 3.1 \times 10^{16}$ m). The evolution of the Universe at large is encoded in a single function, the scale factor $a(t)$, which determines the cosmological geometry through the so-called Friedmann-Lemaître-Robertson-Walker metric. The evolution of $a(t)$ depends on the density of matter and energy following Einstein's equations, which, for this particular setting, are called Friedmann equations. It has been known for almost a century that the Universe is expanding and therefore the time derivative of the scale factor is positive. The value of $\dot{a}(t)/a(t)$ today is called the Hubble constant $H_0 \approx 70$ (km/s)/Mpc. Another important notion is that of the cosmological redshift z, which, roughly speaking, is the general relativistic generalization of the Doppler effect $z = \Delta\lambda/\lambda$. The light arriving at Earth today from distant galaxies was emitted in the far past and had to surmount the gravitational potential, getting redshifted to smaller frequencies. The value of z from a distant source can be related to the value of the scale factor at the time of emission $a(t) = 1/(1+z)$ (the scale factor today is taken to be one by definition). Therefore, the value of z for a given observation indicates its time of emission and the distance from the source.

A fundamental quantity defining the geometry of the Universe is the density ρ of matter-energy, as compared to the critical density that has the value $\rho_c \approx 9 \times 10^{-27}$ kg/m^3. If $\rho > \rho_c$, the curvature of space would be positive and, therefore, we would live in a hypersphere. On the contrary, $\rho < \rho_c$ the space would be a hyperboloid of negative curvature. However, it is known today that $\rho \approx \rho_c$ to a very high accuracy and therefore a fixed time slice of spacetime looks like a Euclidean three dimensional space. This means that the Universe is flat.

A crucial aspect of the standard model of cosmology is the combination of different forms of matter and energy that adds up to the density $\rho \approx \rho_c$. The model takes its name, ΛCDM (cosmological constant-cold dark matter), from this composition. The most up-to-date estimates [1] are based on observations of the Planck mission and are as follows: Around 68% of the total energy budget is in the form of a cosmological constant, also called dark energy, which pushes the Universe outwards (a sort of "anti-gravity") and is responsible for the accelerated expansion ($\ddot{a}(t) > 0$, a shocking discovery that led to the Nobel Prize in Physics 2011). Another 27% is in the form of cold dark matter (CDM), where the word "cold" means non-relativistic. This CDM gravitates and behaves like ordinary matter but, apart from gravity, it does not interact with standard matter or only does so in an extremely weak way. In fact, in the standard ΛCDM modeling, it only interacts gravitationally with baryonic matter and with itself. As of today, its nature is unknown although there are of course many possible theories, see below. Finally, a meager 5% is in the form of baryons (say, ordinary protons and neutrons) of which only a fraction are packed in stars and emit light. Other forms of known matter or energy like photons and neutrinos are also present but only contribute to a negligible fraction, well below 1%. It should be said that these are the fractions today. They have changed during the universal expansion since they scale differently with the size of the Universe.

A non-trivial cosmological evolution has taken place and can be summarized in the following history of the Universe. The modeling starts with a Big Bang at an extremely large temperature. Since then until today, the Universe is expanding and cooling down. After the Big Bang, there was an epoch of rapid accelerated expansion, the so-called inflationary period. When inflation ended, and as the temperature dropped, the known forms of matter sequentially assembled. Shortly after inflation, protons and neutrons formed from the cooling down of the quark-gluon plasma. Later light nuclei appeared, as described by the successful theory of Big Bang nucleosynthesis. At a later time, around $z = 1100$, and with a much smaller temperature, the protons and electrons combined to form atoms. This is usually called recombination. At this point, with the absence of free electrons, the Universe became mostly transparent. Most of the photons that were present at that time have freely propagated until today, getting redshifted by the cosmological expansion. They constitute the cosmic microwave background (CMB), whose detailed observation is one of the fundamental sources of information about the evolution of the Universe. A few hundred million years after that, gravitational attraction started forming structures, ultimately producing the stars, galaxies, clusters, etc., that we observe today. The estimated age of the Universe since the Big Bang is around 14 Gyr (14×10^9 years).

Weird as it may seem, the Λ CDM roughly introduced above is extremely successful in the explanation of many independent sets of observations. As the author of [88] puts it: "what we do know is now so well confirmed by diverse data that it is likely to be true". In the same Ref. [88], there is a brief historical account of how theoretical predictions and observations interplayed since the first half of the 20th century. The observations leading to the present cosmological model include, among others, the details of the CMB spectrum and its anisotropies [1], the formation of structure [10], the formation of galaxies at different redshifts [19], the abundance of different nuclei [27] or the redshift of light from distant supernovae [91]. Moreover, evidence for dark matter does not come only from cosmological considerations, but also from the physics at galactic scales. Indeed, the suggestion that it could exist came originally from the rotation of stars within galaxies, that cannot be explained with standard dynamics if only ordinary matter is taken into account. The dynamics of galactic clusters needs also some form of dark matter. Maybe the most spectacular confirmation comes from strong lensing, from which the existence of extremely massive clumps of matter can be inferred for places where not enough ordinary matter is present. Most of the aforementioned data stem from the detection of photons from distant sources. In the future, new windows to the Universe like the detection of cosmic neutrinos or gravitational waves might get opened.

Despite its successes, the picture presented above leads to an embarrassing conclusion: we do not understand the nature of 95% of the Universe. Quantum field theory predicts the existence of a cosmological constant, but the predicted value is around sixty order of magnitude larger than the Λ inferred from cosmological observations. This is the cosmological constant problem, see e.g., [18], which certainly calls for a better understanding. Concerning dark matter, there is the obvious question: what is it made off? The observables described above only require that it is non-relativistic and weakly interacting but do not constrain, for instance, the mass of its basic constituents. A great number of possibilities have been proposed and thoroughly studied [32]. Just to mention some of the most popular models, the elementary dark matter particle could be an axion (with mass $m\,c^2 \approx 10^{-5}$ eV), a right-handed neutrino ($m\,c^2 \approx 10^3$ eV), a weakly interacting massive particle (WIMP, with $m\,c^2 \approx 10^{12}$ eV) or even a primordial black hole heavier than the Sun. Many experimental possibilities have been explored for the direct or indirect detection of these hypothetical particles. For instance, one of the motivations of the Large Hadron Collider at CERN is to discover supersymmetric particles that could play the role of WIMPS. But neither the LHC nor any other experiment has given a clear signature to be identified with a dark matter particle. Several experiments are projected to scan the space of possible theories and parameters, as e.g., the International Axion Observatory [6]. It could well be that the discovery is around the corner but, at present, it is fair to say that nothing is known for sure. In fact, a serious alternative is that dark matter does not exist but that the conventional dynamics based on Newton and Einstein equations fails at large scales. That is the idea of Modified Newtonian Dynamics (MOND) [74], proposed more than thirty years ago but which accurately fits recent data [72] (although it fails to explain other phenomena attributed to dark matter). In summary, despite many efforts, the nature, the dynamics and even the existence of dark matter remain a mystery whose understanding is one of the most important open problems for fundamental physics today. Other unsolved issues of ΛCDM concerning the early Universe are the nature of the field that drove inflation or the origin of matter-antimatter asymmetry (why do we observe baryons but not antibaryons?).

Apart from these theoretical conundrums, ΛCDM also faces some observational problems whose solution might be useful in the resolution of the explained dilemmas. Among them, there are the so-called small-scale crises [107], which are differences between CDM simulations and observations at galactic or sub-galactic scales. For instance, the cusp-core problem [30] is that CDM predicts cusp profiles with large densities at galactic centers while observations seem to favor smoother distributions. This might be because CDM phenomenology is not sufficiently well understood or could be pointing to other kind of physics. That is the motivation for the scalar field dark matter scenario to which we turn now.

3 Scalar Field Dark Matter and Schrödinger-Poisson Equation

Let us now focus on one of the proposed scenarios and assume that the elementary dark matter particle is a spin-0 ultra-light boson (sometimes called axion-like particle) of mass around $m_a\,c^2 \approx 10^{-22}$ eV or a similar order of magnitude [60, 101]. This is the basis of a model that has been described under different names: scalar field dark matter (SFDM) [102], wavelike dark matter (ψ DM) [97], fuzzy dark matter (FDM)

[49, 50], ultra-light axion dark matter (ULA DM) [68] or Bose-Einstein condensate dark matter (BEC-DM) [17, 23, 44]. In this section, we will introduce the essentials of the model, its motivations and its formalism. Necessarily because of its limited extent, this introduction will be somewhat superficial. For much more complete presentations and lists of references, we refer the reader to three comprehensive reviews which have been written in recent years under different perspectives [66, 68, 102] and to the appealing papers [14, 50], that also include useful review material and general considerations.

We start by discussing the motivation for the model in terms of cosmological and astrophysical observations. As described in Sect. 2, ΛCDM is extremely successful in many aspects, including the description of large structures and the CMB. But it suffers from some problems at small scales, where small means sizes of several kiloparsecs, comparable to galaxies. In particular, there is the cusp-core problem mentioned in Sect. 2. There is also the missing satellite problem [57, 75] which consists in that, typically, CDM simulations give rise to a number of satellite galaxies for the Milky Way well above the number of observed ones. Thus, it seems that ΛCDM over-predicts the amount of structure at these scales. It could be that this only happens because observations are not precise enough or because the modeling is not perfect and could, for instance, be missing important effects from baryonic physics. Maybe the discordance means that dark matter is not completely cold and there are fractions of hot (relativistic) or warm (semi-relativistic) dark matter. Or it could be that it is not collisionless, as suggested by self-interacting dark matter theories. There are numerous works studying all these hypothesis, see e.g., [65, 71, 86] and references therein.

SFDM considers a rather simple possibility: that dark matter satisfies a wave equation, where self-gravitation appears as a nonlinear nonlocal term. If this is the case, it is natural to expect that all the CDM phenomenology is recovered above some length scale but differences are found at shorter distances. This can be heuristically understood with an optical analogy: when all other scales are much larger than the wavelength, the dynamics of light can be easily interpreted as that of a bunch of photons or in terms of geometrical optics. The wavelike nature of light becomes manifest when probing scales comparable to λ. In the dark matter framework, the length scale for which the wavelike underlying nature start playing a role grows with $1/m_a$, and that is the reason why such tiny masses $m_a c^2 \approx 10^{-22}$ eV are needed to affect the physics of galaxies. In this context, the formation of structure is impeded below the axion Jeans scale [54], related to m_a (in the optics language, one would say that the diffraction term opposes the modulation instability). Thus, SFDM can address the missing satellite problem in a natural way. On the other hand, the nonlinear Schrödinger-Poisson equation supports stable solitons [31], which can be identified with galactic cores to address the cusp-core problem [43, 69]. Recently, a remarkable numerical simulation has borne out the SFDM expectations of solving small scale problems while keeping the large scale successes of CDM [97].

It is also interesting to briefly present the motivation for such a light boson from the particle physics point of view. This paragraph just wants to state that a value of $m_a c^2 \approx 10^{-22}$ eV is not crazy, even if it is so many orders of magnitude below any known massive form of matter. A motivation comes from string theory, in which the standard model for particles should emerge from a compactification of a higher-dimensional theory. At the classical level, those compactifications typically produce many massless bosonic fields which do not transform under the standard model gauge group and therefore are, at most, very weakly coupled to it. And moreover, the masses are protected by symmetry from acquiring perturbative quantum corrections. Only non-perturbative corrections can give a mass which, accordingly, is naturally tiny compared to other scales. A mass of around $m_a c^2 \approx 10^{-22}$ eV appears naturally in this context [50], although there is a large model dependence that can shift this value by several orders of magnitude. These considerations have led to the hypothesis of the string axiverse [7], dominated by many light scalar fields. However, the essential motive for having light scalars is not particularly linked to string theory, as it comes from the mass protection due to symmetry and can naturally arise in field theory, see e.g., [4]. (Pseudo-)Goldstone bosons resulting from symmetry breaking are typically light (that is why pions are much lighter than protons and might be the reason why the Higgs mass is not much larger). Ultra-light axions are assumed to appear with non-relativistic initial conditions in the early Universe from some symmetry breaking of this sort. For instance, Ref. [56] proposes that the ultraviolet completion of the standard model is a random quantum field theory with a large gauge group. It is shown that light pseudoscalars are a rather general consequence of the postulate. In [56], they are interpreted in terms of the quantum chromodynamic (QCD) axion, but the argument can be readily generalized. In fact, the light scalars we are discussing here are expected to emerge from the symmetry breaking mechanism proposed in the QCD axion context [82], and that is why they are sometimes called axion-like particles (ALPs) or ultra-light axions (ULAs) [68]. The difference is that symmetry breaking is not linked to QCD dynamics but to some other hidden theory.

We now turn to the basic mathematical formalism. The Schrödinger-Poisson system arises as the non-relativistic limit of the equations governing a massive scalar coupled to gravity [95]. This implies that all relevant velocities are well below c and that gravitational fields are not large. This is a good approximation for most of the dynamics related to dark matter, although it breaks down near black holes, where the gravitational field becomes strong and the full general relativistic formalism must be used. The Einstein-Hilbert action coupled to a real massive scalar reads:

$$S = \int d^4x \sqrt{-g} \left[\frac{c^3}{16\pi G} R + \frac{\hbar^2}{2c^2} (\partial\phi)^2 + \frac{1}{2} m_a^2 \phi^2 \right] \tag{1}$$

Many works consider more general forms for the scalar potential but we do not include them here. See [14] for references on variants of the model. The Euler-Lagrange equation for the scalar is:

$$\frac{\hbar^2}{c^2\sqrt{-g}} \partial_\mu \left(g^{\mu\nu} \sqrt{-g} \partial_\nu \right) \phi = m_a^2 \phi \tag{2}$$

In order to continue, we will use a Newtonian gauge to describe small perturbations of the Friedmann-Lemaître-Robertson-Walker metric:

$$ds^2 = -c^2\left(1 + 2\frac{\Phi}{c^2}\right)dt^2 + \left(1 - 2\frac{\Phi}{c^2}\right)a(t)^2 dx_i dx^i \tag{3}$$

where $\Phi \ll c^2$ is the gravitational potential for Newton's equations. In order to take the non-relativistic limit, we insert in Eq. (2) the following ansatz for the scalar:

$$\phi = \sqrt{\frac{c}{2m_a}}\left(\psi e^{-i\frac{m_a c^2}{\hbar}t} + \psi^* e^{i\frac{m_a c^2}{\hbar}t}\right) \tag{4}$$

A straightforward derivation leads to:

$$i\hbar\partial_t\psi + \frac{3}{2}i\hbar\frac{\dot{a}}{a}\psi = -\frac{\hbar^2}{2m_a a^2}\nabla^2\psi + m_a\Phi\,\psi \tag{5}$$

We have taken the non-relativistic limit with $\partial_t^2\psi \ll \frac{m_a c^2}{\hbar}\partial_t\psi$ (the equivalent of paraxial approximation in optics) and $\frac{\dot{a}}{a} \ll \frac{m_a c^2}{\hbar}$. On the other hand, the usual weak field limit of Einstein equations leads to Poisson equation for the gravitational potencial sourced by the scalar:

$$\nabla^2\Phi = 4\pi G m_a a^2|\psi|^2 \tag{6}$$

The quantity $\int a^3|\psi|^2 d^3x$ is a constant of motion that can be identified with the number of bosons in a given volume. It is worth remarking that these equations have been written in the literature using multiple conventions for how to include the factors of a in the fields and coordinates. For instance, Eqs. (5) and (6) coincide with the conventions of [39] but those of [97] are obtained by $\tilde{\psi} = a^{\frac{3}{2}}\psi$, $d\tau = a^{-2}dt$, $V = a\,\Phi$. Schrödinger-Poisson Eqs. (5) and (6) constitute the foundation of the SFDM scenario. Formally, the nonlinearity is given by a nonlocal term that has been studied in other fields of physics, see Sect. 7. Local nonlinearities of the Kerr form $|\psi|^2\psi$ in Eq. (5) have been studied in many contributions, e.g., [12, 37, 60]. Depending on the underlying theory leading to the light scalar, they could be either attractive [50] or repulsive [4], but we will not include them in the following.

Undoubtedly, Eq. (5) resembles the mean field description of a gas of Bose condensed cold atoms [28] and, certainly, if dark matter is made of ultra-light particles, the typical distance between particles is orders of magnitudes smaller than the de Broglie wavelength. Therefore, it is natural to interpret this kind of dark matter as a cosmological Bose-Einstein condensate [17, 23, 44]. However, even if the quantum nature of the underlying theory is an interesting issue, it should be clear that Eqs. (5) and (6) are just classical equations for the scalar field. In fact, $\hbar$ enters in the equations through the combination $\hbar/m_a$ and $\hbar$ itself plays a role when interpreting the matter wave in terms of particles.

The consequences of Eqs. (5) and (6) have been thoroughly analyzed in the literature. We present an upshot here, mentioning that detailed descriptions and lists of references can be found in the reviews [50, 68, 102]. The model gives a compelling solution for small scale ($\sim$few kpc) crises. We have already mentioned the cusp-core problem and the missing satellite problem, but SFDM also addresses the "too big to fail" problem or the survival of globular clusters around the Fornax dwarf galaxy. For the moment, there is no obvious observational evidence against the model and typical mass estimates are in the range $m_a \approx 1\text{–}10 \times 10^{-22}$ eV. If the mass is smaller, the missing satellite problem would be over-solved, meaning that the amount of predicted structure would be smaller than the observed one. If the mass is larger, the model cannot be relevant to the cusp-core problem (we will come back to this question in Sect. 5). Of course, this is based on the simple form of the model presented here and for the ULA comprising all or a large fraction of dark matter. Possible variations of the modeling or a better understanding of baryonic physics could open a little bit the allowed parameter space. In any case, there are two points that are worth mentioning: first, this is a testable theory and the increasing precision of the observations can disprove it or favor it in the coming years. Second, there is a need for massive and precise numerical studies of the nonlinear Schrödinger-Poisson system or variations thereof. CDM simulations have been developed for many years based on N-body methods, but SFDM requires a change of computational paradigm, at least for some observables. In words of [68] "the field of study of (SFDM) simulations is simply young compared to that of CDM N-body simulations". In our opinion, this is interesting for researchers in nonlinear science and opens the possibility of collaboration between experts in simulations of nonlinear Schrödinger equation, widely studied in many physical contexts for decades, and cosmologists. Methods and insights inherited from other fields might be useful and result in interdisciplinar cross-fertilization.

The most spectacular signature of any kind of dark matter particle would be its direct detection at Earth. There are many experiments designed with that goal, trying to prove the existence of WIMPs or QCD axions. Nevertheless, if ultra-light axions constitute dark matter, it does not seem possible to verify their presence that way. It is therefore important to understand which indirect evidence can be sought for to confirm or disprove the hypothesis. That can be accomplished by comparing the increasingly precise data concerning observations as those described in the previous paragraph with increasingly precise computations from the theoretical model, as it was done in [98] for galaxy formation at high z. Apart from that, there is a remarkable possibility put forward in [55]: the rapid oscillation of frequency $m_a c^2/\hbar \approx 1.5\,(m_a c^2/10^{-22}\,\text{eV}) \times 10^{-7}$ Hz, see Eq. (4), can induce an oscillating delay in arrival time of pulsar signals that, despite being tiny, could be measured in the near future. Which other astrophysical signatures would be clearly distinctive for the ULA scenario is an interesting open question.

We close this section with an important remark. Up to this point, we have discussed the scalar field dark matter scenario. The same equations are applicable to other light scalars like the quantum chromodynamic (QCD) axions of mass $m\,c^2 \approx 10^{-5}$ eV. For ultra-light axions or QCD axions, the Schrödinger-Poisson equation is a fundamental equation describing the non-relativistic limit of the elementary degrees of freedom. But we want to stress here that the applicability of Schrödinger-Poisson is much more general. Numerical computations for the evolution of other cold dark matter types typically rely on the so-called N-body simulations. It was pointed out long ago [29, 109] that the Schrödinger-Poisson system reproduces, above some scale, the results of N-body computations and can be competitive in terms of computational cost. In this setting, the mass entering Eqs. (5) and (6) is not the property of any elementary particle, but just an effective parameter that can be fitted to data. The coincidence between N-body simulations and Schrödinger-Poisson equations has been recently based on firmer grounds by appealing to Nelson quantization and the Calogero conjecture [21]. Therefore, we emphasize that, even if it turned out that scalar field dark matter is not the correct theory, efficient numerical computations of Schrödinger-Poisson would be relevant for astrophysics and cosmology anyway.

4 Numerical Methods

The goal of this section is to provide an introduction to the numerical methods that can be used to deal with Eqs. (5) and (6). It is addressed to non-experts in numerical computation that are interested in getting started in performing simulations of the partial differential equations that govern the evolution in time. With that aim, we propose and describe the split-step Fourier method (also called beam propagation method), which is widely used for the NLSE in nonlinear optics [2] because of its good properties of stability and accuracy [2, 103]. It is possible to perform fully three-dimensional simulations as those shown in Sect. 5 in a desktop computer in reasonable time. Moreover, its actual implementation is rather simple and provides a benchmark and a good starting point if, afterwards, one wants to apply more advanced methods. This split-step method has been applied in the context of SFDM to describe structure formation [111] and galactic collisions [81]. At the end of this section we also provide a short, biased and incomplete discussion of state-of-the-art computing methods for the problem at hand.

The first step of course is to write the equations in dimensionless form. For simplicity, we will restrict ourselves to the case where cosmological evolution is negligible ($a = 1$) and write:

$$i\partial_t \psi = -\frac{1}{2}\nabla^2 \psi + \Phi\, \psi \tag{7}$$

$$\nabla^2 \Phi = 4\pi |\psi|^2 \tag{8}$$

This form of the equations is obtained by taking:

$$\hat{t} = \zeta\, t, \quad \hat{\mathbf{x}} = \sqrt{\frac{\hbar\zeta}{m_a}}\mathbf{x}, \quad \hat{\psi} = (G m_a \zeta^2)^{-1/2}\psi, \quad \hat{\Phi} = \frac{\hbar}{m_a\zeta}\Phi \tag{9}$$

where for clarity we have written the dimensionful quantities of (5) and (6) with hats. We have introduced an arbitrary constant ζ with units of time. This freedom means that for any solution of (7) and (8), one can build a one-parameter family of solutions by appropriate rescaling.

The split-step Fourier method relies on the fact that the separate numerical evolution associated to each of the two terms on the right hand side of (7) is rather simple. The Laplacian is readily integrated in Fourier space while the term without derivatives is integrated in spatial coordinates. For a time interval Δt, one can compute in turn both contributions by transforming to Fourier space and back. This induces an error related to the commutator of the operators, that is of order Δt^2. Therefore, we compute:

$$\psi(t+\Delta t) = \mathscr{F}^{-1}\left[e^{-i\frac{1}{2}(2\pi\mathbf{k})^2\Delta t}\mathscr{F}[e^{-i\,\Phi(t)\,\Delta t}\psi(t)]\right] + \mathscr{O}(\Delta t^2) \tag{10}$$

where $\mathscr{F}$ stands for the three-dimensional discrete Fourier transform that can be calculated with a standard fast Fourier transform algorithm. Notice that the form of (10) ensures the exact conservation of the norm $N = \int |\psi|^2 d^3\mathbf{x}$. The step (10) is repeated until the desired value of t is reached. The numerical integration depends on Δt and also on the widths of the computational window ($w_1 \times w_2 \times w_3$) and the number of grid points $N_1 \times N_2 \times N_3$. Obviously, the method cannot resolve distances smaller than $\Delta x_i = w_i/N_i$. The grid in Fourier space has a spacing $\Delta k_i = 1/w_i$. Taking smaller values of Δt increases the accuracy but also the computational time. If Δt is too large the algorithm can become unstable. The convergence of the method can be checked by comparing a computation performed more than once with different values of Δt and Δx_i. The choice of Δt, Δx_i has to be made carefully in order to preserve the desired accuracy and not to unnecessarily increase the computational cost. The precision of the method can also be tested, for instance, by tracking the evolution of a soliton, whose form should be preserved. The soliton profile will be discussed in detail in the next section. In Fig. 1, we plot the result of a check, by depicting the maximum value of $|\psi|$ for a soliton moving with a certain velocity.

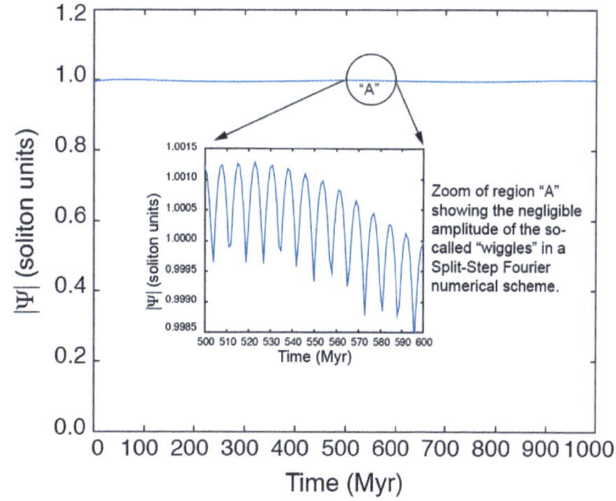

Fig. 1 An example of a numerical check of the split-step method. The maximum value of $|\psi|$ should remain exactly constant during propagation. Some wiggles appear in the actual computation but the size of the errors can be kept under control. Reproduced with permission from [81]. Creative Commons Attribution License (CC BY)

It is also important to comment on boundary conditions. Since (10) relies on Fourier transformation, it automatically introduces periodic boundary conditions for ψ. Typically, this is not a good approximation to reality. In some simulations, e.g., soliton collisions, it is possible to avoid that a non-negligible fraction of the energy reaches the edge of the computational box, making irrelevant the boundary conditions for ψ. However, in other situations, this problem has to be dealt with. The usual solution is to introduce a sponge, namely a term $-i\,V(\mathbf{x})\psi$ in (7) that produces an extra factor of $e^{-V(\mathbf{x})\Delta t}$ in (10). $V(\mathbf{x})$ has support in the vicinity of the boundary and artificially absorbs the energy that escapes the window. This procedure is customarily used in optics [2] and it was introduced in the SFDM context in [13, 41, 42].

We still have to discuss how to solve (8) in order to compute Φ at each time step. One possibility would be to write $\Phi(t,\mathbf{x}) = -\int \frac{|\psi(t,\mathbf{x}')|^2}{|\mathbf{x}-\mathbf{x}'|}d^3\mathbf{x}'$, but performing this integral with a standard trapezoidal method is prohibitive. There are different numerical methods to deal with Poisson equation and here we will mention two. The simplest one is just to use again discrete Fourier transform and write $\Phi(t) = -\frac{1}{4\pi}\mathscr{F}^{-1}\left[\frac{1}{\mathbf{k}^2+\varepsilon}\mathscr{F}[|\psi(t)|^2]\right]$ where we have introduced $\varepsilon \ll w_i^{-2}$ to avoid the infinity at $\mathbf{k}=0$. Notice that $\mathbf{k}=0$ corresponds to an additive constant for Φ that only contributes as a global phase to ψ. This method for computing Φ is simple and fast, but it implements periodic boundary conditions for Φ. This is not bad for a cosmological simulation where energy is not confined to a region in space, but it is not realistic, for instance, in the computation of a galactic merger. In that case, it is more accurate to utilize monopolar boundary conditions, where the boundary values of Φ are fixed as if all the mass was concentrated at the center of the computational box $\Phi|_{boundary} = N/|\mathbf{x}|$ for all times t. The procedure makes sense if the center of mass coincides with $\mathbf{x}=0$ and the total linear momentum is zero (conditions that can always be met by a translation and a Galilean transformation). This leads to an elliptic equation with Dirichlet boundary conditions that can be solved in a finite difference scheme where the derivatives are approximated in terms of nearest neighbor stencils. The problem gets reduced to a linear system $A\cdot\tilde{\Phi} = B$ where $\tilde{\Phi}$ is a $(N_1-2)(N_2-2)(N_3-2)$ vector with the unknown values of Φ at the grid points. A is a square heptadiagonal matrix and B includes the boundary conditions and the values of the source $4\pi|\psi(t,\mathbf{x})|^2$. This is a huge linear system but, since A is sparse, it can be efficiently solved with standard techniques of linear algebra. For instance, we propose to use an iterative symmlq algorithm that is included in typical linear algebra packages. $\Phi(t,\mathbf{x})$ can be used as a good starting point for the iteration to solve for $\Phi(t+\Delta t,\mathbf{x})$. Notice that the two kinds of boundary conditions mentioned above (periodic and monopolar) tend to each other and to the exact solution when the computational window becomes much larger than the region where $|\psi|^2$ has support.

Surely, there are more powerful methods than the one described above to deal with (7) and (8). The state-of-the-art simulation of the cosmological evolution of wavelike dark matter is [97], where a highly optimized adaptive mesh refinement algorithm and computation in graphics processing units were used. The adaptive mesh refinement allowed the authors to resolve very disparate length scales. There are also recent developments in computing science that might find application in further improving the algorithms used in the dark matter context, as for instance the method for evaluation of the Coulomb potential using non-uniform fast Fourier transform [52]. Finally, let us mention that [58, 64] provide codes for the efficient integration of three-dimensional NLSE with nonlocal interactions. The second of these papers uses parallelization and acceleration with graphic processing units.

5 Soliton Dynamics Confronted with Galaxies and Clusters

In this section, we first discuss the spherically symmetric ground state soliton solution of Eqs. (7) and (8). In the SFDM scenario, this can be related to the mass distribution at the core of galaxies, potentially solving the cusp-core problem. We review this issue in Sect. 5.1. In Sect. 5.2, we present the results of simulations of soliton collisions, which are relevant for the encounter of galaxies. Wave interference during those collisions is a plausible explanation for a mysterious observation, see Sect. 5.3. This section is mostly based on [81].

5.1 Soliton Solutions and Galactic Cores

The nonlinear system (7) and (8) supports a one-parameter family of robust, self-trapped, spherically symmetric, stationary and stable solitary waves (which are usually called solitons in an abuse of language). They have the form $\psi(t, \mathbf{x}) = e^{i\beta t} f(r)$, $\Phi(t, \mathbf{x}) = \varphi(r)$ where $r = |\mathbf{x}|$. With this ansatz, the equations read:

$$0 = -\frac{1}{2}\frac{d^2 f(r)}{dr^2} - \frac{1}{r}\frac{df(r)}{dr} + \varphi(r) f(r) + \beta f(r) \qquad (11)$$

$$0 = \frac{d^2\varphi(r)}{dr^2} + \frac{2}{r}\frac{d\varphi(r)}{dr} - 4\pi f(r)^2 \qquad (12)$$

The propagation constant β can be included as an additive constant in the Poisson potential $\tilde{\varphi}(r) = \varphi(r) + \beta$. Defining $\lim_{r\to\infty} \varphi(r) = 0$, the parameter β is associated to the asymptotic value of $\tilde{\varphi}(r)$ In view of the residual rescaling in (9), it is enough to look for the regular and normalizable solution with $f(r = 0) = 1$, which can be found by standard algorithms (e.g., a shooting technique). We find $\beta = 2.454$. The solution is plotted in Fig. 2.

The most general soliton solution is found with $|\psi(r = 0)| = \alpha$ for any positive α and with any constant velocity $\mathbf{v}$:

$$\psi(t, \mathbf{x}) = \alpha f(\sqrt{\alpha}|\mathbf{x} - \mathbf{v}t|) e^{i(\alpha\beta t + \mathbf{v}\cdot\mathbf{x} - \frac{1}{2}|\mathbf{v}|^2 t)}, \quad \Phi(t, \mathbf{x}) = \alpha\varphi(\sqrt{\alpha}|\mathbf{x} - \mathbf{v}t|) \qquad (13)$$

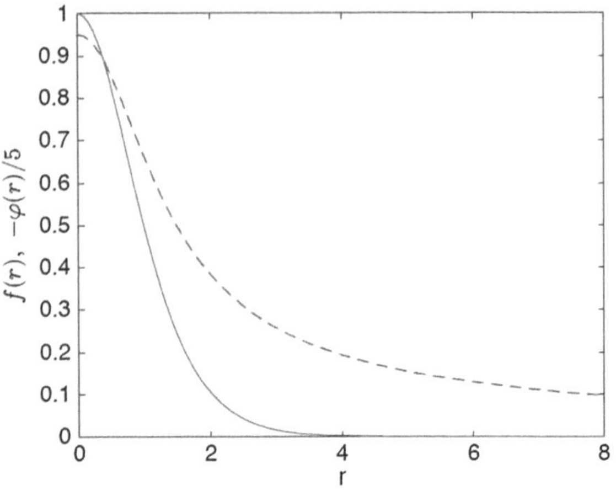

Fig. 2 Soliton profile (blue solid line) and its associated gravitational potential (red dashed line). We are plotting $-\varphi(r)/5$ in order to refer both curves to the same axis. The potential at the origin takes the value $\varphi(0) = -4.76$. Reproduced with permission from [81]. Creative Commons Attribution License (CC BY)

The norm of the soliton is $N_{sol} = \int_0^\infty |\psi|^2 d^3\mathbf{x} = 3.883\sqrt{\alpha}$ and its radius, defined as the half width at half of the maximum of $|\psi|^2$ is $r_{sol} = 0.69/\sqrt{\alpha}$. Multiplying these quantities and going back to the dimensionful quantities using (9) and $M_{sol} = m_a \int |\hat{\psi}|^2 d^3\hat{\mathbf{x}}$, we find:

$$M_{sol} r_{sol} \approx \frac{2.68\hbar^2}{m_a^2 G} \approx 2.3 \times 10^{10} \left(\frac{m_a c^2}{10^{-23}\,\text{eV}}\right)^{-2} \text{kpc} M_\odot, \tag{14}$$

where $M_\odot \approx 1.99 \times 10^{30}$ kg is the mass of the Sun. This soliton solution, and its generalization to the case of having a cubic local nonlinearity, has been rediscovered a number of times in different contexts, including quantum mechanics [76], cold atoms [79] and dark matter [24, 39]. In the SFDM context, it is identified with the dark matter distribution in a galaxy [60]. More precisely, such a soliton lives at the center of the galaxy. Around it, with smaller density, there is a gravitationally trapped pseudo-stationary DM distribution evolving incoherently, that constitutes most of the galactic halo. It has been shown by numerical simulation that that is end state of soliton mergers [99, 100]. The incoherent distribution density falls as power law, like the Navarro-Frenk-White profile deduced in the CDM scenario from N-body simulations. Thus, the DM distribution in SFDM looks like the one for CDM except near the center where it is smoother, and therefore, it can naturally solve the cusp-core problem.

Different authors have tried to estimate the ultra-light axion mass by (indirectly) matching this profile to observations, see e.g., [35, 69, 97] and [14] for more references. Being based on different sets of data, observables and methodologies, the results are not always consistent, but all estimates lie in the approximate range 10^{-24} eV $\lesssim m_a c^2 \lesssim 10^{-22}$ eV. For larger masses, the diffraction term is not strong enough to solve the cusp-core problem. On the other hand, from cosmological observables, the bound $m_a c^2 \gtrsim 10^{-22}$ eV has been obtained, we will come back to this point in Sect. 5.3.

5.2 Soliton Collisions and Galactic Mergers

Since the solitons can be associated with galactic cores, soliton collisions play a role for galactic collision and mergers. This possibility was first studied in [12], where it was demonstrated than interference patterns appear and this can lead to observational differences when compared with other fluid models of dark matter [36]. In this subsection, we will present simulations which include also the luminous matter modeled as a test particle governed by a classical equation of motion [81]. This is interesting because how dark and luminous matter get displaced from each other is a very important observable in galactic clusters, as we will explain in detail in Sect. 5.3. It is interesting to notice that a more elaborate model for the luminous matter, based on N-body simulations, was put forward in [40]. The collisional dynamic of this kind of solitons was also studied in [26].

Figures 3, 4 and 5 present results for six simulations of Eqs. (7) and (8) with initial conditions:

$$\psi(t=0,\mathbf{x}) = \alpha f(\sqrt{\alpha}|\mathbf{x}-\mathbf{x_0}|)e^{i(v\cdot x)} + \alpha f(\sqrt{\alpha}|\mathbf{x}+\mathbf{x_0}|)e^{-i(v\cdot x-\Delta\phi)} \tag{15}$$

We have taken two solitons of equal mass colliding head-on for simplicity. We introduce a relative phase $\Delta\phi$ between both. For a collision within a cluster, this value is essentially random. The relative velocity is $2|\mathbf{v}|$. In the captions of the figures, we display the dimensionful parameters associated to each collision. We fix $m_a c^2 = 0.2 \times 10^{-23}$ eV in all the cases.

The left column of Fig. 3 corresponds to a collision with both solitons in phase. Due to constructive interference a blob of matter appears at the center when the test particles have not arrived yet. Continuing the evolution, the dark matter solitons cross each other and recapture the luminous matter. The right column corresponds to an encounter in phase opposition. The solitons bounce back from each other. While they are getting stopped, the luminous matter goes ahead due to its own inertia. After the bounce, it gets trapped again by the gravitational potential of the soliton.

The fact that luminous matter moves ahead of dark matter does not need a fine tuned phase opposition. We show that in Fig. 4, where we repeat the simulations with relative phases of $3\pi/4$ and $7\pi/8$. In these cases, the symmetry is lost and matter can be transferred from one of the blobs to the other one. Notwithstanding, the dynamics is not very different from the $\Delta\phi = \pi$ case: the dark matter concentrations bounce back and, during the process, the test particles move away from the DM maxima.

In Fig. 5, we show an example with a larger initial relative velocity. In the case of phase coincidence, the velocity is enough to produce an interference pattern, similar to [12, 36]. The collision in phase opposition can be described again as a bounce due to the repulsion induced by destructive interference.

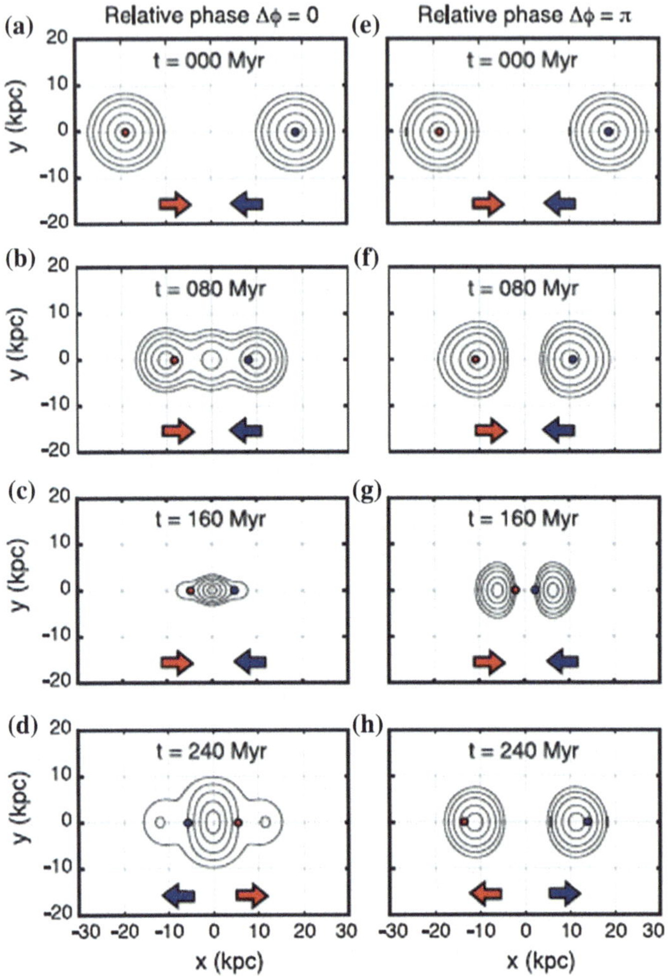

Fig. 3 Simulation of the collision of two solitons in phase coincidence (panels **a–d**) and in phase opposition (panels **e–h**). The lines of the contour plot represent the dark matter density (integrated along the z-direction). The dots are test particles moving according to classical mechanics in the gravitational potential generated by dark matter, and the arrows are an indication of the direction of their motion. They are a toy representation of the stars in galactic collisions, see Sect. 5.3 for more details. Initially, the center of the solitons are separated by 40 kpc, their initial relative velocity is 200 km/s and the mass is $10^{11} M_{\odot}$ for each soliton. The mass of the ultra-light axions has been taken to be $m_a c^2 = 0.2 \times 10^{-23}$ eV. The computation was done as described in Sect. 4 considering a computational box of $(100 \text{ kpc})^3$ discretized in a $360 \times 360 \times 128$ grid. The time step Δt corresponds to 0.1 Myr. Reproduced with permission from [81]. Creative Commons Attribution License (CC BY)

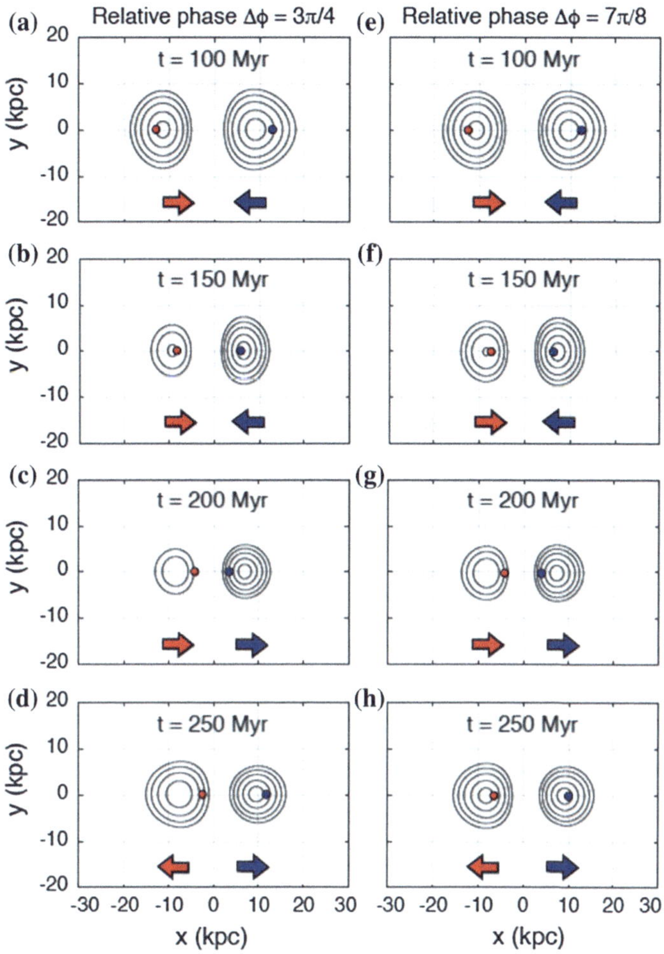

Fig. 4 Simulation of the collision of two solitons with relative phases $\Delta\phi = 3\pi/4$ (panels **a**–**d**) and $\Delta\phi = 7\pi/8$ (panels **e**–**h**). The rest of parameters are as in Fig. 3

The conclusion is that, even if the dark matter axions only feel the gravitational force, their wavy nature makes a difference with respect to the point particles, which only feel gravity too. The difference is due to classical wave interference and produces, as a natural result, relative displacements between both. We explore the physical consequences of this statement in Sect. 5.3.

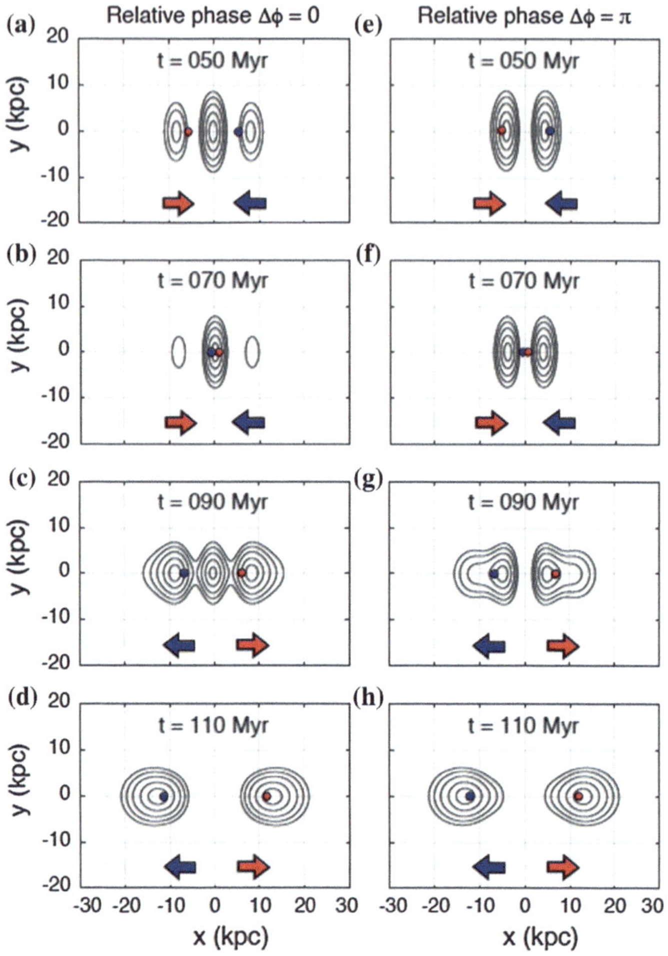

Fig. 5 Simulation of the collision of two solitons in phase coincidence (panels **a**–**d**) and in phase opposition (panels **e**–**h**). The initial relative velocity is 600 km/s. The rest of parameters are as in Fig. 3 except for the Δt that has been reduced to 0.033 Myr

5.3 A Discussion on Galactic Clusters: The Abell 3827 Puzzle

The observation of galactic collisions can provide non-trivial information about how dark matter interacts with itself and with ordinary matter. They provide non-trivial checks for any model of dark matter. The time scales involved are much larger than the human scale and therefore deductions are made based on a fixed picture. The total mass can be divided in three groups: dark matter, which can be mapped by gravitational lensing; stars, that emit in the visible spectrum; and gas, that can be mapped with X-ray telescopes. The gas interacts electromagnetically and is subject to ram pressure. The stars are essentially point-like, do not collide with each other, and only interact gravitationally. In the standard cold dark matter model, the DM only feels gravity as well. Thus, the expectation is to have stars and DM distributions together while the gas can be displaced from them. This is confirmed by most of the observations, including the famous Bullet cluster [67], but also many others [46]. This has been used to set upper limits in the DM self-interaction cross-section [46].

A notable exception is the Abell 3827 cluster [22]. It presents some lucky peculiarities: it is nearby (in cosmic scale) and its casual alignment with a bright source has allowed the astronomers to deduce from strong gravitational lensing a rather detailed map of the dark matter density, resolving structures of the order of some kiloparsecs. This contrasts with other clusters, where this level of detail is typically not achievable. It came as a surprise that there is an offset of around 1.6 kpc between the stars and the dark matter clump associated to one of the merging galaxies [110]. Despite careful analysis, no plausible explanation in terms of standard physics was found [70], and it was proposed that the cluster was the first evidence of DM self-interaction [70, 96, 110]. This interpretation, however, is not free of problems [53] and the puzzle still requires a better explanation. Since the involved length scales are of about a few kiloparsecs, it could potentially be a new small scale crisis of ΛCDM and it is natural to wonder whether SFDM can solve it. That was the goal of [81], where we showed that DM-stars offsets can be naturally expected because of effective forces due to interference during soliton collisions, see Figs. 3, 4 and 5. This is of course very well known in many nonlinear coherent systems, where the importance of relative phases has been emphasized many times, see e.g., [78, 80]. Without any need of fine tuning, it is easy to propose some initial conditions that generate dynamically an offset similar to the one observed in Abell 3827, see Fig. 6 [81].

We have shown that the Schrödinger wave equation naturally generates offsets between DM and stars due to interference. We now discuss in more detail whether it can possibly explain the of Abell 3827 cluster. Loosely speaking, the strong lensing analysis of that observation tells the following [70]: there is an offset of 1.6 kpc for the stars of a galaxy which is about 10 kpc away from another one. The masses in the DM clumps are of the order of $10^{11} M_{\odot}$. In order to have a natural explanation, we would need cores of radii of a few kpc since, otherwise, they would not affect each other at these distances. If we insert $r_{sol} = 5$ kpc, $M_{sol} = 10^{11} M_{\odot}$ into Eq. (14) we get $m_a c^2 \approx 0.2 \times 10^{-23}$ eV. This is a rough order of magnitude estimate of the axion mass needed to produce the pattern. The value is consistent with the most recent estimates from rotation curves [14] and velocity dispersions in dwarf galaxies [35]. It is also consistent with CMB constraints [48]. On the other hand, cosmological probes like formation of high-z galaxies [20, 98] or the Lyman-α forest [5] point towards a moderately larger mass $m_a c^2 \gtrsim 10^{-22}$ eV [50]. This seems to rule out the simplest explanation by about one order of magnitude in the axion mass. This discrepancy is not huge and it could be alleviated by departing from the simplest modeling. For instance, axions could be just a fraction of dark matter, there could be a local interaction term of the form $|\psi|^2\psi$, modeling the core in terms of the ground state soliton could be too naive, there could be several axions at play with

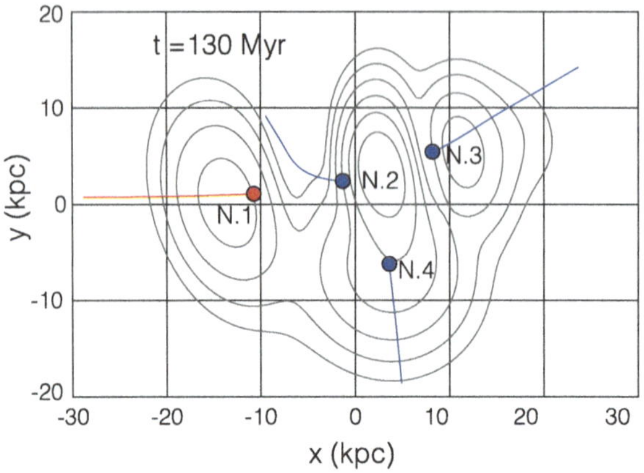

Fig. 6 Result of a simulation that generates dynamically an offset between stars and dark matter similar to the one observed in Abell 3827. The initial conditions consist of four separated solitons representing the cores of the merging galaxies. Initially, the stars, represented by the dots, are at their center, but the interference during the collision generates non-trivial displacements. The importance of relative phases and wave interference in soliton collisions is well known in many coherent nonlinear systems. Reproduced with permission from [81]. Creative Commons Attribution License (CC BY)

different masses, etc. Thus, whether the Abell 3827 offset might possibly come from soliton interference at the galactic scale remains an open question and an undoubtedly appealing possibility.

Finally, let us comment on another yet unexplained observation. Gravitational lensing analysis suggests that the Abell 520 cluster presents a large concentration of dark matter at its center, which seems to coincide with the gas and not with the stars and, therefore, presents a problem for standard cold dark matter [51]. In [62, 105], it was suggested that this could be explained in SFDM due to wave effects. In fact, a matter distribution of that sort can appear naturally in that context, see panels c and d of Fig. 3. Nevertheless, Abell 520 consists of a collision of hundreds of galaxies and therefore it is not clear that the effect can be explained in terms of individual soliton collisions. Moreover, the length scales involved are much larger than in Abell 3827, of the order of hundreds of kpc. For a natural modeling, we would need cores of those sizes, which, regarding the discussion above, are not expected to exist. Thus, it seems unlikely that the Abell 520 and the Abell 3827 puzzle can be explained in terms of the same physics of wave interference.

6 Scalar Field Dark Matter and Supermassive Black Holes

At the center of most of the massive galaxies, there is a so-called supermassive black hole (SMBH). Typical SMBHs have masses in the range $10^6 - 10^{10} M_\odot$. Question about their formation and their relation to galactic evolution are under intense research in astrophysics. In this section, we will comment on some intriguing questions connecting the scalar field dark matter model with SMBHs, more with the goal of arousing the curiosity of the reader than of presenting rigorous results. We overview some progress that has been made in the literature and include some further remarks.

As we have discussed, in the SFDM model the inner part of the galactic dark matter distribution is a self-trapped soliton. A question of concern is whether it can coexist with the SMBH at its center [104]. In general relativity, there is a "no-hair" theorem, roughly stating that there cannot be stationary distributions of energy surrounding a black hole (notice that Eqs. (5) and (6) lose their validity in the vicinity of a black hole because the Newtonian approximation to gravity breaks down for large gravitational fields). This means that, eventually, the scalar field must either be absorbed by the SMBH or get dispersed to infinity. But the theorem does not say how long does that take and, if the absorption rate is low [104], the SMBH + soliton configuration can persist for cosmological time. In that case, it is said that, instead of having hair, the black hole has a "wig" [8]. The authors of [8] performed the relevant numerical computations with the general relativistic equations and found a relation between the black hole mass and m_a for the wig to stay in place for times comparable to the age of the Universe. Indeed, that is what happens for the typical SMBH masses and the typical ULA mass. This is a non-trivial self-consistency check of the SFDM model.

Another interesting question is whether the collapse of a scalar field distribution might be at the origin of the SMBHs themselves. In order to provide an estimate, we may wonder whether the soliton of Schrödinger-Poisson equation can be subject to a collapse instability. That would happen is the soliton mass is confined within its Schwarzschild radius $R_{Sch} = 2\,G\,M/c^2$. Inserting $R_{sol} \lesssim R_{Sch}$ in Eq. (14) we find a condition for black hole collapse.

$$\frac{m_a M}{M_P^2} \gtrsim 1 \qquad (16)$$

where M is to be associated with the black hole mass and $M_P = \sqrt{\frac{\hbar c}{G}} \approx 2.2 \times 10^{-8}$ kg is the Planck mass. This order of magnitude estimate is confirmed by more elaborate computations as a variational approach to the breathing of a dark matter clump [38] or computations in numerical relativity [47]. Inserting benchmark values $m_a = 10^{-22}$ eV/c^2, $M_{SMBH} = 10^8 M_\odot$, we find $m_a M/M_P^2 \approx 10^{-4}$. It is curious that this value is not so far from one, taking into account the immensely disparate masses included in the formula (16). In this context, [38, 47] have proposed that axions a few orders of magnitude heavier than $m_a = 10^{-22}$ eV/c^2 could have collapsed in SMBHs, potentially solving some difficulties of other models concerning the formation of massive objects in the far past (with redshift $z \gtrsim 6$). The authors of [38] also point out that an eventual observation of a SMBH with $10^{12} M_\odot$ could be an indirect indication of the existence of ultra-light axions. In any case, modeling the coevolution of galaxies and SMBHs, including the accretion of matter by the black hole and other important issues is an extremely complicated problem which should be addressed before reaching solid conclusions. In any case, the scalar field dark matter model opens interesting possibilities in this framework.

A separate interesting issue is that of the $M - \sigma$ relation, an empirical finding that states that $M_{SMBH} \approx \alpha\sigma^\beta$, where σ is the velocity dispersion of the galactic bulge and α, β are constants that can be estimated from data. Initially, it was assumed that $\beta = 4$ but more recent analysis give $\beta \approx 5.1$. The authors of [61] have suggested that the scalar field halos can naturally explain this observation. The essence of the argument is that the black hole singularity affects the whole wave-like dark matter distribution. More massive black holes produce tighter bulges and therefore larger velocity dispersions. From the corrected soliton profiles, the authors of [61] derive a version of the $M - \sigma$ relation for an axion mass $m_a = 5 \times 10^{-22}$ eV. In our opinion, the argument of [61] is somewhat heuristic but indeed interesting. We thus find convenient to present the scalar field profiles in the presence of a central black hole. They have been discussed in [61] (and previously in [104]), but we will do it here in a more systematic way, clearly extracting the dependence of the size of the dark matter clump in terms of the rest of physical quantities. We write:

$$i\hbar\partial_t\psi = -\frac{\hbar^2}{2m_a}\nabla^2\psi + m_a\Phi\,\psi - G\frac{m_a M_{SMBH}}{|\mathbf{x}|}\psi, \qquad \nabla^2\Phi = 4\pi G m_a|\psi|^2 \quad (17)$$

where we have included the Newtonian potential for a point-like mass at the origin, to be associated to the black hole. This equation breaks down near the black hole horizon and general relativistic corrections impede the existence of stationary solutions for the scalar. However, as said above, the absorption rate can be extremely small and (17) constitutes a good approximation for most of the space and for long times. It is transformed into:

$$i\,\partial_t\psi = -\frac{1}{2}\nabla^2\psi + \Phi\psi - \frac{1}{|\mathbf{x}|}\psi, \qquad \nabla^2\Phi = 4\pi|\psi|^2 \quad (18)$$

by the rescaling (9) with $\zeta = \hbar^3/(G^2 m_a^3 M_{SMBH}^2)$. We look for ground state stationary solutions of the system (18) by writing $\psi = e^{i\beta t} f(r)$, $\Phi = \varphi(r) - \beta$:

$$0 = -f'' - \frac{2}{r}f' + 2\varphi f - \frac{2}{r}f, \qquad 0 = -\varphi'' - \frac{2}{r}\varphi' + 4\pi f^2 \quad (19)$$

Requiring that f and φ are finite at $r = 0$, we find the following behaviour in terms of two integration constants f_0, φ_0:

$$\frac{f}{f_0} = 1 - r + \frac{r^2}{3}(\varphi_0 + 1) - \frac{r^3}{18}(4\varphi_0 + 1) + \cdots$$
$$\varphi = \varphi_0 + \frac{2\pi}{3}f_0^2 r^2 - \frac{2\pi}{3}f_0^2 r^3 + \cdots \quad (20)$$

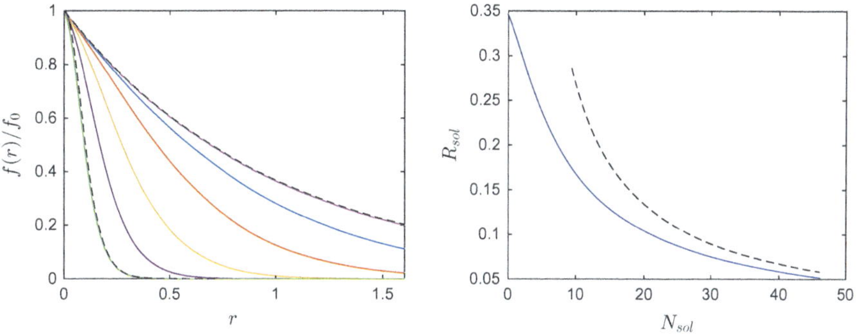

Fig. 7 On the left $f(r)/f_0$ profiles for the stationary ground state solutions of (19) for different values of f_0. From top to bottom: $f_0 = 0.1$ (magenta), $f_0 = 1$ (blue), $f_0 = 3$ (red), $f_0 = 10$ (yellow), $f_0 = 30$ (purple), $f_0 = 100$ (green). The black dashed lines correspond to the asymptotic forms of the profile as mentioned in the text. On the right, we plot the radius of the dark matter distribution (defined as half width at half of the maximum for f^2) as a function of the norm $N_{sol} = 4\pi \int r^2 f(r)^2 dr$. The curve interpolates between $\log 2$ at $N_{sol} \to 0$ and $r_{sol} \approx 2.68/N_{sol}$ (plotted as a dashed black line), and is well approximated by Eq. (21)

Requiring $\lim_{r\to\infty} f(r) = 0$ and $f'(r) < 0$ for all r, there is a unique solution for every $f_0 > 0$. The corresponding value of φ_0 can be found by a shooting technique. The family of solutions interpolates between the ground state of the Coulomb potential $f(r) = f_0 e^{-r}$ for $f_0 \ll 1$ and the profile without Coulomb potential of Fig. 2 ($f_0 \gg 1$). Some results are depicted in Fig. 7.

By fitting the curve for the radius of the dark matter distribution, we find:

$$r_{sol} \approx \frac{9 + 2.68 N_{sol}}{26 + 8.65 N_{sol} + N_{sol}^2} \tag{21}$$

Taking into account that the unit of length is $\hbar^2 G^{-1} m_a^{-2} M_{SMBH}^{-1}$ and that the total dark matter mass in the stationary solution is $M_{sol} = M_{SMBH} N_{sol}$, Eq. (21) provides an approximation for the size of the ground state of a dark matter clump with mass M_{sol} surrounding a black hole of mass M_{SMBH}.

As a final remark, let us comment on another empirical relation for SMBHs, which states that $M_{SMBH} \propto M_{tot}^{1.6}$, where M_{tot} is, essentially, the mass of the halo, see e.g., [16] for more precise definitions. If we assume that the $M - \sigma$ relation holds $M_{SMBH} \propto \sigma^{\beta}$ and that the bulge velocity dispersion is related to the soliton mass and radius by the virial theorem $\frac{GM_{sol}}{R_{sol}} \approx \sigma^2$, we find that $M_{SMBH} \propto M_{sol}^{\beta}$, where we have used $M_{sol} \propto 1/R_{sol}$, Eq. (14). Now, results of galaxy formation with Schrödinger-Poisson equation suggest $M_{sol} \propto M_{halo}^{\frac{1}{3}}$ [99], leading to $M_{SMBH} \propto M_{halo}^{\beta/3}$ that, taking into account $\beta \approx 5.1$, comes really close to the empirical relation. Plainly, this paragraph is just heuristic wishful thinking, but this coincidence might deserve a more serious consideration.

7 Physical Analogues

In this section, we briefly comment on the different uses that the Schrödinger-Poisson system of equations has found in physics. This digression is particularly suitable for a multidisciplinary book in nonlinear science. We also discuss the possibility of implementing tabletop experiments that (partially) mimic aspects of cosmological or astrophysical evolution like the formation of structure or the collision of galaxies.

In previous sections, we have introduced the usage of SPE as a classical equation for gravity coupled to a scalar which applies to boson stars [95], QCD axion DM [39], ultra-light axion DM [60, 101] or as an approximation to any kind of cold dark matter [21]. Long ago, it was also proposed that the SPE can have implications for the foundations of quantum mechanics [31], since including the gravitational term in the quantum Schrödinger equation can help in interpreting the collapse of the wave-function [84]. The concept can be defined as a gravitization of quantum mechanics rather than a quantization of gravity [85]. It should hinder the spread of a free wave-function that is unavoidable in the linear picture leading to measurable consequences whose observation constitutes a serious technological challenge [34, 73]. This line of research led to efforts in numerical implementations that have an obvious overlap with those in the DM context [45].

The SPE has also appeared in contexts where gravity plays no role and where the local nonlinearity typically appears in relation to a transport process. For instance, in one dimension it has been widely used for the modeling of electrons in semiconductors, e.g., [59].

In nonlinear optics, nonlocal nonlinearities have been subject of comprehensive studies (e.g., [3]) and the SPE has appeared in relation to liquid nematic crystals [25] and thermo-optical media [94]. This last setting has already been used with the goal of simulating gravitational effects [11, 92]. The dynamics in this case is two-dimensional, with the propagation distance playing the role of time and the laplacian acts on the plane transverse to propagation. $|\psi|^2$ is associated to the optical intensity and the Poisson potential Φ is related to temperature. In [77], it has been pointed out that the setup provides an opportunity for a tabletop experiment with many analogies with the phenomena discussed in the DM framework. Even if the equations differ in the number of dimensions, there are many qualitative similarities as, for instance, the formations of solitonic cores surrounded by extended incoherent halos [77].

Another interesting framework is that of Bose-Einstein condensed cold atoms. The impressive experimental control developed in the last decades paves the way for the simulation of gravitational phenomena like event horizons [33] or the nonlinear dynamics underlying black hole collapse [15]. In particular, the SPE can be engineered by averaging out the anisotropic nonlinear interactions leaving just Newtonian attraction [79] and leading to another possible tabletop simulation of gravity, Eqs. (5) and (6). More recently, it has also been shown that certain cold atoms systems in microwave cavities are governed by a kind of SPE in one [89] or two dimensions [90].

8 Summary and Outlook

The goal of this chapter has been to emphasize the relevance of (versions of) the nonlinear Schrödinger equation for cosmology and, in particular, for the modeling of dark matter. In our opinion, there are possibilities for collaboration and cross-fertilization with other areas of physics, where the NLSE is an essential tool. For instance, other chapters of this book deal with its usage in optics and cold atom condensates. We have tried to provide a first introduction of the cosmological setting for non-experts who might be interested in this long-standing problem of fundamental physics. We have also given a succinct presentation of a simple numerical method for the integration of Eqs. (5) and (6). This could be useful for people interested in getting started in the algorithms and computational schemes that are useful for dealing with the NLSE. We envisage the possibility of cross-disciplinary implications along two different lines: from the formal point of view, insight and methods developed in one physical framework might be transferred to a different one with related governing equations. On the other hand, it is interesting to implement laboratory experiments that, to some extent, reproduce cosmological dynamics, see Sect. 7.

Several factors underscore the timeliness of research in the topic covered in this chapter. First, astrophysical and cosmological observation are getting more and more precise and they call for a more accurate theoretical understanding and more powerful computations including, in particular, dark matter dynamics. Let us quote for instance the words of [87]: "The interpretation of cosmological observations increasingly requires a precise understanding of non-linear structure formation". Also, Earth-based experiments will soon be helpful in confirming or setting more stringent limits in order to discriminate between the many hypothesis that have been formulated for DM. At the same time, hardware and software are rapidly developing and it is possible to perform computations that were out of reach one decade ago.

In this context, we have mostly discussed the scalar field dark matter scenario, that uses the Schrödinger-Poisson system as a non-relativistic description of a cosmological matter wave. This model been studied for some time, starting with the seminal papers [49, 60, 101]. In the last few years, it has experienced an upsurge, partly motivated by more powerful numerical techniques and by the persistent lack of evidence for other forms of dark matter in facilities like the LHC. We have presented some examples of astrophysical phenomena that can be interpreted with the cosmic matter wave hypothesis. In particular, in Sect. 5, we have exposed the work of [81], that shows that interference between solitons can produce large effective forces that can be relevant for galactic encounters and suggests that the mysterious behavior of the Abell 3827 cluster could be a natural consequence of this phenomenon. In Sect. 6, we have reviewed the relation between black holes and the dark matter condensate and presented the ground state eigenfunctions of Schrödinger-Poisson in the presence of the Newtonian potential generated by a point-like mass concentration.

The NLSE is also useful in other scenarios. The more standard cold dark matter theory heavily relies on N-body simulations. Above some length scale, the Schrödinger formalism gives the same results [21, 97, 109] and, thus, the NLSE can provide an alternative computational scheme. Roughly speaking, the differences are the "quantum pressure" (namely the diffraction term, that avoids the concentration of energy in small volumes) and the appearance of interference. An intermediate possibility is that of a fluid equation with the quantum pressure term but without displaying interference [106]. Discriminating between these sometimes subtle characteristics might shed light on our understanding of the Universe.

Dark matter, ultra-light axions, solitons and nonlinear wave equations are exciting topics. We hope that this contribution has been able to convey their importance and the timeliness of active research in these related areas.

Acknowledgements We acknowledge financial support from Ministerio de Economía y Competitividad (MINECO) through grants FIS2014-58117-P, FIS2014-61984-EXP, and from Xunta de Galicia through grant GPC2015/019.

References

1. Ade, P., Aghanim, N., Arnaud, M., Ashdown, M., Aumont, J., Baccigalupi, C., Banday, A., Barreiro, R., Bartlett, J., Bartolo, N., et al.: Planck 2015 results-XIII. Cosmological parameters. Astron. Astrophys. **594**, A13 (2016)
2. Agrawal, G.P.: Nonlinear Fiber Optics. Academic press (2007)
3. Alberucci, A., Jisha, C.P., Smyth, N.F., Assanto, G.: Spatial optical solitons in highly nonlocal media. Phys. Rev. A **91**(1), 013841 (2015)
4. Alexandre, J.: Dynamical mechanism for ultralight scalar dark matter. Phys. Rev. D **92**(12), 123524 (2015)
5. Amendola, L., Barbieri, R.: Dark matter from an ultra-light pseudo-Goldsone-boson. Phys. Lett. B **642**(3), 192–196 (2006)
6. Armengaud, E., Avignone, F., Betz, M., Brax, P., Brun, P., Cantatore, G., Carmona, J., Carosi, G., Caspers, F., Caspi, S., et al.: Conceptual design of the International Axion Observatory (IAXO). J. Instrum. **9**(05), T05002 (2014)
7. Arvanitaki, A., Dimopoulos, S., Dubovsky, S., Kaloper, N., March-Russell, J.: String axiverse. Phys. Rev. D **81**(12), 123530 (2010)
8. Barranco, J., Bernal, A., Degollado, J.C., Diez-Tejedor, A., Megevand, M., Alcubierre, M., Núñez, D., Sarbach, O.: Schwarzschild black holes can wear scalar wigs. Phys. Rev. Lett. **109**(8), 081102 (2012)
9. Bauer, D., Buckley, J., Cahill-Rowley, M., Cotta, R., Drlica-Wagner, A., Feng, J.L., Funk, S., Hewett, J., Hooper, D., Ismail, A., Kaplinghat, M., Kusenko, A., Matchev, K., McKinsey, D., Rizzo, T., Shepherd, W., Tait, T.M., Wijangco, A.M., Wood, M.: Dark matter in the coming decade: complementary paths to discovery and beyond. Phys. Dark Universe **7–8**, 16–23 (2015)
10. Behroozi, P.S., Wechsler, R.H., Conroy, C.: The average star formation histories of galaxies in dark matter halos from z= 0–8. Astrophys. J. **770**(1), 57 (2013)
11. Bekenstein, R., Schley, R., Mutzafi, M., Rotschild, C., Segev, M.: Optical simulations of gravitational effects in the Newton-Schrodinger system. Nat. Phys. **11**, 872–878 (2015)
12. Bernal, A., Guzman, F.S.: Scalar field dark matter: head-on interaction between two structures. Phys. Rev. D **74**(10), 103002 (2006)
13. Bernal, A., Guzman, F.S.: Scalar field dark matter: nonspherical collapse and late-time behavior. Phys. Rev. D **74**(6), 063504 (2006)
14. Bernal, T., Fernández-Hernández, L.M., Matos, T., Rodríguez-Meza, M.A.: Rotation curves of high-resolution LSB and SPARC galaxies in wave (fuzzy) and multistate (ultra-light boson)

scalar field dark matter. Mon. Not. R. Astron. Soc. **475**(2), 1447–1468 (2017). https://doi.org/10.1093/mnras/stx3208, arXiv:1701.00912. 1 April 2018

15. Biasi, A.F., Mas, J., Paredes, A.: Delayed collapses of Bose-Einstein condensates in relation to anti-de sitter gravity. Phys. Rev. E **95**, 032216 (2017)
16. Bogdán, Á., Goulding, A.D.: Connecting dark matter halos with the galaxy center and the supermassive black hole. Astrophys. J. **800**(2), 124 (2015)
17. Böhmer, C., Harko, T.: Can dark matter be a Bose-Einstein condensate? J. Cosmol. Astropart. Phys. **06**, 025 (2007)
18. Bousso, R.: The cosmological constant. Gen. Relat. Gravit. **40**(2–3), 607–637 (2008)
19. Bower, R., Benson, A., Malbon, R., Helly, J., Frenk, C., Baugh, C., Cole, S., Lacey, C.G.: Breaking the hierarchy of galaxy formation. Mon. Not. R. Astron. Soc. **370**(2), 645–655 (2006)
20. Bozek, B., Marsh, D.J., Silk, J., Wyse, R.F.: Galaxy UV-luminosity function and reionization constraints on axion dark matter. Mon. Not. R. Astron. Soc. **450**(1), 209–222 (2015)
21. Briscese, F.: Theoretical foundations of the Schrödinger method for LSS formation. Eur. Phys. J. C. **2017**(77), 623 (2016). https://doi.org/10.1140/epjc/s10052-017-5209-7, arXiv:1612.04572
22. Carrasco, E., Gomez, P., Verdugo, T., Lee, H., Diaz, R., Bergmann, M., Turner, J., Miller, B., West, M.: Strong gravitational lensing by the super-massive cD galaxy in Abell 3827. Astrophys. J. Lett. **715**(2), L160 (2010)
23. Chavanis, P.H.: BEC dark matter, Zeldovich approximation, and generalized Burgers equation. Phys. Rev. D **84**(6), 063518 (2011)
24. Chavanis, P.H., Delfini, L.: Mass-radius relation of Newtonian self-gravitating Bose-Einstein condensates with short-range interactions. II. Numerical Results. Phys. Rev. D **84**(4), 043532 (2011)
25. Conti, C., Peccianti, M., Assanto, G.: Route to nonlocality and observation of accessible solitons. Phys. Rev. Lett. **91**(7), 073901 (2003)
26. Cotner, E.: Collisional interactions between self-interacting nonrelativistic boson stars: effective potential analysis and numerical simulations. Phys. Rev. D **94**(6), 063503 (2016)
27. Cyburt, R.H., Fields, B.D., Olive, K.A., Yeh, T.H.: Big bang nucleosynthesis: present status. Rev. Mod. Phys. **88**(1), 015004 (2016)
28. Dalfovo, F., Giorgini, S., Pitaevskii, L.P., Stringari, S.: Theory of Bose-Einstein condensation in trapped gases. Rev. Mod. Phys. **71**(3), 463 (1999)
29. Davies, G., Widrow, L.M.: Test-bed simulations of collisionless, self-gravitating systems using the Schrödinger method. Astrophys. J. **485**(2), 484 (1997)
30. De Blok, W.: The core-cusp problem. Adv. Astron. **2010**, 789293 (2010)
31. Diósi, L.: Gravitation and quantum-mechanical localization of macro-objects. Phys. Lett. A **105**, 199 (1984)
32. Feng, J.L.: Dark matter candidates from particle physics and methods of detection. Annu. Rev. Astron. Astr. **48**, 495–545 (2010)
33. Garay, L., Anglin, J., Cirac, J., Zoller, P.: Sonic black holes in dilute Bose-Einstein condensates. Phys. Rev. A **63**(2), 023,611 (2001)
34. Giulini, D., Großardt, A.: Gravitationally induced inhibitions of dispersion according to the Schrödinger-Newton equation. Class. Quant. Grav. **28**(19), 195026 (2011)
35. Gonzáles-Morales, A.X., Marsh, D.J., Peñarrubia, J., Ureña-López, L.: Unbiased constraints on ultralight axion mass from dwarf spheroidal galaxies. Mon. Not. R. Astron. Soc. **472**(2), 1346–1360 (2017). https://doi.org/10.1093/mnras/stx1941, arXiv:1609.05856. 1 December 2017
36. González, J., Guzmán, F.: Interference pattern in the collision of structures in the Bose-Einstein condensate dark matter model: comparison with fluids. Phys. Rev. D **83**(10), 103513 (2011)
37. Goodman, J.: Repulsive dark matter. New Astron. **5**(2), 103–107 (2000)
38. Gupta, P.D., Thareja, E.: Supermassive black holes from collapsing dark matter Bose-Einstein condensates. Class. Quant. Grav. **34**(3), 035006 (2017)

39. Guth, A.H., Hertzberg, M.P., Prescod-Weinstein, C.: Do dark matter axions form a condensate with long-range correlation? Phys. Rev. D **92**(10), 103513 (2015)
40. Guzmán, F., González, J., Cruz-Pérez, J.: Behavior of luminous matter in the head-on encounter of two ultralight BEC dark matter halos. Phys. Rev. D **93**(10), 103535 (2016)
41. Guzmán, F.S., Urena-López, L.A.: Evolution of the Schrödinger-Newton system for a self-gravitating scalar field. Phys. Rev. D **69**(12), 124033 (2004)
42. Guzman, F.S., Urena-Lopez, L.A.: Gravitational cooling of self-gravitating Bose condensates. Astrophys. J. **645**(2), 814 (2006)
43. Harko, T.: Bose-Einstein condensation of dark matter solves the core/cusp problem. J. Cosmol. Astropart. P. **2011**(05), 022 (2011)
44. Harko, T.: Cosmological dynamics of dark matter Bose-Einstein condensation. Phys. Rev. D **83**(12), 123515 (2011)
45. Harrison, R., Moroz, I., Tod, K.: A numerical study of the Schrödinger-Newton equations. Nonlinearity **16**(1), 101 (2002)
46. Harvey, D., Massey, R., Kitching, T., Taylor, A., Tittley, E.: The nongravitational interactions of dark matter in colliding galaxy clusters. Science **347**(6229), 1462–1465 (2015)
47. Helfer, T., Marsh, D.J., Clough, K., Fairbairn, M., Lim, E.A., Becerril, R.: Black hole formation from axion stars. J. Cosmol. Astropart. Phys. **03**, 055 (2017). https://doi.org/10.1088/1475-7516/2017/03/055, arXiv:1609.04724
48. Hlozek, R., Grin, D., Marsh, D.J., Ferreira, P.G.: A search for ultralight axions using precision cosmological data. Phys. Rev. D **91**(10), 103512 (2015)
49. Hu, W., Barkana, R., Gruzinov, A.: Fuzzy cold dark matter: the wave properties of ultralight particles. Phys. Rev. Lett. **85**(6), 1158 (2000)
50. Hui, L., Ostriker, J.P., Tremaine, S., Witten, E.: Ultralight scalars as cosmological dark matter. Phys. Rev. D **95**, 043541 (2017)
51. Jee, M., Mahdavi, A., Hoekstra, H., Babul, A., Dalcanton, J., Carroll, P., Capak, P.: A study of the dark core in A520 with the Hubble Space Telescope: The mystery deepens. Astrophys. J. **747**(2), 96 (2012)
52. Jiang, S., Greengard, L., Bao, W.: Fast and accurate evaluation of nonlocal Coulomb and dipole-dipole interactions via the nonuniform FFT. SIAM J. Sci. Comput. **36**(5), B777–B794 (2014)
53. Kahlhoefer, F., Schmidt-Hoberg, K., Kummer, J., Sarkar, S.: On the interpretation of dark matter self-interactions in Abell 3827. Mon. Not. R. Astron. Soc. Lett. **452**(1), L54–L58 (2015)
54. Khlopov, M.Y., Malomed, B.A., Zeldovich, Y.B.: Gravitational instability of scalar fields and formation of primordial black holes. Mon. Not. R. Astron. Soc. **215**(4), 575–589 (1985)
55. Khmelnitsky, A., Rubakov, V.: Pulsar timing signal from ultralight scalar dark matter. J. Cosmol. Astropart. P. **2014**(02), 019 (2014)
56. Kiritsis, E.: Gravity and axions from a random UV QFT. In: EPJ Web of Conferences, vol. 71, p. 00068. EDP Sciences (2014)
57. Klypin, A., Kravtsov, A.V., Valenzuela, O., Prada, F.: Where are the missing galactic satellites? Astrophys. J. **522**(1), 82 (1999)
58. Kumar, R.K., Young-S, L.E., Vudragović, D., Balaž, A., Muruganandam, P., Adhikari, S.: Fortran and C programs for the time-dependent dipolar Gross-Pitaevskii equation in an anisotropic trap. Comput. Phys. Commun. **195**, 117–128 (2015)
59. Laux, S.E., Stern, F.: Electron states in narrow gate-induced channels in Si. Appl. Phys. Lett. **49**(2), 91–93 (1986)
60. Lee, J.W., Koh, I.G.: Galactic halos as boson stars. Phys. Rev. D **53**(4), 2236 (1996)
61. Lee, J.W., Lee, J., Kim, H.C.: The M-sigma relation of super massive black holes from the scalar field dark matter (2015). arXiv:1512.02351
62. Lee, J.W., Lim, S., Choi, D.: BEC dark matter can explain collisions of galaxy clusters (2008). arXiv:0805.3827
63. Liddle, A.: An Introduction to Modern Cosmology. Wiley (2015)

64. Lončar, V., Young-S, L.E., Škrbić, S., Muruganandam, P., Adhikari, S.K., Balaž, A.: OpenMP, openMP/MPI, and CUDA/MPI C programs for solving the time-dependent dipolar Gross-Pitaevskii equation. Comput. Phys. Commun. **209**, 190–196 (2016)
65. Lovell, M.R., Frenk, C.S., Eke, V.R., Jenkins, A., Gao, L., Theuns, T.: The properties of warm dark matter haloes. Mon. Not. R. Astron. Soc. **439**, 300–317 (2014)
66. Magana, J., Matos, T.: A brief review of the scalar field dark matter model. J. Phys. Conf. Ser. **378**, 012012 (2012) (IOP Publishing)
67. Markevitch, M., Gonzalez, A., Clowe, D., Vikhlinin, A., Forman, W., Jones, C., Murray, S., Tucker, W.: Direct constraints on the dark matter self-interaction cross section from the merging galaxy cluster 1E 0657–56. Astrophys. J. **606**(2), 819 (2004)
68. Marsh, D.J.: Axion cosmology. Phys. Rep. **643**, 1–79 (2016)
69. Marsh, D.J.E., Pop, A.R.: Axion dark matter, solitons and the cusp-core problem. Mon. Not. R. Astron. Soc. **451**, 2479 (2015)
70. Massey, R., Williams, L., Smit, R., Swinbank, M., Kitching, T.D., Harvey, D., Jauzac, M., Israel, H., Clowe, D., Edge, A., et al.: The behaviour of dark matter associated with four bright cluster galaxies in the 10 kpc core of Abell 3827. Mon. Not. R. Astron. Soc. **449**(4), 3393–3406 (2015)
71. Mavromatos, N.E., Argüelles, C.R., Ruffini, R., Rueda, J.A.: Self-interacting dark matter. Int. J. Mod. Phys. D **26**, 1730007 (2017)
72. McGaugh, S.S., Lelli, F., Schombert, J.M.: Radial acceleration relation in rotationally supported galaxies. Phys. Rev. Lett. **117**(20), 201101 (2016)
73. van Meter, J.R.: Schrödinger-Newton "collapse" of the wavefunction. Class. Quantum Gravity **28**(21), 215013 (2011)
74. Milgrom, M.: A modification of the Newtonian dynamics as a possible alternative to the hidden mass hypothesis. Astrophys. J. **270**, 365–370 (1983)
75. Moore, B., Ghigna, S., Governato, F., Lake, G., Quinn, T., Stadel, J., Tozzi, P.: Dark matter substructure within galactic halos. Astrophys. J. Lett. **524**(1), L19 (1999)
76. Moroz, I.M., Penrose, R., Tod, P.: Spherically-symmetric solutions of the Schrödinger-Newton equations. Class. Quantum Gravity **15**(9), 2733 (1998)
77. Navarrete, A., Paredes, A., Salgueiro, J.R., Michinel, H.: Spatial solitons in thermo-optical media from the nonlinear Schrödinger-Poisson equation and dark matter analogs. Phys. Rev. A **95**, 013844 (2017)
78. Nguyen, J.H., Dyke, P., Luo, D., Malomed, B.A., Hulet, R.G.: Collisions of matter-wave solitons. Nat. Phys. **10**(12), 918–922 (2014)
79. O'Dell, D., Giovanazzi, S., Kurizki, G., Akulin, V.: Bose-Einstein condensates with 1/r interatomic attraction: electromagnetically induced "gravity". Phys. Rev. Lett. **84**(25), 5687 (2000)
80. Paredes, Á., Feijoo, D., Michinel, H.: Coherent cavitation in the liquid of light. Phys. Rev. Lett. **112**(17), 173901 (2014)
81. Paredes, A., Michinel, H.: Interference of dark matter solitons and galactic offsets. Phys. Dark Universe **12**, 50–55 (2016). https://creativecommons.org/licenses/by/4.0/, https://doi.org/10.1016/j.dark.2016.02.003
82. Peccei, R.D.: The strong CP problem and axions. In: Axions, pp 3–17. Springer (2008)
83. Peebles, P.J.E.: Principles of Physical Cosmology. Princeton University Press (1993)
84. Penrose, R.: On gravity's role in quantum state reduction. Gen. Relat. Gravit. **28**(5), 581–600 (1996)
85. Penrose, R.: On the gravitization of quantum mechanics 1: Quantum state reduction. Found. Phys. **44**(5), 557–575 (2014)
86. Pontzen, A., Governato, F.: Cold dark matter heats up. Nature **506**(7487), 171–178 (2014)
87. Pontzen, A., Slosar, A., Roth, N., Peiris, H.V.: Inverted initial conditions: exploring the growth of cosmic structure and voids. Phys. Rev. D **93**(10), 103519 (2016)
88. Primack, J.R.: Precision cosmology. New Astron. Rev. **49**(2), 25–34 (2005)
89. Qin, J., Dong, G., Malomed, B.A.: Hybrid matter-wave-microwave solitons produced by the local-field effect. Phys. Rev. Lett. **115**(2), 023901 (2015)

90. Qin, J., Dong, G., Malomed, B.A.: Stable giant vortex annuli in microwave-coupled atomic condensates. Phys. Rev. A **94**(5), 053611 (2016)
91. Riess, A.G., Filippenko, A.V., Challis, P., Clocchiatti, A., Diercks, A., Garnavich, P.M., Gilliland, R.L., Hogan, C.J., Jha, S., Kirshner, R.P., et al.: Observational evidence from supernovae for an accelerating universe and a cosmological constant. Astron. J. **116**(3), 1009 (1998)
92. Roger, T., Maitland, C., Wilson, K., Westerberg, N., Vocke, D., Wright, E.M., Faccio, D.: Optical analogues of the Newton-Schrödinger equation and boson star evolution. Nat. Commun. **7**, 13492 (2016)
93. Roos, M.: Introduction to Cosmology. Wiley (2015)
94. Rotschild, C., Cohen, O., Manela, O., Segev, M., Carmon, T.: Solitons in nonlinear media with an infinite range of nonlocality: first observation of coherent elliptic solitons and of vortex-ring solitons. Phys. Rev. Lett. **95**(21), 213904 (2005)
95. Ruffini, R., Bonazzola, S.: Systems of self-gravitating particles in general relativity and the concept of an equation of state. Phys. Rev. **187**(5), 1767 (1969)
96. Schaller, M., Robertson, A., Massey, R., Bower, R.G., Eke, V.R.: The offsets between galaxies and their dark matter in Λ cold dark matter. Mon. Not. R. Astron. Soc. Lett. **453**(1), L65–L69 (2015)
97. Schive, H.Y., Chiueh, T., Broadhurst, T.: Cosmic structure as the quantum interference of a coherent dark wave. Nat. Phys. **10**(7), 496–499 (2014)
98. Schive, H.Y., Chiueh, T., Broadhurst, T., Huang, K.W.: Contrasting galaxy formation from quantum wave dark matter, ψDM, with ΛCDM, using Planck and Hubble data. Astrophys. J. **818**(1), 89 (2016)
99. Schive, H.Y., Liao, M.H., Woo, T.P., Wong, S.K., Chiueh, T., Broadhurst, T., Hwang, W.P.: Understanding the core-halo relation of quantum wave dark matter from 3D simulations. Phys. Rev. Lett. **113**(26), 261302 (2014)
100. Schwabe, B., Niemeyer, J.C., Engels, J.F.: Simulations of solitonic core mergers in ultralight axion dark matter cosmologies. Phys. Rev. D **94**(4), 043513 (2016)
101. Sin, S.J.: Late-time phase transition and the galactic halo as a Bose liquid. Phys. Rev. D **50**(6), 3650 (1994)
102. Suárez, A., Robles, V.H., Matos, T.: A review on the scalar field/Bose-Einstein condensate dark matter model. In: Accelerated Cosmic Expansion, pp. 107–142. Springer (2014)
103. Taha, T.R., Ablowitz, M.I.: Analytical and numerical aspects of certain nonlinear evolution equations. II. Numerical, nonlinear schrödinger equation. J. Comput. Phys. **55**(2), 203–230 (1984)
104. Urena-Lopez, L.A., Liddle, A.R.: Supermassive black holes in scalar field galaxy halos. Phys. Rev. D **66**(8), 083005 (2002)
105. Valle, D.C., Mielke, E.W.: Solitonic axion condensates modeling dark matter halos. Ann. Phys. **336**, 245–260 (2013)
106. Veltmaat, J., Niemeyer, J.C.: Cosmological particle-in-cell simulations with ultralight axion dark matter. Phys. Rev. D **94**(12), 123523 (2016)
107. Weinberg, D.H., Bullock, J.S., Governato, F., de Naray, R.K., Peter, A.H.: Cold dark matter: controversies on small scales. P. Natl. Acad. Sci. USA **112**(40), 12249–12255 (2015)
108. Weinberg, S.: Gravitation and cosmology: principles and applications of the general theory of relativity, vol. 1. Wiley, New York (1972)
109. Widrow, L.M., Kaiser, N.: Using the Schrödinger equation to simulate collisionless matter. Astrophys. J. **416**, L71 (1993)
110. Williams, L.L., Saha, P.: Light/mass offsets in the lensing cluster Abell 3827: evidence for collisional dark matter? Mon. Not. R. Astron. Soc. **415**(1), 448–460 (2011)
111. Woo, T.P., Chiueh, T.: High-resolution simulation on structure formation with extremely light bosonic dark matter. Astrophys. J. **697**(1), 850 (2009)

Adiabatic Invariants of Second Order Korteweg-de Vries Type Equation

Piotr Rozmej and Anna Karczewska

Abstract In this chapter we analyze the existence and forms of invariants of the extended Korteweg-de Vries equation (KdV2). This equation appears when the Euler equations for shallow water are extended to the second order, beyond Korteweg-de Vries (KdV). We show that contrary to KdV for which there is an infinite number of invariants, for KdV2 there exists only one, connected to mass (volume) conservation of the fluid. For KdV2 we found only so-called *adiabatic invariants*, that is, functions of the solutions which are constants neglecting terms of higher order than the order of the equation. In this chapter we present two methods for construction of such invariants. The first method, a direct one, consists in using constructions of higher KdV invariants and eliminating non-integrable terms in an approximate way. The second method introduces a near-identity transformation (NIT) which transforms KdV2 into equation (asymptotically equivalent) which is integrable. For the equation obtained by NIT, exact invariants exist, but they become approximate (adiabatic) when the inverse NIT transformation is applied and original variables are restored. Numerical tests of the exactness of adiabatic invariants for KdV2 in several cases of initial conditions are presented. These tests confirm that relative changes in these approximate invariants are small indeed. The relations of KdV invariants and KdV2 adiabatic invariants to conservation laws are discussed, as well.

Keywords Shallow water waves · Nonlinear equations · Invariants of KdV2 equation · Adiabatic invariants

P. Rozmej (✉)
Faculty of Physics and Astronomy, Institute of Physics, University of Zielona Góra, Szafrana 4a, 65-246 Zielona Góra, Poland
e-mail: p.rozmej@if.uz.zgora.pl

A. Karczewska
Faculty of Mathematics, Computer Science and Econometrics, University of Zielona Góra, Szafrana 4a, 65-246 Zielona Góra, Poland
e-mail: a.karczewska@wmie.uz.zgora.pl

V. Carmona et al. (eds.), *Nonlinear Systems, Vol. 1*,
Understanding Complex Systems, https://doi.org/10.1007/978-3-319-66766-9_6

1 Introduction

The celebrated Korteweg-de Vries equation (KdV) [31], whose origin is the set of Euler's shallow water and long wavelength equations, now enjoys a paradigmatic status in the field of nonlinear partial differential equations. There is a huge number of research papers concerned with weakly nonlinear, dispersive and long wavelength problems in which KdV is shown as the lowest approximation of wave motion in a number of fields of physics, see, e.g., monographs [8, 22, 36, 39, 41] and references therein.

It is accepted fact that KdV gives an infinite number of invariants or conservation laws also referred to as integrals of motion [4, 8, 35, 37]. The two first KdV invariants concern the preservation of mass (volume) of the fluid and conservation of its total momentum. The next one is related to energy conservation. The higher KdV invariants have no simple interpretation. KdV is, however, the result of an approximation of the set of the Euler equations within the perturbation approach, limited to the first order in expansion with respect to parameters assumed to be small. KdV has been extended to the second order (KdV2) by a number of authors, e.g., [6, 21, 25, 33, 34, 42]. In [23, 24, 26, 28] the authors have derived the KdV2 equation for an uneven bottom, introducing an additional small parameter related to bottom variation. Here the term *second order* is defined as the order of perturbation expansion with respect to small parameters. However, this advanced form is lacking in exactly conserved quantities except for the ubiquitous mass law.

Many papers, e.g., [4, 7, 9–11, 13, 16–19, 29, 30, 44] assert integrability of second order KdV type equations and existence of higher invariants. Specifically Benjamin and Olver [4] have discussed Hamiltonian structure, symmetries and conservation laws in respect of water waves. A near-identity transformation (NIT), first published by Kodama [29, 30] and since used by many other authors, e.g., [7, 9–11, 13, 16–19, 44], makes it possible to transform the second order KdV type equations into Hamiltonian forms which are asymptotically equivalent. The existence of the Hamiltonian form for the transformed equation supplies the full hierarchy of invariants, which appear to be adiabatic invariants in respect of the original equation.

The lack of exact invariants in the system forces one to look for adiabatic (approximate) ones, as in [5]. Recently we developed a simple method to calculate such adiabatic invariants, allowing us to derive them directly using the original 'physical' equation (equally applicable to equations expressed in dimensional variables) [21]. Our method is as follows: one constructs the KdV2 in a similar fashion as one does for KdV invariants and then applies the addition of KdV, multiplied by a small parameter, to cancel the non-integrable terms. In [21] we focused on this direct method mentioning NIT-based derivation of adiabatic invariants rather briefly. In this chapter the NIT method is discussed more broadly with particular attention paid to energy conservation law.

It is shown in [40] that KdV2 for uneven bottom [23, 26] is not symmetry-integrable since it admits no genuinely generalized symmetries.

The chapter substantially extends results published recently in [25]. In order to introduce the reader to higher order nonlinear equations beyond KdV several earlier achievements [23, 24, 26, 28] are recalled in Sect. 2. The set of Euler's equations for the inviscid and incompressible fluid and irrotational motion is introduced and the perturbation technique leading to KdV and KdV2 equations is described. Then analytic solutions for KdV and the recently obtained ones for KdV2, solitonic [23] and periodic [21], are presented and their properties compared.

In Sect. 3 we recall derivations of lowest invariants of KdV and their relations to conservation laws. In Sect. 4 a direct extension of the methods used in Sect. 3 for the KdV2 equation is presented. Particular forms of second and third adiabatic invariants for KdV2 are obtained.

In Sect. 5 near-identity transformation is introduced and applied to find general forms of lowest adiabatic invariants for KdV2. Relations of adiabatic invariants of KdV2 to formulas for the momentum and energy of the system are discussed, as well. The quality of adiabatic invariants is tested in numerics in Sect. 6. The main results are summarised in Sect. 7.

2 KdV and KdV2 Equations

First, we will recall briefly the derivation of KdV and KdV2 equations.

The natural assumptions in the shallow water wave problem are the following. Since water viscosity and compressibility are very small the fluid is assumed to be inviscid and incompressible. For gravity waves velocities of fluid particles are small, as well, therefore the motion can be considered as irrotational. This property allows us to introduce velocity potential ϕ. The velocity potential fulfils the Laplace equation for entire fluid volume. The set of Euler's equations contains also the kinematic and dynamic boundary conditions at the free surface and the kinematic boundary condition at the impenetrable bottom. The full set of equations for the velocity potential $\phi(x, y, z, t)$, as well as its derivation, is published in many textbooks, for instance, see [39, Chap. 5]. A typical procedure consists in introducing two small parameters $\alpha = a/h$ and $\beta = (h/l)^2$ and in application of perturbation approach with respect to these parameters. Here a is the amplitude of a surface wave η, h is water depth and l is a typical surface waves wavelength.

An approximation in deriving KdV and higher order nonlinear wave equations is correct when two small parameters α and β are of the same order of magnitude. The definitions of small parameters α and β and the geometry of the problem are shown in Fig. 1. The parameters α, β have the same meaning as the parameters ε, δ^2 in [39], respectively. These notations follow those in the paper [6], in which a systematic method for the derivation of wave equations of different orders is given. In [23, 26] we have introduced a third parameter $\delta = a_h/h$, where a_h denotes the amplitude of bottom changes. This new parameter allowed us to derive second order equation for surface waves over a non-flat bottom using the same perturbative approach as for derivation of KdV or higher-order KdV-like equations.

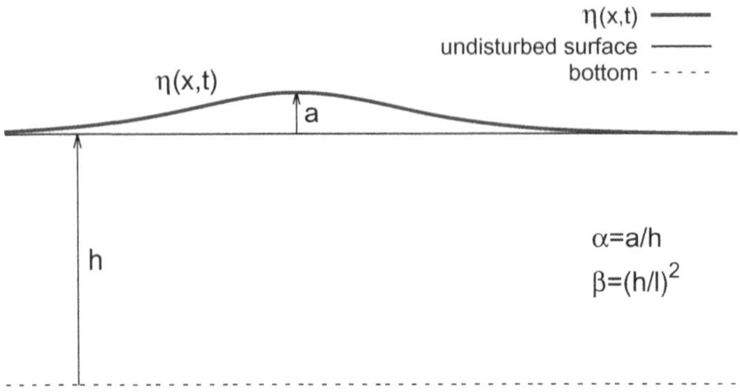

Fig. 1 Schematic view of the geometry of the problem

In what follows we restrict discussion to the 2-dimensional flow, $\phi(x, z, t)$ (which means translational symmetry with respect to y axis). Here, the horizontal coordinate is denoted by x and the vertical one is denoted by z.

A convenient way of studying the problem is introducing non-dimensional variables. They are defined as follows

$$\begin{aligned} \tilde{\eta} &= \eta/a, \quad \tilde{\phi} = \phi/(l\frac{a}{h}\sqrt{gh}), \\ \tilde{x} &= x/l, \quad \tilde{z} = z/h, \quad \tilde{t} = t/(l/\sqrt{gh}). \end{aligned} \tag{1}$$

The set of hydrodynamic equations for 2-dimensional flow in the non-dimensional variables takes the simpler form (henceforth all tildes have been omitted)

$$\beta\phi_{xx} + \phi_{zz} = 0, \tag{2}$$

$$\eta_t + \alpha\phi_x\eta_x - \frac{1}{\beta}\phi_z = 0, \quad \text{for} \quad z = 1 + \alpha\eta \tag{3}$$

$$\phi_t + \frac{1}{2}\alpha\phi_x^2 + \frac{1}{2}\frac{\alpha}{\beta}\phi_z^2 + \eta = 0, \quad \text{for} \quad z = 1 + \alpha\eta \tag{4}$$

$$\phi_z = 0, \quad \text{for} \quad z = 0. \tag{5}$$

The Laplace equation (2) is valid for the entire volume of the fluid. The Eq. (3) represents so called kinematic boundary condition at the (unknown) surface whereas the Eq. (4) is so called dynamic boundary condition at the surface. The Eq. (5) expresses the boundary condition at the impenetrable flat bottom. For abbreviation the partial derivatives with respect to corresponding variables are denoted by subscripts, i.e. $\phi_{nx} \equiv \frac{\partial^n \phi}{\partial x^n}$ and so on.

Next, the velocity potential is postulated in the form of power series

$$\phi(x, z, t) = \sum_{m=0}^{\infty} z^m \phi^{(m)}(x, t). \tag{6}$$

The Laplace equation (2) permits the expression of all $\phi^{(2m)}$ functions by the derivatives $\phi^{(0)}_{2mx}$, and all $\phi^{(2m+1)}$ functions by the derivatives $\phi^{(1)}_{2mx}$. Since the boundary condition at the bottom (5) sets $\phi^{(1)} = 0$, all $\phi^{(2m+1)}$ vanish and one obtains the following velocity potential

$$\phi = \phi^{(0)} - \frac{1}{2}\beta z^2 \phi^{(0)}_{2x} + \frac{1}{24}\beta^2 z^4 \phi^{(0)}_{4x} + \frac{1}{720}\beta^3 z^6 \phi^{(0)}_{6x} + \dots . \tag{7}$$

The presence of small parameters α, β in the set of hydrodynamic equations (2)–(5) and (7) allows us to apply the perturbation technique and to derive equations in first and second order with respect to these parameters. Next we insert $\phi(x, z, t)$ given by (7) into (3) and (4) retaining only terms up to second order in small parameters α, β. The Eq. (4) is then differentiated with respect to x and finally $w(x, t)$ is substituted in place of $\phi^{(0)}_x(x, t)$ in both equations. By this procedure one obtains a set of two coupled nonlinear differential equations which, in general, can be studied at different orders of approximation. This is a second order Boussinesq system

$$\eta_t + w_x + \alpha(\eta w)_x - \frac{1}{6}\beta w_{3x} - \frac{1}{2}\alpha\beta(\eta w_{2x})_x + \frac{1}{120}\beta^2 w_{5x} = 0, \tag{8}$$

$$w_t + \eta_x + \alpha w w_x - \frac{1}{2}\beta w_{2xt} + \frac{1}{24}\beta^2 w_{4xt} + \frac{1}{2}\alpha\beta[-2(\eta w_{xt})_x + w_x w_{2x} - w w_{3x}] = 0. \tag{9}$$

Burde and Sergyeyev [6] show a way of eliminating sequentially the $w(x, t)$ variable and deriving a single equation for $\eta(x, t)$ using the higher order perturbative approach. In their method the properties of solutions to lower order equations for w and η are used in derivations of corrections to equations in the next order. In theory this method can be used up to an arbitrary order. For the reader's convenience we will present it briefly below.

2.1 KdV Equation

Limitation of the Boussinesq system (8), (9) to first order in α, β

$$\eta_t + w_x + \alpha(\eta w)_x - \frac{1}{6}\beta w_{3x} = 0, \tag{10}$$

$$w_t + \eta_x + \alpha w w_x - \frac{1}{2}\beta w_{2xt} = 0 \tag{11}$$

results in the derivation of a KdV equation. First, notice that in zeroth order the above equations

$$\eta_t + w_x = 0, \qquad w_t + \eta_x = 0 \tag{12}$$

hold when $\eta = w$ and $\eta_x = w_x$. It follows that $\eta_t = -\eta_x$ and $w_t = -w_x$.

Next, one seeks solutions of the first order set (10), (11) requiring that w, η fulfil (12) and introducing first order corrections $C^{(\alpha)}$, $C^{(\beta)}$

$$w = \eta + \alpha C^{(\alpha)} + \beta C^{(\beta)}. \tag{13}$$

Insertion of (13) into (10), (11) and neglection of higher order terms gives

$$\alpha\left(C_x^{(\alpha)} + 2\eta\eta_x\right) + \beta\left(C_x^{(\beta)} - \frac{1}{6}\eta_{3x}\right) = 0 \tag{14}$$

$$\alpha\left(C_t^{(\alpha)} + \eta\eta_x\right) + \beta\left(C_t^{(\beta)} - \frac{1}{2}\eta_{2xt}\right) = 0. \tag{15}$$

Because of the correction functions appearing already in first order, it is enough to use a zeroth order formula relating their space and time derivatives. Therefore we use $C_t^{(\alpha)} = -C_x^{(\alpha)}$, $C_t^{(\beta)} = -C_x^{(\beta)}$ (like $\eta_t = -\eta_x$, $w_t = -w_x$) in (14), (15). (Otherwise, if one takes, for instance, $C_t^{(\alpha)} = -C_x^{(\alpha)} + \alpha C_1 + \beta C_2$, then terms with C_1, C_2 appear in second order and consequently are neglected). Inserting these relations, subtracting (14) form (15) and equating separately to zero terms with coefficients α and β one obtains

$$C_x^{(\alpha)} = -\frac{1}{2}\eta\eta_x \qquad \text{and} \qquad C_x^{(\beta)} = \frac{1}{3}\eta_{3x}. \tag{16}$$

Integration gives

$$C^{(\alpha)} = -\frac{1}{4}\eta^2 \qquad \text{and} \qquad C_x^{(\beta)} = \frac{1}{3}\eta_{2x}. \tag{17}$$

Then Eqs. (11) and (10) take the final form

$$w = \eta - \frac{1}{4}\alpha\eta^2 + \frac{1}{3}\beta\eta_{3x}, \tag{18}$$

$$\eta_t + \eta_x + \frac{3}{2}\alpha\eta\eta_x + \frac{1}{6}\beta\eta_{3x} = 0. \tag{19}$$

Equation (19) is the famous Korteweg-de Vries (KdV) equation in fixed reference frame (and scaled dimensionless variables). There are several forms of KdV equation in the literature. Transformation $\bar{x} = x - t,\ \ \bar{t} = t$ converts (19) into

$$\eta_{\bar{t}} + \frac{3}{2}\alpha\eta\eta_{\bar{x}} + \frac{1}{6}\beta\eta_{3\bar{x}} = 0. \tag{20}$$

Additional scaling by $\tilde{x} = \sqrt{\frac{3}{2}}\bar{x},\ \ \tilde{t} = \frac{1}{4}\sqrt{\frac{3}{2}}\alpha\bar{t}$ transforms (20) into

$$\eta_{\tilde{t}} + 6\eta\eta_{\tilde{x}} + \frac{\beta}{\alpha}\eta_{3\tilde{x}} = 0 \quad \text{or} \quad \eta_{\tilde{t}} + 6\eta\eta_{\tilde{x}} + \eta_{3\tilde{x}} = 0, \quad \text{when} \quad \beta = \alpha. \tag{21}$$

Equation (20) gives the form of KdV in the reference frame moving with a characteristic velocity (equal to 1 in dimensionless variables, which corresponds to $\sqrt{gH}$ in original variables). Forms like (21) or similar are preferred in mathematical papers. Sometimes the inverse transform to dimensional variables is applied yielding

$$\eta_t + c\eta_x + \frac{3}{2}\frac{c}{H}\eta\eta_x + \frac{cH^2}{6}\eta_{3x} = 0, \tag{22}$$

where $c = \sqrt{gH}$ is the limiting long wave speed [3]. Then, solutions of (22) can be directly compared to experimental data.

2.2 *Extended KdV (KdV2)*

Extending considerations of the Boussinesq equations (8), (9) to second order we make use of first order solutions (18) and (19). So, applying the perturbation technique described by [6], we postulate w in the form (18) plus higher order corrections, that is

$$w = \eta - \frac{1}{4}\alpha\eta^2 + \frac{1}{3}\beta\eta_{3x} + \alpha^2 C^{(\alpha^2)} + \alpha\beta C^{(\alpha\beta)} + \beta^2 C^{(\beta^2)}, \tag{23}$$

where $C^{(\alpha^2)}$, $C^{(\alpha\beta)}$, $C^{(\beta^2)}$ are yet unknown functions of η and its derivatives. Proceeding similarly as in first order equations and using the same properties of relations for time and space derivatives of correction functions, that is, $C_t^{(\cdot)} = -C_x^{(\cdot)}$ one obtains them in the form

$$C^{(\alpha^2)} = \frac{1}{8}\eta^3, \qquad C^{(\alpha\beta)} = \frac{3}{16}\eta_x^2 + \frac{1}{2}\eta\eta_{2x}, \qquad C^{(\beta^2)} = \frac{1}{10}\eta_{4x}. \tag{24}$$

Then the final form of second order equations is

$$w = \eta - \frac{1}{4}\alpha\eta^2 + \frac{1}{3}\beta\eta_{3x} + \frac{1}{8}\alpha^2\eta^3 + \alpha\beta\left(\frac{3}{16}\eta_x^2 + \frac{1}{2}\eta\eta_{2x}\right) + \frac{1}{10}\beta^2\eta_{4x}, \tag{25}$$

$$\eta_t + \eta_x + \frac{3}{2}\alpha\eta\eta_x + \frac{1}{6}\beta\eta_{3x} - \frac{3}{8}\alpha^2\alpha\beta\left(\frac{23}{24}\eta_x\eta_{2x} + \frac{5}{12}\eta\eta_{3x}\right) + \frac{19}{360}\beta^2\eta_{5x} = 0. \tag{26}$$

Equation (26) was derived by Marchant and Smyth [33] (directly from the set of Euler equations and alternatively from Luke's Lagrangian [32]) and called by the authors the *extended KdV*. This is a second order extension of KdV in dimensionless variables and fixed reference frame. We call it in short **KdV2**. In principle KdV2 solutions should be a better approximation of the solutions to the Boussinesq set than

KdV solutions. They should be, as well, reasonable approximations in a wider range of small parameters α, β.

2.3 Analytic Solutions of KdV and KdV2

KdV gained enormous success as an approximation common for many problems in nonlinear physics. KdV is integrable and has solutions exhibiting a rich variety of properties. The standard derivation of analytic solutions in the form of single solitonic functions (in terms of hyperbolic functions) and periodic functions (Jacobi elliptic functions) is presented in many textbooks or monographs, see, e.g., [1, 8, 20, 36, 39, 41]. It consists in the introduction of the new variable $\xi = x - vt$. Then KdV is transformed to a nonlinear ordinary differential equation (ODE) which can be integrated two times leading to the equation

$$\frac{\beta}{3\alpha}\left(\eta_{\xi}\right)^{2} = -\eta^{3} + 2c_{1}\eta^{2} + r\eta + s, \tag{27}$$

where $c_1 = \frac{v-1}{\alpha}$, r and s are integration constants. The particular case $r = s = 0$ leads to the soliton solution

$$\eta(x,t) = A\,\mathrm{Sech}^{2}\left(\sqrt{\frac{3A}{4}\frac{\alpha}{\beta}}\left[x - t\left(1 + \frac{\alpha}{2}\right)\right]\right). \tag{28}$$

When one is interested only in mathematical properties of KdV solutions A can be an arbitrary positive constant. However, if physical properties are considered A should be close to one, otherwise the resulting solution contradicts the basic assumption for the derivation ($\frac{A}{H} = \alpha \ll 1$).

When integration constants are nonzero a thorough analysis shows the existence of periodic solutions in terms of Jacobi elliptic functions cn^2 (or equivalently dn^2). The solutions have the form (cnoidal wave)

$$\eta(x,t) = A\,\mathrm{cn}^{2}\left[B\left(x - vt\right), m\right] + D, \tag{29}$$

where A, B, D, v are constants and $m \in [0, 1]$ is the elliptic parameter. Constant $D < 0$ is necessary in order to ensure that the volumes of water elevations and depressions with respect to the undisturbed water level are the same (volume conservation condition). When the elliptic parameter $m \to 1$ the distance between crests of cnoidal wave increases to infinity resulting in a soliton solution as the limit. When $m \to 1$ the limiting profile is the usual cosine wave.

KdV possesses one more important property. There exist exact n-soliton solutions which can be derived from the inverse scattering theory, see, e.g., [1, 2, 12, 14, 38].

Not much was known about analytic solutions to KdV2 till recently. In [23] we showed that KdV2 has an exact single soliton solution of the same form as KdV (28) but with different coefficients. The derivation is following. Proceeding similarly as in the KdV case, that is, introducing $\xi = x - vt$ one transforms (26) into ODE. Postulating the solution in the form $\eta(\xi) = A\,\mathrm{Sech}^2(B\,\xi)$ results in an equation of the form

$$C_2\,\mathrm{Sech}^2(B\,\xi) + C_4\,\mathrm{Sech}^4(B\,\xi) + C_6\,\mathrm{Sech}^6(B\,\xi) = 0, \tag{30}$$

where C_2, C_4, C_6 are functions of unknowns A, B, v and coefficients of the KdV2 equation. Equation (30) holds when all C_i vanish simultaneously. Then, solving the set $C_2 = 0$, $C_4 = 0$ and $C_6 = 0$ one obtains formulas for the coefficients A, B, v which determine the solution. Condition $C_6 = 0$ implies a quadratic equation for $z = \dfrac{B^2\beta}{A\,\alpha}$ with solutions

$$z_1 = \frac{43 - \sqrt{2305}}{152} \approx -0.033, \qquad z_2 = \frac{43 + \sqrt{2305}}{152} \approx 0.599. \tag{31}$$

Then the final formulas are

$$A = \frac{z - \frac{3}{4}}{\alpha\, z(\frac{11}{4} - \frac{19}{3}z)}, \quad B = \sqrt{\frac{z - \frac{3}{4}}{\beta(\frac{11}{4} - \frac{19}{3}z)}}, \quad v = 1 + \frac{z - \frac{3}{4}}{(\frac{11}{4} - \frac{19}{3}z)}\left(\frac{2}{3} + \frac{38}{45}\frac{z - \frac{3}{4}}{(\frac{11}{4} - \frac{19}{3}z)}\right). \tag{32}$$

Solutions obtained with $z = z_1$ have to be rejected. In this case B is imaginary, $B = i\bar{B}$. Then $\mathrm{Sech}^2[B(x - vt)] = \left(\cos^2[\bar{B}(x - vt)]\right)^{-1}$. The solution has poles for some arguments, so it has no physical sense.

There is an important difference between solitonic solutions to KdV2 and KdV. There is no freedom for the former ones, and for a given α, β three equations $C_i = 0$ completely determine the coefficients A, B, v of the solutions. For derivation of KdV coefficients the equation analogous to (30) contains only two lower order terms. Then there are only two equations for three coefficients A, B, v. This means that there is one parameter family of solutions. Usually B, v are expressed as functions of positive A which can be arbitrary within some interval (as long as it does not contradict the basic assumption $\frac{A}{h} \ll 1$). Moreover, for KdV2 solitons the ratio $\frac{B^2}{A} = \frac{\alpha}{\beta} z \approx 0.6\frac{\alpha}{\beta}$ and $v \approx 1.1145$ whereas for KdV $\frac{B^2}{A} = 0.75\frac{\alpha}{\beta}$ and $v = 1 + \frac{\alpha}{2}$.

The exact periodic solutions of KdV2 obtained by us in [21] are very fresh. Encouraged by the success of the method used in [23] to derive the soliton solution to KdV2 we postulated periodic solutions to KdV2 in the same form as periodic solutions to KdV (29). Then with a similar procedure as that described above for the solitonic case one arrives at an equation analogous to (30)

$$F_0 + F_2\,\mathrm{cn}^2(B\,\xi) + F_4\,\mathrm{cn}^4(B\,\xi) = 0, \tag{33}$$

where $F_i = F_i(A, B, D, v)$. Then the set of equations $F_i = 0$ supplemented by the volume conservation condition allows us to determine all four unknown coefficients A, B, D, v of the solution as functions of the elliptic parameter m. The condition $F_4 = 0$ gives the same quadratic equation for $z' = \frac{B^2\beta}{A\alpha}\frac{1}{m}$ as the equation $C_6 = 0$ for the solitonic case. Therefore roots z_1', z_2' are the same as z_1, z_2 in (31).

The periodicity of cn function ensures $\lambda/2 = 2K(m)/B$, where λ is the wavelength and $K(m)$ is the complete elliptic integral of the first kind. Then the volume conservation condition

$$\int_0^{\lambda/2} [A\,\mathrm{cn}^2(B\xi, m)]d\xi = 0 \tag{34}$$

yields relations between A, D and m

$$D = -\frac{A}{m}\left(\frac{E(m)}{K(m)} + m - 1\right), \tag{35}$$

where $E(m)$ is the complete elliptic integral. In explicit formulas for coefficients A, B^2, D the factor $\mathrm{EK}(m) = 3\frac{E(m)}{K(m)} + m - 2$ appears. The function $\mathrm{EK}(m)$, $m \in [0, 1]$ has the root $m_s \approx 0.96115$ and is positive when $m < m_s$ and negative when $m > m_s$. Then because two z' roots have different signs there are two branches of KdV2 solutions.

1. The branch with $z' = z_2$. $B^2 > 0$ and then the real B is obtained only when $\mathrm{EK}(m) < 0$, that is when $m > m_s$. Therefore $A > 0$, $D < 0$, and the solution is a 'normal' cnoidal wave with amplitude of crests larger than depressions.
2. Branch with $z' = z_1$. B is real-valued when $m < m_s$. This implies $A < 0, D > 0$, and the solution is an inverted cnoidal profile. Such solutions do not exist for KdV.

For both branches there exist such intervals of m that $B^2 < 0$. However, these solutions (after transforming them to functions of real arguments) exhibit singularities for some arguments and therefore have no physical sense.

For detailed derivation of the analytic periodic solutions of KdV2 and discussion of their properties, see [21].

3 KdV Invariants

It is widely known, see, e.g., [8, Chap. 5], that an equation with a form analogous to the form of the continuity equation

$$\frac{\partial T}{\partial t} + \frac{\partial X}{\partial x} = 0, \tag{36}$$

corresponds to some *conservation law*. In (36) T and X are analogs to density and flux, respectively. Functions T and X may depend on $x, t, \eta, \eta_x, \eta_{2x}, \ldots,$ but not on η_t. The Eq. (36) can be applied, in particular, to KdV and to the equations of the KdV type, such as (47). If functions T and X_x are integrable on $x \in (-\infty, \infty)$ and $\lim_{x\to\pm\infty} X = \text{const}$ (this case corresponds to soliton solutions), then integration of Eq. (36) gives

$$\frac{\mathrm{d}}{\mathrm{d}t}\left(\int_{-\infty}^{\infty} T\,dx\right) = 0 \quad \text{or} \quad \int_{-\infty}^{\infty} T\,dx = \text{const.} \tag{37}$$

$$\text{since} \qquad \int_{-\infty}^{\infty} X_x\,dx = X(\infty, t) - X(-\infty, t) = 0. \tag{38}$$

The same conclusion can be drawn for periodic solutions (cnoidal waves). In this case limits in the integrals (37), (38) have to be replaced by $(a, a + \Lambda)$, where Λ is the wave length of the cnoidal wave and a is arbitrary.

For the KdV equation (20) the first two invariants are easily obtainable. When (20) is presented in the form

$$\frac{\partial \eta}{\partial t} + \frac{\partial}{\partial x}\left(\eta + \frac{3}{4}\alpha\eta^2 + \frac{1}{6}\beta\eta_{xx}\right) = 0, \tag{39}$$

the conservation of mass (volume) law is immediately obtained

$$I^{(1)} = \int_{-\infty}^{\infty} \eta\,dx = \text{const.} \tag{40}$$

Multiplication of (20) by η leads to

$$\frac{\partial}{\partial t}\left(\frac{1}{2}\eta^2\right) + \frac{\partial}{\partial x}\left(\frac{1}{2}\eta^2 + \frac{1}{2}\alpha\eta^3 - \frac{1}{12}\beta\eta_x^2 + \frac{1}{6}\beta\eta\eta_{xx}\right) = 0, \tag{41}$$

which results in the following form of the second invariant

$$I^{(2)} = \int_{-\infty}^{\infty} \eta^2\,dx = \text{const.} \tag{42}$$

Invariants $I^{(1)}$ (40) and $I^{(2)}$ (42) have the same form both in fixed and moving reference frames.

Denoting by $\text{KDV}(x, t)$ the left hand side of (20) and taking

$$3\eta^2 \times \text{KDV}(x, t) - \frac{2}{3}\frac{\beta}{\alpha}\eta_x \times \frac{\partial}{\partial x}\text{KDV}(x, t) = 0 \tag{43}$$

one obtains

$$\frac{\partial}{\partial t}\left(\eta^3-\frac{1}{3}\frac{\beta}{\alpha}\eta_x^2\right)+ \tag{44}$$
$$\frac{\partial}{\partial x}\left(\frac{9}{8}\alpha\eta^4+\frac{1}{2}\beta\eta_{2x}\eta^2-\beta\eta_x^2\eta+\eta^3+\frac{1}{18}\frac{\beta^2}{\alpha}\eta_{2x}^2-\frac{1}{9}\frac{\beta^2}{\alpha}\eta_x\eta_{3x}-\frac{1}{3}\frac{\beta}{\alpha}\eta_x^2\right)=0.$$

This gives the third invariant for KdV (20) in the fixed reference frame

$$I^{(3)}=\int_{-\infty}^{\infty}\left(\eta^3-\frac{1}{3}\frac{\beta}{\alpha}\eta_x^2\right)dx=\text{const.} \tag{45}$$

The same formula is obtained for the third KdV invariant in the moving frame [24].

In the subject literature, see, e.g., [3, 8], $I^{(2)}$ is attributed to conservation of momentum and $I^{(3)}$ to conservation of energy. However, as pointed out in [24] they are not exact momentum and energy, respectively.

For any solutions of KdV preserving their shapes during the motion, that is, for single soliton solutions and cnoidal solutions, integrals of any power of the solution $\eta(x,t)$ and any power of its arbitrary derivative with respect to x are invariants. That is,

$$I^{(p,k)}=\int_{-\infty}^{\infty}(\eta_{kx})^p dx=\text{const}, \tag{46}$$

where $p\in\mathbb{R}$ is an arbitrary real number, and $k=0,1,2,\ldots$. An arbitrary linear combination of $I^{(p,k)}$ is an invariant, as well.

4 KdV2 Adiabatic Invariants—Direct Method

We now consider the KdV2 equation [24, Eq. (1)]

$$\eta_t+\eta_x+\frac{3}{2}\alpha\,\eta\eta_x+\frac{1}{6}\beta\,\eta_{3x} \tag{47}$$
$$-\frac{3}{8}\alpha^2\eta^2\eta_x+\alpha\beta\left(\frac{23}{24}\eta_x\eta_{2x}+\frac{5}{12}\eta\eta_{3x}\right)+\frac{19}{360}\beta^2\eta_{5x}=0,$$

named as the *extended KdV* by Marchant and Smyth [33, Eq. (2.8)]. They derived (47) both from Euler's hydrodynamic equations and Luke's Lagrangian [32]. The equation has been considered by several authors, see, e.g., [6, 21, 23–26, 28, 33, 34]. As stated above, we call it KdV2.

In [24], we note that $I^{(1)}=\int_{-\infty}^{\infty}\eta\,dx$ is the exact invariant of (47) representing the conservation of mass as it does for KdV.

4.1 Second Invariant

The second invariant of KdV, $I^{(2)} = \int_{-\infty}^{\infty} \eta^2\, dx$ is **not** an invariant of KdV2, because, see [24, Sec. III B], after multiplication of Eq. (47) by η one obtains

$$\frac{\partial}{\partial t}\left(\frac{1}{2}\eta^2\right) + \frac{\partial}{\partial x}\left[\frac{1}{2}\eta^2 + \frac{1}{2}\alpha\eta^3 + \frac{1}{6}\beta\left(-\frac{1}{2}\eta_x^2 + \eta\eta_{2x}\right) - \frac{3}{32}\alpha^2\eta^4 \right. \tag{48}$$
$$\left. + \frac{19}{360}\beta^2\left(\frac{1}{2}\eta_{xx}^2 - \eta_x\eta_{3x} + \eta\eta_{4x}\right) + \frac{5}{12}\alpha\beta\,\eta^2\eta_{2x}\right] + \frac{1}{8}\alpha\beta\,\eta\eta_x\eta_{2x} = 0\,.$$

It is not possible to express the final term in (48) as $\frac{\partial}{\partial x}X(\eta, \eta_x, \ldots)$. Then contrary to KdV case the quantity $\int_{-\infty}^{+\infty} \eta^2 dx$ is not conserved. There are no exact higher order invariants of (47) as well.

It is possible, though, to determine approximate invariants of (47), whose terms which violate the invariance are of the third order in α, β. Our simple method allows us to determine such approximate invariants without big effort. It works by forming an equation which contains functions T and X by means of some manipulations with KdV2. In this equation there are terms in X which are non-integrable with respect to x similarly as the last term in (48). By adding a linear combination of the form $(c_1\alpha + c_2\beta) \times KdV2(x, t)$ to that equation, dropping the third order terms we can determine c_1 and/or c_2 such that the non-integrable terms cancel. (Equivalently, we add a linear combination of the form $(c_1\alpha + c_2\beta) \times KdV(x, t)$ without dropping any term.) This action yields a new T' function and an approximate conservation law for $\int_{-\infty}^{\infty} T' dx$.

The first approximate invariant can be obtained by adding to (48) Eq. (47) multiplied by $c_1\alpha\eta^2$, neglecting terms of third-order and selecting a proper value of c_1 in order to cancel the term $\frac{1}{8}\alpha\beta\,\eta\eta_x\eta_{2x}$. When this is done we are left with the expression

$$c_1\alpha\eta_t\eta^2 + c_1\alpha\eta^2\eta_x + c_1\frac{3}{2}\alpha^2\eta^3\eta_x + c_1\frac{1}{6}\alpha\beta\eta^2\eta_{3x}. \tag{49}$$

In integration over x of (49), terms $c_1\alpha\eta^2\eta_x$ and $c_1\frac{1}{6}\alpha\beta\eta^2\eta_{3x}$ are integrable with respect to x and then can be included into the flux function X.

The last term in (49) can be transformed to $-\frac{1}{3}c_1\alpha\beta\eta\eta_x\eta_{2x}$. It cancels with $\frac{1}{8}\alpha\beta\,\eta\eta_x\eta_{2x}$ when $c_1 = \frac{3}{8}$. Then the first term in (49) yields

$$c_1\alpha\eta_t\eta^2 = \frac{\partial}{\partial t}\left(\frac{1}{8}\alpha\eta^3\right). \tag{50}$$

Due to integrability of the other terms the approximate invariant of KdV2 is obtained as ($\frac{1}{2}$ is omitted)

$$I_{\rm ad}^{(2\alpha)} = \int_{-\infty}^{\infty} \left(\eta^2 + \frac{1}{4}\alpha\,\eta^3\right) dx \approx \text{const.} \tag{51}$$

However, there is another way to remove the last term in (48) and get an alternative form of the second approximate invariant. This goal can be achieved by adding to (48) Eq. (47) multiplied by $c_2\beta\eta_{2x}$, dropping again third-order terms and selecting a proper value of c_2 to remove the term $\frac{1}{8}\alpha\beta\,\eta\eta_x\eta_{2x}$. Then new terms are

$$c_2\beta\eta_t\eta_{2x} + c_2\beta\eta_x\eta_{2x} + c_2\frac{3}{2}\alpha\beta\eta\eta_x\eta_{2x} + c_2\frac{1}{6}\beta^2\eta_{2x}\eta_{3x}. \tag{52}$$

In integration over x of (52), the terms $c_2\beta\eta_x\eta_{2x}$ and $c_2\frac{1}{6}\beta^2\eta_{2x}\eta_{3x}$ are integrable with respect to x and then can be included into X. The cancellation of non-integrable terms

$$c_2\frac{3}{2}\alpha\beta\eta\eta_x\eta_{2x} + \frac{1}{8}\alpha\beta\,\eta\eta_x\eta_{2x} = 0$$

implies $c_2 = -\frac{1}{12}$.

Integration of the first term in (52) over x gives

$$\int_{-\infty}^{\infty} c_2\beta\eta_t\eta_{2x}dx = c_2\beta\left(\eta_t\eta_x|_{-\infty}^{\infty} - \int_{-\infty}^{\infty}\eta_{tx}\eta_x\right) = -c_2\beta\int_{-\infty}^{\infty}\frac{\partial}{\partial t}\left(\frac{1}{2}\eta_x^2\right). \tag{53}$$

Since terms with $\eta_x\eta_{2x}$ and $\eta_{2x}\eta_{3x}$ can be expressed as $\left(-\frac{1}{2}\eta_x^2\right)_x$ and $\left(-\frac{1}{2}\eta_{2x}^2\right)_x$, respectively, one gets finally

$$\frac{\partial}{\partial t}\int_{-\infty}^{\infty}\frac{1}{2}\left(\eta^2 + \frac{1}{12}\beta\eta_x^{\,2}\right)dx + F(\eta,\eta_x,\eta_{2x})|_{-\infty}^{\infty} = O(\alpha^3), \tag{54}$$

where $F(\eta,\eta_x,\eta_{2x})$ results from the integration of the flux term. Since solutions to KdV2 are either solitonic or periodic then this term vanishes.

This gives an adiabatic (approximate) invariant of KdV2 (47) in the form

$$I_{\rm ad}^{(2\beta)} = \int_{-\infty}^{\infty}\left(\eta^2 + \frac{1}{12}\beta\,\eta_x^{\,2}\right)dx \approx \text{const.} \tag{55}$$

The existence of two independent adiabatic invariants $I_{\rm ad}^{(2\alpha)}$ and $I_{\rm ad}^{(2\beta)}$ means also that

$$I_{\rm ad}^{(2)} = \varepsilon\, I_{\rm ad}^{(2\alpha)} + (1-\varepsilon)I_{\rm ad}^{(2\beta)} = \int_{-\infty}^{\infty}\left(\eta^2 + \varepsilon\frac{1}{12}\alpha\eta^3 + (1-\varepsilon)\frac{1}{12}\beta\,\eta_x^{\,2}\right)dx \tag{56}$$

is an adiabatic invariant for any ε, that is, there exists one parameter family of adiabatic second invariant of KdV2.

4.2 Third Invariant

In order to find the third invariant for KdV2 one can follow the procedure described in Sect. 3, Eqs. (43)–(45), but for KdV2 equation. Take

$$3\eta^2 \times \mathrm{KDV2}(x,t) - \frac{2}{3}\frac{\beta}{\alpha}\eta_x \times \frac{\partial}{\partial x}\mathrm{KDV2}(x,t) = 0 \tag{57}$$

and consider a simpler case, when $\beta = \alpha$. The result is

$$\begin{aligned}
&\frac{\partial}{\partial t}\left(\eta^3 - \frac{1}{3}\eta_x^2\right) + \frac{\partial}{\partial x}\left(\eta^3 - \frac{1}{3}\eta_x^2 + \alpha\frac{9}{8}\eta^4 - \alpha^2\frac{9}{40}\eta^5\right) \\
&+\alpha\left(-\eta_x^3 - \eta\eta_x\eta_{2x} + \frac{1}{2}\eta^2\eta_{3x}\right) + \alpha^2\left(\frac{1}{2}\eta\eta_x^3 + \frac{25}{8}\eta^2\eta_x\eta_{2x} - \frac{23}{36}\eta_x\eta_{2x}^2\right. \\
&\qquad\left. + \frac{5}{4}\eta^3\eta_{3x} - \frac{11}{12}\eta_x^2\eta_{3x} - \frac{5}{18}\eta\eta_x\eta_{4x} + \frac{19}{120}\eta^2\eta_{5x}\right).
\end{aligned} \tag{58}$$

In (58) we omitted terms which vanish under integration over x. All terms in the second and third rows of (58) are non-integrable. However, taking an integral of the form $\int_{-\infty}^{\infty}\ldots dx$ and integrating by parts they can be reduced to two types of non-integrable terms. All terms in the bracket with α become proportional to $\eta\eta_x\eta_{2x}$. All terms in the bracket with α^2 reduce to $\eta\eta_x\eta_{2x}$ and $\eta_x\eta_{2x}^2$. Then using procedures described above for second adiabatic invariant, that is, by adding to (58) the KdV multiplied by proper factors one can cancel these non-integrable terms. The added terms supply additional terms in the T function. As in the case of second invariant this action is not unique and there is some freedom in the form of final adiabatic invariant. One of admissible forms is

$$I_{\mathrm{ad}}^{(3)} = \int_{-\infty}^{\infty}\left(\eta^3 - \frac{1}{3}\eta_x^2 - \alpha\eta^4 + \frac{7}{12}\alpha\,\eta\eta_x^2\right)dx. \tag{59}$$

It should be noted that the first two terms in (59) are the same as the third KdV invariant.

The method presented enables us to derive higher order adiabatic invariants, as well.

5 Near-Identity Transformation for KdV2 in Fixed Frame

Our research was performed in the fixed reference frame. It was motivated by two facts. First, already pointed out in [24, Eq. (39)], even for KdV energy has **noninvariant form** (Ali and Kalisch [3] showed this fact in dimension variables). Second, our purpose is to study invariants, and approximate invariants not only for KdV and KdV2, but also for the KdV2 equation with non-flat bottom, derived in [23, 26]. For this equation it is only the fixed reference frame that makes sense.

Second order versions of KdV type equations are not unique since there exist transformations which transform the given equation into an equation of the same form but with some coefficients altered. These equations are asymptotically equivalent, that is, their solutions converge to the same form when small parameters tend to zero. Therefore such transformation, called *near-identity transformation* (NIT), is often used to convert higher order nonlinear differential equations to their asymptotically equivalent forms which can be integrable. Such NIT was first introduced by Kodama [29, 30] and then used and generalized by many authors, see, e.g., [9–11, 13, 15, 16, 19, 34]. Below we apply NIT in the form suitable for the KdV2 equation.

Introduced below is the near-identity transformation in the form used by the authors of [9]

$$\eta = \eta' + \alpha a \eta'^2 + \beta b \eta'_{xx} + \ldots, \tag{60}$$

where a, b are some constants. (Here, we choose $+$ sign. The inverse transformation, up to terms of second order, is $\eta' = \eta - \alpha a \eta^2 - \beta b \eta_{xx} + \ldots$).

NIT should preserve the form of the KdV2 (47), at most altering some coefficients. Then it is possible to choose coefficients a, b of NIT such that the transformed equation possesses a Hamiltonian (see the consequences in the Sect. 5.2).

Insertion (60) into (47) yields (terms of order higher than the second in α, β are neglected)

$$\begin{aligned} \eta'_t + \eta'_x + \alpha\left[\left(\frac{3}{2}+2a\right)\eta'\eta'_x + 2a\eta'\eta'_t\right] + \beta\left[\left(\frac{1}{6}+b\right)\eta'_{3x} + b\eta'_{xxt}\right] \\ + \alpha\beta\left\{\left[\left(\frac{23}{24}+a+\frac{3}{2}b\right)\eta'_x\eta'_{2x}\right] + \left[\left(\frac{5}{12}+\frac{1}{3}a+\frac{3}{2}b\right)\eta'\eta'_{3x}\right]\right\} \\ + \alpha^2\left(-\frac{3}{8}+\frac{9}{2}a\right)\eta'^2\eta'_x + \beta^2\left[\left(\frac{19}{360}+\frac{1}{6}b\right)\eta'_{5x}\right] = 0. \end{aligned} \tag{61}$$

Since terms with time derivatives (η'_t, η'_{xxt}) appear in first order with respect to small parameters we can replace them by appropriate expressions obtained from KdV2 (47) limited to first order, that is from KdV (20)

$$\eta'_t = -\eta'_x - \frac{3}{2}\alpha\eta'\eta'_x - \frac{1}{6}\beta\eta'_{3x} \tag{62}$$

and

$$\eta'_{xxt} = \partial_{xx}\left(-\eta'_x - \frac{3}{2}\alpha\eta'\eta'_x - \frac{1}{6}\beta\eta'_{3x}\right) = -\eta'_{3x} - \frac{3}{2}\alpha(3\eta'_x\eta'_{2x} + \eta'\eta'_{3x}) - \frac{1}{6}\beta\eta'_{5x}. \tag{63}$$

Inserting (62) and (63) into (61) one obtains

$$\begin{aligned}\eta'_t + \eta'_x + \frac{3}{2}\alpha\eta'\eta'_x + \frac{1}{6}\beta\eta'_{3x} + \alpha^2\left(-\frac{3}{8} + \frac{3}{2}a\right)\eta'^2\eta'_x \\ + \alpha\beta\left[\left(\frac{23}{24} + a - 3b\right)\eta'_x\eta'_{2x} + \frac{5}{12}\eta'\eta'_{3x}\right] + \frac{19}{360}\beta^2\eta'_{5x} = 0.\end{aligned} \tag{64}$$

Equation (64) for η' has the same form as KdV2 (47) with only two coefficients altered. The coefficient in front of the term with $\alpha^2\eta^2\eta_x$ is changed from $-\frac{3}{8}$ to $-\frac{3}{8} + \frac{3}{2}a$ and the coefficient in front of the term with $\alpha\beta\eta_x\eta_{2x}$ is changed from $\frac{23}{24}$ to $\frac{23}{24} + a - 3b$.

5.1 NIT—Second Adiabatic Invariant

For the NIT-transformed KdV2 equation (64) one can find the second invariant in the same way as previously, that is multiplying (64) by η' and requiring that the coefficient in front of the non-integrable term vanishes. This gives

$$\int_{-\infty}^{\infty}\eta'\left[\frac{5}{12}\eta'\eta'_{3x} + \left(\frac{23}{24} + a - 3b\right)\eta'_x\eta'_{2x}\right]dx = 0. \tag{65}$$

Since

$$\int_{-\infty}^{\infty}\eta'^2\eta'_{3x}\,dx = -2\int_{-\infty}^{\infty}\eta'\eta'_x\eta'_{2x}\,dx \tag{66}$$

one obtains

$$\left(-2\frac{5}{12} + \frac{23}{24} + a - 3b\right)\int_{-\infty}^{\infty}\eta'\eta'_x\eta'_{xx}\,dx = 0 \implies a - 3b + \frac{1}{8} = 0. \tag{67}$$

Then under the condition

$$a - 3b = -\frac{1}{8} \tag{68}$$

the integral $\int_{-\infty}^{\infty}\eta'^2 dx$ is the exact invariant of the Eq. (64).

Using inverse NIT

$$\eta' = \eta - \alpha a\eta^2 - \beta b\eta_{xx} + \ldots, \tag{69}$$

and neglecting higher order terms, one gets

$$\int_{-\infty}^{\infty} \eta'^2 dx \approx \int_{-\infty}^{\infty} \left[\eta^2 - 2\alpha a \eta^3 - 2\beta b \eta \eta_{xx}\right] = \int_{-\infty}^{\infty} \left[\eta^2 - 2\alpha a \eta^3 + 2\beta b \eta_x^2\right] dx, \quad (70)$$

where the last term was obtained through integration by parts. The r.h.s. of (70) is the most general form of the second adiabatic invariant of KdV2 under the condition (68), that is, one parameter family of adiabatic invariants

$$I_{\rm ad}^{(2)} = \int_{-\infty}^{\infty} \left[\eta^2 - 2\alpha a \eta^3 + 2\beta b \eta_x^2\right] dx \approx \text{const.} \quad (71)$$

In particular, with $a = 0$, $b = \frac{1}{24}$

$$I_{\rm ad}^{(2)} = \int_{-\infty}^{\infty} \left(\eta^2 + \frac{1}{12}\beta \eta_x^2\right) dx = I_{\rm ad}^{(2\beta)} \quad (72)$$

and with $b = 0,\ a = -\frac{1}{8}$

$$I_{\rm ad}^{(2)} = \int_{-\infty}^{\infty} \left(\eta^2 + \frac{1}{4}\alpha \eta^3\right) dx = I_{\rm ad}^{(2\alpha)}. \quad (73)$$

These adiabatic invariants are the same as those obtained in the direct way in (51) and (55).

The above formulas come from NIT (60) in which the sign + was used. However, if in (60) the sign − is chosen then the condition (68) is replaced by $a - 3b = \frac{1}{8}$. The signs of the inverse NIT become opposite and then the final forms of adiabatic invariants remains the same as in (71)–(73).

5.2 NIT—Third Adiabatic Invariant

NIT-transformed KdV2 (64) describes waves in the fixed frame. In order to determine its Hamiltonian form let us convert (64) to a moving frame by transformation

$$\bar{x} = x - t, \quad \bar{t} = t, \quad \partial_x = \partial_{\bar{x}}, \quad \partial_t = -\partial_{\bar{x}} + \partial_{\bar{t}}. \quad (74)$$

Then (64) can be written in more general form as

$$\eta_{\bar{t}} + \alpha A \eta \eta_{\bar{x}} + \beta B \eta_{3\bar{x}} + \alpha^2 A_1 \eta^2 \eta_{\bar{x}} + \beta^2 B_1 \eta_{5\bar{x}} + \alpha\beta \left(G_1 \eta \eta_{3\bar{x}} + G_2 \eta_{\bar{x}} \eta_{2\bar{x}}\right) = 0, \quad (75)$$

where

$$A = \frac{3}{2}, \quad B = \frac{1}{6}, \quad A_1 = -\frac{3}{8} + \frac{3}{2}a, \quad B_1 = \frac{19}{360}, \quad G_1 = \frac{5}{12} \quad G_2 = \frac{23}{24} + a - 3b. \tag{76}$$

In the following we drop bars over t and x, remembering that now we work in the moving reference frame.

In particular, the parameters a, b of NIT can be chosen such that

$$G_2 = 2G_1. \tag{77}$$

In this case the Hamiltonian for the Eq. (75) exists. The condition (77) with (76) gives

$$\frac{23}{24} + a - 3b = 2\frac{5}{12} \qquad \Longrightarrow \qquad a - 3b = -\frac{1}{8}.$$

This is the same condition as (68). This condition supplies one parameter family of NIT, assuring Hamiltonian form of the NIT-transformed KdV2 (75) in the moving frame.

This Hamiltonian form is

$$\eta'_t = \frac{\partial}{\partial x}\left(\frac{\delta \mathcal{H}}{\delta \eta'}\right), \tag{78}$$

where the Hamiltonian $H = \int_{-\infty}^{\infty} \mathcal{H}\, dx$ has density

$$\mathcal{H} = -\frac{1}{6}\alpha A \eta'^3 + \frac{1}{2}\beta B \eta_x'^2 - \frac{1}{12}\alpha^2 A_1 \eta'^4 - \frac{1}{2}\beta^2 B_1 \eta_{xx}'^2 + \frac{1}{2}\alpha\beta G_1 \eta' \eta_x'^2. \tag{79}$$

Since $\mathcal{H} = \mathcal{H}(\eta', \eta'_x, \eta'_{xx})$, then the *functional derivative* in (78) is

$$\begin{aligned} \frac{\delta \mathcal{H}}{\delta \eta'} &= \frac{\partial \mathcal{H}}{\partial \eta'} - \frac{\partial}{\partial x}\frac{\partial \mathcal{H}}{\partial \eta'_x} + \frac{\partial^2}{\partial x^2}\frac{\partial \mathcal{H}}{\partial \eta'_{xx}} \\ &= -\frac{1}{2}\alpha A \eta'^2 - \beta B \eta'_{xx} - \frac{1}{3}\alpha^2 A_1 \eta'^3 - \alpha\beta G_1\left(\frac{1}{2}\eta_x'^2 + \eta'\eta'_{xx}\right) - \beta^2 B_1 \eta'_{4x}. \end{aligned} \tag{80}$$

Insertion (80) into (78) yields

$$\eta'_t = -\alpha A \eta' \eta'_x - \beta B \eta'_{3x} - \alpha^2 A_1 \eta'^2 \eta'_x + \beta^2 B_1 \eta'_{5x} - \alpha\beta G_1 (2\eta'_x \eta'_{xx} + \eta' \eta'_{xx}). \tag{81}$$

We see that the Hamiltonian form of KdV2 in the moving frame exists under the condition that the coefficient at the term $\eta'_x \eta'_{xx}$ is two times larger that the coefficient at the term $\eta' \eta'_{xxx}$. This is obtained by a proper choice of a, b parameters of NIT, which is the condition (68).

Now, the Hamiltonian is the exact constant of motion for the NIT-transformed equation (75) under the condition (68)

$$\int_{-\infty}^{\infty}\left[-\frac{1}{6}\alpha A \eta'^3 + \frac{1}{2}\beta B \eta_x'^2 - \frac{1}{12}\alpha^2 A_1 \eta'^4 - \frac{1}{2}\beta^2 B_1 \eta_{xx}'^2 + \frac{1}{2}\alpha\beta G_1 \eta' \eta_x'^2\right] dx = \text{const}. \tag{82}$$

In order to obtain the adiabatic invariant of the original Eq. (47) it is necessary to perform the inverse NIT, that is

$$\eta' = \eta - \alpha a \eta^2 - \beta b \eta_{xx} \tag{83}$$

and then to neglect in the Hamiltonian density all higher order terms. This yields

$$\begin{aligned}\mathscr{H} = & -\frac{1}{6}\alpha A \eta^3 + \frac{1}{2}\beta B \eta_x^2 + \alpha^2\left(\frac{1}{2}aA - \frac{1}{12}A_1\right)\eta^4 \\ & + \beta^2\left(-\frac{1}{2}B_1\eta_{2x}^2 - bB\eta_x\eta_{3x}\right) + \alpha\beta\left[\left(\frac{1}{2}G_1 - 2aB\right)\eta\eta_x^2 + \frac{1}{2}bA\eta^2\eta_{2x}\right],\end{aligned} \tag{84}$$

with the condition (68).

Now, we restore the original notation $A = \eta$ and numerical values of coefficients (76). Using relations which come from integration by parts

$$\int_{-\infty}^{\infty} \eta_x \eta_{3x}\, dx = -\int_{-\infty}^{\infty} \eta_{2x}^2\, dx, \quad \int_{-\infty}^{\infty} \eta^2 \eta_{2x}\, dx = -2\int_{-\infty}^{\infty} \eta\eta_x^2\, dx$$

and changing irrelevant sign one obtains finally

$$\begin{aligned}I_{\text{ad}}^{(3)} = \int_{-\infty}^{\infty} \Bigg[& \frac{1}{4}\alpha\eta^3 - \frac{1}{12}\beta\eta_x^2 - \alpha^2\left(\frac{1}{32} + \frac{5}{8}a\right)\eta^4 + \beta^2\left(\frac{19}{720} - \frac{1}{6}b\right)\eta_{2x}^2 \\ & + \alpha\beta\left(\frac{1}{3}a + \frac{3}{2}b - \frac{5}{24}\right)\eta\eta_x^2\Bigg].\end{aligned} \tag{85}$$

We obtain one parameter family (68) of adiabatic invariants related to energy.

In a particular case, when in (68), we set $a = 0, \quad b = \frac{1}{24}$ and then

$$I_{\text{ad}}^{(3)} = \int_{-\infty}^{\infty} \left[\frac{1}{4}\alpha\eta^3 - \frac{1}{12}\beta\eta_x^2 - \frac{1}{32}\alpha^2\eta^4 + \frac{7}{720}\beta^2\eta_{2x}^2 - \frac{7}{48}\alpha\beta\eta\eta_x^2\right] dx. \tag{86}$$

When, in (68), we set $a = -\frac{1}{8}, \quad b = 0$, then we obtain

$$I_{\text{ad}}^{(3)} = \int_{-\infty}^{\infty} \left[\frac{1}{4}\alpha\eta^3 - \frac{1}{12}\beta\eta_x^2 + \frac{3}{64}\alpha^2\eta^4 + \frac{19}{720}\beta^2\eta_{2x}^2 - \frac{1}{4}\alpha\beta\eta\eta_x^2\right] dx. \tag{87}$$

Another particular form of (85) can be obtained when one sets

$$\frac{19}{720} - \frac{1}{6}b = 0 \implies b = \frac{19}{120}, \quad a = \frac{7}{20}.$$

Then, (85) reduces to

$$I_{\mathrm{ad}}^{(3)} = \int_{-\infty}^{\infty} \left[\frac{1}{4}\alpha\eta^3 - \frac{1}{12}\beta\eta_x^2 - \frac{1}{4}\alpha^2\eta^4 + \frac{7}{240}\alpha\beta\eta\eta_x^2 \right] dx. \tag{88}$$

In a similar way one can set

$$\frac{1}{3}a + \frac{3}{2}b - \frac{5}{24} = 0 \quad \Longrightarrow \quad b = \frac{1}{10}, \quad a = \frac{7}{40}.$$

In this case the adiabatic invariant has the form

$$I_{\mathrm{ad}}^{(3)} = \int_{-\infty}^{\infty} \left[\frac{1}{4}\alpha\eta^3 - \frac{1}{12}\beta\eta_x^2 - \frac{9}{64}\alpha^2\eta^4 + \frac{7}{720}\beta^2\eta_{2x}^2 \right] dx. \tag{89}$$

5.3 *Momentum and Energy for KdV2*

Relations between invariants and conservation laws are not as simple as might be expected, even for KdV. In this subsection we present these relations for motion in a fixed reference frame. Expressions of energy for KdV and KdV2 in the moving frame can be found in [24, 28].

5.3.1 KdV Case

The first KdV invariant, $\int_{-\infty}^{\infty} \eta\, dx = \text{const}$, represents volume (mass) conservation of the incompressible fluid.

When components of momentum are calculated as integrals over the fluid volume from momentum density the results are as follows.

$$p_x = p_0 \int_{-\infty}^{\infty} \left[\eta + \frac{3}{4}\alpha\eta^2 \right] dx = p_0 \left[I_1 + \frac{3}{4}\alpha\, I_2 \right] \quad \text{and} \quad p_y = 0, \tag{90}$$

where p_0 is a constant in units of momentum. Since the vertical component of the momentum is zero and the horizontal component is expressed by the two lowest invariants we have the conservation of momentum law.

The total energy in the fixed frame is, see, e.g., [24, Eq. (39)] (E_0 is a constant in energy units)

$$\begin{aligned} E_{\mathrm{tot}} &= E_0 \int_{-\infty}^{\infty} \left(\alpha\eta + (\alpha\eta)^2 + \frac{1}{4}(\alpha\eta)^3 \right) dx \\ &= E_0 \left(\alpha I^{(1)} + \alpha^2 I^{(2)} + \frac{1}{4}\alpha^2 I^{(3)} + \frac{1}{12}\alpha^2\beta \int_{-\infty}^{\infty} \eta_x^2\, dx \right). \end{aligned} \tag{91}$$

The energy (91) in the fixed reference frame **has noninvariant form**. The last term in (91) generates tiny deviations from energy conservation only when η_x changes in time in the soliton frame of reference, which occurs during soliton collisions only.

5.3.2 KdV2 Case

Volume conservation, $I_1 = \int_{-\infty}^{\infty} \eta \, dx = \text{const}$, is fulfilled for KdV2, too.

Calculation of momentum components within second order approximation of Euler's equations gives also a vanishing vertical component $p_y = 0$. For a horizontal component one gets

$$p_x = p_0 \int_{-\infty}^{\infty} \left(\eta + \frac{3}{4}\alpha\eta^2 - \frac{1}{8}\alpha^2\eta^3 - \frac{7}{48}\alpha\beta\eta_x^2 \right) dx \tag{92}$$
$$= p_0 \left[I_1 + \frac{3}{4}\alpha \int_{-\infty}^{\infty} \left(\eta^2 - \frac{1}{6}\alpha\eta^3 - \frac{7}{36}\beta\eta_x^2 \right) dx \right].$$

The total momentum of the fluid is composed of two terms. The first is proportional to the volume. The second, an integral in the lower row of (92), contains the same functional terms η^2, η^3, η_x^2 as the expressions for the second adiabatic invariants (56) and (71) but with slightly different coefficients. Analogously to the KdV case (90) one can write

$$p_{x(\mathrm{ad})} \approx p_0 \left[I_1 + \frac{3}{4}\alpha I_{\mathrm{ad}}^{(2)} \right]. \tag{93}$$

We will see in Sect. 6 that $p_{x(\mathrm{ad})}$, given by (93) has much smaller deviations from a constant value than p_x given by (92).

Energy, $E_{\mathrm{tot}} = T + V$, for the system governed by KdV2 is, see, e.g., [24, Eq. (91)],

$$E_{\mathrm{tot}} = E_0 \int_{-\infty}^{\infty} \left(\alpha\eta + (\alpha\eta)^2 + \frac{1}{4}(\alpha\eta)^3 - \frac{3}{32}(\alpha\eta)^4 - \frac{7}{48}\alpha^3\beta\eta\eta_x^2 \right) dx. \tag{94}$$

This expression can be written as

$$E_{\mathrm{tot}} = E_0 \left[\alpha I_1 + \alpha^2 I_{\mathrm{ad}}^{2\beta} + \alpha^2 \int_{-\infty}^{\infty} \left(\frac{1}{4}\alpha\eta^3 - \frac{1}{12}\beta\eta_x^2 - \frac{3}{32}\alpha^2\eta^4 - \frac{7}{48}\alpha\beta\eta\eta_x^2 \right) dx \right]$$
$$\approx E_0\,\alpha \left[I_1 + \alpha \left(I_{\mathrm{ad}}^{2\beta} + \alpha I_{\mathrm{ad}}^{3} \right) \right], \tag{95}$$

where $I_{\mathrm{ad}}^{2\beta}$ is given by (55) and I_{ad}^{3} was chosen in the form (88). Equation (95) shows that the energy of the system described by KdV2 in a fixed frame is approximately given by the sum of exact first invariant and combination of second and third adiabatic invariants. Since there is one parameter freedom in these adiabatic invariants other particular approximate formulas for the energy are admissible, as well. Because of the approximate character of adiabatic invariants the energy of the system is not a conserved quantity. When motion of several solitons is considered the largest changes in the energy occur when solitons change their shapes during collisions, see, e.g., [24, Fig. 4].

6 Numerical Tests

One might question how good these invariants are. The calculations given below offer some insight.

To start with we calculated the time evolution, governed by Eq. (47), for three particular waves. The finite difference method (FDM) of Zabusky [43], generalized for precise calculation of higher derivatives [23, 26] was used. The finite element method (FEM) used for the same problems in [27] give the same results for soliton's motion. 1-, 2- and 3-soliton solutions of KdV were chosen as initial conditions. For the 3-soliton solution the amplitudes were chosen to be 1.0, 0.6 and 0.3, for the 2-soliton solution the amplitudes were chosen as 1.0 and 0.3 and for the single soliton the chosen amplitude was 1.0. The profiles of these waves evolving according to (47) at some instants are presented in Fig. 2. Vertical shifts by 0.2 and horizontal shifts by 30 were used on the figure to avoid overlaps. All these results were obtained for small parameters $\alpha = \beta = 0.1$.

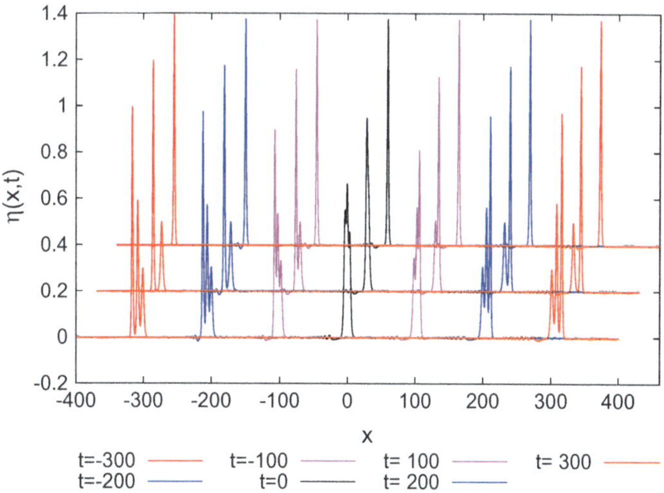

Fig. 2 Time evolution of initially 1-, 2- and 3-soliton KdV solution according to KdV2 (47). Reproduced with permission from [21]. Copyright (2017) by Elsevier

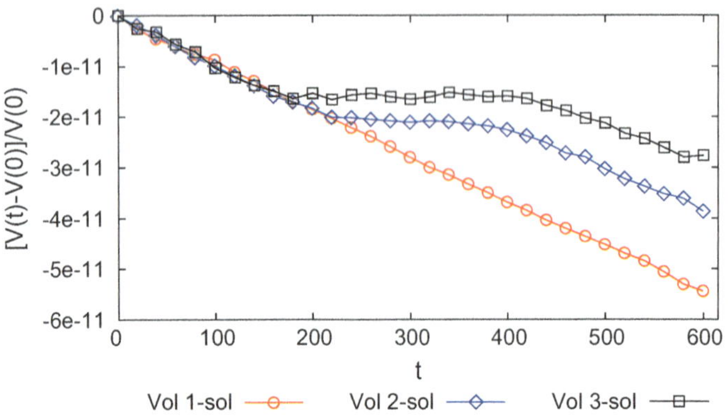

Fig. 3 Numerical precision of the volume conservation law for the three waves displayed in Fig. 2. Reproduced with permission from [21]. Copyright (2017) by Elsevier

Since the volume should be conserved exactly its presentation can verify the precision of numerical evolution. The numerical values of this invariant shown in Fig. 3 are constant up to 10 digits.

6.1 Momentum (Non)Conservation and Adiabatic Invariant $I_{ad}^{(2)}$

To study approximate invariants $I_{\mathrm{ad}}^{(2\beta)}$ and $I_{\mathrm{ad}}^{(2\alpha)}$ we write each of them as the sum of two terms

$$I_{\mathrm{ad}}^{(2\alpha)} = \int_{-\infty}^{\infty} \eta^2\, dx + \int_{-\infty}^{\infty} \frac{1}{4}\alpha\, \eta^3\, dx =: Ie(t) + Ia(t), \tag{96}$$

$$I_{\mathrm{ad}}^{(2\beta)} = \int_{-\infty}^{\infty} \eta^2\, dx + \int_{-\infty}^{\infty} \frac{1}{12}\beta\, \eta_x^2\, dx =: Ie(t) + Ib(t). \tag{97}$$

The first terms in (96) and (97) are the same as the exact KdV invariant.

The changes of adiabatic invariants $I_{\mathrm{ad}}^{(2\alpha)}$ (96) and $I_{\mathrm{ad}}^{(2\beta)}$ (97) presented in Fig. 4 correspond to waves displayed in Fig. 2. In this scale both adiabatic invariants look perfectly constant. To verify how good these invariants are we show how they change with respect to the initial values.

Figure 5 shows changes in the quantities Ie, Ia and Ib for all three 1-, 2-, and 3-soliton waves presented in Fig. 2. These relative changes are defined as

$$Ie = \frac{Ie(t) - Ie(0)}{Ie(0) + Ia(0)}, \quad Ia = \frac{Ia(t) - Ia(0)}{Ie(0) + Ia(0)}, \quad Ib = \frac{Ib(t) - Ib(0)}{Ie(0) + Ia(0)}.$$

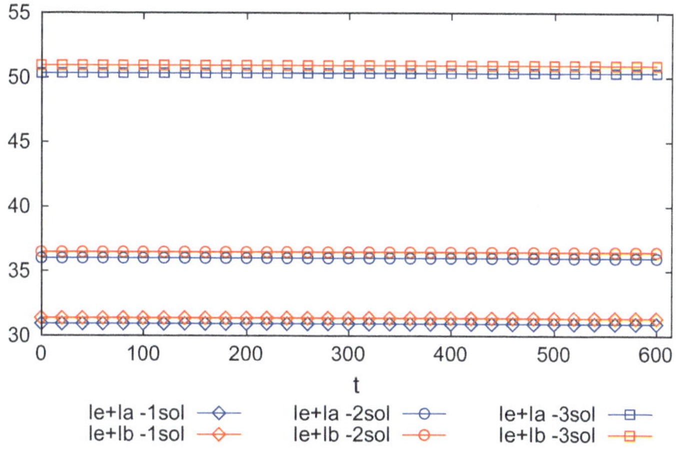

Fig. 4 Absolute values of the adiabatic invariants (96) and (97) for the time evolution shown in Fig. 2. Reproduced with permission from [21]. Copyright (2017) by Elsevier

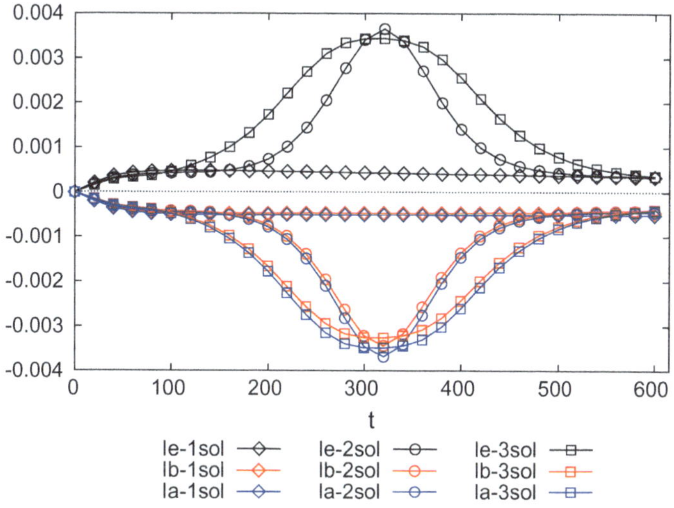

Fig. 5 Relative changes of *Ia* and *Ib* as a functions of time for the three waves presented in Fig. 2. Reproduced with permission from [21]. Copyright (2017) by Elsevier

The figure shows that the corrections *Ia*, *Ib* to the KdV invariant *Ie* have almost the same absolute values as *Ie* but with opposite sign. Therefore one can expect that their summations with *Ie* should only produce small variations of approximate invariants $I^{(2\alpha)}_{\rm ad}$ and $I^{(2\beta)}_{\rm ad}$.

Figure 6 confirms this expectation. For long term evolution, the relative variations of presented approximate invariants are less than the order of 0.00025.

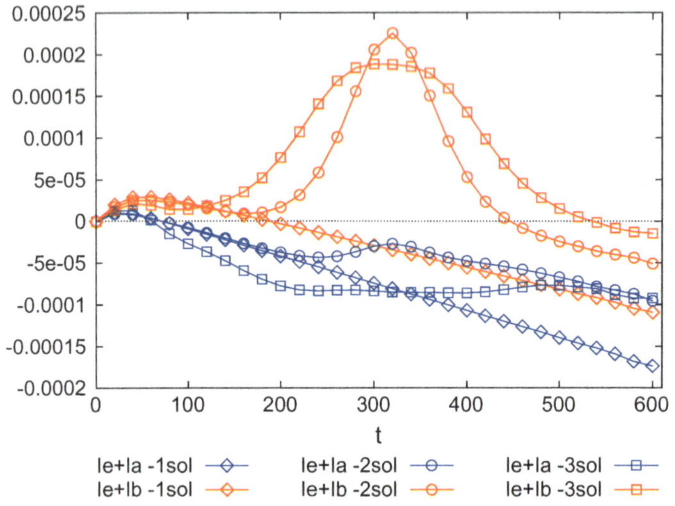

Fig. 6 Relative changes of the approximate invariants: $I_{\text{ad}}^{(2\alpha)}$, denoted as $Ie + Ia$ and $I_{\text{ad}}^{(2\beta)}$ denoted as $Ie + Ib$ for the three waves displayed in the Fig. 2. Reproduced with permission from [21]. Copyright (2017) by Elsevier

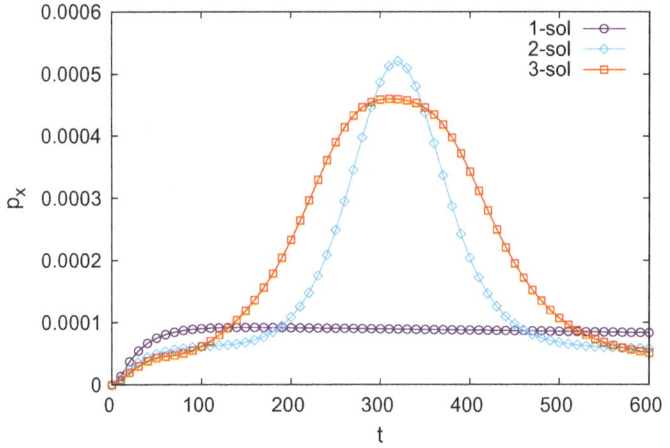

Fig. 7 Relative changes of p_x (92) as a function of time for the three waves presented in Fig. 2

As we have already mentioned the fluid momentum is related to the adiabatic invariant $I_{\text{ad}}^{(2)}$. Let us compare the momentum given by definition (92) with its approximation expressed by adiabatic invariant (93). The former is presented in Fig. 7. In the latter, displayed in Fig. 8, for $I_{\text{ad}}^{(2)}$ we used (56) with $\varepsilon = \frac{1}{2}$. It is clear that the approximate momentum expressed by exact first invariant and adiabatic invariant $I_{\text{ad}}^{(2)}$ suffers much smaller fluctuations than the exact momentum (92).

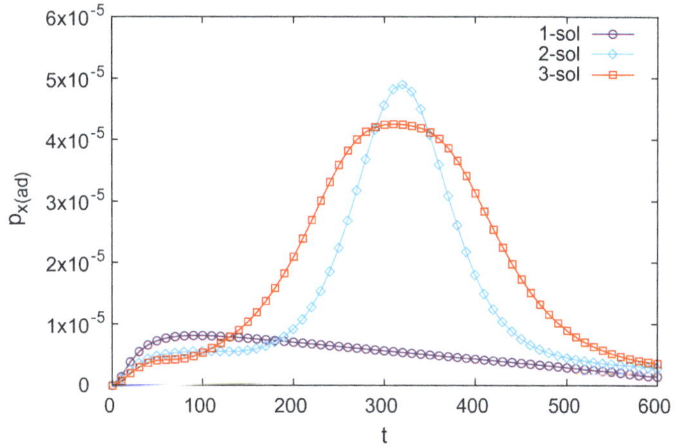

Fig. 8 Relative changes of $p_{x(\mathrm{ad})}$ (93) as a function of time for the three waves presented in Fig. 2

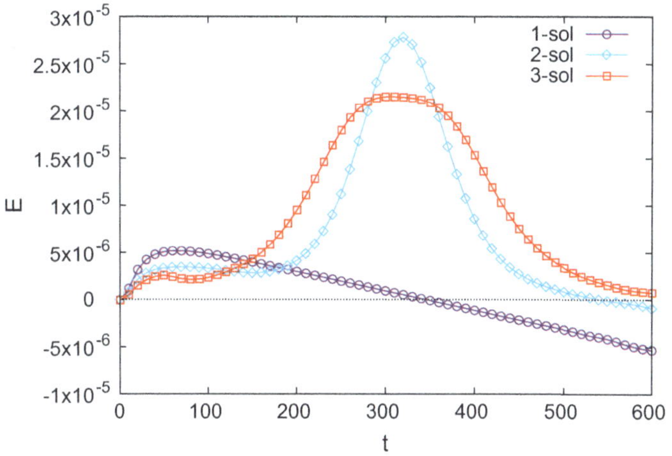

Fig. 9 Relative changes of energy as a function of time for the three waves presented in Fig. 2

6.2 *Energy (Non)Conservation and Adiabatic Invariant $I^{(3)}_{ad}$*

How close to constant values are adiabatic invariants $I^{(2)}_{\mathrm{ad}}$ and $I^{(3)}_{\mathrm{ad}}$? The relative changes of the energy (94) for three waves shown in Fig. 2 are displayed in Fig. 9.

The energy (94) can be approximated by a linear combination of three terms (95), exact invariant I_1 and adiabatic invariants $I^{(2)}_{\mathrm{ad}}$ and $I^{(3)}_{\mathrm{ad}}$. Relative changes of that approximate energy (95) are displayed in Fig. 10. Comparing Figs. 9 and 10 we see, that, as in the case of momentum, the approximate energy expressed by adiabatic invariants varies less than the exact one.

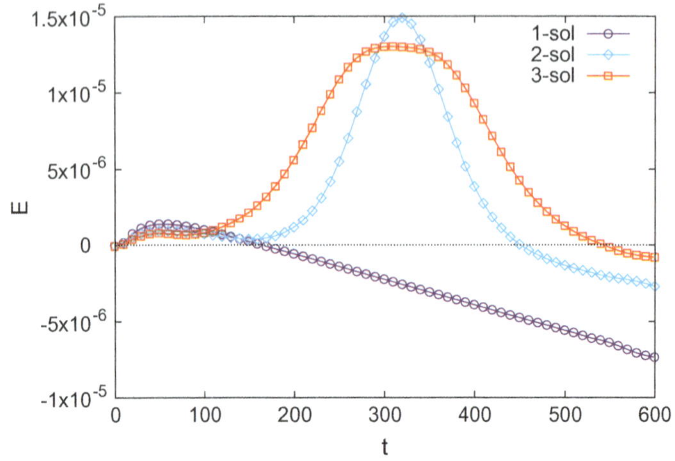

Fig. 10 Relative changes of energy approximated by adiabatic invariants (95) for the three waves presented in Fig. 2

Other than volume conservation, which maintains virtually numerical precision (see, Fig. 3), the adiabatic invariants presented in Figs. 6 and 10, and the energy shown in Fig. 9 over longer periods slowly decrease with time. In our assessment the reason can be found in the fact that 1-, 2-, 3-soliton solutions of the KdV equation, taken as initial conditions, **are not exact** solutions of the KdV2 equation. The 1-soliton analytic solution of KdV2 equation derived in [23] preserves exactly its shape and then possesses the infinite number of invariants. The same is true for recently found [21] exact analytic periodic solutions of KdV2. However, we doubt the existence of exact n-soliton solutions for KdV2 as it does not belong to a hierarchy of integrable equations. Additionally the 2- or 3-soliton solutions of an integrable equation like those obtained through NIT are likewise not exact solutions of (47). Therefore the deviations from exact solutions will lead to dissipation.

7 Summary and Conclusions

In this chapter several properties of solutions to the KdV2 equation (extended KdV equation) are discussed. First, the shallow water problem is formulated within the framework of the motion of ideal fluid under gravitational force with proper boundary conditions. After introducing appropriate scaled variables this model can be considered at different stages of approximation. Next, a general method for derivation of the wave equation is described which can be applied up to arbitrary order in the perturbative approach. Limitation to the first order results in the KdV equation (19), while a second order approach gives KdV2 (26). Then, solutions to KdV are referred to and compared to analytic solutions to KdV2 found by us recently (solitonic [23] and periodic [21] ones). Next, in the main part of the paper, invariants of KdV and adiabatic invariants of KdV2 are described in detail.

Presented is a means of direct calculation of adiabatic invariants for KdV2 which was developed in [25]. This method can be applied directly to equations expressed in the fixed reference frame. Small parameters of $\alpha \neq \beta$ should be of similar order but not necessarily equal.

The method does not depend on a transformation to a particular moving frame, nor on a near-identity transformation. This makes calculations of second invariant simpler. It can be used also to calculate invariants of higher order.

The NIT-based method, developed in Sect. 5, seems to be more suitable for the adiabatic invariant related to energy since it gives directly the most general form of this invariant.

Numerical tests have verified that the second and third adiabatic invariants related to momentum and energy, respectively, have indeed almost constant values. The small deviations from these almost constant values are largest during soliton collisions.

Because the KdV2 equation has non-integrable form, the energy is not an exact constant (see, e.g., Fig. 9).

There is, however, an intriguing kind of paradox with KdV2 invariants. On the one hand, exact invariants related to momentum and energy do **not** exist, only adiabatic ones are found. On the other hand, despite the non-integrability of KdV2, there exist exact analytic solutions of KdV2. The form of the single soliton solution of KdV2 was found in [23, Sect. IV]. Recently, in [21], we found analytic periodic solutions of KdV2 known as *cnoidal waves*. These KdV2 solutions have the same form as corresponding KdV solutions, but with different coefficients. Both of these solutions preserve their shapes during motion, so for such initial conditions the infinite number of invariants like those given by (46) exist. When initial conditions have the form different from analytic solutions only adiabatic invariants are left.

Acknowledgements The authors thank Prof. Eryk Infeld and Prof. George Rowlands for inspiring discussions.

References

1. Ablowitz, M.J., Clarkson, P.A.: Solitons, Nonlinear Evolution Equations and Inverse Scattering. Cambridge University Press, Cambridge (1991)
2. Ablowitz., M.J., Segur, H.: Solitons and Inverse Scattering Transform. SIAM, Philadelphia (1981)
3. Ali, A., Kalisch, H.: On the formulation of mass, momentum and energy conservation in the KdV equation. Acta Appl. Math. **133**, 113–131 (2014)
4. Benjamin, B.T., Olver, P.J.: Hamiltonian structure, symmetries and conservation laws for water waves. J. Fluid Mech. **125**, 137–185 (1982)
5. Bullough, R.K., Fordy, A.P., Manakov, S.V.: Adiabatic invariants theory of near-integrable systems with damping. Phys. Lett. A **91**, 98–100 (1982)

6. Burde, G.I., Sergyeyev, A.: Ordering of two small parameters in the shallow water wave problem. J. Phys. A: Math. Theor. **46**, 075501 (2013)
7. Dodd, R.: On the integrability of a system of coupled KdV equations. Phys. Lett. A **89**, 168–170 (1982)
8. Drazin., P.G., Johnson, R.S.: Solitons: An Introduction. Cambridge University Press, Cambridge (1989)
9. Dullin, H.R., Gottwald, G.A., Holm, D.D.: An integrable shallow water equation with linear and nonlinear dispersion. Phys. Rev. Lett. **87**, 194501 (2001)
10. Dullin, H.R., Gottwald, G.A., Holm, D.D.: Camassa-holm, Korteweg-de Vries-5 and other asymptotically equivalent equations for shallow water waves. Fluid Dyn. Res. **33**, 73–95 (2003)
11. Dullin, H.R., Gottwald, G.A., Holms, D.D.: On asymptotically equivalent shallow water equations. Physica D **190**, 1–14 (2004)
12. Eckhaus, W., van Harten, A.: The inverse scattering method and the theory of solitons. An introduction. In: North-Holland Mathematics Studie, vol. 50. North Holland, Amsterdam (1981)
13. Fokas, A.S., Liu, Q.M.: Asymptotic integrability of water waves. Phys. Rev. Lett. **77**, 2347–2351 (1996)
14. Gardner, C.S., Greene, J.M., Kruskal, M.D., Miura, R.M.: Method for solving the Korteweg de Vries equation. Phys. Rev. Lett. **19**, 1095–1097 (1967)
15. Grimshaw, R.: Internal solitary waves. In: Presented at the International Conference Progress in Nonlinear Science, Held in Nizhni Novogrod, July 2001. Dedicated to the 100-th Anniversary of Alexander A. Andronov (2001)
16. Grimshaw, R., Pelinovsky., E., Talipova, T.: Modelling internal solitary waves in the costal ocean. Surv. Geophys. **28**, 273–298 (2007)
17. He, Y.: New exact solutions for a higher order wave equation of KdV type using multiple G'/G-expansion methods. Adv. Math. Phys. 148132 (2014)
18. He, Y., Zhao, Y.M., Long, Y.: New exact solutions for a higher-order wave equation of KdV type using extended F-expansion method. Math. Prob. Eng. 128970 (2013)
19. Hiraoka, Y., Kodama, Y.: Normal forms for weakly dispersive wave equations. Lect. Notes Phys. **767**, 193–196 (2009)
20. Hirota, R.: The Direct Method in Soliton Theory. Cambridge University Press, Cambridge (2004) (First published in Japanese (1992))
21. Infeld, E., Karczewska, A., Rowlands, G., Rozmej, P.: Exact cnoidal solutions of the extended KdV equation. Acta Phys. Polon. A **133**, 1191–1199 (2018)
22. Infeld, E., Rowlands, G.: Nonlinear Waves, Solitons and Chaos, 2nd edn. Cambridge University Press, Cambridge (2000)
23. Karczewska, A., Rozmej, P., Infeld, E.: Shallow-water soliton dynamics beyond the Korteweg de Vries equation. Phys. Rev. E **90**, 012907 (2014)
24. Karczewska, A., Rozmej, P., Infeld, E.: Energy invariant for shallow-water waves and the Korteweg-de Vries equation: doubts about the invariance of energy. Phys. Rev. E **92**, 053202 (2015)
25. Karczewska, A., Rozmej, P., Infeld, E., Rowlands, G.: Adiabatic invariants of the extended KdV equation. Phys. Lett. A **381**, 270–275 (2017)
26. Karczewska, A., Rozmej, P., Rutkowski, L.: A new nonlinear equation in the shallow water wave problem. Phys. Scr. **89**, 054026 (2014)
27. Karczewska, A., Rozmej, P., Rutkowski, L.: A finite element method for extended KdV equations. Annal. UMCS Sectio AAA Phys. **70**, 41–54 (2015)
28. Karczewska, A., Rozmej, P., Rutkowski, L.: Problems with energy of waves described by Korteweg-de Vries equation. Int. J. Appl. Math. Comp. Sci. **26**, 555–567 (2016)
29. Kodama, Y.: Normal forms for weakly dispersive wave equations. Phys. Lett. A **112**, 193–196 (1985)
30. Kodama, Y.: On integrable systems with higher order corrections. Phys. Lett. A **107**, 245–249 (1985)
31. Korteweg, D., de Vries, G.: On the change of form of long waves advancing in a rectangular canal, and on a new type of long stationary waves. Philos. Mag. **39**, 422–443 (1895)

32. Luke, J.C.: A variational principle for a fluid with a free surface. J. Fluid Mech. **27**, 395–397 (1967)
33. Marchant, T.R., Smyth, N.F.: The extended Korteweg-de Vries equation and the resonant flow of a fluid over topography. J. Fluid Mech. **221**, 263–288 (1990)
34. Marchant, T.R., Smyth, N.F.: Soliton interaction for the extended Korteweg-de Vries equation. IMA J. Appl. Math. **56**, 157–176 (1996)
35. Miura, R.M., Gardner, C.S., Kruskal, M.D.: KdV equation and generalizations II. Existence of conservation laws and constants of motion. J. Math. Phys. **9**, 1204–1209 (1968)
36. Newell, A.C.: Solitons in Mathematics and Physics. Society for Industrial and Applied Mathematics, Philadelphia, PAS, New York (1985)
37. Olver, P.J.: Applications of Lie groups to differential equations. Springer, New York (1993)
38. Osborne, A.R.: Nonlinear ocean waves and the inverse scattering transform. Elsevier, Academic Press, Amsterdam (2010)
39. Remoissenet, M.: Waves Called Solitons: Concepts and Experiments. Springer, Berlin (1999)
40. Sergyeyev, A., Vitolo, R.F.: Symmetries and conservation laws for the Karczewska-Rozmej-Rutkowski-Infeld equation. Nonlinear Anal. Real World Appl. **32**, 1–9 (2016)
41. Whitham, G.B.: Linear and Nonlinear Waves. Wiley, New York (1974)
42. Yang, J.: Dynamics of embedded solitons in the extended Korteweg-de Vries equations. Stud. Appl. Math. **106**, 337–365 (2001)
43. Zabusky, N.J.: Solitons and bound states of the time-independent Schrödinger equation. Phys. Rev. **168**, 124–128 (1968)
44. Zhao, Y.M.: New exact solutions for a higher-order wave equation of KdV type using the multiple simplest equation method. J. Appl. Math. 848069 (2014)

Nonlinear Gravitational Waves and Solitons

Francisco R. Villatoro

"Ladies and gentlemen. We have detected gravitational waves! We did it!"

—David Reitze, aLIGO Executive Director, February 11, 2016.

Abstract Gravitational wave astronomy is born in 2016. The laser interferometers of Advanced LIGO have detected two gravitational waves, each one generated by two black hole pairs. The observed wave profiles result from the fusion of two stellar-mass black holes into a single rotating black hole, with the emission of gravitational radiation with energy in the solar-mass scale. Indeed, they are the most violent astrophysical events recorded to date. Since gravitational waves solve the weak-field approximation of the Einstein equations in vacuum, in this limit, they evolve as linear waves. However, gravitational waves are intrinsically nonlinear waves; in fact, the chirp of the signal, the change in frequency observed by Advanced LIGO detectors, is due to the nonlinearity at the sources, even being negligible far away from them. Both cylindrical and planar nonlinear gravitational waves can be interpreted as soliton solutions of Einstein's equations outside the sources. Actually, even black holes, the main sources of gravitational radiation, are two-soliton solutions of Einstein's equations in vacuum. Gravitational solitons differ from standard nonlinear solitons in several aspects, including new phenomena such as multi-soliton coalescence, a phenomenon that emits low-amplitude radiation. Indeed, the pair-of-pants solution for the fusion of two black holes can be interpreted in such a way. In conclusion, although gravitational waves propagate in spacetime like linear waves, at their sources they are nonlinear gravitational waves and, even, gravitational solitons.

Keywords Gravitational waves · Black holes · Gravitational solitons · Inverse scattering method

---

F. R. Villatoro (✉)
Department Lenguajes y Ciencias de la Computación, Escuela de Ingenierías Industriales, Ampliación del Campus de Teatinos, Universidad de Málaga, 29071 Málaga, Spain
e-mail: villa@lcc.uma.es; frvillatoro@uma.es

V. Carmona et al. (eds.), *Nonlinear Systems, Vol. 1*,
Understanding Complex Systems, https://doi.org/10.1007/978-3-319-66766-9_7

1 Introduction

The birth of a new black hole was the most violent astrophysical event ever recorded by Humanity. GW150914, The Event, results from the collision of a pair of stellar-mass black holes with about 30 and 35 solar masses, that merges into a Kerr black hole with about 62 solar masses [6]. Three solar masses in energy were radiated in less than two tenths of a second; this is more than fifty times more energy than the combined light power from all the stars in the observable Universe, in terms of radiated power during the same time interval. This major scientific breakthrough was announced on Febrary 11, 2016, but the signal was recorded on September 14, 2015 by the two detectors of the Advanced Laser Interferometer Gravitational-wave Observatory (aLIGO).

A new window on the Universe has been open, the field of gravitational-wave astronomy. During the aLIGO Run O1 not one but two signals has been observed. The second one, GW151226, was produced by the coalescence of two black holes with about 8 and 14 solar masses, resulting in one with 21 solar masses and the emission of 1 solar mass in energy during one second [4]. The evidence for the existence of gravitational waves was strongly supported by studying the motions of tightly orbiting pairs of neutron stars in our Galaxy. Their orbits shrink exactly as predicted by Einstein's theory for the emission of gravitational wave energy. Two astronomers, Russel Hulse and Joseph Taylor were awarded with the 1993 Nobel Prize in Physics for this indirect confirmation using the binary star system known as PSR 1913 + 16.11, the orbit of a pulsar and a neutron star.

The direct detection of gravitational waves by aLIGO provides new and more stringent ways to test general relativity under the most extreme conditions. Several predictions of Einstein's theory of gravitation have been confirmed at once by aLIGO. In fact, the previous evidence for the existence of black holes was circumstantial, being based on the effects of black hole candidates on their immediate surroundings. Gravitational waves GW150914 and GW151226 are the first evidence for the existence of event horizons, Kerr black holes, and binary black hole systems. Moreover, black holes can collide and merge to create a new one, emitting gravitational waves with a waveform that exactly matches aLIGO observations, as previously shown by computer simulations using numerical relativity.

This chapter summarizes my plenary talk at NoLineal 2016, Seville, 07–10th June, 2016. My goal was to review the current status of gravitational waves from the point of view of nonlinear physics. The contents of this chapter are as follows. The history of gravitational waves is summarized in Sect. 2; in more detail, after a presentation of the theory for linear waves in Sect. 2.1, the problem of the existence of nonlinear waves is discussed in Sect. 2.2, and the current definition of nonlinear gravitational waves in Sect. 2.3. The birth of gravitational wave astronomy is reviewed in Sect. 3; more precisely, the working operation of interferometric gravitational detectors is recapitulated in Sect. 3.1, the estimated astrophysical parameters of the signal GW150914 are discussed in Sect. 3.2, and those of GW151226 and LVT151012 in Sect. 3.3. Section 4 is devoted to the fusion (or coalescence) of two black holes; the history of black holes is briefly summarized in Sect. 4.1; the two-body problem in general relativity is very different from Newtonian theory, as surveyed in Sect. 4.2; during the fusion of two blakck holes regions with nontrivial topology could temporally appear as shown in Sect. 4.3. Nonlinear gravitational waves and black holes could be understood as gravitational solitons as reviewed in Sect. 5; after a brief summary of soliton theory in nonlinear physics presented in Sect. 5.1, its application to gravitation is sketched in Sect. 5.2, with emphasis in planar and cylindrical gravitational waves in Sect. 5.3, and in black holes in Sect. 5.4. Finally, Sect. 6 summarizes this chapter and focuses in the next future of gravitational wave astronomy.

2 Brief History of Gravitational Waves

The cosmological constant was considered Einstein's biggest blunder. However, it was not his only one. The existence of gravitational waves and black holes bothered him during his whole life. He thought that they are unphysical phenomena because, in both cases, the known exact solutions of the fully nonlinear equations of gravitation have singularities. The answers of these conundrums were clarified by a younger generation of physicists after Einstein's death. The turning point in the history was the famous Conference on the *Role of Gravitation in Physics*, held at the University of North Carolina at Chapel Hill on January 18–23, 1957. But everything started two years before, in the first conference dealt almost exclusively with general relativity; a jubilee celebration of the 50th anniversary of the theory of special relativity, held in Berne (unfortunately, Einstein could not attend either). After these two conferences several scientific breakthroughs convince the majority of the relativists that both, gravitational waves and black holes, exist in the Universe.

Let me briefly review the historic milestones on gravitational waves. For the readers interested in a detailed presentation, I recommend the books by Weinstein [37] and Kennefick [24]; for a briefer summary, my recommendation is the papers by Thorne [33], and Hill and Nurowski [19]. In this section, for brevity, I will not cite the major original references; they can be consulted in the bibliographies of the recommended sources. My main intention is to highlight the most significant ideas and issues from the point of view of nonlinear waves.

2.1 *Linear Gravitational Waves*

Gravitation is instantaneous in Newton's theory. Hence, it predicts the nonexistence of gravitational waves. The gravitational field is governed by a scalar potential $\phi(\mathbf{x}, t)$ that solves a linear, elliptic, partial differential equation, the widely known Poisson's equation given by

$$-\nabla^2\phi(\mathbf{x}, t) = 4\pi\, G\rho(\mathbf{x}, t)\,,$$

where G is Newton's universal gravitation constant, ρ is the mass density (the source of gravitation in Newton's theory), ∇^2 is the Laplacian, $\mathbf{x}$ is the position in space, and t is time. In this elliptic equation time is a dummy variable.

The wave equation was introduced by Euler in 1756, as an extension to three spatial dimensions of the one-dimensional wave equation proposed by d'Alembert in 1746. Curiously, Poisson's equation was first introduced in 1813 as the limit of the wave equation when the speed of propagation is infinite. At the epoch it was natural to consider the case of a finite, but very large, speed for gravitation. This idea was stated by Laplace in the volume IV of his *Treatise in Celestial Mechanics* (1805); moreover, using the orbit of the Earth, Laplace's estimates that c_g is at least seven million larger than the speed of light in vacuum, c. But Laplace did not write explicitly the wave equation for gravitational waves; it was first written by Riemann in 1858 (published posthumously in 1867) and Lorenz in 1861. The resulting linear, hyperbolic, partial differential is given by

$$\Box\,\phi(\mathbf{x},t) \equiv \frac{1}{c_g^2}\,\frac{\partial^2\phi(\mathbf{x},t)}{\partial t^2} - \nabla^2\phi(\mathbf{x},t) = 4\,\pi\,G\,\rho(\mathbf{x},t)\,. \qquad (1)$$

Weber, Gauss, Riemann, Maxwell, Lévy, and Gerber, among others 19th century physicists, independently estimated the speed of gravitation [24]. The best published result was $c_g \approx 305{,}000$ km/s, nearly the current value of c. It was obtained by Gerber in 1898 by using the perihelion shift of Mercury's orbit (although his calculation contains an error that makes this result doubtful).

The theory of relativity introduced by Einstein in 1905 postulates that the maximum speed of a signal coincides with c, a limit that the speed of gravitational waves cannot exceed. Obviously, the most natural possibility is $c_g = c$, the case considered by Poincaré, Abraham, Nördstrom, and other physicists; they propose Eq. (1) as a relativistic scalar field theory that reconciles Newton's gravitation with special relativity. However, in 1907 Einstein suggested that gravitation must affect time, the so-called gravitational time dilation, by using a beautiful and indisputable *gendanken* problem. Since time can be measured by a light clock, a flash of light bouncing between two mirrors, light must gravitate, in contradiction with Newton's theory (because light has no mass).

In order to incorporate Maxwell's electromagnetism as a source of gravitation in Eq. (1), a scalar magnitude built from the electromagnetic energy-momentum tensor $T_{\alpha\beta}$ should be used instead of the density of mass. The most obvious possibility is the trace of such a tensor, so Eq. (1) should be written as $\Box\,\phi = 4\,\pi\,G\,T$, where $T = T^{\alpha}_{\alpha} = \eta^{\alpha\beta}\,T_{\alpha\beta}$, with $\eta_{\alpha\beta}$ is the Minkowski metric. However, this idea has a big problem, the electromagnetic energy-momentum tensor is traceless ($T = 0$); in modern language, the photon is massless. In order to solve this problem the gravitational field should be governed by equations more complicated than a scalar wave equation. At the epoch it is widely known that a vector field equation similar to Maxwell's electromagnetism cannot be used, since Maxwell himself showed that, in such a case, masses should repel instead of attract.

Einstein accepted in 1912 the concept of space-time introduced by Minkowski in 1908. He understood then that the gravitational field, beyond the *curvature in time* suggested by gravitational time dilation, should be the result of the curvature of both space and time. Several mathematicians and physicists started the search for the space-time field equations, including Nördstrom and Hilbert. Einstein initiates his own path by asking for help from his friends Grossmann and Besso. Let me emphasize that, at the time, Einstein was aware that the future field equations should allow the propagation of gravitational waves. In fact, in a congress in Vienna in 1913, Born asked Einstein about the speed of propagation of gravitation [37]. He answers Born that, in the weak field approximation, the linear field equations can be used to approximate a small disturbance propagating in a flat space-time. The resulting gravitational wave should propagate exactly at the speed of light in vacuum.

General relativity, the current theory of gravitation, was developed by Einstein between 1907 and 1915. The field equations were presented in November 1915, a set of 16 coupled hyperbolic-elliptic nonlinear partial differential equations for the ten components of the metric $g_{\alpha\beta}$ of space-time in 3+1 dimensions given by

$$R_{\alpha\beta} - \frac{1}{2}\, g_{\alpha\beta}\, R = \frac{8\,\pi\,G}{c^4}\, T_{\alpha\beta}\,, \tag{2}$$

where $R_{\alpha\beta}$ is the Ricci tensor, the contraction of the Riemann curvature tensor, and R is the scalar curvature, the contraction of the Ricci tensor. The mathematical explanation and physical interpretation of Eq. (2) is outside the scope of this chapter. Usually, Wheeler is quoted: “mass tells space-time how to curve, and space-time tells mass how to move.” But let me highlight here that the most important difference with Newton’s theory is that the source of gravitation is not mass density, but energy and momentum densities, incorporated into the energy-momentum tensor $T_{\alpha\beta}$. Since energy is always positive, like mass, it results in a positive curvature. Notice that negative curvature cannot be the result of negative energy (or mass), which do not exist in Nature. However, negative pressure do exist (the so-called dark energy), resulting in a negative curvature (the acceleration of cosmic expansion).

Einstein’s field Eq. (2) is highly redundant; the Bianchi identities throw out four components and the diffeomorphism invariance (also referred to as the general covariance) of the metric throw out another four, so only two degrees of freedom have physical meaning. These nonlinear equations are elliptic (gravitational field equations) or hyperbolic (gravitational wave equations) depending on the *gauge* used to deal with their redundancy. In the last case, Einstein’s Principle of Equivalence (locally, space-time follows special relativity) ensures that gravitational perturbations propagate exactly at the speed of light in vacuum.

Gravitational waves were considered by Einstein in June 1916. He linearized the full non-linear field equations using the so-called harmonic condition, a suggestion to him by de Sitter. Einstein’s paper contains a mathematical error (something understandable since he wrote 15 publications in 1916). This error leads him to conclude that there are three different types of gravitational waves: longitudinal, transverse, and mixed. Apparently, no energy is transported by the longitudinal and transverse waves, so they were fictitious waves reflecting vibrations of the reference system. Only the mixed gravitational waves should be physical, being produced by a point mass. Einstein's error was found by Nordström in 1917 and by Schrödinger in 1918, after carefully repeating the original calculations [37]. They communicate it to Einstein, who finally accepted their results and published in 1918 a corrected version of his 1916 paper. Like electromagnetic waves, gravitational waves are transverse waves; mixed waves do not exist and longitudinal waves are an artifact (they are equivalent to flat space-time after a change of coordinates). However, several physicists, like Eddington in 1922, considered that even transversal gravitational waves are ripples in the coordinates with no physical meaning.

Einstein's paper studies a very small perturbation of the Minkowski metric $\eta_{\alpha\beta}$ given by the metric tensor $g_{\alpha\beta} = \eta_{\alpha\beta} + h_{\alpha\beta}$, where $|h_{\alpha\beta}| \ll 1$. Equation (2) after neglecting the quadratic and higher-order terms in $h_{\alpha\beta}$, reduces to the set of decoupled linear wave equations

$$\Box\, \bar{h}_{\alpha\beta} \equiv \left(-\frac{1}{c^2}\frac{\partial^2}{\partial t^2} + \nabla^2\right) \bar{h}_{\alpha\beta} = \frac{16\pi\, G}{c^4} T_{\alpha\beta}\,, \tag{3}$$

where $\bar{h}_{\alpha\beta} = h_{\alpha\beta} - \eta_{\alpha\beta} h/2$ is the so-called *trace-reversed* metric perturbation, and the Lorentz gauge condition $\partial \bar{h}_{\alpha\beta}/\partial x_\beta = 0$ has been used. In vacuum, the wave equation $\Box\, \bar{h}_{\alpha\beta} = 0$ has *plane wave solutions* given by the real part of the expression

$$\bar{h}_{\alpha\beta} = A_{\alpha\beta}\, \exp(\mathrm{i}\, k_\gamma\, x^\gamma)\,,$$

where the components of the amplitude polarization tensor $A_{\alpha\beta}$ and the wave vector $k^\gamma = (-\omega/c, k^i)$ are all constants. The gauge condition implies that this gravitational wave is transversal, $A_{\alpha\beta}\, k^\beta = 0$, i.e., the amplitude and the wave vector are orthogonal. Moreover, gravitational waves propagate exactly at the speed of light in vacuum, since the wave vector is light-like, $k^\gamma\, k_\gamma = 0$ (equivalent to $\omega^2 = c^2\, |\mathbf{k}|^2$).

The Lorentz gauge condition does not fully fix the metric, since only two components of the polarization tensor are independent. Usually, the *transverse-traceless* (TT) gauge $A_{0\alpha} = 0$ is used; in this gauge $A_{ij}\, k^j = 0$ (spatially *transverse* wave), $A^i_i = 0$ (*traceless* wave), and $\bar{h}_{\alpha\beta} = h_{\alpha\beta}$. For a gravitational wave moving in the z-direction $k^0 = k_z = \omega/c$, and $k_x = k_y = 0$, the TT gauge conditions lead to $A_{0\alpha} = A_{z\alpha} = 0$, $A_{xx} = -A_{yy}$, and $A_{xy} = A_{yx}$. The two independent components of the polarization tensor are A_{xx} and A_{xy}, corresponding to the pure polarization waves

$$h_+ = A_{xx}\, e^{-\mathrm{i}\omega\,(t - z/c)}, \qquad \text{and} \qquad h_\times = A_{xy}\, e^{-\mathrm{i}\omega\,(t - z/c)},$$

and referred to as plus- and cross-polarizations, respectively. Both polarizations can be illustrated by using an animation of the movement of a circular ring of free particles in the (x, y) plane during the passage of the gravitational wave. The tidal deformation of the circular ring deforms into an elliptical ring of the same area alternatively both in the horizontal and in the vertical directions for plus-polarization, and both in the diagonal and in the counterdiagonal directions for cross-polarization.

Gravitational waves are generated by time-varying quadrupole moments, due to the conservation of mass and linear momentum forbid the emission of either monopolar or dipolar radiation. Let us recall that in acoustics, radiation is dominated by monopolar emission, and that in electromagnetism, it is dominated by dipolar radiation, due to charge conservation. Already in 1918, Einstein studied the generation of gravitational waves by *slowly varying* sources. He obtained the so-called quadrupole formula, the solution of Eq. (3) when the energy-momentum tensor corresponds to a ball of perfect fluid in linearized gravity. Specifically, the spatial components of $\bar{h}_{\alpha\beta}$ are given by

$$\bar{h}_{ij} = \frac{2}{r}\frac{G}{c^4}\frac{d^2}{dt^2}Q_{ij}(t - r/c)\,, \tag{4}$$

where $r^2 = x_i\, x_i$ is the radial distance, Q_{ij} is the quadrupole moment that in the TT gauge is written as

$$Q_{ij}(t) = \int \rho(t, x_i)\left(x_i\, x_j - \frac{1}{3}\delta_{ij}\, r^2\right) d^3x\,.$$

The quadrupole formula (4) shows that spherical bodies cannot emit gravitational waves. The non-spherical part of an emitter is characterized by an asymmetry fudge factor $0 \leq s \leq 1$, such that $s = 0$ corresponds to a spherically symmetric source. The order-of-magnitude of the amplitude h of gravitational waves generated by a non-spherical source of mass M, with typical size R, quadrupole moment $Q \sim s\, M\, R^2$, and typical internal velocity v is given by

$$h \sim \frac{2\,G}{c^4\, r}\left(\frac{v}{R}\right)^2 s\, M\, R^2 \sim \frac{R}{r}\frac{R_S}{R}\left(\frac{v}{c}\right)^2 s\,, \tag{5}$$

where the Schwarzschild radius is $R_S \equiv 2\,G\,M/c^2$. Hence, the main sources of gravitational waves are nonspherical ($s \lesssim 1$), compact sources ($R \gtrsim R_S$), moving at relativistic speeds ($v \lesssim c$). For example, for the coalescence of two neutron stars in a distant galaxy, the estimate gives $h \lesssim 10^{-22}$. Such a tiny value is the main barrier that current observatories have had to overcome.

2.2 *Nonlinear Gravitational Waves*

The existence problem for nonlinear gravitational wave solutions of the fully nonlinear Eq. (2) always burdened Einstein [19]. If such solutions do not exists, it makes no sense the use of the linearized theory, being their solutions pure artifacts of the linearization. In the 1920s Brinkmann, a mathematician, published several papers on a class of Ricci flat metrics having radiative properties. Nowadays they are called *pp-waves*, but at that time these papers were unnoticed by physicists, including Einstein. Indeed, their physical interpretation has to wait until 1957 when Bondi, Pirani and Robinson discovered nonlinear, plane gravitational waves, noticing that they are a special case of pp-waves. Perhaps, if Einstein had noticed Brinkmann's papers the history of gravitational waves could have been completely different.

In June 1936, Einstein and Rosen submitted a paper on gravitational waves to the journal *Physical Review* [37]. They found a family of exact solutions of the non-linear field equations, but they not satisfy the harmonic condition and contained singularities. Einstein changed his mind with regard to gravitational waves and claimed they did not exist. However, the paper was rejected after peer review (the anonymous referee was Robertson). Einstein becomes furious and preferred to publish the paper elsewhere. In 1936 Infeld replaced Rosen as Einstein's assistant. Infeld met Robertson who clarified to him the mistake in Einstein's paper (without revealing his role as anonymous reviewer). In Cartesian coordinates (t, x, y, z) the metric has singularities, but after a change to cylindrical coordinates (t, ρ, ϕ, z), the singularities disappear, only remaining that at $\rho = 0$, which is associated with the point source of the waves. Hence, Einstein–Rosen metric describes cylindrical gravitational waves rather than plane gravitational waves. Infeld told to Einstein, who replied that he also had found the problem (in fact he refused to read the reviewer's report). Einstein corrected the paper with Rosen that appeared in 1937 in the *Journal of the Franklin Institute* with another title ("On Gravitational Waves" instead of "Do Gravitational Waves Exist?").

Einstein was not completely convinced on the existence of gravitational waves, neither Infeld, due to his influence [19]. In 1954 Infeld suggested to his Ph.D. student Trautman the rigourous study of this matter. In 1957 Trautman extended Sommerfeld's outgoing radiation conditions from electromagnetism to gravitation, proving that gravitational radiation exists, propagates at the speed of light, and carries energy. Also in 1957, but independently, Bondi published in the journal *Nature* a singularity-free solution of a plane gravitational wave. A subsequent paper by Bondi, Pirani and Robinson, published in 1958, introduces a rigourous definition of a *plane* gravitational wave in the full Einstein theory [16]: a space-time which satisfies vacuum Einstein's equations with a 5-dimensional group of isometries (isomorphic to the corresponding group for the electromagnetic field); this group acts transitively on the space-time, so it is a homogeneous space (nonsingular at every point).

In the early 1960s, after the works of Trautman, Bondi, Pirani, Robinson, and others the theoretical arguments for the existence of gravitational waves were accepted by the whole community of relativists. However, its observation is extremely difficult because they pass through surrounding matter with impunity, by contrast with electromagnetic waves and even neutrinos [33]. The first experiments to detect gravitational waves began with Weber and his resonant mass detectors in the 1960s. In 1969 he announces the first candidate detection. During a six-year period of excitement, 15 other research groups around the world tried to reproduce the detection by using similar bar-detectors. Sadly, no convincing evidence of a direct detection was reached.

The unsuccessful experimental efforts of the early 1970s pointed the way toward new detectors based on laser interferometers. M. Gerstenshtein and V. I. Pustovoit were the pioneers who published in 1962 a paper with this idea, but in Russian [31]. Independently, Weber and Forward also consider the idea in 1964. Weiss analyzed the noise sources and sensitivity in 1969, determining that scale-kilometre arm lengths are required. Weiss, today considered the father of laser interferometers for gravitational wave detection, effectively *invented LIGO* in his famous unpublished report in MIT's Lincoln Research Laboratory of Electronics in 1972. The National Science Foundation (NSF) funded several research proposals of Weiss from 1974 to 1983. After meeting Thorne and Drever in the late 1970s, the general features of LIGO were defined, including the Fabry-Pérot design and the technique of laser power recycling. The Troika (Thorne, Drever, and Weiss) managed the project until it requires Big Science management. Vogt, Barish, and Reitze were the subsequent executive directors.

By the early 2000s, a set of interferometric detectors was completed, including TAMA 300 in Japan, GEO 600 in Germany, LIGO in the United States, and Virgo in Italy. Combinations of these detectors made joint observations from 2002 through 2011 into a global network [17]. However, only upper limits on a variety of gravitational-wave sources were set. Fortunately, after a long road to success, in September 2015, Advanced LIGO directly detects a gravitational wave, the first direct observation of a binary black hole system merging to form a single black hole.

2.3 The Definition of Gravitational Waves

In 1916 Einstein posed the following problem: when a metric $g_{\alpha\beta}$ solving the Einstein equations (2) can be interpreted as gravitational waves on a background? He knew that the answer is nontrivial. For example, in Cartesian coordinates (t, x, y, z), with $c = 1$, the metric $g_{\alpha\beta} = \eta_{\alpha\beta} + h_{\alpha\beta}$ with

$$\begin{aligned} h_{\alpha\beta}\, dx^\alpha dx^\beta =& \cos(t-x)\,(2+\cos(t-x))\, dt^2 + \cos^2(t-x)\, dx^2 \\ & - 2\cos(t-x)\,(1+\cos(t-x))\, dt\, dx\,, \end{aligned}$$

is a solution of both Eqs. (3) and (2). Apparently, it represents a gravitational wave in spacetime moving at the speed of light. However, it was created from the flat Minkowski metric in Cartesian coordinates $(\bar{t}, x, y, z)$ by the change of variables $\bar{t} = t + \sin(t-x)$. Hence, it is equivalent to flat spacetime and it does not represent any gravitational wave.

Currently, there are several mathematical definitions of the concept of gravitational wave. For physicists, the most useful is the asymptotic definition developed by Trautman [36]. For simplicity of the exposition, let us take a spacetime in cylindrical coordinates (t, r, θ, z), with a metric $g_{\alpha\beta} = \eta_{\alpha\beta} + h_{\alpha\beta}$. It satisfies Trautman's *radiative boundary conditions* if, as $r \to \infty$, the metric satisfy

$$
\begin{aligned}
g_{\alpha\beta} &= \eta_{\alpha\beta} + O(1/r)\,, & \partial_r g_{\alpha\beta} &= h_{\alpha\beta}\, k_r + O(1/r^2)\,,\\
h_{\alpha\beta} &= O(1/r)\,, & h_{\alpha\beta}\, k^\beta &= \frac{h}{2}\,\eta_{\alpha\beta}\, k^\beta + O(1/r^2)\,,
\end{aligned}
\tag{6}
$$

where $h = \eta^{\alpha\beta}\, h_{\alpha\beta}$. These equations generalize the Sommerfeld's radiative condition for electromagnetic waves.

The energy-momentum tensor for linear gravitational waves is usually calculated by the second-order quantity introduced by Landau and Lifshitz [25] given by

$$
t_{\alpha\beta} = \frac{c^4}{32\,\pi\, G}\,\left(\partial_\alpha\, h_{\alpha\beta}\right)\,\left(\partial_\beta\, h^{\alpha\beta}\right)\,,
$$

where $h_{\alpha\beta}$ solves the linear wave Eq. (3). But, essentially, gravitational waves are a nonlinear phenomenon, so the physical interpretation of this tensor is not straightforward. Instead, Trautman defines a gravitational radiation energy-momentum tensor given by

$$
t_\alpha{}^\beta = \tau\, k_\alpha k^\beta + O(1/r^3)\,, \qquad \tau = \frac{c^4}{32\,\pi\, G}\,\left(h^{\alpha\beta}\, h_{\alpha\beta} - \frac{h^2}{2}\right)\,,
$$

where τ is non-negative. The total energy momentum tensor $T_{\alpha\beta} + t_{\alpha\beta}$ is conserved and

$$
p_\alpha = \int_\Sigma (T_\alpha{}^\beta + t_\alpha{}^\beta)\, d S_\beta\,,
$$

taken over a *time-like* "cylindrical" hypersurface Σ at spatial infinity, is the total energy and momentum radiated through Σ. It can be shown that the total energy $p_0 \geq 0$, as expected.

Following the ideas of Trautman, the existence of gravitational radiation is characterized by $p_\alpha \neq 0$. Linear electromagnetic waves cannot transport is own source, the electric charge, but their energy-momentum tensor is not restricted to be linear. Similarly, linear gravitational waves cannot transport its own source, the energy and momentum [12]; in fact, their energy-momentum tensor is null at first-order, with a non-null result requiring that nonlinear, higher-order terms be taken into account. Hence, the propagation of gravitational energy results from the nonlinearity of the field equations, i.e., gravitational waves are intrinsically a nonlinear phenomenon. However, the analysis of the detection of gravitational waves using laser interferometers in the Earth is based on the linear (weak field) approximation.

3 Gravitational Wave Astronomy

The direct detection of gravitational waves was Science's 2016 Breakthrough of the Year. The discovery of gravitational waves was already recognized with the 1993 Nobel Prize in Physics, thanks to the discovery and analysis of the binary pulsar PSR B1913+16 by Hulse and Taylor. However, in my opinion, a new prize will be awarded for the birth of gravitational wave astronomy. It not only coincides with the centenary of the prediction of gravitational waves by Einstein and the discovery of the black hole solution by Schwarzschild; it is also the culmination of over forty years of observational effort.

The Troika, the three physicists who conceived of LIGO, Rainer Weiss of the Massachusetts Institute of Technology (MIT) in Cambridge, and Ronald Drever and Kip Thorne of the California Institute of Technology (Caltech) in Pasadena, were in the red carpet for the 2017 Nobel Prize in Physics. Unfortunately, Drever passed away on March 7, 2017, in Edinburgh, Scotland. Hence, the Nobel Prize committee must decide to award Weiss and Thorne alone, like in the case of the 2013 Nobel to the Higgs boson, that excluded the deceased Brout, or to include a third candidate; in the last case, the most promising researcher is Barry Barish of Caltech, the father of LIGO's hardware.[1]

Let me briefly review how aLIGO detectors work and the main properties of the gravitational waves observed to date [18]. I will cite the original references, where further details are presented. The most relevant fact to take into account is that new signals from the coalescence of stellar-mass black holes are expected in the next years. Such signals will revolutionize our current knowledge of the nonlinear dynamics of compact astrophysical objects.

3.1 *The aLIGO Detectors*

Advanced LIGO is the world's largest gravitational wave observatory [1]. Its design goal is to reach a factor of ten increase in sensitivity over a broad frequency band than (Initial) LIGO. However, in the aLIGO Run O1, which spanned September 12, 2015 through January 12, 2016, and the aLIGO Run O2 started in November 30, 2016, ending on August 25, 2017, only a factor of three has been achieved. LIGO sensitivity is limited by photon shot noise at frequencies above 150 Hz, and by a superposition of multiple noise sources. The volume of the region of the universe explored by the detectors increases as the cubic power of the sensitivity; so aLIGO Run O1 has explored a volume nine times larger than the one explored in the last run of LIGO, being is future goal to reach a factor of one hundred in volume.

Figure 1 shows a simplified scheme of the LIGO detector. They are composed of two giant laser interferometers located three thousands of kilometres apart, one in Livingston, Louisiana, and the other one in Hanford, Washington. Each detector has

[1] Weiss, Thorne and Barish were awarded the Nobel Prize in Physics 2017.

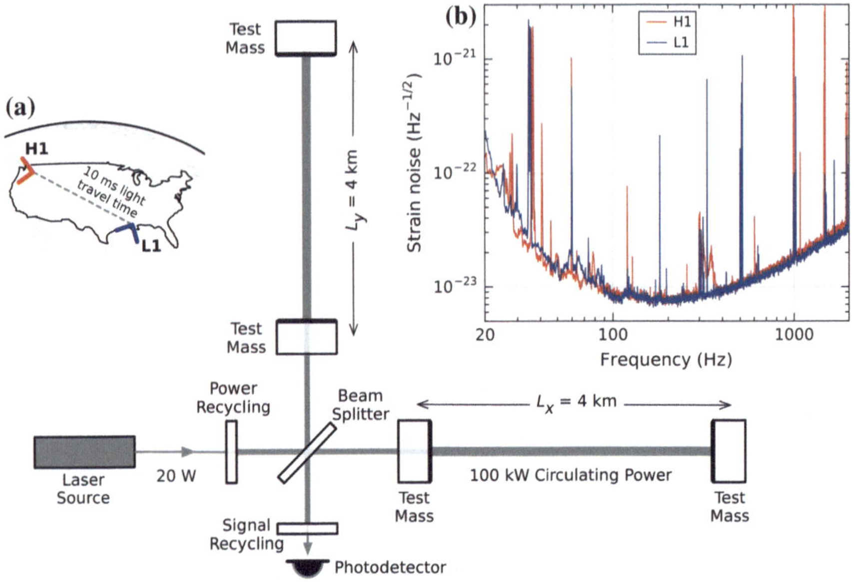

Fig. 1 Simplified diagram of an advanced LIGO detector (not to scale). Inset **a** location and orientation of the LIGO detectors at Hanford, WA (H1) and Livingston, LA (L1). Inset **b** the instrument noise for each detector during aLIGO Run O1. Reproduced from [6]; published by the American Physical Society with a Creative Commons Attribution 3.0 License

two orthogonal *arms* with a length of 3994.5 m (4 km). They are illuminated with the light of a Nd:YAG laser at wavelength 1064 nm (282 THz) in continuous-wave mode; its power is increased from 20 to 700 W by using a power-recycling mirror before it enters through a beam splitter into the two arms (1.2-m diameter ultra-vacuum tubes with a pressure below 1 μPa = 10^{-9} torr; in volume, the second largest ultra-vacuum after that of LHC tubes at CERN). Each arm contains two test mass mirrors acting as a Fabry–Pérot resonant cavity, where the light power increases up to 100 kW; this value is the major limitation of the sensibility of Runs O1 and O2, being the plan to go up to 750 kW in future runs. The test masses are ultrapure cylinders of fused silica with 34 cm diameter, 20 cm height, and 40 kg weight, polished to nanometre smoothness; they are suspended in a quadruple-pendulum system for seismic noise isolation. At the output the two light beams are recombined by using a partially transmissive signal-recycling mirror, which optimizes the gravitational wave signal at the photodetector.

Let me clarify two common misconception related to laser interferometers. First, they do not act as rulers, but as clocks. Usually it is stated that they directly measure the difference in length between two perpendicular arms. However, for a detector with an arm length $L = 4$ km and a gravitational wave with a maximum amplitude of $h = 10^{-21}$ the change in the length of one arm is about $h\,L/2 \sim 2 \times 10^{-18}$ m, a lengths smaller than a thousand the width of a proton. Such an extremely small distance cannot be directly measured by a ruler using the interference of laser pulses with a wavelength of $\sim 10^{-6}$ m; the measurement must be indirect. Specifically, LIGO's interferometers act as clocks that measure the travel time variation of light-signals between the two freely falling mirrors. They are observed as phase shifts in the returning light. In fact, the changes in the arm length do not enter into the equation for the phase difference responsible of the interference pattern at the detector.

The second misconception is the so-called interferometer paradox [21]. The gravitational wave changes the length of arms in exactly the same amount that the wavelength of the light resonating between the mirrors at their extremes, so these effects might cancel and there would be no measurable effects. The solution of the apparent paradox is as simple as the first one. The gravitational wave has a wavelength in the km scale, do not affecting the wavelength and properties of the light pulses inside the arms that are in the μm scale. There is no light resonating between the mirrors; the observable magnitude in the detectors is the time of flight between the test masses, thanks to the constancy of the speed of light in vacuum. The golden rule to remember is that weak gravitational fields always can be understood as time *curvature*, being the spatial curvature usually negligible.

Let me clarify the above misconceptions by using a simple calculation. Let us assume that the arms of the interferometer with a common length L are in the x and y directions, aligned with the polarization of the incident gravitational wave [20, 32]. A gravitational wave propagating along the z-axis, with its polarizations aligned with the x- and y-axes, is described by the space-time metric given by

$$ds^2 = -c^2\,dt^2 + (1 + h(t))\,dx^2 + (1 - h(t))\,dy^2 + dz^2\,,$$

where the small wave amplitude $h_{\alpha\beta}$ is such that $\sqrt{1 \pm h(t)} \approx 1 \pm h(t)/2$, the metric $g_{\alpha\beta} = \eta_{\alpha\beta} + h_{\alpha\beta}$, with $\eta_{\alpha\beta}$ is Minkowski's flat space-time, and $h(t) = h_{11}(t) = -h_{22}(t)$ is a oscillatory function of local time. The physical interpretation of this metric is that the wave of amplitude $h(t)$ causes that the distance between freely falling test masses to change by factors of $1 + h(t)/2$ and $1 - h(t)/2$ in the x and y arms of the interferometer, respectively.

The frequency of the gravitational wave $h(t)$ is between 10 Hz and 1 kHz, which is very tiny compared to that of the laser light beam (about 282 THz). Hence, during the travel time of a light wave crest through the arms the function $h(t)$ is approximately constant. The path of a light ray between the test masses follows $ds^2 = 0$, then

$$c^2\,dt^2 = (1 + h(t))\,dx^2 + (1 - h(t))\,dy^2 + dz^2\,.$$

The round-trip travel time for two pulses separated at the beam splitter and returning to it after travelling each arm is given by

$$\Delta t_x = \left(1 + \frac{h(t)}{2}\right)\frac{2\,L}{c}\,, \qquad \Delta t_y = \left(1 - \frac{h(t)}{2}\right)\frac{2\,L}{c}\,,$$

such that their time difference yields

$$\Delta t = \Delta t_x - \Delta t_y = \frac{2L}{c}\, h(t)\,.$$

Hence, the phase difference between the two beams of light is given by

$$\Delta\phi = \frac{4\pi\, L}{\lambda}\, h(t)\,. \tag{7}$$

It depends only on the arrival times of wave crests (the return time). The changes of the length L of the interferometer arm do not enters directly into Eq. (7). The quotient between the arm length and the wavelength of the light amplifies the amplitude of the gravitational wave in order to allow its detection. The phase shift for $\lambda = 1\ \mu$m, $L = 4$ km and $h = 10^{-21}$ is $\Delta\phi \sim 5 \times 10^{-11}$ rad. For an interferometer with Fabry-Pérot cavities there is an additional gain factor $2\mathscr{F}/\pi$, where $\mathscr{F}$ is the so-called *finesse* of the cavity [27]. For a typical value of $\mathscr{F} = 200$ the measured phase shift is $\Delta\phi \sim 10^{-8}$ rad, accessible to current metrology technologies.

Gravitational wave astronomy requires the determination of the direction on the sky of the source. Since, the detectors are essentially omni-directional, like microphones for sound instead of conventional telescopes for electromagnetic waves, they have nearly all-sky sensitivity. The sky localization of the source is largely determined by the time delay between different observatories. Ideally, all ground-based interferometers should operate as a global network. However, the signals observed during aLIGO Run O1 have only been recorded by its two detectors. Hence, current uncertainty in their location is very large. During aLIGO Run O2 in 2017 a third detector, Virgo, a single 3-km detector at Cascina, near Pisa, Italy, will start its operation. And in 2018, KAGRA, in the Kamioka Observatory, Gifu, Japan, a single 3-km detector like Virgo, is expected to enter in operation as well. The future global network of interferometric detectors during the 2020 s promise a brilliant future for gravitational wave astronomy [26].

3.2 The Event: GW150914

The Event, GW150914, was a textbook-like gravitational wave signal [6]. It was observed by the two detectors of the aLIGO on September 14, 2015 at 09:50:45 UTC, arriving first at Livingston detector, and $6.9^{+0.5}_{-0.4}$ ms later at Hanford detector. Serendipity accompanied the discovery, since the official start of aLIGO Run O1 was scheduled to 18 September 2015. Fortunately, the signal was observed during the engineering run ER8 (a commissioning phase not intended to detect astrophysical signals, but that does not preclude this possibility). About 8 cycles of the signal was observed during 0.2 s. It is a chirp, a signal with increasing frequency and amplitude from 35 to 250 Hz. The peak gravitational-wave strain was 1.0×10^{-21}. The p-value,

Table 1 Parameters of the three most significant events observed by aLIGO Run O1. Quoted with 90% credible intervals. The parameters have been taken from Refs. [2] and [5]

Event	GW150914		GW151226	LVT151012
Signal-to-noise ratio r	23.7		13.0	9.7
False alarm rate	$<6.0 \times 10^{-7}$		$<6.0 \times 10^{-7}$	0.37
p-value	7.5×10^{-8}		7.5×10^{-8}	0.045
Significance	$>5.3\sigma$		$>5.3\sigma$	1.7σ
Reference source	Abbott et al. [2]	Abbott et al. [5]	Abbott et al. [2]	Abbott et al. [2]
Primary mass $m_1/M_\odot$	$36.2^{+5.2}_{-3.8}$	$35.4^{+5.0}_{-3.4}$	$14.2^{+8.3}_{-3.7}$	23^{+18}_{-6}
Secondary mass $m_2/M_\odot$	$29.1^{+3.7}_{-4.4}$	$29.8^{+3.3}_{-4.3}$	$7.5^{+2.3}_{-2.3}$	13^{+4}_{-5}
Effective inspiral spin ξ_{eff}	$-0.06^{+0.14}_{-0.14}$	$-0.04^{+0.14}_{-0.16}$	$0.21^{+0.20}_{-0.10}$	$0.0^{+0.3}_{-0.2}$
Final mass $M_f/M_\odot$	$62.3^{+3.7}_{-3.1}$	$62.2^{+3.7}_{-3.4}$	$20.8^{+6.1}_{-1.7}$	35^{+14}_{-4}
Final spin a_f	$0.68^{+0.05}_{-0.06}$	$0.68^{+0.05}_{-0.06}$	$0.74^{+0.06}_{-0.06}$	$0.66^{+0.09}_{-0.10}$
Radiated energy $E_{\text{rad}}/(M_\odot c^2)$	$3.0^{+0.5}_{-0.4}$	$3.0^{+0.5}_{-0.4}$	$1.0^{+0.1}_{-0.2}$	$1.5^{+0.3}_{-0.4}$
Luminosity distance D_L/Mpc	410^{+160}_{-180}	440^{+160}_{-180}	440^{+180}_{-190}	1000^{+500}_{-500}
Source redshift z	$0.09^{+0.03}_{-0.04}$	$0.093^{+0.030}_{-0.036}$	$0.09^{+0.03}_{-0.04}$	$0.20^{+0.09}_{-0.09}$

called false alarm probability by LIGO Collaboration, the probability that random noise fluctuation was confused as a signal, was estimated $<2 \times 10^{-7}$, corresponding to a significance greater than 5σ, as shown in Table 1. Moreover, the estimation of the signal-to-noise ratio (SNR) was 24; let us recall that the detection of a signal by aLIGO requires SNR $>$ 10.

The interpretation of the signal, shown in the top left plot in Fig. 2, was obtained by using Bayesian inference, based on waveform models for binary black hole fusions [8]. The models are calibrated using full numerical solutions of Einstein's equations. The initial analysis was done using two methods [7]: A non-precession EOBNR technique, based on an effective-one-body (EOB) analytical approximation with parameters tuned by numerical relativity (NR), and a precession IMRPhenom technique, that incorporates the effects of spins aligned with the orbital angular momentum of the black hole binary into the EOBNR model. Both methods agree, so the overall results come from averaging the two. An improved analysis of GW150914 using a precessing EOBNR model was published later [5]. Furthermore, a direct comparison with the results from numerical relativity solutions of Einstein's equations for binary black hole coalescence has also been published [3]. The last process is computationally very expensive, but all the methods yield very similar results.

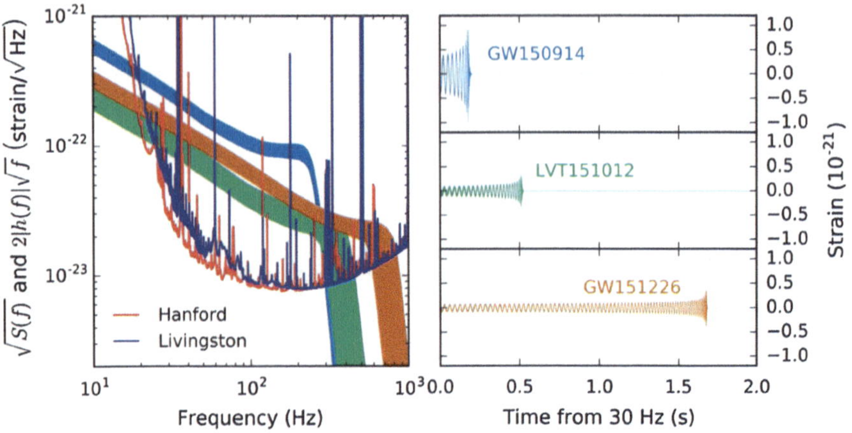

Fig. 2 The waveforms of the three signals observed in aLIGO Run O1 (right) and the corresponding sensitivity of the detector (left). Reproduced from [2]; published by the American Physical Society with a Creative Commons Attribution 3.0 License

Let me explain the notation used in Table 1 for the newcomers. The statistical results are written as the median value plus the range needed to safely enclose the 90% of the probability, therefore, the resulting interval contains the true value with equal probability above or below the median with a 10% chance to be outside this range. For example, the mass of the bigger black hole shown in Table 1 is given by $36^{+5}_{-4}M_{\odot}$, where $M_{\odot}$ indicates the mass of our Sun; the meaning of this notation is that there is a 5% chance that the mass is below $32M_{\odot} = (36 - 4)M_{\odot}$, and a 5% chance that it is above $41M_{\odot} = (36 + 5)M_{\odot}$.

The waveform of the signal GW150914 corresponds to the inspiral and merger of a pair of black holes, and the ringdown of the resulting single black hole. Table 1 shows that the initial black hole masses at 90% C.L. are $m_1 = 36^{+5}_{-4}M_{\odot}$ (the primary), and $m_2 = 29^{+4}_{-4}M_{\odot}$ (the secondary). The final black hole (the remnant) has a mass of $M_f = 62^{+4}_{-4}M_{\odot}$, with $3.0^{+0.5}_{-0.4}M_{\odot}c^2 \simeq 5.3^{+0.9}_{-0.8} \times 10^{47}$ J of energy radiated away as gravitational waves. Notice that the improved analysis [5] prefers $m_1 = 35^{+5}_{-3}M_{\odot}$, $m_2 = 30^{+3}_{-4}M_{\odot}$, and $M_f = 62^{+4}_{-3}M_{\odot}$; this result is consistent with the original one.

Astrophysical black holes are described by their mass and their spin (how much they rotate). The three black holes responsible for the signal GW150914 are the largest observed stellar-mass black holes to date. Their (dimensionless) spins a_i are defined to be between 0 (no spin) and 1 (the maximum value); the final black hole's spin is $a_f = 0.67^{+0.05}_{-0.07}$. The spins of the initial black holes cannot be measured precisely because the signal is too short, and it is not known the orientation of the binary. For these reasons, the estimated spins are compatible with spinless black holes, specifically, $a_1 = 0.3^{+0.5}_{-0.3}$, and $a_2 = 0.5^{+0.5}_{-0.4}$. Another interesting parameter related to the spin of the black hole binary is the so-called effective inspiral spin ξ_{eff}; it has a value of +1 if both black holes have maximal spin values, and are rotating the same way as the binary is orbiting, and a value of -1 if the black holes have maximum spin values and are both rotating exactly the opposite way to the binary's orbit. It is found that the effective spin of the signal GW150914 is compatible with a null value, $-0.06^{+0.17}_{-0.18}$. This could mean that both black holes have small spins, or that they have larger spins that are not aligned with the orbit (or between each other).

The analysis of the signal GW150914 estimates that the source lies at a luminosity distance of 410^{+160}_{-180} Mpc (a megaparsec is a unit of length equal to about 3 million light-years). The corresponding redshift $z = 0.09^{+0.03}_{-0.04}$ is small, so the effect on the source parameters is negligible. Therefore, the merger happened sometime between 700 million years and 1.6 billion years ago. Note also that it is very difficult to localize the position in the sky of the source by using only the two aLIGO detectors. A possibility to obtain a precise localization is by means of a neutrino or electromagnetic counterpart, but none is expected from a binary black hole. However, searches for astrophysical counterparts of the signal GW150914 have been prosecuted. As expected they were fruitless. No high-energy neutrinos were observed by ANTARES (Mediterranean sea), IceCube (Antarctica), KamLAND (Japan), the Pierre Auger Observatory (Argentina) and Super-Kamiokande (Japan) [11]. Neither electromagnetic signals in gamma-ray, X-ray, optical, infra-red, or radio observatories [10]. The exception is a weak transient source above 50 keV, observed by the Fermi gamma-ray telescope about 0.4 s after the GW150914 event was detected by aLIGO, with a significance of 2.9 σ [10]. However, the non-detection of this event by INTEGRAL/SPI-ACS suggests that this counterpart is consistent with a background fluctuation; hence, the most plausible explanation is a chance coincidence.

3.3 *The Christmas Gift and the Candidate*

Serendipity is part of the scientific endeavour. The signal GW150914 could be easily missed if aLIGO Run O1 start a little later as scheduled. The signal GW151226, too, might be missed if the Run O1 were not continued during Christmas holidays (an stop already done during aLIGO Run O2). Let me summarizes the main features of this second gravitational wave [4] that arrived at Earth 03:38:53 GMT on 26 December 2015. First observed in Livingston and 1.1 ± 0.3 ms later in Hanford. About 55 cycles of the signal was observed during a whole second. The signal is a chirp with increasing frequency and amplitude from 35 up to 450 Hz. Specifically, 45 cycles for the inspiral phase with frequency from 35 up to 100 Hz, and about 7 cycles for the merger and 3 for the ringdown. The peak gravitational-wave strain was $h_{\max} = 3.4 \times 10^{-22}$. The signal-to-noise ratio (SNR) was 13, with a p-value similar to that of GW150914, so the significance was greater than 5σ, as shown in Table 1.

The analysis of the signal GW151226 results in an estimation of the masses of the initial black holes of $m_1 = 14.2^{+8.3}_{-3.7} M_\odot$ and $m_2 = 7.5^{+2.3}_{-2.3} M_\odot$; the two black holes merge to form a final black hole of mass $M_f = 20.8^{+6.1}_{-1.7} M_\odot$. Across the entire coalescence, gravitational waves carry away $1.0^{+0.1}_{-0.2} M_\odot c^2 \simeq 1.8^{+0.2}_{-0.4} \times 10^{47}$ J of energy.

Note that usually the black hole masses are rounded to 14 and 8 solar masses, resulting in one with 21, emitting 1 solar mass. The spin measurements for GW151226, allows the determination of at least one of the black holes, although it cannot be said which; it had to have a spin of greater than 0.21 with 99% probability (as shown in Table 1 for the effective inspiral spin). The estimation of the luminosity distance of the signal was 440^{+160}_{-180} Mpc, and its redshift $z = 0.09^{+0.03}_{-0.04}$, both very similar to those of GW150914.

During the aLIGO Run O1 there were another signal, LVT151012, but only a candidate (the name "LVT" is short for *LIGO–Virgo trigger*). LVT151012 was seen first by Hanford detector then at Livingston. Table 1 shows an estimation its signal-to-noise ratio of 9.6 (hence, it is a candidate because it is smaller than 10), with a signal significance of only 1.7σ. Under the assumption that it is astrophysical in origin, it corresponds to another binary black hole merger. Table 1 shows that the component masses are $m_1 = 23^{+18}_{-6} M_\odot$ and $m_2 = 13^{+4}_{-5} M_\odot$, merging into a final black hole with mass $M_f = 35^{+14}_{-4} M_\odot$, emitting $1.5^{+0.3}_{-0.4} M_\odot$ in energy. The spin of the final black holes was estimated in $a_f = 0.66^{+0.09}_{-0.10}$, but the spin of the initial ones is unknown (the effective spin $\chi_{\rm eff} = 0.0^{+0.3}_{-0.2}$ is compatible with zero). The luminosity distance was larger than the other two signals, specifically 1000^{+500}_{-500} Mpc, corresponding to a redshift $z = 0.20^{+0.09}_{-0.09}$, meaning it is about twice as far away as GW150914 and GW151226 sources.

4 Black Hole Binary Coalescence

The existence of black holes is currently accepted by the majority of astrophysicists [23]. Even if until 2016 the only evidence was indirect, like the movement of stars around a supermassive black hole or the electromagnetic emissions from the accretion disk in binaries. Nowadays, aLIGO has attained the first direct observation of black holes thanks to the detection of gravitational waves. Obviously, isolated black holes do not radiate gravitationally, but they emit radiation during their interaction with other black holes or neutron stars in black hole binaries.

Let me briefly review the current knowledge on the dynamics of black hole binaries obtained from numerical general relativity simulations [30]. First, the history of black holes is summarized here following Ref. [23]. Then, I stress a very interesting feature of the event horizon during the fusion of two black holes that it is not discussed in the current reviews on the topic: the appearance of an unstable toroidal topology, a ringhole, in the classical pants solution for black hole coalescence; it is very unstable, strongly emitting gravitational waves [15].

4.1 Black Holes

Black holes as compact massive objects with an escape velocity equal to the speed of light in vacuum were first discussed by Michell in 1784, although the mathematical calculations in Newton's gravitation were first made by Laplace in 1796. In Einstein's general relativity the concept appear in the first exact (spherically symmetric) solution of Einstein's equations in vacuum obtained by Schwarzschild in 1916, given by

$$ds^2 = -\left(1 - \frac{r_s}{r}\right) c^2\, dt^2 + \left(1 - \frac{r_s}{r}\right)^{-1} dr^2 + r^2\, d\theta^2 + r^2 \sin^2\theta\, d\theta^2\,,$$

where $r_S = 2\,G\,M/c^2$ is the Schwarzschild radius, M is the mass and spherical coordinates $(x^0, x^1, x^2, x^3) = (t, r, \theta, \phi)$ has been used. Black holes are extremely compact astrophysical objects. The Earth has a Schwarzschild radius of 9 mm, stellar-mass black holes are just a few kilometres across, and Sagittarius A*, the supermassive black hole at the centre of the Milky Way, with four million solar masses is smaller than 10% of Earth's orbital radius.

Let me stress that in general relativity a black hole is not a massive object, but just vacuum, pure curved space-time. They do not have a surface or physical boundary, being their size associated to its event horizon, where the scape velocity equals the speed of light in vacuum; hence, in this point of no return not even light can scape of the black hole's gravitational field. Since the black hole is an asymptotically flat solution, the total energy associated to its curvature, or equivalently to its gravitational field, can be properly defined. This total energy is referred to as mass, although it has no relation with the mass of a physical body. Arnowitt, Deser and Misner in 1959 clarified this point introducing the so-called ADM mass M, associated with the energy of the curvature field calculated asymptotically from infinity.

Einstein and other physicists refuse black hole solutions due to the presence of two singularities, one at the center of symmetry $r = 0$, and another one on the event horizon $r = r_S$. The general idea in the 1920 s is that the *Schwarzschild singularity* do not exist in physical reality because matter cannot be concentrated arbitrarily. But during the 1930 s, the Chandrasekhar limit for the mass of a compact massive star results in the study of the gravitational collapse by Oppenheimer and Snyder in 1939. Using interior co-moving coordinates, the collapse proceeds to zero radius in a finite proper time; however, for the external viewer the contraction slows down and *freezes* exactly at the gravitational radius.

Officially, Wheeler coined the term *black hole* in a lecture in late 1967, in order to avoid the longest term *gravitationally completely collapsed object*. However, the term was already used in December 1963 at the Texas Symposium on Relativistic Astrophysics in Dallas. The term first appears in print in the January 18, 1964 issue of Science News Letter, bandied in Cleveland at a meeting of the American Association for the Advancement of Science (the author was the reporter Ann Ewing). A few days later, it also appears in the January 24, 1964, issue of Life magazine (the author was the editor Al Rosenfeld). Both reporters listen the term in the Dallas meeting, but they do not remember the original author.

Black holes are not considered astrophysical objects until the 1960 s. One of the fathers of the soliton concept, Kruskal, introduced new coordinates that remove the apparent singularity at the event horizon in 1960. Kerr discovered in 1963 the solution for a rotating black hole, physically more appropriate for an astrophysical object. Novikov and Zel'dovich, Shklovsky, Burbidge, among others, proposed the hypothesis that some X-ray emitting binary systems are black holes with accretion disk from a close star. The discovery of the first astrophysical black hole is associated to such an interpretation for the X-ray binary Cygnus X-l, discovered in 1964; today it is widely accepted that it is a black hole with more than ten solar masses accreting matter from a blue supergiant star.

The strongest evidence for the existence of black holes comes from Sagittarius A*, the bright and very compact astronomical radio source at the center of the Milky Way. Discovered in 1974 and named in 1982, its mass has been determined by monitoring stellar orbits around it during the last twenty years. In particular, the star S2 has an elliptical orbit with a period of 15.2 years and a closest distance of 17 light hours from the center of the central object. From its motion, Sgr A* mass is estimated in about four million solar masses.

4.2 The Two-Body Problem in General Relativity

In general relativity the two-body problem is *unstable*, eventually resulting in the decay of the orbit and the collision of the two bodies. This fact is in contrast with Newtonian gravitation, where the two-body problem is stable and exactly solvable in closed form by using conic sections. Fortunately, the time scale of the instability in general relativity is larger than the Hubble time, except for very close compact objects, like neutron stars and black holes. In the last case, black hole binaries, the final result will always be a single rotating black hole described by the Kerr metric.

Figure 3 shows a reconstruction of the signal GW150914 observed by aLIGO at Hanford and its interpretation by the best-matching waveform computed by the numerical solution of Einstein field equations [6]. The signal starts when the two black holes are separated by only five Schwarzschild radii, with a relative velocity over a 30% of the speed of light in vacuum. The speed increases as both black holes approach each other until they merge; the result is not a black hole, but an unstable gravitational object that decays into one in a few tenths of a second.

The dynamics of black hole binaries can be broken down into four stages: *Newtonian, inspiral, plunge/merger*, and *ringdown* (see the top plots in Fig. 3). The gravitational wave emission is too weak to be detected in the *Newtonian* stage, where the two black holes are very far apart. In the *inspiral* regime the gravitational wave emission dominates driving the black holes to closer separation. This phase is well-modelled by post-Newtonian methods resulting in the so-called *effective one body* (EOB) theory. This theory yields waveforms surprisingly close to full numerical

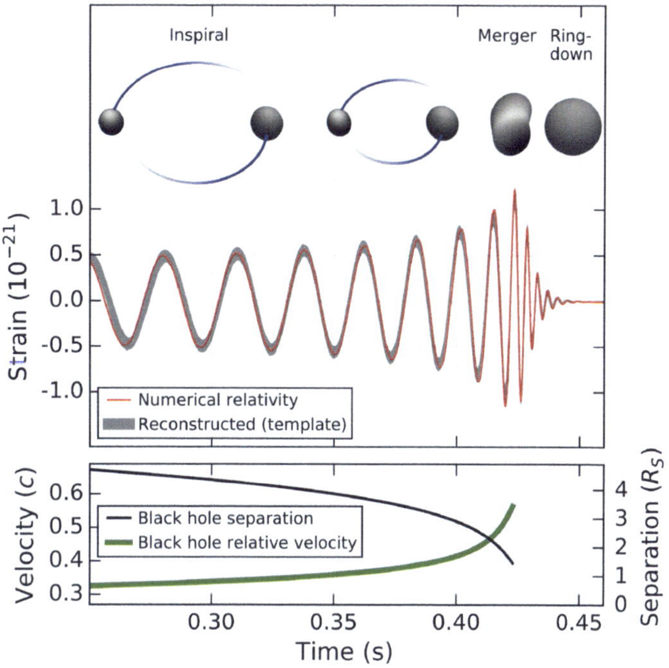

Fig. 3 The interpretation of the signal GW150914 as seen at Hanford over the three stages of the event: inspiral, merger and ringdown. The separation and velocity of the black holes are illustrated in the lower plot. Reproduced from [6]; published by the American Physical Society with a Creative Commons Attribution 3.0 License

results even until very close to the *plunge/merger* phase. The black holes plunge together to form a single object, the merger, with a strong emission of gravitational waves; the luminosity is on the order of one-hundredth of the Planck luminosity (about 10^{59} ergs/s), making black hole mergers by far the most energetic events in the post-big-bang era of the Universe. This phase is very short, lasting on the order of one to two gravitational wave cycles. For equal mass black holes upwards of 3% of the rest mass energy of the system is radiated away. A single black hole results as the consequence of a merger in the *ringdown* stage. This phase could be calculated using perturbative techniques, since the remnant black hole is a perturbed Kerr spacetime with *quasi-normal modes* in its event horizon. The ringdown frequency is several times higher than the orbital frequency in the last few inspiral cycles, and the decay time is quite short, so the majority of the energy lost during ringdown (1–2% of the rest mass) is emitted quite rapidly.

The collision of two black holes is usually referred to as either the *trousers* or the *pair of pants* solution, illustrated in Fig. 4. It cannot be treated analytically, although topological arguments can yield a general picture [22, 29]. The analysis by using numerical relativity, that solves Einstein's equations for general relativity, begins with the works of Hahn and Lindquist in 1964. However, the first calculations of

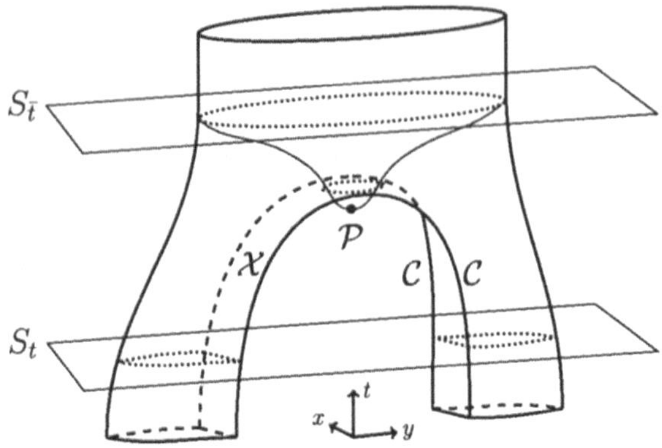

Fig. 4 A (2+1)-dimensional *pair of pants* representation of slices of constant time S_t and $S_{\bar{t}}$ through a binary black hole event horizon. The spatial hypersurface of crossover points $\mathscr{X}$ is surrounded on both sides by lines of caustics $\mathscr{C}$. The event horizon is toroidal, being $\mathscr{P}$ the center of the hole in the torus. Reproduced with permission from [15]. Copyrighted by the American Physical Society

the gravitational radiation during the collision of two black holes was first obtained by the Binary Black Hole Alliance in 1995, published by Matzner et al. [28]. Of course, the precise track of the coalescence and merger in a circular orbit as they spiral together was not solved until recent teraflop supercomputers were available.

There are two methods of formulating Einstein's equations amenable to stable numerical integration of binary black hole spacetimes [30]. The first one is based on the use of *generalized harmonic coordinates* to re-express Einstein's equations into a set of wave-like equations with constraint-preserving boundary conditions. The second one is the *Baumgarte-Shapiro-Shibata-Nakamura* formalism with *moving punctures*, based on the use of a conformal decomposition of the spatial metric to separate the extrinsic curvature into "radiative" versus "non-radiative" degrees of freedom. It is beyond the scope of this chapter to discuss either method in detail. But let me emphasize that both methods yield similar results in practice, so the choice between them is a matter of personal preference or aesthetics [30].

4.3 Ringholes in the Pair of Pants Solution

Numerical relativity has a prohibitive cost, so the interpretation of the waveform signals observed by gravitational wave interferometers are regularly done by using post-Newtonian approaches that only require a fine adjustment of their parameters from numerical simulations. Nevertheless, there are some features that a post-Newtonian approach cannot deal with, like nontrivial topology horizons that appear before the merger formation, as shown in Fig. 4 and the numerical simulations of Ref. [38]. Indeed, a numerical simulation of the signal GW150914 shows that a toroidal topology, a ringhole, appears in the event horizon, as shown in Fig. 5 from Ref. [15].

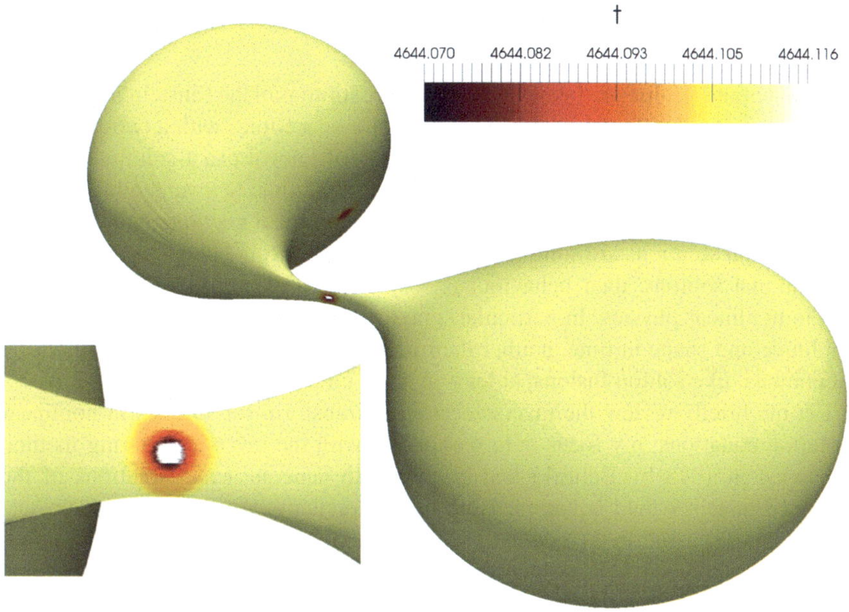

Fig. 5 Toroidal event horizon found in a GW150914 consistent SpEC simulation. The inset figure in the bottom left corner shows a zoomed in and slightly rotated viewpoint of the hole in the event horizon. Reproduced with permission from [15]. Copyrighted by the American Physical Society

The topological censorship theorem states that the ringhole must collapse faster than it would take light to traverse it. Hence, they emit a large amount of gravitational radiation.

In 1998 Siino, and in 1999 Husa and Winicour predicted that the event horizon of a generic binary black hole system should develop a brief toroidal event horizon due to the distribution of caustics and crossover points during the merger phase. Recently, Bohn, Kidder, and Teukolsky [15] has simulated with the Spectral Einstein Code (SpEC) a binary black hole with a mass ratio of 1.25 and spin parameters consistent with the source of the signal GW150914. Figure 5 shows that main result of their simulation, the emergence in a spacelike foliation of the spacetime of a toroidal topology inside the event horizon of the merger. The event horizon is a $2+1$-dimensional hypersurface whose localization requires a proper coordinate system with a specific time slicing, i.e., an event horizon finding code. A new one has been developed in Ref. [15] to be incorporated into SpEC in order to resolve fine-scale features of the event horizon. Figure 5 is the result of this new feature in the last version of SpEC; in fact, when using the standard SpEC time coordinate slicing such a ringhole is not observed. In the context of this chapter, the most important feature of the toroidal event horizon is that it transforms into a spherical event horizon in a few milliseconds, accompanied by an intense emission of gravitational radiation.

5 Gravitational Solitons

Black holes and nonlinear gravitational waves are solitons of the Einstein's field equation. Gravitational solitons are exact solutions for spacetimes with a two-parameter group of isometries. Surprisingly, the whole set of known exact solutions to Einstein's equations are particular cases of soliton solutions. Schwarzchild and Kerr black holes, and their generalizations, exact and cylindrical gravitational waves, and cosmological solutions are solitons. Despite the term *soliton* being used to describe gravitational solitons, their behaviour is very different from other (classical) solitons in nonlinear physics. In particular, gravitational solitons do not preserve their amplitude and shape in time, neither their collisions are elastic, admitting inelastic phenomena, like soliton fusions, at least in some circumstances.

Let me briefly review the inverse scattering transform for solitons in nonlinear evolution equations; my focus is to compare it with the inverse scattering method developed in 1978 by Belinski and Zakharov to generate exact solutions of the vacuum Einstein's field equation. Both exact gravitational waves and black hole metrics are discussed in the context of gravitational solitons with emphasis in their fusion or coalescence, a phenomenon not observed in classical solitons. This section is based on the book by Belinski and Verdaguer [14], highly recommended to the reader interested in a detailed presentation; a briefer one can be found in Ref. [13].

5.1 Solitons

Solitons are nonlinear waves with particle-like behaviour [9]. Specifically, they have finite and localized energy, a finite velocity of propagation, and a persistent shape even after they collide. Its prehistory starts in the XIX-th century, but the soliton concept was introduced by Zabusky and Kruskal in 1965. These researchers made a numerical analysis of the Korteweg–de Vries (KdV) equation as a continuous model of the Fermi–Pasta–Ulam–Tsingou problem. The particle-like behaviour observed in their simulations was understood around 1967 thanks to the work of Gardner, Greene, Kruskal, and Miura (GGKM). They develop the so-called Inverse Scattering Transform (IST) to obtain the general solution of the Cauchy problem for the KdV equation. In 1968, Lax showed that the equations solvable with the IST of GGKM correspond to the compatibility condition of a pair of linear operators, the so-called pair of Lax operators.

The IST developed by GGKM applies to first-order evolution equations. Each nonlinear, one-dimensional, partial differential equation (PDE) solvable by the IST is the compatibility condition of the pair of Lax operators $(\mathscr{L}, \mathscr{A})$. Let us write the equation as

$$\mathscr{P}_1(u) \equiv \partial_t u - F(u, \partial_x u, \partial_x^2 u, \ldots) = 0 ,$$

where t is the time variable, x is the space coordinate, and F is a nonlinear function. The IST interprets this equation as the compatibility condition

$$(\partial_t \mathscr{L} + \mathscr{L}\mathscr{A} - \mathscr{A}\mathscr{L})\,\psi \equiv \mathscr{P}_1(u)\,\psi\,,$$

of a pair of linear differential operators $\mathscr{L}(\partial_x)$ and $\mathscr{A}(\partial_x)$, both with coefficients depending on $u(x,t)$ and its derivatives, associated to the linear problems

$$\mathscr{L}\psi = \lambda\,\psi\,, \qquad \partial_t\psi = \mathscr{A}\psi\,.$$

The IST requires that the eigenvalues of the problem $\mathscr{L}\psi = \lambda\,\psi$, do not change in time, i.e., $\partial_t\lambda = 0$, and that $\tilde{\psi} \equiv \partial_t\psi - \mathscr{A}\psi$ be an eigensolution, i.e., $\mathscr{L}\tilde{\psi} = \lambda\,\tilde{\psi}$. For example, for the KdV equation

$$\partial_t u - 6\,u\,\partial_x u + \partial_x^3 u = 0\,,$$

the Lax operators discovered by GGKM are

$$\mathscr{L} = -\partial_x^2 + u\,, \qquad \mathscr{A} = -4\,\partial_x^3 + 6\,u\,\partial_x + 3\,u_x\,.$$

In 1972 Zakharov and Shabat showed that the IST of GGKM is applicable to the IVP for the cubic nonlinear Schrödinger equation. Wadati showed in 1972 that the method extends to the modified Korteweg–de Vries (mKdV) equation. In 1973 Ablowitz, Kaup, Newell, and Segur (AKNS) showed that the IVP for the sine-Gordon equation can be solved in the same way, but instead of using a Lax pair, they introduce the so-called pair of AKNS operators. Since then, many other nonlinear evolution equations have been discovered to be solvable by the Inverse Scattering Transform which is currently interpreted as a nonlinear generalization of the Fourier Transform.

For second-order evolution equations the IST developed by AKNS should be used, which is applicable to first-order ones too. The nonlinear PDE solvable by the IST is the compatibility condition of the AKNS operators $(\mathscr{U}, \mathscr{V})$. The equation

$$\mathscr{P}_2(u) \equiv \partial_\tau^2 u - G(u, \partial_z u, \partial_z^2 u, \ldots) = 0\,,$$

must be the compatibility condition

$$(\partial_\tau \mathscr{U} - \partial_z \mathscr{V} + \mathscr{U}\mathscr{V} - \mathscr{V}\mathscr{U})\,\Psi \equiv \mathscr{P}_2(u)\,\Psi\,,$$

of a pair of linear differential operators $\mathscr{U}(\partial_z;\lambda)$ and $\mathscr{V}(\partial_z;\lambda)$, both with coefficients depending on $u(z,\tau)$ and its derivatives, associated to the problems

$$\partial_z\Psi = \mathscr{U}\,\Psi\,, \qquad \partial_\tau\Psi = \mathscr{V}\,\Psi\,.$$

The AKNS method requires that the spectral parameter λ in the problem $\partial_z\Psi = \mathscr{U}\,\Psi$ do not change in time, i.e., $\partial_\tau\lambda = 0$, and that $\tilde{\Psi} \equiv \partial_\tau\Psi - \mathscr{V}\psi$ be a solution of $\partial_z \tilde{\Psi} = \mathscr{U}\, \tilde{\Psi}$. For example, for the sine-Gordon equation $\partial_t^2 u - \partial_x^2 u = \sin(u)$, written as $\partial_z \partial_\tau u = \sin u$, after using light-cone coordinates $t = \tau - z$ and $x = \tau + z$, the AKNS pair is given by

$$\mathscr{U} = \begin{pmatrix} -\mathrm{i}\,\lambda & -\partial_z u/2 \\ \partial_z u/2 & \mathrm{i}\,\lambda \end{pmatrix}, \quad \mathscr{V} = \frac{\mathrm{i}}{4\lambda} \begin{pmatrix} \cos u & \sin u \\ \sin u & -\cos u \end{pmatrix}, \quad \Psi = \begin{pmatrix} \partial_z \psi \\ \psi \end{pmatrix}.$$

5.2 *Gravisolitons*

The inverse scattering method (ISM) developed by Belinski and Zakharov (BZ) in 1978 was inspired in the IST of AKNS. The gravisolitons, or gravitational solitons, are metrics $g_{\alpha\beta}$, for $\alpha, \beta = 0, 1, 2, 3$, admitting two commuting Killing vectors, i.e., two Lie symmetries. Basically, there are four spacetimes admitting these symmetries: cylindrical gravitational waves, exact plane waves, inhomogeneous cosmologies, and stationary axisymmetric spacetimes. In the first three cases, both Killing vectors are space-like, while in the last case, one Killing vector is space-like and the other one is time-like. In order to compare the IST with the ISM, in this section only the last case will be considered, i.e., gravisolitons representing black hole solutions over a Minkowski spacetime [14].

Let us take a four-dimensional metric in cylindrical coordinates $(x^0, x^1, x^2, x^3) = (t, r, \theta, z)$, using relativistic units with $c = 1$. The metric is stationary and axisymmetric if it has the two Killing vectors ∂_t and ∂_θ, that generate the time flow and, locally, the axial symmetry. Such a metric can be written as

$$ds^2 = g_{ab}(r, z)\, dx^a\, dx^b + f(r, z)\, (dr^2 + dz^2)\,, \tag{8}$$

where g_{ab}, for $a, b = 0, 2$, is a Lorentz metric with determinant $\det g_{ab}(r, z) = -r^2$ (the last condition can be taken without loss of generality). Notice that this metric is a generalization of the Schwarzschild metric for a spherical, stationary black hole, first published in 1916; it also generalizes the Kerr metric for an axisymmetric, rotating black hole, first published in 1963.

The Einstein equations in vacuum $R_{\alpha\beta} = 0$ with metric (8) transform into simple equations for $g_{ab}(r, z)$ and $f(r, z)$. Specifically, a 2×2 matrix, elliptic, partial differential equation for the metric $g_{ab}(r, z)$, and two scalar, hyperbolic, partial differential equations for $f(r, z)$. The metric $g_{ab}(r, z)$ solves

$$\partial_r U + \partial_z V = 0\,, \tag{9}$$

where U and V are the 2×2 matrices

$$U = r(\partial_r\, g_{ab})\, g^{ab}\,, \qquad V = r(\partial_z\, g_{ab})\, g^{ab}\,, \tag{10}$$

and g^{ab} is the inverse of the matrix g_{ab}. And the function f solves

$$\partial_r f = -\frac{f}{r} + \frac{f}{4r}\mathrm{Tr}\left(U^2 - V^2\right), \qquad \partial_z f = \frac{f}{2r}\mathrm{Tr}\,(U\,V)\,, \tag{11}$$

where Tr is the trace operator. In order to obtain a soliton solution, first Eqs. (9) and (10) are solved for g_{ab}, and subsequently Eq. (11) for f. Notice that the application of the trace operator to Eq. (9) results in a linear wave equation for the absolute value of the square root of the determinant of the metric g_{ab}, whose solution is $\vartheta_1(r-z) + \vartheta_2(r+z)$.

Apparently, Eqs. (9)–(11) are not related to the IST developed by AKNS. However, Belinski and Zakharov interpreted Eq. (9) as the compatibility condition for an AKNS pair of linear differential equations given by

$$D_r\psi = \frac{r\,U + \lambda\,V}{\lambda^2 + r^2}\,\psi\,, \quad D_z\psi = \frac{r\,V - \lambda\,U}{\lambda^2 + r^2}\,\psi\,, \tag{12}$$

where λ is a complex constant independent of the coordinates r and z, $\psi(r, z; \lambda)$ is the so-called generating matrix, such that $g(r, z) = \psi(r, z; 0)$, and the differential operators D_r and D_z are defined as

$$D_r = \partial_r + \frac{2\,\lambda\,r}{\lambda^2 + r^2}\,\partial_\lambda\,, \quad D_z = \partial_z - \frac{2\,\lambda^2}{\lambda^2 + r^2}\,\partial_\lambda\,. \tag{13}$$

The determination of the general solution of Eqs. (9)–(11) is outside the scope of this review; the interested reader is referred to Ref. [14]. Let me only highlight the main difference between the ISM of BZ and the IST of both GGKM and AKNS. The IST, when applicable to a given nonlinear evolution equation, can be used to obtain the general solution of its Cauchy problem; this solution is composed of a multisoliton solution (a finite number of solitons in pairwise interaction) over a radiation background. However, the ISM cannot be used to obtain the general solution of Einstein field equation with two Killing vectors, but only to obtain gravitational multisoliton solutions. Indeed, the ISM is reminiscent of the Bäcklund transformation or the dressing operator methods that allow the construction of multisoliton solutions by explicit combination of individual solitons.

The ISM allows the construction of multisoliton solutions over a given background metric, the so-called "seed solution," by combining individual gravisolitons. From a seed g_0, the matrices U_0 and V_0 are computed by solving Eqs. (9) and (10), and the generating matrix $\psi_0(r, z; \lambda)$ by solving Eqs. (12) and (13). For the Minkowski metric $g_0 = \mathrm{diag}(-1, r^2)$, $f_0 = 1$, and $\psi_0 = \mathrm{diag}(-1, r^2 - 2\,z\,\lambda - \lambda^2)$. Next, a new generating matrix $\psi = \chi\,\psi_0$ is determined, where the *dressing matrix* $\chi(r, z; \lambda)$ is a 2×2 matrix admitting, for multisoliton solutions, the form

$$\chi = I + \sum_{k=1}^{n} \frac{R_k}{\lambda - \mu_k}\,, \tag{14}$$

where I stands for the identity matrix, and the residue matrices R_k and the poles μ_k are independent of λ, i.e., they depend only on the coordinates r and z. The n poles are usually referred to as "pole trajectories" since they are nonconstant functions $\mu_k(r, z)$ that satisfies the pair of differential equations given by

$$\partial_z \mu_k = \frac{-2\,\mu_k^2}{\mu_k^2 + r^2}, \quad \partial_r \mu_k = \frac{2\,r\,\mu_k}{\mu_k^2 + r^2}, \tag{15}$$

whose solutions are the roots of the quadratic algebraic equation

$$\mu_k^2 + 2\,(z - \omega_k)\,\mu_k - r^2 = 0\,, \tag{16}$$

where ω_k are arbitrary, generally complex, constants. Hence, the two pole trajectories for each ω_k are given by

$$\mu_k = \omega_k - z \pm \sqrt{(w_k - z)^2 + r^2}\,. \tag{17}$$

The matrices R_k in Eq. (14) are degenerate and determined by two vectors, i.e., $(R_k)_{ab} = n_a^{(k)}\,m_b^{(k)}$; the vector $m^{(k)}$ depends only on the seed metric g_0, but the vector $n^{(k)}$ has arbitrary values, the so-called BZ parameters of each soliton. After some algebra [14], the metric (8) can be written as

$$g = \left(I - \sum_{k=1}^{n} \frac{R_k}{\mu_k} \right) g_0\,. \tag{18}$$

The determinant of the matrix g is

$$\det g = (-1)^n\, r^{2n} \left(\prod_{k=1}^{n} \frac{1}{\mu_k^2} \right) \det g_0\,, \tag{19}$$

where the condition $\det g_0 = -r^2$ implies that the number of gravisolitons n must always be even; an odd number leads to an unphysical metric signature with the opposite sign (which is equivalent to flat space without gravisolitons). All stationary axisymmetric solitons in Eq. (18) appear in pairs forming two-soliton bound states.

The simplest case for the metric (8) is the two-soliton solution, i.e., two poles on a flat spacetime background. For two complex poles, i.e., with ω_k complex, the metric presents a naked singularity without a horizon, so it is forbidden by the cosmic censorship hypothesis. In the case of two real poles, i.e., with ω_k real, the singularity in the metric (8) is hidden behind an event horizon, so it corresponds to a black hole solution. Specifically, the Kerr–NUT (Newman–Unti–Tamburino) black hole is obtained as the most general two-soliton solution with real-valued pole trajectories μ_1 and μ_2 in Eq. (14). Let me omit the detailed mathematical expression, but only emphasize that it has three real parameters m, a, and b. In the Kerr metric ($b = 0$) and the Scharzchild metric ($a = b = 0$) the parameter m corresponds to the ADM mass of the black hole. The Taub–NUT metric ($a = 0, b \neq 0$) and the general Kerr–NUT metric ($a \neq 0$, $b \neq 0$) are not asymptotically flat so they cannot be interpreted as physical black holes.

For further details on the construction and the analysis of the general properties of gravitational n-soliton solutions obtained by the ISM my best recommendation for the reader is the book by Belinski and Verdaguer [14].

Let me clarify why, in physical applications of general relativity, gravisolitons do not have the key role that solitons have in nonlinear physics. From the mathematical point is understandable that gravitational solitons are localized perturbations on a gravitational background described by the ISM that behave like solitons described by the IST. However, their physical interpretation have several drawbacks. Specifically, it is difficult to define concepts like amplitude, velocity of propagation, or even shape for gravisolitons in analogy to those concepts in solitons. Even there is no notion of energy (or mass) for the gravisolitons; although the root of this problem is that energy cannot be defined locally in general relativity, but only globally in *asymptotically flat* spacetimes (isolated systems). Moreover, for real-pole trajectories, the field of the gravisolitons is not smooth in spacetime, presenting discontinuities in the first derivatives at some null hypersurfaces. In fact, there is no time evolution of the gravisolitons from a free state at $t = -\infty$ to a free state at $t \to \infty$, without the appearance of singularities in some regions.

Gravisolitons share some properties with classical solitons in nonlinear physics but they cannot be considered *true* solitons, since the poles in the ISM are not constant, but pole trajectories. In the collision of two gravisolitons their *shapes* change, instead of the shape preservation property of solitons. The usual shift in the positions of solitons after their collisions, in the case of gravisolitons, corresponds to a change in shape and a shift in time. Furthermore, gravisolitons show surprising properties, from the point of view of classical solitons, like soliton coalescence, or pole fusion [14]. Concretely, two pole trajectories can coincide at some point in spacetime and fuse together transforming into a single pole trajectory, i.e., a single pole. The fusion of two solitons into only one is not observed in other physical systems having solitons.

5.3 *Cylindrical and Planar Gravitational Waves*

Exact, nonlinear, gravitational waves are referred to as either cylindrical waves or planar waves. The first ones are cylindrically symmetric gravisolitons, referred to as cylindrical waves by Kompaneets in 1958, who introduced the metric

$$ds^2 = e^{2(\gamma-\psi)}\,(-dt^2 + dr^2) + r^2\,e^{-2\psi}\,d\theta^2 + e^{2\psi}\,(dz + \omega\,d\theta)^2\,, \qquad (20)$$

where the functions γ, ψ, and ω depend on t and r only; this metric, for $\omega = 0$, reduces to the Einstein–Rosen metric published in 1937. The second ones are plane-fronted gravitational waves with parallel rays (pp-waves), first studied by Brinkmann in 1925, and Baldwin and Jeffery in 1926, but currently referred to as planar waves. Both cylindrical and planar waves have two polarizations, like linear gravitational waves, reducing to them asymptotically.

Cylindrical and planar gravitational waves are gravisolitons for a dynamical metric with two commuting, space-like, Killing vectors. Cylindrical solitons are given by

$$ds^2 = f(t,r)\,(-dt^2 + dr^2) + g_{ab}(t,r)\,dx^a\,dx^b\,, \tag{21}$$

where g_{ab}, for $a, b = 2, 3$, is an Euclidean metric with determinant $\det g_{ab}(t,r) = r^2$. This dynamical and cylindrical metric has the Killing vectors ∂_θ and ∂_z, that generate, locally, the cylindrical symmetry. The Einstein equations in vacuum $R_{\alpha\beta} = 0$ with the metric (22) transform into simple differential equations for $g_{ab}(t,r)$ and $f(t,r)$; concretely, a 2×2 matrix, hyperbolic (instead of elliptic like in black hole solitons), partial differential equation for the metric $g_{ab}(t,r)$, and two scalar, hyperbolic, partial differential equations for $f(t,r)$. The ISM proceeds similarly to the case presented in the previous section, but with small and significant differences; let me omit further details and, again, refer to the book by Belinski and Verdaguer [14].

Planar solitons are given by

$$ds^2 = f(u,v)\,du\,dv + g_{ab}(u,v)\,dx^a\,dx^b\,, \tag{22}$$

where $u = t - z$, $v = t + z$, and g_{ab}, for $a, b = 1, 2$, is an Euclidean metric with determinant $\det g_{ab}(u,v) = \vartheta(u,v)^2$. This dynamical metric has the Killing vectors ∂_r and ∂_θ, that generate, locally, the planar symmetry (in cylindrical coordinates).

The exact gravitational waves obtained by the ISM of BZ generally have the problems already found by Einstein and other physicists, the appearance of singularities during their mutual interactions. When two of such nonlinear waves collide, due to mutual focusing, the interaction region becomes locally isometric to the interior of the Schwarzchild metric, with a singularity or Cauchy horizon where $\det g = 0$. The current interpretation of this phenomenon is that a black hole forms when two *pp*-waves, or two cylindrical waves, collide. This interpretation has resulted from the analysis of multisoliton gravitational solutions obtained by the ISM; although, it was pioneered by Tomimatsu and Sato in 1972 before the development of the ISM.

5.4 *The Fusion of Black Hole Solitons*

The source of the gravitational waves observed by aLIGO, the coalescence of two stellar-mass black holes, in principle, can be described by an exact solution using the ISM of BZ applied to the metric (8). Astrophysical black holes are described by either Schwarzchild or Kerr gravisolitons, that corresponds to two pole solitons. A reasonable proposal is that black hole binary systems be interpreted as four pole solitons (14). However, the physical interpretation of these metrics, already discovered before the development of the ISM by Tomimatsu and Sato [34], has been a subject of controversy.

The double Kerr solution is a four soliton solution with two pairs of pole trajectories (μ_1, μ_2) and (μ_3, μ_4). The fusion or coalescence of both Kerr black holes into a single one results from taking the limit $\omega_3 \to \omega_1$ and $\omega_4 \to \omega_2$, keeping the NUT parameter $b = 0$ for asymptotic flatness; after this limit process, the pole trajectories coincide at some point (r^*, z^*), i.e. when $\mu_1(r^*, z^*) = \mu_3(r^*, z^*)$ and $\mu_2(r^*, z^*) = \mu_4(r^*, z^*)$, where a final Kerr-like solution results. Apparently, the process is straightforward, however, there is a physical problem, the appearance of a naked ring singularity and closed-time curves near the horizon in the final solution; the cosmic censorship hypothesis forbidden this kind of black hole gravisoliton.

Let me conjecture a new physical interpretation of the pole fusion. The 2-soliton Kerr solution has two non-null parameters, the mass m and the spin a (recall that $b = 0$). However, the asymptotically flat, 4-soliton, double Kerr solution has five parameters; additionally to the masses (m_1 and m_2) and the spins (a_1 and a_2), there is a positive distortion (or deformation) parameter δ. Originally, Tomimatsu and Sato interpreted this parameter as an integer equal to half of the number of solitons ($\delta = 1$ for the Kerr solution and $\delta = 2$ for the double Kerr solution). The physical difficulty with this parameter is the appearance of a ring naked singularity outside the horizon for $\delta > 1$; note that this singularity is similar to the ringhole discussed in Sect. 4.3.

The fusion of the double Kerr solution results in a single Kerr solution, but with $\delta = 2$ instead of $\delta = 1$. Hence, for the majority of physicists, the Tomimatsu–Sato fusion solution is unphysical [14]. However, there is nothing that forbids non-integer values for δ. In fact, the distortion parameter is related to the total quadrupole momentum Q of the 4-pole gravisoliton, given by

$$Q = M^3 \left(a^2 + (1 - a^2) \frac{\delta^2 - 1}{3\,\delta^2} \right),$$

where $M = m_1 + m_2$ and $a = J/M$ is the total angular momentum of the binary system. The stability analysis of the Tomimatsu–Sato has not been published, but the cosmic censorship hypothesis suggests that the $\delta = 2$ Kerr solution resulting from the fusion must be unstable; in such a case, I conjecture that it can dynamically change into a $\delta = 1$ Kerr solution. In this process the quadrupole momentum reduces its initial value in an amount of $\Delta Q = M^3\,(1 - a^2)/4$ by the emission of gravitational waves.

The detailed analysis of the continuous change of the distortion parameter from $\delta = 2$ to $\delta = 1$ for the double Kerr solution during pole fusion has not been published yet. The process requires numerical simulations similar to those presented by Tomizawa and Mishima for the fusion of cylindrical waves [35]. They have shown that the fusion of two ingoing solitons at past null infinity result in a single outgoing soliton at near future null infinity, including the loss of some energy in the form of linear gravitational waves. Further research in this topic is welcome.

6 Conclusions

The direct detection of gravitational waves generated by the fusion of two black holes by both interferometers of Advanced LIGO was Science's 2016 Breakthrough of the Year. Gravitational waves solve the weak-field approximation of the Einstein equations in vacuum, a limit in which they evolve as linear waves. However, exact gravitational waves can be obtained by means of using the inverse scattering method developed by Belinski and Zakharov. These nonlinear gravitational waves behave as gravitational solitons, or gravisolitons, propagating energy and momentum like an intrinsically nonlinear phenomenon. Moreover, Kerr black holes are also soliton solutions of Einsteins equation in vacuum. Surprisingly, there is an exact solution for the fusion of two black holes; its physical interpretation is controversial, since ring singularities appear outside the event horizon. However, numerical simulations of the collision of two black holes in general relativity also show a ringhole in the pair of pants solution. It is very unstable, radiating energy in the form of gravitational waves and transforming into a standard horizon. This author has conjectured that a similar interpretation can be done for the exact multisoliton solution modelling the coalescence of two black holes. The confirmation of this conjecture requires gravisoliton numerical simulations.

The future of gravitational wave astronomy is brilliant. In 2017, the three-kilometre laser interferometer Advanced Virgo, in Italy, has joined the two four-kilometre interferometers of Advanced LIGO, making easy the precise localization of the sources in the sky. In 2018, the three-kilometre interferometer Kamioka Gravitational Wave Detector (KAGRA), in Japan, is expected to enter operation. And in 2020, another four-kilometre Advanced LIGO interferometer must be listening gravitational waves in India. Moreover, a consortium of European partners is planning a 10 Km laser interferometer, the Einstein Telescope, for the late 2020 s. Finally, the European Laser Interferometer Space Antenna (eLISA), a constellation of three satellites forming a triangular interferometer with one-million-kilometre arms, is currently in development by the European Space Agency, to be deployed in space in the mid-2030s. What new astrophysical phenomena will gravitational wave astronomy give us in the future? Nobody knows, but the field is really exciting.

Acknowledgements The author acknowledges financial support from project TIN2014-56494-C4-1-P from Programa Estatal de Fomento de la Investigación Científica y Técnica de Excelencia del Ministerio de Ciencia e Innovación (MICINN) of Spain.

References

1. Aasi, J., et al.: Advanced LIGO. Class. Quant. Grav. **32**, 074001 (2015)
2. Abbott, B.P., et al.: Binary black hole mergers in the first advanced LIGO observing run. Phys. Rev. X **6**(4), 041015 (2016)
3. Abbott, B.P., et al.: Directly comparing GW150914 with numerical solutions of Einsteins equations for binary black hole coalescence. Phys. Rev. D **94**(6), 064035 (2016)

4. Abbott, B.P., et al.: GW151226: observation of gravitational waves from a 22-solar-mass binary black hole coalescence. Phys. Rev. Lett. **116**(24), 241103 (2016)
5. Abbott, B.P., et al.: Improved analysis of GW150914 using a fully spin-precessing waveform model. Phys. Rev. X **6**(4), 041014 (2016)
6. Abbott, B.P., et al.: Observation of gravitational waves from a binary black hole merger. Phys. Rev. Lett. **116**(6), 061102 (2016)
7. Abbott, B.P., et al.: Properties of the binary black hole merger GW150914. Phys. Rev. Lett. **116**(24), 241102 (2016)
8. Abbott, B.P., et al.: Effects of waveform model systematics on the interpretation of GW150914. Class. Quant. Grav. **34**(10), 104002 (2017)
9. Ablowitz, M.J., Segur, H.: Solitons and the Inverse Scattering Transform. SIAM, Philadelphia, USA (1981)
10. Ackermann, M., et al.: Fermi-LAT observations of the LIGO event GW150914. Astrophys. J. **823**(1), L2 (2016)
11. Adrian-Martinez, S., et al.: High-energy Neutrino follow-up search of gravitational wave event GW150914 with ANTARES and IceCube. Phys. Rev. D **93**(12), 122010 (2016)
12. Aldrovandi, R., Pereira, J.G., da Rocha, R., Vu, K.H.: Nonlinear gravitational waves: their form and effects. Int. J. Theor. Phys. **49**, 549–563 (2010)
13. Beheshti, S., Tahvildar-Zadeh, S.: Dressing with control: using integrability to generate desired solutions to Einstein's equations. In: Cuevas-Maraver, J., Kevrekidis, P.G., Williams, F. (eds.) The Sine-Gordon Model and Its Applications: From Pendula and Josephson Junctions to Gravity and High-Energy Physics, pp. 207–231. Springer (2014)
14. Belinski, V., Verdaguer, E.: Gravitational Solitons. Cambridge University Press, Cambridge, UK (2005)
15. Bohn, A., Kidder, L.E., Teukolsky, S.A.: Toroidal horizons in binary black hole mergers. Phys. Rev. D **94**(6), 064009 (2016)
16. Bondi, H., Pirani, F.A.E., Robinson, I.: Gravitational waves in general relativity iii. exact plane waves. Proc. R. Soc. London **A251**, 519–533 (1959)
17. Braginsky, V.B., Bilenko, I.A., Vyatchanin, S.P., Gorodetsky, M.L., Mitrofanov, V.P., Prokhorov, L.G., Strigin, S.E., Khalili, F.: Ya.: The road to the discovery of gravitational waves. Phys. Usp. **59**(9), 879–885 (2016)
18. Cherepashchuk, A.M.: Discovery of gravitational waves: a new chapter in black hole studies. Phys. Usp. **59**(9), 910–917 (2016)
19. Denson Hill, C., Nurowski, P.: How the green light was given for gravitational wave search. arXiv:1608.08673 [physics.hist-ph] (2016)
20. Faraoni, V.: A Common misconception about LIGO detectors of gravitational waves. Gen. Rel. Grav. **39**, 677–684 (2007)
21. Garfinkle, D.: Gauge invariance and the detection of gravitational radiation. Am. J. Phys. **74**, 196–199 (2006)
22. Hawking, S.W., Ellis, G.F.R.: The Large Scale Structure of Space-Time. Cambridge Monographs on Mathematical Physics. Cambridge University Press, Cambridge, UK (1973)
23. Israel, W.: Dark stars: the evolution of an idea. In: Hawking, S. W., Israel, W. (eds.) Three Hundred Years of Gravitation, pp. 199–276. Cambridge University Press (1987)
24. Kennefick, D.: Traveling at the Speed of Thought: Einstein and the Quest for Gravitational Waves. Princeton University Press, Princeton, US (2007)
25. Landau, L.D., Lifshitz, E.M.: The Classical Theory of Fields, Course of Theoretical Physics, vol, vol. 2. Pergamon Press, Oxford, UK (1975)
26. Lipunov, V.M.: Astrophysical meaning of the discovery of gravitational waves. Physics-Uspekhi **59**(9), 918–928 (2016)
27. Maggiore, M.: Gravitational Waves. Vol. 1: Theory and Experiments. Oxford Master Series in Physics. Oxford University Press, New York, USA (2007)
28. Matzner, R.A., Seidel, H.E., Shapiro, S.L., Smarr, L., Suen, W.M., Teukolsky, S.A., Winicour, J.: Geometry of a black hole collision. Science **270**, 941–947 (1995)

29. Misner, C.W., Thorne, K.S., Wheeler, J.A.: Gravitation. W. H. Freeman, San Francisco, USA (1973)
30. Pretorius, F.: Binary black hole coalescence. In: Colpi, M., Casella, P., Gorini, V., Moschella, U., Possenti, A. (eds.) Physics of Relativistic Objects in Compact Binaries: From Birth to Coalescence, pp. 305–369. Springer (2009)
31. Pustovoit, V.I.: On the direct detection of gravitational waves. Physics-Uspekhi **59**(10), 1034–1051 (2016)
32. Saulson, P.R.: If light waves are stretched by gravitational waves, how can we use light as a ruler to detect gravitational waves? Am. J. Phys. **65**, 501–505 (1997)
33. Thorne, K.S.: Gravitational radiation. In: Hawking, S.W., Israel, W. (eds.) Three Hundred Years of Gravitation, pp. 330–458. Cambridge University Press (1987)
34. Tomimatsu, A., Sato, H.: New exact solution for the gravitational field of a spinning mass. Phys. Rev. Lett. **29**, 1344–1345 (1972)
35. Tomizawa, S., Mishima, T.: Nonlinear effects for a cylindrical gravitational two-soliton. Phys. Rev. D **91**(12), 124058 (2015)
36. Trautman, A.: Radiation and boundary conditions in the theory of gravitation. Bull. Acad. Polon. Sci. **6**, 407–412 (1958)
37. Weinstein, G.: General Relativity Conflict and Rivalries: Einstein's Polemics with Physicists. Cambridge Scholars Publishing, Cambridge, UK (2015)
38. Winicour, J.: Characteristic evolution and matching. Living Rev. Rel. **12**, lrr-2009-3 (2009)

Part III
Differential and Difference Equations

Local Integrability for Some Degenerate Nilpotent Vector Fields

Antonio Algaba, Isabel Checa and Cristóbal García

Abstract This work is about the analytic integrability problem around the origin in a family of degenerate nilpotent vector fields. The integrability problem for planar vector fields with first Hamiltonian component having simple factors in its factorization on $\mathbb{C}[x, y]$ is solved in Algaba et al. (Nonlinearity 22:395–420, 2009) [5]. Nevertheless, when the Hamiltonian function has multiple factors on $\mathbb{C}[x, y]$ is an open problem. In this second case our problem is framed. More concretely, we study the following degenerate systems:

$$\dot{x} = -y(x^{2n} + ny^2) + \cdots, \quad \dot{y} = x^{2n-1}(x^{2n} + ny^2) + \cdots,$$

with $n \in \mathbb{N}$, where its first quasi-homogeneous component has Hamiltonian function given by $(x^{2n} + ny^2)^2/(2n)$. The analytic integrability of the above system is not completely solved and only partial results are obtained. The results are applied to some particular families of degenerate vector fields for which the integrability problem is completely solved.

Keywords Integrability problem · Degenerate center problem · First integral
Orbital normal form · Blow-up · Conservative-dissipative splitting
Nilpotent systems

A. Algaba · I. Checa (✉) · C. García
Departamento de Matemáticas, Centro de Investigación de Física Teórica y
Matemática FIMAT, Universidad de Huelva, 21071 Huelva, Spain
e-mail: isabel.checa@dmat.uhu.es

A. Algaba
e-mail: algaba@uhu.es

C. García
e-mail: cristoba@uhu.es

V. Carmona et al. (eds.), *Nonlinear Systems, Vol. 1*,
Understanding Complex Systems, https://doi.org/10.1007/978-3-319-66766-9_8

1 Introduction

The study of the integrability is to find the existence of first integrals and finding the functional class which this first integral of a given differential system must belong to. This is one of the main open problems in the qualitative theory of differential systems in $\mathbb{R}^2$ (see [5, 13, 15, 22, 38] and references therein).

A method for determining the phase plane of a given planar system around an equilibrium point consists in obtaining a first integral; that is, a function which is nonconstant, defined in some nonempty open subset of $\mathbb{R}^2$, which is constant along the solution curves of the system.

Another open problem is the center problem, which consists in the distinction between a center and a focus. This problem is closely related to the integrability and the reversibility problem (see [3, 4, 6, 7, 16, 21, 22, 34, 39] and references therein).

In this work, we will denote the system as

$$\dot{\mathbf{x}} = \mathbf{F}(\mathbf{x}), \tag{1}$$

where $\mathbf{x} = (x, y)^T \in \mathbb{R}^2$ and $\mathbf{F} = (P, Q)^T$ with P and Q analytic functions that vanish at the origin.

The above system is formally (analytically) integrable if there exists a formal (analytical) first integral. Clearly, H is a $\mathscr{C}^1$ first integral if, and only if, it satisfies $\nabla H \cdot \mathbf{F} = 0$ (integrability equation).

With regard to the center problem, a system has a center at a singular point only if it has linear part of center type, i.e. with imaginary eigenvalues, or nilpotent linear part or null linear part and it also is monodromic. Any nondegenerate center has always a local analytic first integral in a neighborhood of its singular point (see [7, 22, 31, 35–38]), that is, the center problem and the integrability problem are equivalent to a nondegenerate singular point. However, there exist nilpotent and degenerate centers that do not have a local analytic first integral (see [23, 27–30, 33, 34, 38, 40] and references therein). There exist methods to determine nondegenerate and nilpotent centers of a given polynomial system (see [1, 10, 12, 17, 25, 26, 30, 33, 34]). Nevertheless there is not any method to determine centers for a general degenerate singular point.

The characterization of a nilpotent center in terms of its normal form is due to Moussu [33] and Berthier and Moussu [19]. Strózyna and Zoładek [39] have obtained the orbital normal form of nilpotent centers with analytic first integral. Any nilpotent center has a local analytic first integral if, and only if, it is analytically equivalent to the Hamiltonian system $\dot{x} = y$, $\dot{y} = -x^{2k-1}$ where $k > 1$. Chavarriga et al. [23] study the analytical integrability for reversible nilpotent centers. The integrability problem has been studied for a few families of degenerate singular points (see [2, 5, 8, 9, 25, 26] and references therein).

In [5], it is studied the integrability problem for a class of planar systems and there are established the following results.

Theorem 1 *Suppose that the first quasi-homogeneous component of the degenerate system* $\mathbf{F}$ *is* $\mathbf{F}_r = \mathbf{X}_h = (-\partial h/\partial y, \partial h/\partial x)^T$*, where* h *has only simple factors in its factorization on* $\mathbb{C}[x, y]$*. Then, the quoted system is formally integrable if, and only if, it is formally conjugated, via dissipative transformations, to a divergence-free system.*

Theorem 2 *Suppose that the first quasi-homogeneous component of the degenerate system* $\mathbf{F}$ *is* $\mathbf{F}_r = \mathbf{X}_h = (-\partial h/\partial y, \partial h/\partial x)^T$*, where* h *has only simple factors in its factorization on* $\mathbb{C}[x, y]$*. Suppose that the quoted system is analytically (or formally) integrable. Then, one of the first integrals is of the form* $I = h + \cdots$*, where the dots denote higher-order quasi-homogeneous terms.*

Using these results, the integrability problem for the next families can be solved. In [8], degenerate systems of the following form were considered

$$\dot{x} = y^3 + 3\mu x^2 y + o(|x, y|^3), \qquad \dot{y} = -x^3 - 3\mu x y^2 + o(|x, y|^3), \qquad \mu \in \mathbb{R}, \tag{2}$$

and the analytic integrability problem for them was analyzed establishing the following result.

Theorem 3 *System* (2) *is analytically integrable if, and only if, it is formally orbital equivalent to* $\dot{x} = y^3 + 3\mu x^2 y$*,* $\dot{y} = -x^3 - 3\mu x y^2$*.*

In [9] it was considered degenerate systems of the form

$$\dot{x} = y^3 + 2ax^3 y + \cdots, \qquad \dot{y} = -x^5 - 3ax^2 y^2 + \cdots, \qquad a \in \mathbb{R}, \tag{3}$$

and the analytic integrability problem for them was also studied where here the dots mean terms of higher order than the first component in the quasi-homogeneous expansion (see definition of quasi-homogeneous expansion below). The next result was obtained in [9].

Theorem 4 *System* (3) *is analytically integrable if, and only if, it is formally orbital equivalent to* $\dot{x} = y^3 + 2ax^3 y - 2\beta_9 x^4 y$*,* $\dot{y} = -x^5 - 3ax^2 y^2 + 4\beta_9 x^3 y^2$*, where* β_9 *is a function of the parameters of the first quasi-homogeneous components of system* (3).

Theorem 1 solves the integrability problem for $\mathbf{F} = \mathbf{F}_r + \cdots$, in the case that $\mathbf{F}_r = \mathbf{X}_h$ where all the irreducible factors of h on $\mathbb{C}[x, y]$ are simple. The case $\operatorname{div}(\mathbf{F}_r) \neq 0$ or $\mathbf{F}_r = \mathbf{X}_h$ where h has multiple factors on $\mathbb{C}[x, y]$ is an open problem.

Nevertheless a necessary condition so that $\mathbf{F}$ be integrable is that $\mathbf{F}_r$ be also integrable. The integrability problem for quasi-homogeneous vector fields, $\mathbf{F}_r = \mathbf{X}_h + \mu \mathbf{D}_0$ with $\mathbf{D}_0 = (t_1 x, t_2 y)^T$, $\mu \neq 0$, is resolved in [13].

It is known, see [2], that the study of the integrability of a system whose quasi-homogeneous vector field has non null divergence is equivalent to study the integrability of a system whose quasi-homogeneous vector field is Hamiltonian where its Hamilton function has multiple factors on $\mathbb{C}[x, y]$.

In [2], the analytic integrability problem around the origin in a family of degenerate centers was studied and we showed the difficulty of the integrability problem even inside this family of degenerate centers. Concretely, there we considered degenerate systems of the form

$$\dot{x} = -y(x^2 + y^2) + \cdots, \quad \dot{y} = x(x^2 + y^2) + \cdots, \tag{4}$$

where the dots signify terms of higher order than the first component in the homogeneous expansion and which corresponds to $\mathbf{X}_h + \cdots$ whose $h = (x^2 + y^2)^2/4$. In [41], this type of systems was studied and it was proved that there are centers inside this family that do not have an analytic first integral.

Recall that if h has not simple factors on $\mathbb{C}[x, y]$ Theorem 1 can not be applied. In [2], we obtain an useful result to the integrability problem for this kind of systems, because it gives us some necessary conditions to have formal integrability and also the value of $n \in \mathbb{N}$ to have first integrals of the form $I = (x^2 + y^2)^n + \cdots$. Nevertheless, there is a particular case which remains open.

Summarizing, the nodes and saddle-nodes are not analytically integrable. A singular point with linear part of center type is analytically integrable if, and only if, it is a center and if, and only if, it is orbitally linearizable. System (1) is not analytically integrable around a non-resonant saddle singular point, but a resonant saddle has an analytic first integral around the singular point if, and only if, it is orbitally linearizable. A nilpotent vector field, with first quasi-homogeneous component non-reducible, is analytically integrable if, and only if, it is orbitally equivalent to its first quasi-homogeneous component, see [5, 11]. The remaining global open case is the case when we have a degenerate singular point. However there are some partial results. The analytic integrability problem when $\mathbf{F}_r = \mathbf{X}_h$, with h having only simple factors on $\mathbb{C}[x, y]$ is completely solved in [5]. The case $\mathbf{F}_r = \mathbf{X}_h$, with h having multiple factors on $\mathbb{C}[x, y]$ is still open, although in [2] there are some partial results.

We have structured this chapter as follows. In the next section, we summarize some preliminary definitions and some technical results. In the third section, we present our system and we give a formal orbital equivalent normal form for it. We transform this system to obtain an irreducible first quasi-homogeneous component and we analyze the different Newton diagrams. In the fourth and fifth sections we provide integrability conditions for our system. First, we study when this system is reducible and then, we suppose it irreducible, give necessary conditions for the integrability. Finally, we finish the chapter with an application to a particular family of degenerate vector fields.

2 Preliminary Definitions and Technical Results

In this section we remember the following concepts.

Given $\mathbf{t} = (t_1, t_2)$ non-null with t_1 and t_2 non-negative integer numbers without common factors, a function f of two variables verifying $f(\varepsilon^{t_1}x, \varepsilon^{t_2}y) = \varepsilon^k f(x, y)$

is called a quasi-homogeneous function of type $\mathbf{t}$ and degree k and we denote by $f \in \mathscr{P}_k^{\mathbf{t}}$. A vector field $\mathbf{F} = (F_1, F_2)^T$ verifying $F_1 \in \mathscr{P}_{k+t_1}^{\mathbf{t}}$ and $F_2 \in \mathscr{P}_{k+t_2}^{\mathbf{t}}$ is called a quasi-homogeneous vector field of type $\mathbf{t}$ and degree k and we denote by $\mathbf{F} \in \mathscr{Q}_k^{\mathbf{t}}$.

A basis for the vector space $\mathscr{P}_k^{\mathbf{t}}$ is given in [13]: $\mathscr{P}_k^{\mathbf{t}} = \text{span}\{x^{k_1+t_2(k_3-j)}y^{k_2+t_1 j}\}_{j=0}^{k_3}$ if there exist k_1, k_2, k_3 integer numbers with $0 \le k_1 < t_2, 0 \le k_2 < t_1, k_3 \ge 0$ where $k = k_1 t_1 + k_2 t_2 + k_3 t_1 t_2$; otherwise $\mathscr{P}_k^{\mathbf{t}} = \{0\}$.

We can expand any vector field $\mathbf{F}$ into quasi-homogeneous terms of type $\mathbf{t}$ of successive degrees. So, we write the vector field $\mathbf{F}$ as

$$\mathbf{F} = \mathbf{F}_r + \mathbf{F}_{r+1} + \cdots,$$

for some $r \in \mathbb{Z}$, with $\mathbf{F}_j = (P_{j+t_1}, Q_{j+t_2})^T \in \mathscr{Q}_j^{\mathbf{t}}$ and $\mathbf{F}_r \not\equiv \mathbf{0}$. Such expansions will be expressed as $\mathbf{F} = \mathbf{F}_r + q - h.h.o.t.$

We will write $\mathbf{D}_0 = (t_1 x, t_2 y)^T \in \mathscr{Q}_0^{\mathbf{t}}$ and $\mathbf{X}_h = (-\partial h/\partial y, \partial h/\partial x)^T$, a dissipative quasi-homogeneous vector field and the Hamiltonian vector field associated to the polynomial h, respectively. $\mathbf{X}_h \in \mathscr{Q}_r^{\mathbf{t}}$, if $h \in \mathscr{P}_{r+|\mathbf{t}|}^{\mathbf{t}}$, where $|\mathbf{t}| = t_1 + t_2$. Even more, any $\mathbf{F}_k \in \mathscr{Q}_k^{\mathbf{t}}$ can be written as

$$\mathbf{F}_k = \mathbf{X}_h + \mu \mathbf{D}_0 \tag{5}$$

where $h = (\mathbf{D}_0 \wedge \mathbf{F}_k)/(k + |\mathbf{t}|)$ and $\mu = \text{div}(\mathbf{F}_k)/(k + |\mathbf{t}|)$, being $\mathbf{D}_0 \wedge \mathbf{F}_k \in \mathscr{P}_{k+|\mathbf{t}|}^{\mathbf{t}}$ the wedge product of both vector fields and $\text{div}(\mathbf{F}_k) \in \mathscr{P}_k^{\mathbf{t}}$ the divergence of $\mathbf{F}_k$, see [5]. This sum is called the conservative-dissipative splitting of a quasi-homogeneous vector field.

The method to obtain a normal form under orbital equivalence for the system (1) is explained in, for example, [14]. However, we will remember some notions.

In the problem of getting a normal form for the system (1), we analyze the effect of a near-identity transformation $\mathbf{x} = \mathbf{y} + \mathbf{P}_k(\mathbf{y})$ and a reparametrization of the time given by $dt/dT = 1 + \mu_k(\mathbf{x})$, where $\mathbf{P}_k \in \mathscr{Q}_k^{\mathbf{t}}$ and $\mu_k \in \mathscr{P}_k^{\mathbf{t}}$, with $k \ge 1$.

In the transformed system $\dot{\mathbf{y}} = \mathbf{G}(\mathbf{y})$, the quasi-homogeneous terms agree with the original ones up to $r + k - 1$ order and for the degree $r + k$ it has

$$\mathbf{G}_{r+k} = \mathbf{F}_{r+k} - \mathscr{L}_k(\mathbf{P}_k, \mu_k)$$

where we have introduced the homological operator under formal orbital equivalence:

$$\begin{aligned} \mathscr{L}_k : \mathscr{Q}_k^{\mathbf{t}} \times \mathscr{P}_k^{\mathbf{t}} &\longrightarrow \mathscr{Q}_{r+k}^{\mathbf{t}} \\ (\mathbf{P}_k, \mu_k) &\to \mathscr{L}_k(\mathbf{P}_k, \mu_k) = [\mathbf{P}_k, \mathbf{F}_r] - \mu_k \mathbf{F}_r. \end{aligned} \tag{6}$$

Following the ideas of the classical normal form theory, we choose $(\mathbf{P}_k, \mu_k) \in \mathscr{Q}_k^{\mathbf{t}} \times \mathscr{P}_k^{\mathbf{t}}$ adequately to simplify the $(r + k)$-degree quasi-homogeneous term in system (1), by annihilating the part belonging to the range of the linear operator $\mathscr{L}_k$. In this case, it is said that this term is in normal form under orbital equivalence. So, by means of a sequence of near identity transformations and time-reparametrizations system (1) can be formally carried out to normal form under orbital equivalence.

We define the following linear operators, to study the homological operator in the case $\mathbf{F}_r = \mathbf{X}_H$ with $H = h^2, h \in \mathscr{P}^{\mathbf{t}}_{\frac{r+|\mathbf{t}|}{2}}$,

$$\begin{array}{rccc} \ell_{k-r} : & \mathscr{P}^{\mathbf{t}}_{k-r} & \longrightarrow & \mathscr{P}^{\mathbf{t}}_{k} \\ & \mu_{k-r} & \rightarrow & \nabla \mu_{k-r} \cdot \mathbf{X}_H \end{array} \qquad \begin{array}{rccc} \tilde{\ell}_{k-r} : & \mathscr{P}^{\mathbf{t}}_{k-r} & \longrightarrow & \mathscr{P}^{\mathbf{t}}_{k-\frac{r+|\mathbf{t}|}{2}} \\ & \mu_{k-r} & \rightarrow & \nabla \mu_{k-r} \cdot \mathbf{X}_h, \end{array}$$

i.e. the Lie derivative of the lowest degree quasi-homogeneous term of $\mathbf{X}_H$ and $\mathbf{X}_h$, respectively.

As in [14], we have the following lemma modified to our purposes.

Lemma 1 *Suppose that the lowest-degree quasi-homogeneous term of system* (1) *is* $\mathbf{F}_r = \mathbf{X}_H \in \mathscr{Q}^{\mathbf{t}}_r$, *where* $H \in \mathscr{P}^{\mathbf{t}}_{r+|\mathbf{t}|}$. *A complementary subspace to the range (co-range) of* $\mathscr{L}_k$, *with* $k \geq 1$, *can be written as*

$$\mathrm{Cor}(\mathscr{L}_k) = \mathbf{X}_{S_{r+|\mathbf{t}|+k}} \oplus \mathrm{Cor}(\ell_k)\mathbf{D}_0,$$

being $S_{r+|\mathbf{t}|+k}$ *a subspace verifying*

$$\mathrm{Cor}(\ell_{k+|\mathbf{t}|}) = S_{r+|\mathbf{t}|+k} \oplus \left(H\mathrm{Cor}(\ell_{k-r}) \cap \mathrm{Cor}(\ell_{k+|\mathbf{t}|})\right),$$

where $\mathrm{Cor}(\ell_{k-r})$ *is a complementary subspace to the* $\mathrm{Range}(\ell_{k-r})$ *in* $\mathscr{P}^{\mathbf{t}}_k$, *such that* $H\mathrm{Cor}(\ell_{k-r}) \cap \mathrm{Cor}(\ell_{k+|\mathbf{t}|})$ *has maximal dimension.*

Proposition 1 *Suppose that the lowest-degree quasi-homogeneous term of system* (1) *is* $\mathbf{F}_r = \mathbf{X}_H \in \mathscr{Q}^{\mathbf{t}}_r$, *with* $H = h^2$, $h \in \mathscr{P}^{\mathbf{t}}_{\frac{r+|\mathbf{t}|}{2}}$. *A complementary subspace of the range of* ℓ_{k-r} *is*

$$\mathrm{Cor}(\ell_{k-r}) = h\mathrm{Cor}(\tilde{\ell}_{k-r}) \oplus \Delta_k,$$

where Δ_k *is such that* $\mathscr{P}^{\mathbf{t}}_k = \Delta_k \oplus h \cdot \mathscr{P}^{\mathbf{t}}_{k-\frac{r+|\mathbf{t}|}{2}}$.

Proof For all $\mu \in \mathscr{P}^{\mathbf{t}}_{k-r}$, we have $\ell_{k-r}(\mu) = \nabla\mu \cdot \mathbf{X}_{h^2} = 2h\nabla\mu \cdot \mathbf{X}_h = 2h\tilde{\ell}_{k-r}(\mu)$, from where we deduce $Range(\ell_{k-r}) \subset h \cdot Range(\tilde{\ell}_{k-r})$.

Reciprocally, $h \cdot Range(\tilde{\ell}_{k-r}) \subset Range(\ell_{k-r})$ because $h \cdot \tilde{\ell}_{k-r}(\mu) = h \cdot \nabla\mu \cdot \mathbf{X}_h = \nabla\frac{\mu}{2} \cdot \mathbf{X}_{h^2} = \ell_{k-r}(\frac{\mu}{2})$.

Hence we have the equality $Range(\ell_{k-r}) = h \cdot Range(\tilde{\ell}_{k-r})$. The result follows from here.

3 A Family of Perturbed Degenerate Centers

The goal of this chapter is to study the formal integrability problem around the origin of an analytic system of the form

$$(\dot{x}, \dot{y})^T = (x^{2n} + ny^2)(-y, x^{2n-1})^T + q - h.h.o.t. \qquad (7)$$

First we remember that if an analytic vector field is formally integrable around an isolated singular point then it is analytically integrable, see [32]. Then, the formal integrability and the analytic integrability are equivalent.

We start computing a formal orbital equivalent normal form for system, to study the formal integrability.

Theorem 5 *A formally orbital equivalent normal form for system* (7) *is*

$$(\dot{x}, \dot{y})^T = (x^{2n} + ny^2)(-y, x^{2n-1})^T + \sum_{j=3n}^{\infty} \mathbf{X}_{a_j x^{j+n+1} + b_j x^{j+1} y} + (c_j x^j + d_j x^{j-n} y)\mathbf{D}_0 + \sum_{i=n}^{2n-2} e_i^{(1)} x^i h \mathbf{D}_0 + \sum_{j=2}^{\infty} \sum_{i=0}^{2n-2} e_i^{(j)} x^i h^j \mathbf{D}_0, \quad (8)$$

where $h = \frac{1}{2n}(x^{2n} + ny^2)$.

Proof In this case, $r = 3n - 1$ and $|\mathbf{t}| = n + 1$.

In order to calculate a co-range of $\mathscr{L}_k$, that is, a complementary subspace of the range of the homological operator $\mathscr{L}_k$, we apply Lemma 1, i.e. we have that $\mathrm{Cor}(\mathscr{L}_k) = \mathbf{X}_{S_{4n+k}} \oplus \mathrm{Cor}(\ell_k)\mathbf{D}_0$ with $k \geq 1$. Therefore, we must compute $\mathrm{Cor}(\ell_k)$ and the subspace S_{4n+k}.

From Proposition 1, we obtain

$$\mathrm{Cor}(\ell_k) = h\mathrm{Cor}(\tilde{\ell}_{k-r}) \oplus \Delta_{k+3n-1},$$

$k \geq 1$, where we can choose $\Delta_j =< x^j, x^{j-n} y >, j \geq n$.

From [14], $\mathrm{Cor}(\tilde{\ell}_{k-r})$ is

$$\begin{cases} < x^{k+n-1} > & \text{if } 1 \leq k \leq n-1, \\ < x^j h^l > & \text{if } k+n-1 = 2nl + j,\ 0 \leq j \leq 2n-2,\ l \in \mathbb{N} \\ 0 & \text{otherwise} \end{cases}$$

So, $\mathrm{Cor}(\ell_k)$ is:

- If $1 \leq k \leq n-1$,

$$\mathrm{Cor}(\ell_k) =< x^{k+3n-1}, x^{k+2n-1} y, x^{k+n-1} h > .$$

- If $k + 3n - 1 = 2nl + j, 0 \leq j \leq 2n-2, l \in \mathbb{N}$,

$$\mathrm{Cor}(\ell_k) =< x^{k+3n-1}, x^{k+2n-1} y, x^j h^{l+1} > .$$

- If $k + 3n - 1 = 2nl + 2n - 1, l \in \mathbb{N} \cup \{0\}$,

$$\mathrm{Cor}(\ell_k) =< x^{k+3n-1}, x^{k+2n-1} y > .$$

The subspace S_{4n+k} is such that

$$\mathrm{Cor}(\ell_{k+n+1}) = S_{4n+k} \oplus (H\mathrm{Cor}(\ell_{k-3n+1}) \cap \mathrm{Cor}(\ell_{k+n+1}))\,.$$

Considering the previous calculations, we can deduce $S_{4n+k} =< x^{4n+k}, x^{3n+k}y >$, with $k \geq 1$.

Therefore, a co-range of the homological operator is:

- If $1 \leq k \leq n-1$,

$$\mathrm{Cor}(\mathscr{L}_k) =< \mathbf{X}_{x^{k+4n}}, \mathbf{X}_{x^{k+3n}y}, x^{k+3n-1}\mathbf{D}_0, x^{k+2n-1}y\mathbf{D}_0, x^{k+n-1}h\mathbf{D}_0 > .$$

- If $k+3n-1 = 2nl + j, 0 \leq j \leq 2n-2, l \in \mathbb{N}$,

$$\mathrm{Cor}(\mathscr{L}_k) =< \mathbf{X}_{x^{k+4n}}, \mathbf{X}_{x^{k+3n}y}, x^{k+3n-1}\mathbf{D}_0, x^{k+2n-1}y\mathbf{D}_0, x^j h^{l+1}\mathbf{D}_0 > .$$

- If $k+3n-1 = 2nl + 2n - 1, l \in \mathbb{N} \cup \{0\}$,

$$\mathrm{Cor}(\mathscr{L}_k) =< \mathbf{X}_{x^{k+4n}}, \mathbf{X}_{x^{k+3n}y}, x^{k+3n-1}\mathbf{D}_0, x^{k+2n-1}y\mathbf{D}_0 > .$$

The following result is obtained as a consequence.

Theorem 6 *System* (7) *is formally integrable if and only if, system* (8) *is formally integrable.*

We will analyze the formal integrability of system (8). In the following result we apply the blow-up technique and we transform system (8) into another one with first quasi-homogeneous component irreducible.

Lemma 2 *The blow up* $x = u$, $y = u^n v$, *the scaling of time* $dT = u^{3n-1}dt$ *and then the translation* $x_2 = \sqrt{n}v + i\sigma$, *where* $\sigma = \pm 1$ *with* $i = \sqrt{-1}$ *transforms system* (8) *into*

$$\begin{aligned}
u' &= \frac{2}{n}ux_2 + \frac{3i\sigma}{n}ux_2^2 - \frac{1}{n}ux_2^3 - \frac{1}{\sqrt{n}}\sum_{j\geq 3n} b_j u^{j-3n+2} + \frac{1}{\sqrt{n}}\sum_{j\geq 3n}(c_j - \frac{i\sigma}{\sqrt{n}}d_j)u^{j-3n+2} \\
&+ \frac{1}{n}\sum_{j\geq 3n} d_j u^{j-3n+2}x_2 + \frac{1}{2n\sqrt{n}}\sum_{i=n}^{2n-2} e_i^{(1)}u^{i-n+2}x_2(x_2 - 2i\sigma) \\
&+ \frac{1}{\sqrt{n}}\sum_{j\geq 2}\sum_{i=0}^{2n-2}\frac{1}{(2n)^j}e_i^{(j)}u^{i+2nj-3n+2}x_2^j(x_2 - 2i\sigma)^j, \\
x_2' &= -4x_2^2 - 4i\sigma x_2^3 + x_2^4 + \sum_{j\geq 3n}(j+n+1)(a_j - \frac{i\sigma}{\sqrt{n}}b_j)u^{j-3n+1} \\
&+ \frac{1}{\sqrt{n}}\sum_{j\geq 3n}(j+n+1)b_j u^{j-3n+1}x_2. \qquad (9)
\end{aligned}$$

To study the integrability problem of system (8), we analyze the different Newton diagrams in function of the parameters of system (9), and we give some definitions that depend on the vector field in orbital normal form.

The Newton diagram of system (9) (see [20, 24]), has an inner vertex $V_0 = (1, 2)$, associated to the vector $(\frac{2}{n}ux_2, -4x_2^2)^T$, from which leaves a non compact edge.

We consider the following non-negative integers

$$m := \min\left\{j \in \mathbb{N}, j \geq 3n : a_j^2 + \frac{1}{n}b_j^2 \neq 0\right\},$$

$$l := \min\left\{j \in \mathbb{N}, j \geq 3n : c_j^2 + \frac{1}{n}d_j^2 \neq 0\right\},$$

$$k := \min\left\{2nj + i : e_i^{(j)} \neq 0, j \in \mathbb{N} \text{ and } \left\{\begin{array}{c} i = 0, \ldots, 2n-2 \;\; if \;\; j \geq 2, \\ \text{or} \\ i = n, \ldots, 2n-2 \;\; if \;\; j = 1 \end{array}\right\}\right\}, \tag{10}$$

where $\min(\emptyset) := +\infty$.

- In the case $m < 2l - 3n + 1$ with $m < +\infty$ the Newton diagram has a exterior vertex $V_2 = (m - 3n + 2, 0)$ associated to the vector field $(0, (m + n + 1)(a_m - \frac{i\sigma}{\sqrt{n}}b_m)u^{m-3n+1})^T$.
- The case $2l - 3n + 1 < m < +\infty$ has also the inner vertex $V_1 = (l - 3n + 2, 1)$ associated to the vector field $(\frac{1}{\sqrt{n}}(c_l - \frac{i\sigma}{\sqrt{n}}d_l)u^{l-3n+2}, 0)^T$ and two compact edges.
- In the case $m = 2l - 3n + 1 < +\infty$, the Newton diagram has only a unique compact edge, because the vector field $(\frac{1}{\sqrt{n}}(c_l - \frac{i\sigma}{\sqrt{n}}d_l)u^{l-3n+2}, 0)^T$ is not associated to any vertex.
- If $m = +\infty$, we will see by Theorem 9 that system (7) is integrable if $l = k = +\infty$. In this case the system is formally orbital equivalent to system $(x^{2n} + ny^2)(-y, x^{2n-1})^T$.
  Hence the system has a first integral of the form $I = (x^{2n} + ny^2) + \cdots$.
- If $m < +\infty$, with $l = k = +\infty$, system (7) is formally integrable because system (8) is Hamiltonian of the form

$$\begin{aligned}(\dot{x}, \dot{y})^T &= (x^{2n} + ny^2)(-y, x^{2n-1})^T + \sum_{j=3n}^{\infty} \mathbf{X}_{a_j x^{j+n+1} + b_j x^{j+1} y} \\ &= n\mathbf{X}_{h^2} + \sum_{j=3n}^{\infty} \mathbf{X}_{a_j x^{j+n+1} + b_j x^{j+1} y}\end{aligned}$$

  where $h = (x^{2n} + ny^2)/(2n)$. Hence system (7) has a first integral of the form $I = (x^{2n} + ny^2)^2 + \cdots$.

Consequently, from now on, we assume $l < +\infty$ or $k < +\infty$; i.e., we suppose that the vector field is not formally equivalent to a Hamiltonian one (Fig. 1).

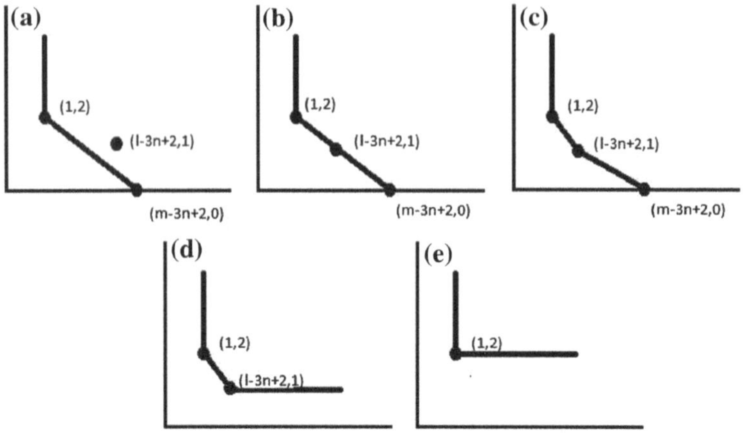

Fig. 1 **a** case $m < 2l - 3n + 1$, **b** case $m = 2l - 3n + 1$, **c** case $m > 2l - 3n + 1$, **d**) case $m < +\infty, l = +\infty$, **e** case $m = l = +\infty$

4 The Main Results

First we consider the case when system (7) is reducible.

Definition 1 The vector field **F** is reducible if there exists a scalar function f, with $f(0) = 0$, such that $\mathbf{F} = f \cdot \mathbf{G}$. In this case, f is the reducibility factor of **F**.

Theorem 7 *Suppose that system* (7) *is formally integrable and suppose that its associated vector field is* $\mathbf{F} = (x^{2n} + ny^2)(-y, x^{2n-1})^T + \cdots$. *System* (7) *has a first integral of the form* $I = (x^{2n} + ny^2) + \cdots$ *if and only if* **F** *is reducible.*

Proof We suppose that the vector field **F** is reducible. Then it can be written

$$\mathbf{F} = (x^{2n} + ny^2 + \cdots)[(-y, x^{2n-1})^T + \cdots] = f \cdot \mathbf{G}.$$

As **F** is formally integrable, the vector field **G** is formally integrable too. Even more, by Theorem 2, a first integral of **G** is of the form $I = (x^{2n} + ny^2) + \cdots$. Therefore, $I = (x^{2n} + ny^2) + \cdots$ is a first integral of system (7).

Now, we assume that $I = (x^{2n} + ny^2) + \cdots$ is a first integral of **F**. By the Arnold Theorem, see [18], there are a change of variables $(x, y) = \phi(u, v)$ with $I(\phi(u, v)) = h(u, v)$, where we denote $h = x^{2n} + ny^2$, being $I(\phi(u, v)) = h(u, v)$ a first integral of the transformed vector field $\mathbf{G} = \phi_* \mathbf{F}$.
We write **G** as $\mathbf{G} = (x^{2n} + ny^2)(-y, x^{2n-1})^T + \sum_{j>3n} \mathbf{G}_j$.
We prove that $\mathbf{G}_j = f_j \cdot (-y, x^{2n-1})^T, \forall j > 3n$.

If h is a first integral of **G**, then it is a first integral for each $\mathbf{G}_j$. So, $\nabla h \cdot \mathbf{G} = 0$ and then $\nabla h \cdot (x^{2n} + ny^2)(-y, x^{2n-1})^T + \sum_{j>3n} \nabla h \cdot \mathbf{G}_j = 0$ and then $\nabla h \cdot \mathbf{G}_j = 0$, for all $j > 3n$.

We write now each $\mathbf{G}_j$ as $\mathbf{G}_j = \mathbf{X}_{g_j} + \mu_j \mathbf{D}_0 + \lambda_j(-y, x^{2n-1})^T$, with $g_j \in \Delta_j$, for all $j > 3n$, (see definition in Proposition 1). Then, $\nabla h \cdot \mathbf{G}_j = \nabla h \cdot \mathbf{X}_{g_j} + 2\mu_j h = 0$; therefore, h is an invariant curve of $\mathbf{X}_{g_j}$ and we can write $g_j = f_j \cdot h, \forall j > 3n$. But this contradicts $g_j \in \Delta_j$; so, we have $g_j = 0, \forall j > 3n$. In consequence, $\nabla h \cdot \mathbf{G}_j = 2\mu_j h = 0$, hence $\mu_j = 0$, for all $j > 3n$.

From here $\mathbf{G}_j = \lambda_j(-y, x^{2n-1})^T, \forall j > 3n$.

Thus we have

$$\mathbf{G} = (x^{2n} + ny^2)(-y, x^{2n-1})^T + \sum_{j>3n} \lambda_j(-y, x^{2n-1})^T =$$

$$(x^{2n} + ny^2 + \sum_{j>3n} \lambda_j)(-y, x^{2n-1})^T.$$

Summary, we can write $\mathbf{G} = f \cdot (-y, x^{2n-1})^T$.

Undoing the change ϕ, we obtain $\mathbf{F} = \tilde{f} \cdot [(-y, x^{2n-1})^T + \cdots]$, where $\tilde{f} = \phi^{-1}(f) = x^{2n} + ny^2 + \cdots$.

Finally, we can write $\mathbf{F}$ as

$$\mathbf{F} = (x^{2n} + ny^2 + \cdots) \cdot [(-y, x^{2n-1})^T + \cdots],$$

thus the vector field $\mathbf{F}$ is reducible.

Next, we suppose that the vector field $\mathbf{F}$ associated to system (8) is irreducible.

We remember the following result proved in [13] that we have modified to our purpose.

Theorem 8 *The quasi-homogeneous system* $\dot{\mathbf{x}} = \mathbf{F}_r = (P, Q)^T$, *with* $\mathbf{F}_r \in \mathcal{Q}_r^{\mathbf{t}}$, P, Q *coprimes,* $PQ \not\equiv 0$ *and* $\mathrm{div}(\mathbf{F}_r) \not\equiv 0$, *is polynomially integrable if, and only if,*

$$(\mathbf{D}_0 \wedge \mathbf{F}_r)(x, y) = cx^{\delta_x} y^{\delta_y} \prod_{i=1}^{m} (y^{t_1} - \lambda_i x^{t_2}), \tag{11}$$

with $r + |\mathbf{t}| = t_1\delta_x + t_2\delta_y + t_1t_2m$, $c \neq 0$, $\delta_x, \delta_y \in \{0, 1\}$, $\delta_x + \delta_y + m \geq 2$ *and* $\lambda_1, \ldots, \lambda_m$ *distinct complex numbers not zero and exist* n_x, n_y, n_i, $i = 1, \ldots, m$ *non-negative integers, not all zero, verifying*

$$\begin{cases} Res[\eta(x, 1), 0] = -\dfrac{1}{t_2} + \dfrac{(n_x + 1)(r + |\mathbf{t}|)}{t_2\widetilde{M}} & \text{if } \delta_x = 1, \\ Res[\eta(1, y), 0] = \dfrac{1}{t_1} - \dfrac{(n_y + 1)(r + |\mathbf{t}|)}{t_1\widetilde{M}} & \text{if } \delta_y = 1, \\ Res[\eta(1, y), \lambda_i^{1/t_1}] = \dfrac{1}{t_1} - \dfrac{(n_i + 1)(r + |\mathbf{t}|)}{t_1\widetilde{M}} & i = 1, \ldots, m, \end{cases} \tag{12}$$

where $\widetilde{M} = t_1(n_x + 1)\delta_x + t_2(n_y + 1)\delta_y + t_1t_2 \sum_{j=1}^{m}(n_j + 1)$ *and* $\eta = \dfrac{\mathrm{div}(\mathbf{F}_r)}{\mathbf{D}_0 \wedge \mathbf{F}_r}$.

Moreover a first integral is of the form

$$I = x^{\delta_x(n_x+1)} y^{\delta_y(n_y+1)} \prod_{i=1}^{m} (y^{t_1} - \lambda_i x^{t_2})^{n_i+1}. \tag{13}$$

We write as $\mathbf{G}$ the vector field associated to system (9). Remember that if system (8) is formally integrable then system (9) is formally integrable too.

Next result we give integrability necessary conditions of system (8), (a formal normal form of system (7)). To prove it we will use Theorem 8.

Theorem 9 *Let* $\mathbf{F}$ *be the associated vector field of system* (8) *and consider* m, l, k *defined in* (10). *We suppose* $\mathbf{F}$ *irreducible. If* $\mathbf{F}$ *is formally integrable then the following conditions are verified.*

- $\min\{l, m\} < k$, $k < +\infty$,
- $m \leq 2l - 3n + 1$, $l < +\infty$,
- *If* $m = 2l - 3n + 1$, $m < +\infty$, *then*

$$(a_m - \frac{i\sigma}{\sqrt{n}} b_m) + \frac{(l - 3n + 1)^2}{16(l - n + 1)^2}(c_l - \frac{i\sigma}{\sqrt{n}} d_l)^2 \neq 0.$$

Proof We will prove the contrary result, that is, if any of the following conditions is verified, then system (8) is not formally integrable.

- $\min\{l, m\} \geq k$, $k < +\infty$,
- $m > 2l - 3n + 1$, $l < +\infty$,
- $m = 2l - 3n + 1$, $m < +\infty$ and $\frac{(l - 3n + 1)^2}{16(l - n + 1)^2}(c_l - \frac{i\sigma}{\sqrt{n}} d_l)^2 + (a_m - \frac{i\sigma}{\sqrt{n}} b_m) = 0$.

We assume that $\min\{l, m\} \geq k$. If $\mathbf{F}$ is the associated vector field of system (8), then we have $\mathbf{F} = \mathbf{F}_{3n-1} + \mathbf{F}_k + \cdots$ with $\mathbf{F}_{3n-1} = \frac{1}{2n} h \mathbf{X}_h$, $h = x^{2n} + ny^2$ and $\mathbf{F}_k = (e_{i_0}^{(j_0)} x^{i_0} h^{j_0} + \mu_l)\mathbf{D}_0 + \mathbf{X}_{g_m}$, where

- $e_{i_0}^{(j_0)} \neq 0$ where $j_0 \geq 2$ and $i_0 = 0, \ldots, 2n - 2$ or $j_0 = 1$ and $i_0 = n, \ldots, 2n - 2$,
- $\mu_l = c_l x^l + d_l x^{l-n} y$ if $k = l$, otherwise $\mu_l = 0$,
- $g_m = a_m x^{m+n+1} + b_m x^{m+1} y$ if $k = m$, otherwise $g_m = 0$.

In the case that the system were integrable, $I = h^p + \sum_{j>2np} I_j$ with $I_j \in \mathscr{P}_j^{(1,n)}$, would be a first integral, by Theorem 7 we have that $p \geq 2$. The integrability condition $(\nabla I \cdot \mathbf{F} = 0)$ is checked and at the lower degree $2np + k$ is

$$\begin{aligned}
0 &= \nabla h^p \cdot \mathbf{F}_k + \nabla I_{2np+k-3n+1} \cdot \mathbf{F}_{3n-1} \\
&= 2npe_{i_0}^{(j_0)} x^{i_0} h^{p+j_0} + 2np\mu_l h^p + ph^{p-1} \nabla h \cdot \mathbf{X}_{g_m} + h \nabla I_{2np+k-3n+1} \cdot \mathbf{X}_{\frac{h}{2n}} \\
&= h[2npe_{i_0}^{(j_0)} x^{i_0} h^{p+j_0-1} + 2np\mu_l h^{p-1} - 2nph^{p-2} \nabla g_m \cdot \mathbf{X}_{\frac{h}{2n}} + \nabla I_{2np+k-3n+1} \cdot \mathbf{X}_{\frac{h}{2n}}]
\end{aligned}$$

which is compatible only if $e_{i_0}^{(j_0)} = 0$; this is a contradiction.

Now, we suppose that $m > 2l - 3n + 1$. The Newton diagram of (9) in this case would have two compact edges (if $m < +\infty$) or one (if $m = +\infty$). In any case, we consider in the edge of type $t = (1, l - 3n + 1)$, because it is common to both situations. The vector field

$$\mathbf{G}_r = (\frac{2}{n}ux_2 + \frac{1}{\sqrt{n}}(c_l - \frac{i\sigma}{\sqrt{n}}d_l)u^{l-3n+2}, -4x_2^2)^T,$$

where $r = l - 3n + 1$, is associated to this edge.

If $\mathbf{F}$ is formally integrable then $\mathbf{G}$ is formally integrable and $\mathbf{G}_r$ is also formally integrable. To apply Theorem 8, first we compute the Hamiltonian function of $\mathbf{G}_r$

$$h = \frac{1}{r + |\mathbf{t}|}(u, (l - 3n + 1)x_2)^T \wedge \mathbf{G}_r = -\frac{2(l - n + 1)}{n(2l - 6n + 3)}ux_2[x_2 - \lambda u^{l-3n+1}],$$

where $\lambda = -\frac{\sqrt{n}(l-3n+1)}{2(l-n+1)}(c_l - \frac{i\sigma}{\sqrt{n}}d_l)$.

On the other hand, the dissipative part of $\mathbf{G}_r$ is

$$\mu = \frac{1}{r + |\mathbf{t}|}div(\mathbf{G}_r) = -\frac{2(4n - 1)}{n(2l - 6n + 3)}\left[x_2 - \frac{\sqrt{n}(l - 3n + 2)}{2(4n - 1)}(c_l - \frac{i\sigma}{\sqrt{n}}d_l)u^{l-3n+1}\right].$$

Applying Theorem 8 and calculating the expressions of the residues, we have

$$\begin{aligned}
Res[\eta(u, 1), 0] &= \frac{4n - 1}{l - n + 1} = \frac{-\widetilde{M} + (2l - 6n + 3)(n_x + 1)}{(l - 3n + 1)\widetilde{M}},\\
Res[\eta(1, x_2), 0] &= -\frac{l - 3n + 2}{l - 3n + 1} = \frac{\widetilde{M} - (2l - 6n + 3)(n_y + 1)}{\widetilde{M}},\\
Res[\eta(1, x_2), \lambda] &= \frac{l^2 + 2l - 9n^2 + 2n + 1}{(l - n + 1)(l - 3n + 1)} = \frac{\widetilde{M} - (2l - 6n + 3)(n_1 + 1)}{\widetilde{M}},
\end{aligned}$$

where $\widetilde{M} = (n_x + 1) + (l - 3n + 1)(n_y + 1) + (l - 3n + 1)(n_1 + 1)$.

The second equation gives us $n_x + 1 = -(l - 3n + 1)(n_1 + 1)$ and considering that $l \geq 3n$, we would have that n_x should be a negative value which implies a contradiction. Hence $\mathbf{G}_r$ is not formally integrable.

Finally we suppose that $m = 2l - 3n + 1$ and

$$\frac{(l - 3n + 1)^2}{16(l - n + 1)^2}(c_l - \frac{i\sigma}{\sqrt{n}}d_l)^2 + (a_m - \frac{i\sigma}{\sqrt{n}}b_m) = 0.$$

In this case, $\mathbf{G}_r = (P, Q)^T$ is

$$\mathbf{G}_r = \begin{pmatrix} \frac{2}{n}ux_2 + \frac{1}{\sqrt{n}}(c_l - \frac{i\sigma}{\sqrt{n}}d_l)u^{l-3n+2} \\ -4x_2^2 + 2(l - n + 1)(a_m - \frac{i\sigma}{\sqrt{n}}b_m)u^{2(l-3n+1)} \end{pmatrix}.$$

Moreover we can write the following dissipative-conservative decomposition:

$$h = \frac{-2(l-n+1)}{n(2l-6n+3)}u(x_2 + \frac{\sqrt{n}(l-3n+1)}{4(l-n+1)}(c_l - \frac{i\sigma}{\sqrt{n}}d_l)u^{l-3n+1})^2 := \alpha\, u \cdot f^2$$

$$\mu = -\frac{2(4n-1)}{n(2l-6n+3)}(x_2 - \frac{\sqrt{n}(l-3n+2)}{2(4n-1)}(c_l - \frac{i\sigma}{\sqrt{n}}d_l)u^{l-3n+1}) \neq 0.$$

We will see that $\mathbf{G}_r$ is irreducible. Otherwise $\mathbf{G}_r = g \cdot \widehat{\mathbf{G}}_s$, with $\widehat{\mathbf{G}}_s \in \mathscr{Q}_s^{\mathbf{t}}$ and $g \in \mathscr{P}_{r-s}^{\mathbf{t}}$.

Thus we have $\mathbf{D}_0 \wedge \mathbf{G}_r = (r + |\mathbf{t}|)\, h = g\mathbf{D}_0 \wedge \widehat{\mathbf{G}}_s$; so, the irreducible factors of g are irreducible factors of h. That is, the possible reducible factors of $\mathbf{G}_r$ are u or f. But u is not a factor of Q and f is not a factor of $P = \frac{2}{n}u(x_2 + \frac{\sqrt{n}}{2}(c_l - \frac{i\sigma}{\sqrt{n}}d_l)u^{l-3n+1})$; otherwise $(x_2 + \frac{\sqrt{n}}{2}(c_l - \frac{i\sigma}{\sqrt{n}}d_l)u^{l-3n+1}) = f$, from where $l = -(n+1)$, which is a contradiction.

By Theorem 8, as $\mathbf{G}_r$ is irreducible and $\mathbf{D}_0 \wedge \mathbf{G}_r$ has not simple factors on $\mathbb{C}[x, y]$, $\mathbf{G}_r$ is not polynomially integrable and then system (9) is not formally integrable.

Next result gives us necessary conditions of integrability for system (7).

Theorem 10 *Let* $\mathbf{F}$ *be the associated vector field of system* (8) *irreducible. Consider* m *and* l *defined in* (10). *If* $\mathbf{F}$ *is formally integrable, then it is verified*

(a) *If* $m < 2l - 3n + 1$, *with* $m - 3n + 1$ *odd, then a first integral of system* (8) *is of the form* $I = (x^{2n} + ny^2)^2 + \cdots$.
(b) *If* $m < 2l - 3n + 1$, *with* $m - 3n + 1$ *even, then a first integral of system* (8) *is of the form* $I = f_1 f_2$ *where* $f_1 \not\equiv f_2$ *y* $f_i = (x^{2n} + ny^2) + \cdots$ *for* $i = 1, 2$.
(c) *If* $m = 2l - 3n + 1$ *and* $(a_m - \frac{i\sigma}{\sqrt{n}}b_m) \neq \frac{n}{2(l-n+1)}\left(c_l - \frac{i\sigma}{\sqrt{n}}d_l\right)^2$ *then exist* $M \in \mathbb{N}$ *and* $N \in \mathbb{N}$, *with* $M \neq N$, *such that*

$$a_m = \frac{1}{16(l-n+1)^2}\left((l+n+1)^2\frac{(M+N)^2}{(M-N)^2} - (l-3n+1)^2\right)(c_l^2 - \frac{1}{n}d_l^2),$$

$$b_m = \frac{1}{8(l-n+1)^2}\left((l+n+1)^2\frac{(M+N)^2}{(M-N)^2} - (l-3n+1)^2\right)c_l d_l.$$

In this case, a first integral of system (8) is of the form $I = f_1^M f_2^N$ *where* $f_1 \not\equiv f_2$ *and* $f_i = (x^{2n} + ny^2) + \cdots$ *for* $i = 1, 2$.

The proof of Theorem 10 is given in the next section.

Remark 1 The case $m = 2l - 3n + 1$ with $(a_m - \frac{i\sigma}{\sqrt{n}}b_m) = \frac{n}{2(l-n+1)}(c_l - \frac{i\sigma}{\sqrt{n}}d_l)^2$, that is the case where the principal part of system (9) is reducible, continues open.

5 Proof of Theorem 10

(i) If $m < 2l - 3n + 1$, in the Newton diagram of system (9), there is a unique compact face of type $(2, m - 3n + 1)$ which is associated to the vector field

$$\mathbf{G}_{m-3n+1} = \left(\frac{2}{n}ux_2, -4x_2^2 + (m+n+1)\left(a_m - \frac{i\sigma}{\sqrt{n}}b_m\right)u^{m-3n+1}\right)^T$$

with the Hamiltonian function h and the dissipative function μ (conservative and dissipative part of $\mathbf{G}_{m-3n+1}$ respectively) given by

$$h = \frac{1}{r+|\mathbf{t}|}\mathbf{D}_0 \wedge \mathbf{G}_{m-3n+1} = -\frac{m+n+1}{n(m-3n+2)}u\left[x_2^2 - n(a_m - \frac{i\sigma}{\sqrt{n}}b_m)u^{m-3n+1}\right],$$
$$\mu = \frac{1}{r+|\mathbf{t}|}div(\mathbf{G}_{m-3n+1}) = -\frac{4n-1}{n(m-3n+2)}x_2. \tag{14}$$

Here we distinguish if $m - 3n + 1$ is odd or even.

- If $m - 3n + 1$ is odd, the Hamiltonian function has the factor $\left[x_2^2 - \lambda u^{m-3n+1}\right]$, where $\lambda = n(a_m - \frac{i\sigma}{\sqrt{n}}b_m)$, which is irreducible. Now, in order to apply Theorem 8, we calculate the residues

$$Res[\eta(u,1),0] = \frac{4n-1}{m+n+1} = \frac{-\widetilde{M} + 2(m-3n+2)(n_x+1)}{(m-3n+1)\widetilde{M}},$$
$$Res[\eta(1,x_2),\lambda^{1/2}] = \frac{4n-1}{2(m+n+1)} = \frac{\widetilde{M} - 2(m-3n+2)(n_1+1)}{2\widetilde{M}},$$

where $\widetilde{M} = 2(n_x+1) + 2(m-3n+1)(n_1+1)$.
Matching the expressions of the residues and considering the value of $\widetilde{M}$, we obtain

$$(n_x+1) = 4n(n_1+1).$$

Then, by Theorem 8, a first integral of $\mathbf{G}_{m-3n+1}$ is of the form

$$I = u^{n_x+1}(x_2^2 - \lambda u^{m-3n+1})^{n_1+1} = u^{4n(n_1+1)}(x_2^2 - \lambda u^{m-3n+1})^{n_1+1} =$$
$$(u^{4n}(x_2^2 - \lambda u^{m-3n+1}))^{n_1+1} = (u^{4n}(x_2^2 - \lambda u^{m-3n+1}))^N,$$

where $N = n_1 + 1$. Then, $I = (u^{4n}(x_2^2 - \lambda u^{m-3n+1}))^N + \cdots$ is a first integral of system (9). By abuse of notation, we use the same letter I to refer to the first integral of the different vector fields.

We can write $I = (u^{4n}(x_2^2 - \lambda u^{m-3n+1}) + \cdots)^N \cdot U$, because the first component of the vector field is integrable, where U is a unity in the ring of the formal series of the variables u, x_2. In consequence, a first integral of system (9) is of the form

$$I = u^{4n}(x_2^2 - \lambda u^{m-3n+1}) + \cdots. \tag{15}$$

On the other hand, if system (7) is formally integrable, a first integral is of the form $I = (x^{2n} + ny^2)^N + \cdots$, with $N \in \mathbb{N}$. Applying the change of variables described in Lemma 2, this first integral becomes $I = u^{2nN}x_2^N(x_2 - 2\sigma i)^N + O(u^{2nN+1})$, that we can write as

$$I = u^{2nN}x_2^N + O(u^{2nN+1}). \tag{16}$$

Considering the type $\mathbf{t} = (3, 1)$ in system (9), the first quasi-homogeneous component of the vector field is $\mathbf{G}_1 = (\frac{2}{n}ux_2, -4x_2^2)^T$. Ordering respect to this type the first integral (15), we get $I = u^{4n}x_2^2 + \cdots$.
Now we consider (16) with respect to the type $\mathbf{t} = (3, 1)$ and we have $I = u^{2nN}x_2^N + \cdots$.
Matching both expressions we conclude that $N = 2$.
Then, for system (7) a first integral is of the form $I = (x^{2n} + ny^2)^2 + \cdots$ and this corresponds to case **(a)**.

- If $m - 3n + 1$ is even, then we can factorize h. From the expression of (14) we have

$$h = -\frac{(m+n+1)}{n(m-3n+2)}u\left[x_2 - \sqrt{n(a_m - \frac{i\sigma}{\sqrt{n}}b_m)}u^{\frac{m-3n+1}{2}}\right]\times$$
$$\left[x_2 + \sqrt{n(a_m - \frac{i\sigma}{\sqrt{n}}b_m)}u^{\frac{m-3n+1}{2}}\right].$$

Consider $R^2 = \sqrt{n^2a_m^2 + nb_m^2}$ and α such that

$$R^2\cos(2\alpha) = na_m,$$

$$R^2\sin(2\alpha) = -\sqrt{n}\sigma b_m.$$

Thereby $\sqrt{n(a_m - \frac{i\sigma}{\sqrt{n}}b_m)} = R(\cos(\alpha) + i\sin(\alpha))$.

Then, we have

$$h = -\frac{m+n+1}{n(m-3n+2)}u[x_2 - \lambda_1 u^{\frac{m-3n+1}{2}}][x_2 - \lambda_2 u^{\frac{m-3n+1}{2}}],$$
$$\mu = -\frac{4n-1}{n(m-3n+2)}x_2,$$

where $\lambda_1 = R\cos(\alpha) + iR\sin(\alpha)$ and $\lambda_2 = -R\cos(\alpha) - iR\sin(\alpha)$.
Considering the type $\mathbf{t} = (1, \frac{m-3n+1}{2})$ and applying Theorem 8 we obtain

$$Res[\eta(u,1),0] = \frac{2(4n-1)}{m+n+1} = \frac{-2\widetilde{M} + 2(m-3n+2)(n_x+1)}{(m-3n+1)\widetilde{M}},$$
$$Res[\eta(1,x_2),\lambda_i] = \frac{4n-1}{m+n+1} = \frac{\widetilde{M} - (m-3n+2)(n_i+1)}{\widetilde{M}}, \quad i = 1,2,$$

where $\widetilde{M} = (n_x+1) + \frac{m-3n+1}{2}(n_1+1) + \frac{m-3n+1}{2}(n_2+1)$.
Defining $M := n_1 + 1$, $N := n_2 + 1$ and solving the equations we get

$$M = N = \frac{1}{4n}(n_x+1),$$
$$\widetilde{M} = \frac{1}{4n}(m+n+1)(n_x+1).$$

Hence by Theorem 8, a first integral of $\mathbf{G}_{m-3n+1}$ is of the form

$$I = u^{n_x+1}(x_2 - \lambda_1 u^{\frac{m-3n+1}{2}})^{n_1+1}(x_2 - \lambda_2 u^{\frac{m-3n+1}{2}})^{n_2+1},$$

that we can write as

$$I = (u^{4n}(x_2 - \lambda_1 u^{\frac{m-3n+1}{2}})(x_2 - \lambda_2 u^{\frac{m-3n+1}{2}}))^N = \left[(u^{2n}x_2 - \lambda_1 u^{m-3n+1})(u^{2n}x_2 - \lambda_2 u^{m-3n+1})\right]^N.$$

Therefore, analogously to the previous case, it can be proved that a first integral of system (9) is of the form

$$I = (u^{2n}x_2 - \lambda_1 u^{m-3n+1} + \cdots)(u^{2n}x_2 - \lambda_2 u^{m-3n+1} + \cdots).$$

As in the previous case, considering the type $\mathbf{t} = (3, 1)$ in system (9), it is easy to prove that system (7) has a first integral of the form

$$I = (x^{2n} + ny^2 + \cdots)(x^{2n} + ny^2 + \cdots).$$

This is case **(b)**.

(ii) If $m = 2l - 3n + 1$, in the Newton diagram of system (9), there is a unique compact edge of type $(1, l - 3n + 1)$ which is associated to the vector field

$$\mathbf{G}_{l-3n+1} = \begin{pmatrix} \frac{2}{n}ux_2 + \frac{1}{\sqrt{n}}(c_l - \frac{i\sigma}{\sqrt{n}}d_l)u^{l-3n+2} \\ -4x_2^2 + 2(l-n+1)(a_m - \frac{i\sigma}{\sqrt{n}}b_m)u^{2(l-3n+1)} \end{pmatrix}.$$

As $(a_m - \frac{i\sigma}{\sqrt{n}}b_m) \neq \frac{n}{2(l-n+1)}(c_l - \frac{i\sigma}{\sqrt{n}}d_l)^2$, the vector field associated to system (9) is irreducible because its first quasi-homogeneous component is irreducible. Then, we can apply Theorem 8, being the Hamiltonian function and the dissipative term of $\mathbf{G}_{l-3n+1}$:

$$h = -\frac{2(l-n+1)}{n(2l-6n+3)}u\Bigg[\left(x_2 + \frac{\sqrt{n}}{4}\frac{(l-3n+1)}{(l-n+1)}(c_l - \frac{i\sigma}{\sqrt{n}}d_l)u^{l-3n+1}\right)^2 - \left(\frac{n(l-3n+1)^2}{16(l-n+1)^2}(c_l - \frac{i\sigma}{\sqrt{n}}d_l)^2 + n(a_m - \frac{i\sigma}{\sqrt{n}}b_m)\right)u^{2(l-3n+1)}\Bigg],$$

$$\mu = -\frac{2(4n-1)}{n(2l-6n+3)}\left[x_2 - \frac{\sqrt{n}(l-3n+2)}{2(4n-1)}(c_l - \frac{i\sigma}{\sqrt{n}}d_l)u^{l-3n+1}\right].$$

By Theorem 9, we have that $\frac{(l-3n+1)^2}{16(l-n+1)^2}(c_l - \frac{i\sigma}{\sqrt{n}}d_l)^2 + (a_m - \frac{i\sigma}{\sqrt{n}}b_m) \neq 0$, hence we can write h as

$$h = \frac{-2(l-n+1)}{n(2l-6n+3)}u\left[x_2 - \lambda_1 u^{l-3n+1}\right]\left[x_2 - \lambda_2 u^{l-3n+1}\right],$$

where

$$\lambda_1 = -\frac{\sqrt{n}(l-3n+1)}{4(l-n+1)}c_l - R\cos(\alpha) + \left(\frac{\sqrt{n}(l-3n+1)}{4(l-n+1)}\sigma d_l - R\sin(\alpha)\right)i,$$

$$\lambda_2 = -\frac{\sqrt{n}(l-3n+1)}{4(l-n+1)}c_l + R\cos(\alpha) + \left(\frac{\sqrt{n}(l-3n+1)}{4(l-n+1)}\sigma d_l + R\sin(\alpha)\right)i,$$

and α is an angle such that

$$R^2\cos(2\alpha) = \left(na_m + \frac{n(l-3n+1)^2}{16(l-n+1)^2}(c_l^2 - \frac{1}{n}d_l^2)\right),$$

$$R^2\sin(2\alpha) = -\left(b_m + \frac{(l-3n+1)^2}{8(l-n+1)^2}c_l d_l\right)\sqrt{n}\sigma,$$

$$R^2 = \sqrt{\left(na_m + \frac{n(l-3n+1)^2}{16(l-n+1)^2}(c_l^2 - \frac{1}{n}d_l^2)\right)^2 + n\left(b_m + \frac{(l-3n+1)^2}{8(l-n+1)^2}c_l d_l\right)^2}.$$

Now, applying Theorem 8, there exist n_1, n_2, n_x non–negative integers, not all zero, verifying

$$Res[\eta(u,1),0] = \frac{4n-1}{l-n+1} = \frac{-\widetilde{M}+(2l-6n+3)(n_x+1)}{(l-3n+1)\widetilde{M}},$$

$$Res[\eta(1,x_2),\lambda_1] = \frac{(4n-1)[\lambda_1 - \frac{\sqrt{n}(l-3n+2)}{2(4n-1)}(c_l - \frac{i\sigma}{\sqrt{n}}d_l)]}{(l-n+1)(\lambda_1-\lambda_2)} = 1 - \frac{(2l-6n+3)(n_1+1)}{\widetilde{M}},$$

$$Res[\eta(1,x_2),\lambda_2] = \frac{(4n-1)[\lambda_2 - \frac{\sqrt{n}(l-3n+2)}{2(4n-1)}(c_l - \frac{i\sigma}{\sqrt{n}}d_l)]}{(l-n+1)(\lambda_2-\lambda_1)} = 1 - \frac{(2l-6n+3)(n_2+1)}{\widetilde{M}},$$

where $\widetilde{M} = (n_x+1) + (l-3n+1)[(n_1+1)+(n_2+1)]$.
From these equations we get

$$c_l \sin(\alpha) + \frac{\sigma}{\sqrt{n}} d_l \cos(\alpha) = 0, \tag{17}$$

$$\frac{1}{4R}\frac{l+n+1}{l-n+1}\sqrt{n}(c_l\cos(\alpha) - \frac{\sigma}{\sqrt{n}}d_l\sin(\alpha)) = -\frac{M-N}{M+N}, \tag{18}$$

where $M = n_1+1$, $N = n_2+1$ and $n_x+1 = 2n(M+N)$
We distinguish four cases:

(1) Case $c_l \neq 0, d_l = 0$.
From (17) we have $\sin\alpha = 0$ and $|\cos(\alpha)| = 1$, hence $\cos(2\alpha) = 1$ and $\sin(2\alpha) = 0$.
From the expression $R^2\sin(2\alpha) = -\left(b_m + \frac{(l-3n+1)^2}{8(l-n+1)^2}c_ld_l\right)\sqrt{n}\sigma$, we deduce $b_m = 0$.

And from the expression $R^2\cos(2\alpha) = \left(na_m + \frac{n(l-3n+1)^2}{16(l-n+1)^2}\left(c_l^2 - \frac{1}{n}d_l^2\right)\right)$, we get $na_m = R^2 - \frac{n(l-3n+1)^2}{16(l-n+1)^2}c_l^2$.
On the other hand, condition (18) can be written as

$$\sqrt{n}(l+n+1)c_l\cos(\alpha)(M+N) = -4R(l-n+1)(M-N).$$

We multiply by $\cos\alpha$ and we isolate $R\cos(\alpha) = -\frac{\sqrt{n}(l+n+1)}{4(l-n+1)}\frac{M+N}{M-N}c_l$ ($M \neq N$).
We have $na_m = R^2 - \frac{n(l-3n+1)^2}{16(l-n+1)^2}c_l^2 = R^2\cos^2(\alpha) - \frac{n(l-3n+1)^2}{16(l-n+1)^2}c_l^2$, and replacing $R\cos\alpha$, we get the expression given in the statement.

(2) Case $c_l = 0, d_l \neq 0$.
Analogously, from (17) we have $\cos\alpha = 0$ and $|\sin(\alpha)| = 1$, hence $\cos(2\alpha) = -1$ and $\sin(2\alpha) = 0$.
From the expression $R^2\sin(2\alpha) - \left(b_m + \frac{(l-3n+1)^2}{8(l-n+1)^2}c_ld_l\right)\sqrt{n}\sigma$, we deduce $b_m = 0$.

And from the expression $R^2\cos(2\alpha) = \left(na_m + \frac{n(l-3n+1)^2}{16(l-n+1)^2}\left(c_l^2 - \frac{1}{n}d_l^2\right)\right)$, we get $na_m = \frac{(l-3n+1)^2}{16(l-n+1)^2}d_l^2 - R^2$.
Condition (18) can be written as

$$(l+n+1)\sigma d_l \sin(\alpha)(M+N) = 4R(l-n+1)(M-N).$$

We multiply by $\sin\alpha$ and we isolate $R\sin(\alpha) = \frac{l+n+1}{4(l-n+1)}\frac{M+N}{M-N}\sigma d_l$ $(M \neq N)$. We have $na_m = \frac{(l-3n+1)^2}{8(l-n+11)^2}d_l^2 - R^2 = \frac{(l-3n+1)^2}{8(l-n+1)^2}d_l^2 - R^2\sin^2(\alpha)$ and replacing $R\sin\alpha$, we get the expression given in the statement.

(3) Case $c_l^2 + d_l^2 \neq 0$, $c_l^2 - \frac{1}{n}d_l^2 = 0$.
From (17) we have $\cos^2\alpha = \sin^2\alpha$.
From the expression $R^2\cos(2\alpha) = \left(na_m + \frac{n(l-3n+1)^2}{16(l-n+1)^2}\left(c_l^2 - \frac{1}{n}d_l^2\right)\right)$, we deduce $a_m = 0$.
In (17) we can isolate $c_l = -\frac{\sigma d_l\cos(\alpha)}{\sqrt{n}\sin(\alpha)}$, and replacing in (18) we get

$$\frac{(l+n+1)\sigma d_l}{4(l-n+11)R\sin(\alpha)} = \frac{M-N}{M+N}.$$

As $M \neq N$, we can isolate $R\sin(\alpha) = \frac{(l+n+1)\sigma d_l}{4(l-n+1)}\frac{M+N}{M-N}$.
On the other hand, if we multiply (17) by $\sin\alpha$, we can isolate $\sin^2\alpha = -\frac{\sigma d_l\cos(\alpha)\sin(\alpha)}{\sqrt{n}c_l}$, and replacing into the last equality we obtain

$$-R^2\frac{\sigma d_l\cos(\alpha)\sin(\alpha)}{\sqrt{n}c_l} = \frac{(l+n+1)^2 d_l^2}{16(l-n+1)^2}\frac{(M+N)^2}{(M-N)^2}.$$

From here we can isolate $R^2\cos(\alpha)\sin(\alpha) = -\frac{(l+n+1)^2}{16(l-n+1)^2}\frac{(M+N)^2}{(M-N)^2}\sqrt{n}\sigma c_l d_l$.
Replacing this last equality in the expression $-\left(b_m + \frac{(l-3n+1)^2}{8(l-n+1)^2}c_l d_l\right)\sqrt{n}\sigma = R^2\sin(2\alpha) = 2R^2\cos(\alpha)\sin(\alpha)$, we can isolate b_m and we obtain the expression given in the statement.

(4) Case $c_l^2 + d_l^2 \neq 0$, $c_l^2 - \frac{1}{n}d_l^2 \neq 0$.
From (17) we have $\tan\alpha = -\frac{\sigma}{n}\frac{d_l}{c_l}$. Moreover

$$\tan(2\alpha) = \frac{2\tan(\alpha)}{1-\tan^2(\alpha)} = \frac{-2\sigma c_l d_l}{\sqrt{n}(c_l^2 - \frac{1}{n}d_l^2)}.$$

On the other hand,

$$\tan(2\alpha) = \frac{\sin(2\alpha)}{\cos(2\alpha)} = \frac{-(b_m + \frac{(l-3n+1)^2}{8(l-n+1)^2}c_l d_l)\sqrt{n}\sigma}{na_m + \frac{n(l-3n+1)^2}{16(l-n+1)^2}(c_l^2 - \frac{1}{n}d_l^2)}.$$

Equating both expressions and isolating a_m, we deduce that

$$a_m = \frac{c_l^2 - \frac{1}{n}d_l^2}{2c_l d_l} b_m.$$

Following the same idea as in the previous case, we get the expression of b_m given in the statement, and replacing it in the last equality we obtain the expression of a_m given in the statement.

Then in this case, system (7) has a first integral of the form

$$I = u^{2n(M+N)}(x_2 - \lambda_1 u^{l-3n+1})^M (x_2 - \lambda_2 u^{l-3n+1})^N.$$

Following the same idea as in previous cases, we can write

$$I = u^{2n(M+N)}(x_2 - \lambda_1 u^{l-3n+1} + \cdots)^M (x_2 - \lambda_2 u^{l-3n+1} + \cdots)^N,$$

i.e.,

$$I = (u^{2n}x_2 - \lambda_1 u^{l-n+1} + \cdots)^M (u^{2n}x_2 - \lambda_2 u^{l-n+1} + \cdots)^N,$$

because the first quasi-homogeneous component of the vector field is polynomially integrable.
Considering the type $\mathbf{t} = (3, 1)$, this first integral becomes

$$I = (u^{2n}x_2 + \cdots)^M (u^{2n}x_2 + \cdots)^N = (u^{2n}x_2)^{M+N} + \cdots.$$

On the other hand, we know that system (7) has a first integral of the form $I = (x^{2n} + ny^2)^{\widetilde{N}} + \cdots$, $\widetilde{N} \in \mathbb{N}$.
After the change of variables described in Lemma 2, it can be written as $I = u^{2n\widetilde{N}} x_2^{\widetilde{N}} + \cdots = (u^{2n}x_2)^{\widetilde{N}} + \cdots$.
Matching both expressions, we deduce that $\widetilde{N} = M + N$.
Then, system (7) has a first integral of the form

$$I = (x^{2n} + ny^2 + \cdots)^M (x^{2n} + ny^2 + \cdots)^N.$$

This is the case **(c)**.

6 Application

We study the following system

$$\begin{pmatrix} \dot{x} \\ \dot{y} \end{pmatrix} = (x^4 + 2y^2)\begin{pmatrix} -y \\ x^3 \end{pmatrix} + \begin{pmatrix} a_{51}x^5y + a_{13}xy^3 \\ b_{80}x^8 + b_{42}x^4y^2 + b_{04}y^4 \end{pmatrix}. \qquad (19)$$

Theorem 11 *System* (19) *has a first integral of the form* $I = x^4 + 2y^2 + \cdots$ *if, and only if, one of the following conditions is satisfied*

(1) $a_{13} = 2a_{51}$, $b_{80} = 0$, $b_{04} = 2b_{42}$.
(2) $b_{80} = -a_{51}$, $b_{42} = -a_{13}$, $b_{04} = 0$.

Proof To obtain the necessary conditions we apply Theorem 7, that is, **F** must be reducible. The second component must be reducible, because the first component is reducible. Thus we obtain:

- $a_{13} = 2a_{51}$, $b_{80} = 0$, $b_{04} = 2b_{42}$; and in this case, $(x^4 + 2y^2)$ is the factor of reducibility,
- or $b_{80} = -a_{51}, b_{42} = -a_{13}$ and $b_{04} = 0$; in this case, $(x^4 + 2y^2 - a_{51}x^5 - a_{13}xy^2)$ is the common factor of both components.

On the other hand, we first suppose that $a_{13} = 2a_{51}$, $b_{80} = 0$ and $b_{04} = 2b_{42}$. In this case system (19) becomes

$$\begin{pmatrix} \dot{x} \\ \dot{y} \end{pmatrix} = (x^4 + 2y^2)\left[\begin{pmatrix} -y \\ x^3 \end{pmatrix} + \begin{pmatrix} a_{51}xy \\ b_{42}y^2 \end{pmatrix}\right], \tag{20}$$

which is integrable with a first integral of the form $I = x^4 + 2y^2 + \cdots$.

We assume now that $b_{80} = -a_{51}$, $b_{42} = -a_{13}$ and $b_{04} = 0$. With these values system (19) yields

$$\begin{pmatrix} \dot{x} \\ \dot{y} \end{pmatrix} = (x^4 + 2y^2 - a_{51}x^5 - a_{13}xy^2)\begin{pmatrix} -y \\ x^3 \end{pmatrix}, \tag{21}$$

which is integrable. Moreover, as it is reducible, applying Theorem 7, a first integral is of the form $I = x^4 + 2y^2 + \cdots$.

Theorem 12 *System* (19), *in generic conditions* ($A \neq 0$), *is analytically integrable if, and only if, one of the following conditions is verified.*

(1) $4a_{51} - 2a_{13} + 4b_{80} - 2b_{42} + b_{04} \neq 0$.
(2) $a_{13} = 2a_{51} + 2b_{80} - b_{42} + \frac{1}{2}b_{04}$, $16a_{51} - 4b_{80} + 10b_{42} - 9b_{04} \neq 0$, $2a_{51} + 4b_{80} - b_{42} \neq 0$, $64a_{51} + 164b_{80} - 50b_{42} + 9b_{04} \neq 0$ *and exist* $M, N \in \mathbb{N}$, *with* $M \neq N$, *such that* $4(-4a_{51} - 4b_{80} + b_{04})N + 5(4b_{80} - 2b_{42} + b_{04})M = 0$.
(3) $a_{13} = 10a_{51} + 4b_{42} - 4b_{04}$, $b_{80} = 4a_{51} + \frac{5}{2}b_{42} - \frac{9}{4}b_{04}$, $2a_{51} + b_{42} - b_{04} \neq 0$,

where $A = -81b_{04}^2 + 1296b_{80}^2 + 324a_{51}b_{04} - 648a_{51}b_{42} + 1296a_{51}b_{80} + 162b_{04}b_{42} - 648b_{42}b_{80}$.

Moreover in the Cases 1 *and* 3 *a first integral of system* (19) *is of the form* $I = (x^4 + 2y^2)^2 + \cdots$ *and in the Case* 2 *is of the form* $I = (x^4 + 2y^2)^{M+N} + \cdots$.

Proof The normal form of system (19) has the following first coefficients:

$$\begin{aligned} a_6 =& \frac{1}{36}(4a_{51} - 2a_{13} + 4b_{80} - 2b_{42} + b_{04}), \ b_6 = 0, \\ c_6 = 0, \ d_6 =& \frac{1}{18}(10a_{51} - a_{13} + 4b_{42} - 4b_{04}). \end{aligned} \tag{22}$$

First we consider the case $a_6 \neq 0$, that is, $4a_{51} - 2a_{13} + 4b_{80} - 2b_{42} + b_{04} \neq 0$, and we apply Theorem 10 statement **(a)**, since $m = 6$ and $m < 2l - 5$ for all $l \geq 6$. In this case we conclude that system (19) is integrable and a first integral is of the form $I = (x^4 + 2y^2)^2 + \cdots$.

We now assume that $a_6 = 0$, that is, $a_{13} = 2a_{51} + 2b_{80} - b_{42} + \frac{1}{2}b_{04}$. The coefficients of the normal form for system (19) are now

$$\begin{aligned} a_6 &= 0, \quad b_6 = 0, \\ c_6 &= 0, \quad d_6 = \frac{1}{36}(16a_{51} - 4b_{80} + 10b_{42} - 9b_{04}), \\ a_7 &= -\frac{1}{6480}(2a_{51} + 4b_{80} - b_{42})(64a_{51} + 164b_{80} - 50b_{42} + 9b_{04}), \ b_7 = 0. \end{aligned}$$

We first consider the case $a_7 \neq 0$ and $d_6 \neq 0$. In this case we are under the conditions of Theorem 10 statement **(c)**, that is, $m = 2l - 5$ with $l = 6$ y $m = 7$. Therefore we have that

$$a_m = \frac{1}{16(l-1)^2}\left((l+3)^2\frac{(M+N)^2}{(M-N)^2} - (l-5)^2\right)(c_l^2 - \frac{1}{2}d_l^2).$$

In our case this implies

$$a_7 = \frac{1}{400}\left[1 - 81\frac{(M+N)^2}{(M-N)^2}\right]\frac{1}{2}d_6^2.$$

If we replace the values of a_7 and d_6, we can write it as $4(-4a_{51} - 4b_{80} + b_{04})N + 5(4b_{80} - 2b_{42} + b_{04})M = 0$. Then, it is satisfied the relation given in statement (2). We now assume that $a_6 = 0$, $d_6 = 0$ and $a_7 \neq 0$. We apply Theorem 10 statement **(b)** and we deduce that system (19) is formally integrable with a first integral of the form $I = (x^4 + 2y^2 + \cdots)(x^4 + 2y^2 + \cdots)$.

Acknowledgements The authors are supported by a MINECO/FEDER grant number MTM2014-56272-C2-02 and by the *Consejería de Educación y Ciencia de la Junta de Andalucía* (projects P12-FQM-1658, FQM-276).

References

1. Algaba, A., Checa, I., García, C., Gamero, E.: On orbital-reversibility for a class of planar dynamical systems. Commun. Nonlinear Sci. Numer. Simulat. **20**, 229–239 (2015)
2. Algaba, A., Checa, I., García, C., Giné, J.: Analytic integrability inside a family of degenerate centers. Nonlinear Anal. Real World Appl. **31**, 288–307 (2016)
3. Algaba, A., Freire, E., Gamero, E., García, C.: Monodromy, center-focus and integrability problems for quasi-homogeneous polynomials systems. Nonlinear Anal. **72**, 1726–1736 (2010)
4. Algaba, A., Fuentes, N., García, C.: Centers of quasi-homogeneous polynomial planar systems. Nonlinear Anal. Real World Appl. **13**, 419–431 (2012)
5. Algaba, A., Gamero, E., García, C.: The integrability problem for a class of planar systems. Nonlinearity **22**, 395–420 (2009)
6. Algaba, A., Gamero, E., García, C.: The reversibility problem for quasi-homogeneous dinamical systems. Discrete Contin. Dyn. Syst. **33**, 3225–3236 (2013)
7. Algaba, A., Gamero, E., García, C.: The center problem. A view from the normal form theory. J. Math. Anal. Appl. **434**(1), 680–697 (2015)
8. Algaba, A., García, C., Giné, J.: Analytic integrability for some degenerate planar systems. Commun. Pure Appl. Anal. **6**, 2797–2809 (2013)
9. Algaba, A., García, C., Giné, J.: Analytic integrability for some degenerate planar vector fields. J. Differ. Equ. **257**, 549–565 (2014)
10. Algaba, A., García, C., Giné, J.: Nilpotent centers via inverse integrating factors. Euro. J. Appl. Math. **27**, 781–795 (2016)
11. Algaba, A., García, C., Giné, J.: The analytic integrability problem for perturbation of non-hamiltonian quasi-homogeneous nilpotent systems. Submited (2017)
12. Algaba, A., García, C., Reyes, M.: The center problem for a family of systems of differential equations having a nilpotent singular point. J. Math. Anal. Appl. **340**, 32–43 (2008)
13. Algaba, A., García, C., Reyes, M.: Integrability of two dimensional quasi-homogeneous polynomial differential systems. Rocky Mt. J. Math. **41**, 1–22 (2011)
14. Algaba, A., García, C., Reyes, M.: Existence of an inverse integrating factor, center problem and integability of a class of nilpotent systems. Chaos Solitons Fractals **45**, 869–878 (2012)
15. Algaba, A., García, C., Reyes, M.: A note on analytic integrability of planar vector fields. Eur. J. Appl. Math. **23**, 555–562 (2012)
16. Algaba, A., García, C., Teixeira, M.: Reversibility and quasi-homogeneous normal form of vector fields. Nonlinear Anal. **73**, 510–525 (2010)
17. Andreev, A., Sadovskii, A.P., Tskialyuk, V.A.: The center-focus problem for a system with homogeneous nonlinearity in the case zero eigenvalues of linear part. J. Differ. Equ. **39**(2), 155–164 (2003)
18. Arnold, V.I.: Local normal forms of functions. Invent. math. **35**, 87–109 (1976)
19. Berthier, M., Moussu, R.: Reversibilit et classification des centres nilpotents. Ann. Inst. Fourier (Grenoble) **44**, 465–494 (1994)
20. Brunella, M., Miari, M.: Topological equivalence of a plane vector field with its principal part defined through newton polyhedra. J. Differ. Equ. **85**, 338–366 (1990)
21. Chavarriga, J., García, I., Giné, J.: Integrability of centers perturbed by quasi-homogeneous polynomials. J. Math. Anal. Appl. **210**, 268–278 (1997)
22. Chavarriga, J., Giacomini, H., Giné, J., Llibre, J.: On the integrability of two-dimensional flows. J. Differ. Equ. **157**, 163–182 (1999)
23. Chavarriga, J., Giacomini, H., Giné, J., Llibre, J.: Local analytic integrability for nilpotent centers. Ergod. Theory Dynam. Syst. **23**, 417–428 (2003)
24. Dumortier, F.: Singularities of vector fields on the plane. J. Differ. Equ. **23**, 53–106 (1977)
25. Giacomini, H., Giné, J., Llibre, J.: The problem of distinguishing between a center and a focus for nilpotent and degenerate analytic systems. J. Differ. Equ. **227**(2), 406–426 (2006)
26. Giacomini, H., Giné, J., Llibre, J.: The problem of distinguishing between a center and a focus for nilpotent and degenerate analytic systems (corrigendum). J. Differ. Equ. **232**, 702 (2007)

27. Giné, J.: Sufficient conditions for a center at a completely degenerate critical point. Int. J. Bifur. Chaos Appl. Sci. Eng. **12**, 1659–1666 (2002)
28. Giné, J.: On the centers of planar analytic differential systems. Int. J. Bifur. Chaos Appl. Sci. Eng. **17**, 3061–3070 (2007)
29. Giné, J.: On the degenerate center problem. Int. J. Bifur. Chaos Appl. Sci. Eng. **21**, 1383–1392 (2011)
30. Giné, J., Llibre, J.: A method for characterizing nilpotent centers. J. Math. Anal. Appl. **413**, 537–545 (2014)
31. Li, J.: Hilbert's 16th problem and bifurcations of planar polynomial vector fields. Int. J. Bifur. Chaos Appl. Sci. Eng. **13**, 47–106 (2003)
32. Mattei, J., Moussu, R.: Holonomie et intégrales premières. Ann. Sci. École Norm. Sup. **4**(13), 469–523 (1980)
33. Moussu, R.: Symetrie et forme normale des centres et foyers degeneres. Ergod. Theory Dynam. Syst. **2**, 241–251 (1982)
34. Pearson, J., Lloyd, N., Christopher, C.: Algorithmic derivation of centre conditions. SIAM Rev. **38**, 691–636 (1996)
35. Poincaré, H.: Mémoire sur les courbes définies par les équations différentielles. J. de Mathématiques **37**, 375–442 (1881)
36. Poincaré, H.: Mémoire sur les courbes définies par les équations différentielles. J. de Mathématiques **8**, 251–296 (1882)
37. Poincaré, H.: Ouvres de Henri Poincaré, vol. I. Gauthier-Villars, Paris (1951)
38. Romanovski, V.G., Shafer, D.: The center and cyclicity problems: a computational algebra approach. Birkhäuser Boston (2009)
39. Strózyna, E., Zoładek, H.: The analytic and normal form for the nilpotent singularity. J. Differ. Equ. **179**, 479–537 (2002)
40. Teixeira, M.A., Yang, J.: The center-focus problem and reversibility. J. Differ. Equ. **174**, 237–251 (2001)
41. Villarini, M.: Algebraic criteria for the existence of analytic first integrals. Differ. Equ. Dynam. Syst. **5**, 439–454 (1997)

A Logistic Non-linear Difference Equation with Two Delays

Francisco Balibrea

Abstract In this chapter we analyze the state of art of logistic first and second order difference equations with two delays. They model the evolution of populations with respect to seasons in time $n \in \mathbb{N}$. Of special interest are those non-linear equations with two delays, particularly due to the effect of food in the evolution of the population. As an adequate tool to understand the behaviors of solutions of the equation, we use an unfolding of it obtaining a discrete dynamical system of dimension two, defined in the unit square. We review some dynamical properties already known like periodic solutions and local linear analysis around the fixed points of the unfolding. Besides we also introduce new results and analysis of behavior of invariant curves. All analysis depend on a parameter a. Our study is mainly devoted to the range $a \in (0, 2]$, the setting where we give some revised results. Some open problems remain for the range when $a \geq 2.27$ where 2.27 is a critical value.

Keywords Non-linear difference equations · Unfolding
Periodic solutions · Invariant curves

1 Introduction

Some phenomena described in Population Dynamics may be modeled using difference equations and systems of them with some delays which means that what happens in one time depends on what has happened several times before. By times we mean seconds, weeks, months, years, seasons, etc, depending on the nature of the considered phenomena.

F. Balibrea (✉)
Group of Dynamical Systems, Facultad de Matemáticas, Universidad de Murcia, Campus de Espinardo, 30100 Murcia, Spain
e-mail: balibrea@um.es

V. Carmona et al. (eds.), *Nonlinear Systems, Vol. 1*,
Understanding Complex Systems, https://doi.org/10.1007/978-3-319-66766-9_9

In the literature on difference equations modeling the evolution of populations, some examples are of the form

$$x_{n+1} = x_n f(x_n) = F(x_n) \tag{1}$$

where f is generally a continuous monotone decreasing map. Such type of equations are called *logistic* because most of known models in Population Dynamics are formulated in this way. If the nonlinearity of f is strong enough, then the sets of *solutions* of the equations, associated to different maps f are composed of sequences of values $(x_n)_{n\geq 0}$ computed using the equations and taking a starting value x_0 or a vector of them, may contain periodic solutions of arbitrary period, aperiodics or chaotics.

The behaviours of solutions can be very complicated as May in [13] and Beddington et al. in [5] mention. They may be formulated also as coupled or independent pairs of equations of the former type to obtain models on two population systems. In these last models, very often it is possible to appreciate more bizarre and chaotic behaviours.

On next sections we are considering some models of populations composed of one or more species and having one or several delays.

2 Logisted Models for One and Several Species

Two of the most employed logistic models for one species are:

$$x_{n+1} = ax_n(1 - x_n)\ a > 0 \tag{2}$$

$$x_{n+1} = x_n e^{a(1-x_n)}\ a > 0 \tag{3}$$

where x_n is measured in terms of a percentage of the carrying capacity of the environment given in numbers of individuals. The model (3) was discussed in [14] and (2) in [8] and in many other references (see for example [7] and [19]). Equation (2) with $0 < a \leq 4$ modelizes the evolution of a single species in a territory and x means the relative population living on it, being always a number belonging to the interval $[0, 1] = I$. For $0 < a < 1$ we have $\lim_{n\to\infty} x_n = 0$ (see [19]) and the population is on its way to extinction. In fact (2) can be seen as the one dimensional discrete dynamical system $(f, [0, 1])$ where $f(x) = ax(1 - x)$. In this case, the sequence $(f^n(x))_{n\geq 0} = Orb_f(x_0)$ is called the orbit of the value x_0 by f where $f^n = f \circ f^{n-1}$ for $n \geq 1$ and f^0 is the *identity* on I. A value x_0 is a periodic value of f is there is a p such that $f^{n+p} = f^n$ for all n. The minimal p holding such condition is called the *period* of x_0. When $p = 1$ then x_0 is a *fixed* value. Instead of using the term value we may use the term point. Further we will use both terminology.

Concerning fixed values or points or further periodic orbits, we give the following notions of attractive or repulsive points for the dynamical system $(\mathbb{X}, f)$ where $\mathbb{X}$ is a compact metric space and f continuous map from $\mathbb{X}$ into itself.

Definition 1 A fixed point $x_0 \in \mathbb{X}$ is said *attractive* if there is a neighborhood $U(x_0)$ such that for any $x \in U(x_0)$ the $Orb_f(x)$ converges to x_0.

Definition 2 A fixed point $x_0 \in \mathbb{X}$ is said *repulsive* if there is a neighborhood $U(x_0)$ such that for every $x \in U(x_0)$ with $x \neq x_0$ we have $f^n(x) \notin U(x_0)$ for some n.

Definition 3 Let $\{x_1, x_2, \ldots, x_k\} = A$ be a periodic orbit of period $k < 1$. Such orbit is said *attractive* if at least one value in A is an attractive value for f^k.

Definition 4 The orbit is repulsive if every value in A is a *repulsive* point for f^k.

In dynamical systems $(\mathbb{I}, f)$ where $f \in C^1(\mathbb{I})$ if p is a fixed point and $|f'(p)| < 1$, then there exists an open interval $U(p)$ such that whenever $x \in U(p)$, then $f^n(x)$ converges to p. If $|f'(p)| > 1$, then there exists an open interval containing p such that all points in the interval $y \neq p$ must leave such interval under interaction of f. These conditions are sufficient for p to be attractive or repulsive. Nothing can be assured if $|f'(p)| = 1$. Such definitions can be extended to periodic orbits of periods greater than one (for more detailed information, see [11]).

When $a \in [0, 3]$, the function f from (2) always has an attractive fixed point, namely $\alpha = 0$ and another $\beta = \frac{a-1}{a} = \overline{x}$ if $a \in (1, 3)$. For the range of parameter $(0, 3)$ the function f has no periodic orbits of higher periods, which means that a population governed by this model with $a \in [0, 3]$ is stable (its behaviour does not change by small perturbations) and therefore the number of individuals approaches in time to an equilibrium (i.e. to a fixed point of f). When $a > 3$ the function f has cycles of higher periods and even chaotic behaviours (see [19]).

Let now consider another example of logistic behaviour by the difference equation

$$x_{n+1} = x_n f(x_n) = a x_n^2 (1 - x_n) \tag{4}$$

It is not difficult to see that when $a \in [0, \frac{27}{4}]$ the function

$$f_a = [ax(1 - x)]x = ax^2(1 - x)$$

describing the population growth, maps continuously $[0, 1]$ into itself. The fixed points of f_a are obtained solving the equation $x(ax^2 - ax + 1) = 0$. Since $\alpha = 0$ is a solution of the former equation, it is a fixed point of f_a (the map f for parameter a). It is immediate that f_a has a zero derivative in α and in [19] it is proved that always α is an attractive fixed point.

In order to find other possible fixed points, we solve the equation $ax^2 - ax + 1 = 0$. We have the following solutions:

1. When $a < 4$ then $\alpha = 0$ is the unique fixed point of f_a. It is immediate to see that for all $x > 0$ we have $f_a(x) < x$ and then any solution $(x_n)_{n \geq 0}$ where $x_0 \in [0, 1]$, converges to α.
2. When $a = 4$ the former second order equation has $\beta = \frac{1}{2}$ as a unique solution. Since is $f_a'(\beta) = 1$ it is difficult to see if such point is attractive or repulsive.

In figure we represent three cases for $a = 3$, $a = 4$ and $a = \frac{16}{3}$. The diagram, as well a direct computation, show that $f_a(x) < x$ for $x < \beta$, $x \neq 0$ and as a consequence, β can not be a repulsive fixed point. In an analogous way, for $x > \beta$ and being x sufficiently close to β, then the sequence generate by x_0 converges to β and it can not be repulsive.

3. For $a > 4$, the former fixed point β splits into two fixed points

$$\beta_1 = \frac{1}{2} - (\frac{1}{4} - \frac{1}{a})^{\frac{1}{2}}$$

$$\beta_2 = \frac{1}{2} + (\frac{1}{4} - \frac{1}{a})^{\frac{1}{2}}$$

it is easy to see that $0 < \beta_1 < \beta_2 < 1$ and $f'_a(\beta_1) > 1$. Thus β_1 is a repulsive fixed point for any value $a \in (4, \frac{27}{4}]$. For the other fixed point β_2 we have

$$f'_a(\beta_2) = 3 - \frac{1}{2}a - \frac{1}{2}(a^2 - 4a)^{\frac{1}{2}}$$

we see that $f'_a(\beta_2) > -1$ if and only if $a < \frac{16}{3}$ and $f'_a(\beta_2) < 1$ for all a. Therefore for $a \in (4, \frac{16}{3})$, β_2 is an attractive fixed point of f_a. Although $f'_a(\beta_2) = -1$ for $a = \frac{16}{3}$, β_2 is an attractive fixed point. To see it we use a Sharkovskii's theorem [18] proving that $f_{\frac{16}{3}}$ has no periodic points of period greater than one.

4. If $a > \frac{16}{3}$ we have $f'_a(\beta_2) < -1$, hence β_2 is a repulsive fixed point. For this range of parameters, f_a has a two cycle (the two different point of it are separated by β_2). For $a > 5.76\ldots$ the first 4-cycle appears and so on. At $a = 5.89\ldots$ the function becomes chaotic (we use such terminology in the sense of Li and Yorke which it is a weak notion of chaos) (see definitions in [19]).

5. For $a > 4$ it is interesting to study and interpret the role of the fixed points β_1 and β_2. As $f_a(1) = 0$ it is evident that there is a point $\gamma \in (\beta_2, 1)$ holding $f_a(\gamma) = \beta_1$. If $(x_n)_{n\geq 0}$ is the sequence starting in x_0, then

$$\lim_{n\to\infty} x_n = 0 \text{ when } x_0 \in [0, \beta_1) \cup (\gamma, 1] \tag{5}$$

$$\lim_{n\to\infty} x_n = \beta_1 \text{ if } x_0 = \beta_1 \text{ or } x_0 = \gamma \tag{6}$$

The property (5) is obtained from the fact that $f_a(x) < \beta_1$ for $x \in [0, \beta_1) \cup (\gamma, 1]$, since f_a is increasing in $[0, \beta_1]$ and deceasing in $[\gamma, 1]$. The condition (6) is evident. If we can back to (β_1, γ), it is immediate that there is a critical value $a_c = 6.6\ldots$ such that

$$f_a(x) \in (\beta_1, \gamma) \text{ whenever and } a \in (4, a_c) \tag{7}$$

After this, and take into account (5), (6) and (7), they imply that β_1 and γ are *threshold* values. If the population would not attain at least β_1, then it will die out.

This situation occurs rather often, a small population is likely to extinct under the influence of various disturbing factors. If the population would attain exactly the value β_1, it would be in equilibrium, in fact theoretically since the equilibrium is very unstable. This means that a small perturbation suffices to push it away from β_1 to one side or the other.

If the population would exceed the second threshold value γ, something similar similar to the first case would happen and the population would die out. This is often the case in reality. We can imagine an over population of some herbivorous animals in a desert with very poor vegetation. The result would be the extinction of the vegetation and the population could not reproduce at a sufficient rate. As a consequence very likely the animals would die out too.

If the population would attain a value in (β_1, γ), it would tend for $a \leq \frac{16}{3}$ to the equilibrium β_2 which is stable. This means that when the population would remove from that state slightly by a small perturbation, it will return to it after some time. In case that $a \in (\frac{16}{3}, a_c)$, the population may behave variously, even chaotically, but can not die out. In last case it varies between β_1 and γ. If a would exceed a_c, the population may die out in this case as well. We can observe that in fact the model given by Eq. (4) is nearer to reality that (2), but its dynamics is much more complicated and difficult to understand.

Frequently in the literature, we may find variants of the former models, which are transformed in variants in the respective difference equations. One of them is the model given by

$$x_{n+1} = ax_n - bx_n^2 \tag{8}$$

with $a, b > 0$ constants. This equation may be written in the form

$$x_{n+1} = x_n[1 + r(1 - \frac{x_n}{K})] \tag{9}$$

with r, $K > 0$ constants. Besides $x = 0$, the other equilibrium value is $\overline{x} = K$. K is interpreted as the capacity of the environment and r is the growth coefficient. With a suitable change of variable, (8) and (9) can be reduced to (4).

Equation (9) may be obtained using an exponential function

$$x_{n+1} = x_n e^{r(1-\frac{x_n}{K})}$$

with a dynamical treatment similar to the previous models.

3 Models for Epidemics

In this subsection we are obtaining different models on spreading of diseases depending on different assumptions. Simple models may be obtained with the following assumption. In every period of time, each individual has the same number of contacts with other individuals, this means to be in equal chance of getting infected with the disease. Additionally we will suppose that the population is constant, i.e. the number of individuals considered does not change in time.The period during which a sick person is contagious is also suppose to be constant in time and its length equals the unit of time (it may be one day, a week, fifteen days, etc). After overcoming the disease, depending on the kind of disease, the individual may either remain permanently immune or return without immunity to the group of susceptible people. Further we will study the second case.

First divide the population in two groups. Let denote by $I(t)$ the number, depending on time, of infected persons and with $S(t)$ the number of individuals who can get infected in time t and T the total population. Now suppose that if two individuals are arbitrarily chosen, one being healthy and the other sick, then the probability of the healthy person of getting infected from the sick within a unit of time is p and does not depend on the choice of the two individuals. Now $q = 1 - p = e^{-\alpha}$ where $\alpha > 0$. The probability, denoted by P, that a given susceptible individual will not get infected in a unit interval of time $(t, t+1)$ depends on the total number of infected persons at the time t. The larger such number is, the less will be P. In fact we will have $P = q^{I(t)}$. Then the probability for one healthy individual to get infected in the chosen unit of time will be $1 - P = 1 - q^{I(t)}$, and the probable number of new cases of disease will be proportional to the total number of healthy persons, that is

$$S(t)(1 - q^{I(t)}) = S(t)(1 - e^{-\alpha I(t)})$$

we have also that

$$I(t+1) = S(t)(1 - e^{-\alpha I(t)})$$

since those individuals who were ill at time t will be well again after a period of a unit time, that is at time $t+1$. Also is evident that

$$S(t) = T - I(t)$$

and introduction the notation

$$x_1(t) = \frac{I(t)}{T}, \; x_2(t) = \frac{S(t)}{T}$$

and

$$\alpha T = a$$

from the last two equations we have

$$x_1(t+1) = x_2(t)(1 - e^{-a x_1(t)})$$

and

$$x_2(t) = 1 - x_1(t)$$

and eliminating $x_2(t)$ we obtain

$$x_(t + 1) = (1 - x_1(t))(1 - e^{-ax_1(t)}),$$

that is

$$y_{n+1} = (1 - y_n)(1 - e^{-ay_n}) \tag{10}$$

where y_n denotes the relative number of sick individuals at time n.

Some variants of (10) may be considered modifying slightly it. Suppose that the infected person is contagious for a period of a unit of time and the next period of unit of time the individual is isolated or immune (depending of the type of disease) and becomes susceptible again. This is taken into account introducing a new function $J(t)$ in the former example, meaning the number of those which are immune at time t, getting the system of equations

$$I(t + 1) = S(t)(1 - e^{-aI(t)})$$

$$J(t + 1) = J(t)$$

$$S(t) + J(t) + I(t) = T$$

introducing a new relative value $x_3 = \frac{I(t)}{T}$ we have

$$x_1(t + 1) = x_2(t)(1 - e^{-ax_1(t)})$$

$$x_3(t + 1) = x_1(t)$$

$$x_1(t) + x_2(t) + x_3(t) = 1$$

eliminating x_3, the second and third equations yield

$$x_1(t) + x_2(t) + x_1(t - 1) = 1$$

and the first equation becomes

$$x_1(t + 1) = (1 - x_1(t) - x_1(t - 1))(1 - e^{-ax_1(t)})$$

or

$$y_{n+1} = (1 - y_n - y_{n-1})(1 - e^{-ay_n})$$

where y_n denotes again the relative number of sick persons at time n. We observe that this equation is the second order, in the sense that has two delays.

In some diseases, the immunity period lasts longer than the disease itself. For example one unit of time for the duration of disease and two for the period of immunity or isolation. The previous equations remain the same except the second which is

$$J(t + 1) = I(t) + I(t - 1)$$

applying the same process we reach

$$y_{n+1} = (1 - y_n - y_{n-1} - y_{n-2})((1 - e^{-ay_n})$$

with the same meaning than in former cases. Of course such procedure may be extended to k delays.

Coming back to the first Eq. (9), we will analyse what happens and interpret the results. For doing this it is necessary to understand the dynamics of the function

$$f_a(x) = (1 - x)(1 - e^{-ax})$$

the proof of next properties can be found, for example in [19]

1. For all values $a \geq 0$, f_a maps $[0, 1] = I$ into itself
2. For $0 \leq a \leq 1$, $f : a$ has a unique fixed point $x = 0$ and every orbit of f_a starting at $x_0 \in I$ converges to 0.
3. If $a > 1$ then f_a has two fixed points, a repulsive fixed point at $x = 0$ and an attractive one $\beta \in (0, \frac{1}{2})$ and every orbit of f_a starting at any $x_0 \in (0, 1)$ converges to β.

The graph of f_a for several values of a can be seen in figure.

These properties allow us to interpret (5). When $a \leq 1$, then $\log q = -\frac{a}{T} \geq -\frac{1}{T}$, thus, if q is large enough which means little probability of getting infected, then the disease will completely disappear after some time. But if q falls under some limit, the disease will not disappear, but the percentage of sick individuals will stabilize after a sufficiently long time at a non-zero value of 100β. Such value is stable. This means that if the percentage of sick persons is changed due to some temporary very unfavourable or very favourable circumstances and it will back to that value after some time.

The analysis of the rest of equations is not easy because we face to difference equations of order bigger than one and will partially do on next sections. For detail account of such analysis can be seen for example in [9].

4 Discrete Population Growth Models for Two Species

The most studied relation between two biological species is that of a predator and its prey where the second serves as food for the former. Under the assumption that both population are homogeneous in age and genetic structure and the same for the environment, that is, natural conditions,it is possible to obtain rather simple systems of equations which model the evolution in time of both populations.

Let us denote by x_i the number of prey and by y_i the number of predator in a given territory at time i. Let us suppose that the reproduction of preys follows a logistic model and thus, the Eq. (5) occurs. Therefore,

$$x_{i+1} = ax_i - bx_i^2$$

with $a > 1$ and $b > 0$. Next, assume that the number of predators is proportional to the number of preys and predators at a unit period of time before, the is,

$$y_{i+1} = cx_i y_i$$

where $c > 0$ and assume that $dx_i y_i$ with $d > 0$ animals become victims of the predators, as a result we have

$$x_{i+1} = ax_i - bx_i^2 - dx_i y_i$$

For simplifying the calculations, we introduce another notation for the constants. The final system of equations is

$$x_{i+1} = (1 + A)x_i - BDx_i^2 - CDx_i y_i$$

$$y_{i+1} = Cx_i y_i$$

where all constants are positive numbers.

To analyse such systems, we start obtaining the fixed points which consists on solving the algebraic system of equations

$$x = (1 + A)x - BDx^2 - DCxy$$

$$y = Cxy$$

We obtain two fixed points $x^1 = 0$ and $x^2 \frac{A}{BC}$. If $y \neq 0$ then the fixed point has coordinates $\overline{x} = \frac{1}{C}, \overline{y} = \frac{A-B}{CD}$ with $A > B$ with is in $y > 0$. Thus the system has three fixed points, $(x^1, 0)$, $(x_2, 0)$ and $(\overline{x}, \overline{y})$. Now it is interesting to test the existence or not of orbit of period two and greater. But these are difficult problems. For more detailed results, see the Refs. [6, 12, 15].

5 Models of Behaviour of One Species Given by Difference Equations of Order Two and Delay

First, let us consider a general difference equation with k delays

$$x_{n+1} = F(x_n, x_{n-1}, \dots, x_{n-k-1}, x_{n-k})$$

with a general function $F : A^k \to A^k$ and $A \subset \mathbb{R}$ which may modelize the dynamics of a population where the population in time $n + 1$ depends on the influence of the same population in previous times until k times. It may be due to several effects delayed in time.

In most models, F is a polynomial P depending on k variables. One example of interest is

$$x_{n+1} = \alpha x_n(\beta - x_{n-k})$$

when $r \in \mathbb{N}$ denotes the delay of the equation and α, β are real positive constant depending of the nature of the population. These examples are called equations of logistic type or simply logistic, since when $r = 0$ we have what is known in the literature as the logistic equation and considered in the former subsection by Eqs. (2) and (4). In particular a great interest has been shown for case $r = 2$, trying to study some variations in the number of individuals of the population provoked by effects which are not effective on next period of time, but after two times.

One of such cases concerns the influence that food of the population in one time has in two times after, (see for example [16] for an interested account of such problem). As a general discrete model of evolution of the population is represented by

$$x_{n+1} = x_n(s_n + b_n)$$

where n is measured as seasons, s_n denotes the probability that members which are alive in n continue being in time $n + 1$ and b_n denotes the birth rate in time n.

If s_n is small in comparison with b_n then we simplify the model and obtain

$$x_{n+1} = b_n x_n$$

which is a non-autonomous discrete difference equation. In general, b_n can depends on $\{x_n, x_{n-1}, x_{n-2}, \dots, x_{n-r}\}$. When we are considering only the effects of birth rate and food we have that $r = 2$. Of course other effects can be considered, not only food.

Further in this paper, we will concentrate to the case where $b_n = g(x_{n-1}, x_{n-2})$. In this setting, x_n will denote and represent the rate of the total number of individuals of the population in time n. Therefore is $0 \leq x_n \leq 1$.

The dynamics of the difference equation is complete when we know the behavior of all solutions. One important property to test is the boundness or unbounded character of solutions. In first case, the *phase space* of the equation is a compact subinterval of $\mathbb{R}$ because all solutions are contained on it. This happens with the logistic equations for some values of parameters α, β. In second case is $\mathbb{R}$. Using a change of variable, the Eq. (1) is obtained in the easiest form

$$x_{n+1} = ax_n(1 - x_{n-1}) \tag{11}$$

where a is a positive real number

Also for some values of a it could be $x_n > 1$. But this is not possible in logistic setting. In this case will take $x_n = 1$ which means that $x_{n+2} = 0$ and from it all component of the solution would be 0 that is the population extincts. In such case we wonder for set of pair of initial points for which the population extincts. The complementary set is the set of *persistent points*, that is, the set of points whose whole orbits remain in a compact interval of the form $[0, K]$.

If x_n denotes the number of members of the population in instant n and fix a compact interval of $[0, K] \subset \mathbb{R}$ where the pairs of initial conditions have to be taken, then x_n can reach any positive value. For (11) we will take $K = 1$ and as a consequence x_n is a rate of population.

In order to understand the dynamics behind this equation we introduce an *unfolding* of it taking

$$x_{n-1} = x,\ x_n = y$$

and instead of considering directly the Eq. (2) we will deal with the planar transformation in order to see in two dimension what happens if the initial conditions are taken in the unit square $Q_1 = Q$.

$$L_a(x, y) = (y, ay(1 - x))$$

where $(x_1, x_0) = P_0$ and $P_0 \in \mathbb{R}^2$, $(x_{n-1}, x_n) = P_n \in \mathbb{R}^2_+$ that means the first quadrant of the plane. In such case, the values of x_n can be obtained by the second projection in the coordinate axes of P_n or the first projection of P_{n-1}. As a consequence, we will consider the two dimensional dynamical system (Q, L_a) where $Q = [0, 1]^2$ and $L : Q \to \mathbb{R}^2$ is given by

$$L_a(x, y) = (y, ay(1 - x)) \tag{12}$$

The role played by Q is that of phase space and the set where the initial conditions are taken. It is immediate that L_a has an inverse in the interior of Q given by

$$L_a^{-1}(x, y) = (\frac{a - y}{ax}, x)$$

Generalizing, it would be considered the same transformation from $\mathbb{R}^2$ into itself given by (12).

5.1 Dynamics of the Equations $x_n = f(x_{n-k})$ with $k \geq 2$

Let

$$x_n = f(x_{n-k}) \tag{13}$$

a delayed difference equation where $f : \mathbb{X} \to \mathbb{X}$ and $\mathbb{X}$ is any set. It is immediate to test that

$$f^r(x_s) = x_{s+rk}$$

We wonder for the solution of the equation generate by the *initial condition* $C_0 = (x_1, x_2, \ldots, x_k) \in \mathbb{X}^k$. The simplest case is when all solutions are periodic, that is, $x_{n+p} = x_n$ for all initial conditions, some $p \in \mathbb{N}$ and all $n \in \mathbb{X}$. The minimum of such p is called the *minimal period or simply period* of $(x_n)_{n\geq 1}$. We will denote by $Per_{de}(f)$, the set of periods of the difference equation and will stated the problem of study its *Periodic Structure*, if possible. This is an important problem in Dynamics, first stated and solved by A.Sharkovsky for interval maps. In the general case of (10), the problem was totally solved for any $k \geq 2$ and in any $\mathbb{X}$. In particular, for $\mathbb{I}^n$ and $\mathbb{S}^n$, where $\mathbb{I} = [0, 1] \in \mathbb{R}$ and $\mathbb{S}^1$ is the unit circle (see [2–4]).

Here we will concentrate in $k = 2$ (difference equations with two delays) in the cases $\mathbb{I}$ which is the main aim of this paper. The results may be extended to $\mathbb{S}^1$. The key point in the proof is study properties of the map

$$F(z_1, z_2, \ldots, z_k) = (z_2, z_3, \ldots, z_k, f(z_1))$$

in particular the general expressions of its iterations

$$F^{nk+j}((x_i)_{i=1}^k) = (f^n(x_{j+1}), f^n(x_{j+2}), \ldots, f^n(x_k), f^{n+1}(x_1), f^{n+1}(x_2), \ldots, f^{n+1}(x_j))$$

for $n \geq 0$ and $1 \leq j \leq k$. With this formula can be proved that

$$Per(F) = Per(f)$$

where with $Per(g)$ we denote the periods of all periodic orbits of a map g.

5.2 Dynamics of (11) for $0 < a \leq 2$

The main aim of this paper is to study some dynamics properties of the dynamical system (Q, L) like existence of periodic points, linear analysis around them and existence in Q of invariant curves.

5.2.1 The Case $0 < a \leq 1$

It is easy to see that the unique fixed point in Eq. (11) is $(0, 0)$ which is a global attractor in Q which means that independently of the initial point $P \in Q$ is $P_{n \to \infty} \to P_0 = (0, 0)$. This can be seen proving that given $\varepsilon \geq 0$ there is $N > 0$ such that for $n \geq N$ is $d(P^n, P_0) < \varepsilon$, where d denotes the euclidean distance. Analytically it is immediate that in $\mathbb{R}^2$ the unique periodic point of L is the fixed point P_0.

5.2.2 The Case $1 < a \leq 2$

To obtain periodic orbits in Eq. (12) we use the analytic expressions of the iterates of L_a which leads to solve algebraics equations the more and more degrees. Unfortunately this procedure is limited in effectiveness.

The equation $L_a(x, y) = (y, ay(1 - x))$ has only two fixed solutions, these starting in th points $P_0 = (0, 0)$ and $(P_c = \frac{a-1}{a}, \frac{a-1}{a})$ which are placed in the diagonal of the first quadrant of the plane contained in Q. In particular, when $a = 2$ we have $(\frac{1}{2}, \frac{1}{2})$.

When we want to obtain analytically the periodic points of period 2, it is necessary to solve the equation

$$L_a^2(x, y) = (x, y)$$

doing thr adequate calculus we reach the equation $(1 - x)(1 - y) = \frac{1}{a^2}$ which apart $(0, 0)$ it has uniquely the solution $(\frac{a-1}{a}, \frac{a-1}{a})$. This means that L_a has not periodic orbits of minimal period 2. Although the difficulties, can be proved (see [16]), that (12) has only an orbit of minimal period 3 in $\mathbb{R}^2$, but since the first coordinate on one of the points in negative, this means that the Eq. (2) has no periodic point of minimal period 3. Since it is difficult the analytic approach, we do not know if there are periodic orbits of minimal period greater than 2. We claim that the answer is partially negative, that is, (2) has no solutions of periods less than 7 except 1. After 7 there are periodic orbits of not all periods only for some values of the parameter a (see for example, [16]) but when $a > 2$.

We conjecture that (11) has at least two solutions of minimal period 4, three of period 5 and in general $n - 2$ of period n.

5.2.3 The Case $a^\star \simeq 2.27$

For approximately $a^\star \simeq 2.27$ the unstable and instable manifolds to $(0, 0)$ (see [20], for such notions) intersect tangentially, the stable manifold is $[0, 1]$ and the instable manifold is a curve tangent to the sides of Q except $\{(0, y) : y \in [0, 1]\}$. Morse-Smale theorem applies and as a consequence in a small neighborhood of such intersection, there are points that are the initial points of solutions periodic of all periods,

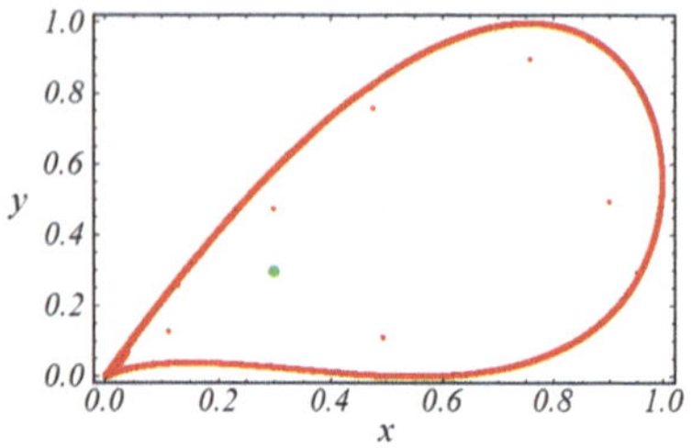

Fig. 1 For $a = 2.27$ the stable and instable manifolds to $(0, 0)$ are tangent. Infinite many periodic orbits appear in a neighborhood of the tangency point

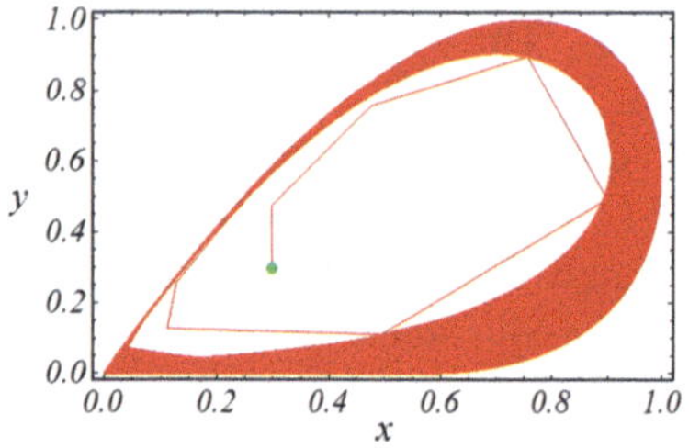

Fig. 2 The curve attracts all orbits started in points inside it

(see [17] and [20]). See Figs. 1 and 2 where it is represented the orbit of the initial point $(0.3, 0.3)$ which it is internally attracted by the drawn curve which it is the image of the segment $\{[0, y] : 0 \leq y \leq 1\}$.

6 A Local Linear Analysis

The difference equation (11) has two *equilibrium* solutions which corresponds to constant sequences $(0)_{n_{-1}}^{\infty}$ and $(\frac{a-1}{a})_{n_{-1}}^{\infty}$. To find solutions of the equation it is needed to know the initial values x_{-1} and x_0. An analytic expression of x_n depending on a and on initial values is very difficult. In fact x_n is a polynomial $P_n(a)$ with coefficients depending of certain factors obtained from the initial conditions. If we denote by d_n the degree of $P_n(a)$, it can be proved by induction on n that

$$d_{n+1} = d_n + 1 + d_{n-1}$$

with $d_1 = 1$ and $d_2 = 2$.

The nature of the fixed solutions can be seen trough a local linear approximation in each of the fixed points of the unfolding transformation L_a and also for the Eq. (11). In fact we will take a neighborhood $W(0, 0) = W$ where the transformation is $L_a(x, y) = (y, ay)$. When $a > 1$ taking any point in W, then the distance of the successive iterations to $(0, 0)$ is increasing, that is $(0, 0)$ is a *repelling* point. For $a < 1$, given a point $P \in W$ the successive iterations have distances to $(0, 0)$ tending to zero, that is, such point is an *attracting* point. If $a = 1$, then the linear analysis does not decide what it is the behavior of any orbit of a point $P \in W$. Nevertheless, the nonlinear analysis made in the previous subsection proves that for such value of a, $(0, 0)$ is also an attracting point.

To study the stable character of the point $P_c = (\frac{a-1}{a}, \frac{a-1}{a})$ we will use the approach of linear stability considering the form and the sign of the eigenvalues of $DL_a(x, y)$ (see [7]). To this end we calculate in P_c the eigenvalues of the jacobian matrix in P_c, that is

$$\det \begin{pmatrix} -\lambda & 1 \\ -ay & a(1-x)-\lambda \end{pmatrix}$$

from which $\lambda_1 = \frac{1+\sqrt{5-4a}}{2}$ and $\lambda_2 = \frac{1-\sqrt{5-4a}}{2}$. The characteristic equation $\lambda^2 - \lambda + (a-1) = 0$ can have two real solutions for $1 \leq \lambda < \frac{5}{4}$, one double real solution in $\lambda = \frac{5}{4}$ and two conjugate complex values for $a \geq \frac{5}{4}$. As a consequence, using again [7], P_c is a *stable node* for $1 < a < \frac{5}{4}$, a *stable focus* since $|\lambda_1|, |\lambda_2| < \sqrt{a-1} < 1$ for $\frac{5}{4} < a < 2$ and an *unstable focus* since $|\lambda_1|, |\lambda_2| > \sqrt{a-1} > 1$ for $a > 2$. Roughly speaking a node is a fixed point attracting in a neighborhood and the orbit starting in it converges monotonically and for a node, the convergence is oscillatory. For values $a = 1, \frac{5}{4}$ and 2 we have phenomena of *bifurcations*. A detailed account can be seen in [1].

To illustrate the former paragraph, consider the equation $\lambda^2 - \lambda + (a-1) = 0$ can be seen as the characteristic equation associated to the linear difference equation

$$X_{n+1} - X_n + (a-1)X_{n-1} = 0$$

which general solution is of the form $X_n = C_1\lambda_1^n + C_2\lambda_2^n$. We will consider the variable X put in the form $X = \rho\cos\theta$. In this case if R_{-1} and R_2 are initial conditions of the equation, chosen in a neighborhood of P_c, and consider the variable X put in the form $X_n = \rho_n\cos\theta_n$ and also the initial conditions are

$$R_{-1}^0 = \rho_{-1}^0\cos\theta_{-1}^0$$

and

$$R_0^0 = \rho_0^0\cos\theta_0^0$$

the solution of the equation is

$$X(n) = C_1\lambda_1^n + C_2\lambda_2^n$$

where $\lambda_{1,2} = \frac{1}{2} + i\frac{\sqrt{4a-5}}{2}$.

The former formulas are interesting since allow us to understand how are the trajectories of points that are attracted by the point P_c. In particular, when $4a - 5 < 0$ then the iterates of P_c spiral out the point, while in the case $4a - 5 > 0$ there is a direct trajectory in the approach.

6.1 Invariant Curves

We are dealing with interesting subsets in the phase space in planar difference equations. They are the *invariant curves* whose dynamics behavior give us an approximate idea of the whole dynamics of the system.

Consider the linear system

$$x_{n+1} = \lambda x_n$$

$$y_{n+1} = \mu y_{n+1}$$

where λ, μ are real numbers. The system can be seen as the two dimensional dynamical system

$$T(x, y) = (\lambda x, \mu y)$$

An *invariant curve* is a plane curve γ holding the following condition: if $(x, y) \in \gamma$, then $(\lambda x, \mu y)$ belongs also to the curve. If the curve is given by the graph of the map $y = F(x)$, then $\mu y = F(\lambda x)$ that is

$$\mu F(x) = F(\lambda x)$$

which is the functional equation of all curves defined by functions F holding the former condition.

It is very easy to check that the general solution of the equation is

$$y = kx^r$$

where

$$r = \log \frac{|\mu|}{|\gamma|}$$

Unfortunately it is difficult or even impossible in most cases of difference and systems of difference equations, to obtain explicit formulas in the say way that the former. That is the case we are dealing in this paper.

We will be devoted with again with the equation

$$x(t+2) = ax(t+1)(1 - x(t))$$

where $t \in \mathbb{C}$. Introducing the change of variable $a^t = z$ we have $x(t) = x(\log_a s) = f(z)$. Then we are looking for solutions of the equation of the form $f(z)$. Then such function verifies

$$f(a^2 z) = af(az)(1 - f(z)) \tag{14}$$

and we have

Theorem 1 *f is an entire function on $\mathbb{C}$ and in such class, it is the unique solution of* (14) if are given the values $c_0, c_1 \in \mathbb{C}$.

Proof Let us suppose that $f(z)$ can be written as the formal power series

$$f(z) = \sum_{j=0}^{\infty} c_j z^j$$

where $c_j \in \mathbb{C}$. Our task is compute such coefficients. It is necessary that

$$\sum_{j=0}^{\infty} c_j a^{2j} z^j = a \sum_{j=0}^{\infty} c_j a^j z^j (1 - \sum_{j=0}^{\infty} c_j z^j) = \sum_{j=0}^{\infty} c_j a^{j+1} z^j - (\sum_{j=0}^{\infty} c_j a^{j+1} z^j)(\sum_{j=0}^{\infty} c_j z^j)$$

where with last product of series we mean the Cauchy product.

We have

$$\sum_{j=0}^{\infty} c_j a^{2j} z^j - \sum_{j=0}^{\infty} c_j a^{j+1} z^j = - \sum_{j=0}^{\infty} \big(c_j a^{j+1} c_0 + c_{j-1} a^j c_1 + c_{j-2} a^{j-1} c_2 + \dots + c_2 a^3 c_{j-2} + c_1 a^2 c_{j-1} + c_0 c_j \big) z^j \tag{15}$$

It is usual take $c_0 = 0$ (the graph of f passes trough 0) and $c_1 = 1$. Identifying coefficients,

$$c_j a^{2j} - c_j a^{j+1} = - \big(c_j a^{j+1} c_0 + c_{j-1} a^j c_1 + c_{j-2} a^{j-1} c_2 + \dots + c_2 a^3 c_{j-2} + c_1 a^2 c_{j-1} + c_0 a c_j \big) \tag{16}$$

where $j \geq 0$. Applying this formula we get in a unique way the coefficients of the power series. Giving $c_0 = 0$, $c_1 = 1$. we have

$$c_2 = -2(a(a-1)^{-1}, c_3 = 4(a^3(a-1))^{-1}, c_4 = -8 \frac{1 + 2a^2}{(a^6(a-1)(a^2 + a + 1))^{-1}}, \dots$$

Proceeding by induction on j we obtain a general expression for the coefficients. It is easy to see that we obtain an immediate evaluation for such coefficients

$$c_j \leq (a^2 - a)^{1-j}$$

for $j \in \mathbb{N}$. As a consequence the radius of the convergence is at least of $(a^2 - a)$ and that the series converges for $|z| < a^2 - a$. Once a is fixed using (14) it is immediate that the series converges also for all $z \in \mathbb{C}$. It is also obtained that the series of second side of (14) converge for $|az| < a(a^2 - a)$ and $|a^2 z| < a^2(a^2 - a)$.

Now we introduce the map h because it has no singularities

$$h(z) = \begin{cases} f(z) & \text{if } |z| < a^2 - a; \\ ah(\frac{z}{a})(1 - h(\frac{z}{a^2})) & \text{if } |z| \geq a^2 - a. \end{cases}$$

The maps f and h coincide in the points where f converges. Using the principle of analytic extensions the two series converges for all $z \in \mathbb{C}$.

Let us consider now that $z = a^t$ is a real positive variable. Introduce now parametrically the curve

$$\gamma(z) = \begin{cases} x = f(z); \\ y = f(az). \end{cases}$$

Then $\gamma(0) = 0$ and it is verified the interesting property

$$L_a(\gamma(z)) = \gamma(z)$$

that is, the curve is *invariant* and it can be used to understand the asymptotic behavior of the original difference equation.

The points of the curve $\gamma(z)$ coincides with the points of the graph of the map $y = F(x)$ up the value where x does not increase, which means until the first local maximum of $x = f(z)$ (see [16]). It is clear that if change the notation z by x, then we can obtain the inverse of the power series denoted by f^{-1} and then

$$y = f(az) = f(af^{-1}(x)) = F(x)$$

6.2 Analysis of the Invariant Curves

We are denoting by Q_n the set $L^n(Q) \cap Q$ for $n \in \mathbb{N}$ that is, the set of points that remains in Q after n-iterations. Such are the *persistent points* after such number of iterations. Denote the closed unit square Q by $OABC$ where $O = (0, 0)$, $A = (0, 1)$, $B = (1, 1)$, $C = (1, 0)$, A *persistent trajectory* is a trajectory $(P_n)_{n=0}^{\infty}$ of points such that for all $n \in \mathbb{N}$, P_n remains in Q and *strictly persistent trajectory* that for which all point including $P_0 = (x_{-1}, x_0)$ remains in the interior of Q. The rest of trajectories are called *non persistent*. It is evident that for having a non-persistent trajectory it is necessary that $P_0 \in OC$ or on some preimage by L_a of such segment and in this case the trajectory must converge to the origin O. The complement of all persistent trajectories composes the *spacing set*.

If $\gamma_1 = \{(x, ax) : x \in [0, 1]\}$ and Δ_1 denotes the triangle bounded by $[y = 0, x = 1, y = 1] = A$ and γ_1, then $Q_1 = Q \cup \Delta_1$. Let $\gamma_i = L_a(\gamma_{i-1}$ for $i = 2, 3, 4$ and

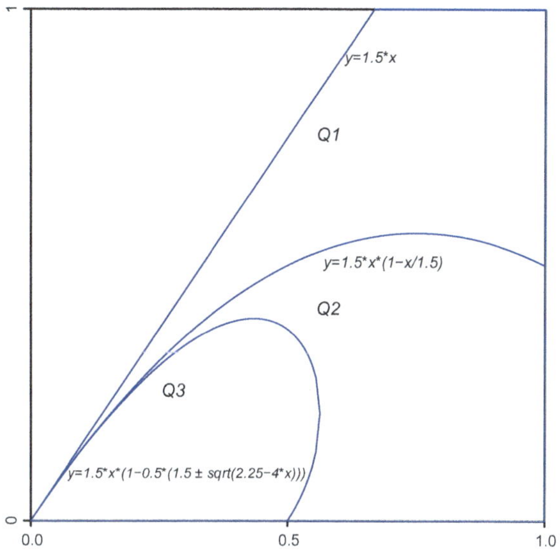

Fig. 3 The three first images of the segment $\{(0, y)\}$ with the maps defining the boundary for $a = \frac{3}{2}$

respectively $Q_i = Q \cup \Delta_i$ where Δ_i is the domain whose boundary is composed of γ_i and A.

It is interesting observe that

$$\gamma_4 = L_a(\gamma_3) = (p_3(v), ap_3(v)(1 - v)) = (u, p^{\pm}(u))$$

where with p_4^+ and p_4^- we denote the concave and convex curves composing the boundary of γ_4. The shape of this set has a drop shape and in fact it is a simply connected Jordan set. The same situation happens from $i \geq 4$ and also is easy to see that

$$Q_1 \supset Q_2 \supset \cdots \supset Q_n \supset \cdots$$

the set Q_n for all $n \leq 4$ is a simply connected Jordan set of a drop shape bounded by the curves $p_n^{\pm}$. It is easy to prove by induction on n that maps p_n^+ are concave. Generally the functions p_n^- have a part of concavity and another of convexity. In Fig. 3 we represent the three first regions Q_i for $i = 1, 2, 3$ and the give the analytic expression of the corresponding curves boundaries.

Lemma 1 *Functions p_n^+ are concave for $n \geq 3$.*

Proof We proceed by induction on $n \geq 3$.

For $n = 3$, calculating the derivative of p_3^+ leads to

$$(p_3^+)'(u) = a[(1 - \frac{a}{2}) + \frac{1}{2}(\sqrt{a^2 - 4u} - \frac{u}{\sqrt{a^2 - 4u,}})]$$

from which we see that the derivative is negative so that the function is decreasing and $(p_3^+)'(a) = a$. Therefore $(p_3^+)''(u) \leq 0$ for all u. Let $j > 3$ and suppose that $p_n^+(u)$ be concave, that is $(p_j^+)'(u)$ is decreasing. Since

$$(p_{j+1}^+)'(v) = a[1 - (p_j^+)'(v) - (p_3^-)'(v)]$$

and

$$(p_j^+)'(u_j^+)(u_j^+)' = 1$$

then we have

$$u_j^+(v) = \frac{1}{(p_j^+)'(u_j(v))'}$$

and

$$p'_{j+1}(v) = a[1 - u_j^+(v) - \frac{v}{(p_j^+)'(u_j(v))'}]$$

Therefore we have that $p'_{j+1}(v)$ is decreasing because u_j^+ it is and also the same is valid using the induction hypothesis for $\frac{v}{(p_j^+)'(u_j(v))}$.

With the previous observations we prove in next result that there is a relationship between the area of Q_n denoted by $|Q_n|$ and the distance between the curves p_n^+ and p_n^- $(d(p_n^+, p_n^-))$.

Theorem 2 *Let $a \in (1, 2]$. Then*

$$lim_{n \to \infty} d(p_n^+, p_n^-) = 0$$

if and only if

$$lim_{n \to \infty} |Q_n| = 0$$

Proof Let suppose that $\lim_{n \to \infty} |Q_n| = 0$ and that $\lim_{n \to \infty} d(p_n^+, p_n^-) \neq 0$. This means that there is $d > 0$ such that for each n there is u_n holding $p_n^+(u_n) - p_n^-(u_n) \geq d$.

Let

$$\Delta = \bigcap_{n=1}^{\infty} Q_n$$

this set is invariant by L, that is, $L(\Delta) = \Delta$ and besides $L|\Delta$ is injective. It is immediate that the point $P_c(\frac{a-1}{a}, \frac{a-1}{a})$ for $a \in (1, 2)$ is a *local attractor* since the eigenvalues of $DL(P_c)$ are $\lambda_{1,2} = \frac{1 \pm \sqrt{5-4a}}{2}$ which verifies $|\lambda_{1,2}| < 1$.

For each Q_n let c_n be the critical point of p_n^+ and $w_n = p_n^+$. Note that $\max\{x : (x, y) \in Q_n\} = w_{n-1}$ and since is $Q_n \supset Q_{n-1}$ we have $w_n < w_{n-1}$. Besides, Δ contains a vertical segment of length greater or equal than d. Let $x_0 \in I$ and $S = \{x_0 \in I$ where $I = [\alpha, \beta]$ such that $\beta - \alpha \geq d$. The set

$$L_a^{-1}(S) = \{(1 - \frac{\beta}{ax_0}, 1 - \frac{\alpha}{ax_0} \times x_0)\}$$

is an horizontal segment since the set

$$L_a^{-2}(S) = \{(x, y) : 1 - \frac{x_0}{ay}, y \in (1 - \frac{\beta}{ax_0}, 1 - \frac{\alpha}{ax_0})\}$$

is a subset of an hyperbola.

Since $p_n^+(x)$ is a concave map, the set M bounded by the subset of hyperbola and the segments joining the extreme points of $L_a^{-2}(S)$ belongs to Δ. Then is $|\Delta| \geq |M| > 0$ that is in contradiction with the fact that $|\Delta|$ can be done as small I wanted since is $\lim_{n\to}|Q_n| = |\Delta| = 0$.

Next result proves that one of the conditions of the theorem always occurs.

Theorem 3 *If* $a \in (1, 2]$, *then* $lim_{n\to}|Q_n| = 0$.

Proof Denote by $J_{L_a}(x, y)$ the jacobian of L_a in (x, y). It is immediate that $J_{L_a}(x, y) = ay$ and

$$J_{L_a^n}(x, y) = \prod_{k=0}^{n-1} ay_k$$

where $L_a^k(x, y) = (x_k, y_k)$. Since it is interesting for what follows, we will compute $J_{L_a 2n+2}(x, y)$. Such jacobian is calculated using the two relations:

1.
$$J_{L_a}^{2n+2}(x, y) = a^{2n+2} \prod_{k=0}^{2n+1} y_k$$

2.
$$(\prod_{k=0}^{2n+1} y_k)^2 = a^{2n} \prod_{k=0}^{2n-1} [y_k(1 - y_k)][\prod_{k=0}^{2n+1} y_k] y_1 y_{2n}$$

from last formulas we obtain

3.
$$\prod_{k=0}^{2n+1} y_k = a^{2n} \prod_{k=0}^{2n-1} [y_k(1 - y_k)] y_1 y_{2n} \leq a^{2n} \frac{1}{4^{2n}} y_1 y_{2n}$$

and finally from (1) and (12)

4.
$$J_{L_a 2n+2}(x, y) \leq a^2 (\frac{a}{2})^{4n}$$

When $a < 2$ we obtain that $\lim_{n\to\infty} J_{L_a^{2n+2}}(P_c) = 0$ and therefore is $\lim_{n\to\infty} J_{L^n}(P) = 0$ for every $P \in Q_4$. From this

$$|Q_n| = \lim_{n\to\infty} \int\int_{Q_4} J_{L_a^{n-4}}(x, y) = 0$$

When $a = 2$, then the set $Q_4 = T \cup R$ where

$$T = \{P \in Q_4 : \lim_{n\to\infty} L_a^n(P) = P_c\}$$

$$R = \{P \in Q_4 : \lim_{n\to\infty} L_a^n(P) \neq P_c\}$$

Then given $\varepsilon > 0$ there exists a subset $T_\varepsilon \subset T$ such that $|T_\varepsilon| \geq |T|(1 - \varepsilon)$ and holding that $d(L_a^n(P), P_c)$ converges uniformly to zero on T_ε. As a consequence if $n \geq N$ is $L_a^n \in B_\varepsilon(P_c)$ for all $P \in T_\varepsilon$. This implies

$$|L_a^n(T)| = \int\int_T J_L^{n-1}(x, y)dxdy \leq \pi\varepsilon^2 + \varepsilon \sup_{P\in Q_4} J_{L_a}^{n-4}(P)$$

for $n > N$. This means that

$$\lim_{n\to\infty} |L_a^n(T)| = 0$$

Analogously, for R we have

$$|L_a^n(R)| = \int\int_R J_{L_a}^{n-4}(x, y)dxdy = 0$$

Therefore

$$|Q_n| \leq |L_a^n(T)| + |L_a^n(R)| \to 0$$

Remark 1 1. For $a \in (1, 2]$ the set Δ is a continuous curve Γ such that $L_a(\Gamma) = \Gamma$ and joining the points P_0 and P_c in the sense of approaching them. Moreless, such approaching to P_c is made in a spiral form for ($a \leq \frac{5}{4}$) or monotonously for ($\frac{5}{4} < a < 2$). Also the set Δ is a local attractor for the system defined by L_a.
2. In general, for $a > 0$ it is easy to see that Δ is also a *global attractor* for L, which means that such set attracts all point of Q. To see it is sufficient to prove that for every subset C such that $C \bigcap \Gamma \neq \emptyset$ it is hold $J_F^n(C) \to 0$ for all points in C.

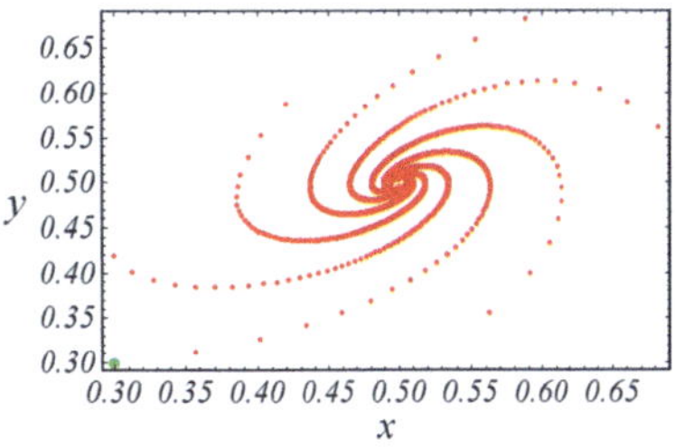

Fig. 4 Orbit when $a = 2$ for the initial point $(0.3, 0.3)$. It is observed that many invariant curves appear tending to the fixed point $(\frac{1}{2}, \frac{1}{2})$

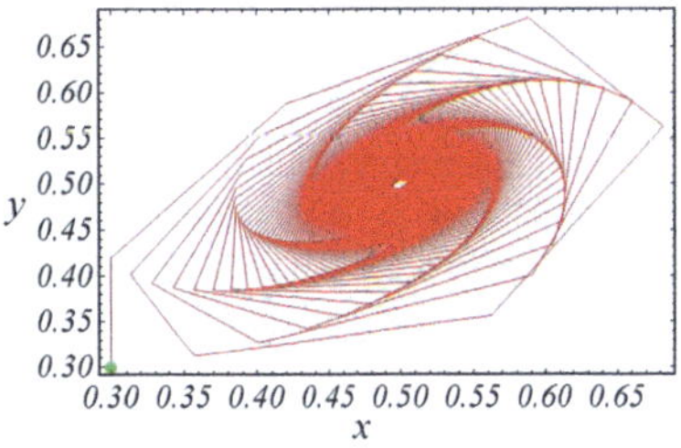

Fig. 5 The same orbit of $(0.3, 0.3)$ joining the dots, being attracted by the fixed point $(\frac{1}{2}, \frac{1}{2})$

3. When $a = 2$, the system has a *supercritic Hopf-Neimark-Sacker bifurcation* (see for example [1] or [10]). From values of a and closed to 2 but greater, the system has another invariant curve around P_c which is a repulsive point while the points of the curve Γ are attracting. The dynamics in this curve is interesting, particularly the appearance of periodic points of many periods (see [1] for a complete analysis of such case). Also in Figs. 4 and 5, observe the graphs of the corresponding invariant curve for $a = 2$.
   After a value of a closed to 2 the curve starts to be highly non-differentiable and appears and disappears after an intermittent process (see a numerical approach also in [1]).
   After the critical value of 2.38 then most of points of Q become scaping points out of $\mathbb{Q}$ and the persistent points (point which stays inside $\mathbb{Q}$ for all iterations of L_a) belong to a Cantor like set (a set homeomorphic to the ternary Cantor set). These facts have not been yet sufficiently clarified in the literature.
4. All results obtained for L_a has to be interpreted in the setting of the Eq. (14).
5. Another open problem is to consider the non-autonomous version of (14)

$$x_{n+1} = q(n)x_n(1 - x_{n-1})$$

   where in particular, $(q(n))_{n=1}^{\infty}$ is a sequence of real numbers. It may be interesting to study those cases where the sequence $q(n)$ be periodic, bounded or convergent.

7 Summary

After a revision of the dynamics of several equations modelling the evolution of populations with one or more species, we have considered a logistic two delays equation and studied some topics of its dynamics.

Periodic solutions of nonlinear difference equations like (11) are difficult to obtain analytically. But the existence of invariant curves in the range of a less or igual to 2 can be obtained analytically and geometrically. There exists a curve joining the two fixed points which is introduced as a power series. The special interest is the case $a = 2.27$.

Acknowledgements This research has been supported by the project MTM2014-51891-P from Spanish Ministerio de Economía (MINECO). This paper is dedicated to the memory of Wieslaw Szlenk, good master and better person.

References

1. Aronson, D.G., Chory, M.A., Hall, G.R., McGehee, R.P.: Bifurcations from an invariant circle for two-parameter families of maps of the plane: a computer-assisted study. Comm. Math. Phys. **83**, 303–354 (1982)
2. Balibrea, F., Linero, A.: Periodic structure of σ-permutations maps on I^n. Aequationes Math. **62**(3), 265–279 (2001)
3. Balibrea, F., Linero, A.: Periodic structure of σ-permutation maps. II. The case $\mathbb{T}^n$. Aequationes Math. **64**, 34–52 (2002)
4. Balibrea, F., Linero, A.: On the periodic structure of delayed difference equations of the form $x_n = f(x_{n-k})$ on $\mathbb{I}$ and $\mathbb{S}^1$. J. Diff. Eq. Appl. **9**, 359–371 (2003)
5. Beddington, J., Free, C., Lawton, J.: Dynamic complexity in predator-prey models framed in difference equations. Nature **255**, 58–60 (1975)
6. Cooke, K.L., Calef, D.F., Level, E.V.: Stability or chaos in discrete epidemic models, pp. 73–93 (1977)
7. Elaydi, S.: An introduction to difference equations. Springer, New York (2005)
8. Guckenheimer, J., Oster, G., Ipaktchi, A.: The dynamics of density dependent population models. J. Math. Biol. **4**, 101–147 (1977)
9. Hadeler, K.: Mathematik für Biologen. Springer (1974)
10. Hale, J.K., Koçak, H.: Dynamics and bifurcations, vol. 3. Springer-Verlag, New York (1991)
11. Holmgren, R.A.: A first course in discrete dynamical systems, 2nd edn. Springer, New York (1996)
12. Marotto, F.R.: The dynamics of a discrete population model with threshold. Math. Biosci. **58**, 123–128 (1982)
13. May, R.M.: Simple mathematical models with very complicated dynamics. Nature **261**, 459–467 (1976)
14. May, R.M., Oster, G.F.: Bifurcations and dynamic complexity in simple ecological models. Am. Natural. **110**, 573–599 (1976)
15. Poluektov, R., Pykh, Y., Shvytov, I.: Dynamical models of ecological systems (in Russian), pp. 101–147. Leningrad, Gidrometeoizdat (1980)
16. Pounder, J., Rogers, T.D.: The geometry of chaos: dynamics of a nonlinear second-order difference equation. Bulletin Math. Biol. **42**, 551–597 (1980)
17. Rogers, T.D., Clarke, B.L.: A continuous planar map with many periodic points. Appl. Math. Comput. **8**, 17–33 (1981)

18. Sharkovskiĭ, A.N.: Coexistence of cycles of a continuous map of the line into itself. In: Proceedings of the Conference "Thirty Years after Sharkovskiĭ's Theorem: New Perspectives" (Murcia, 1994), vol. 5, pp. 1263–1273 (1995)
19. Smí tal, J.: On functions and functional equations. Adam Hilger, Ltd., Bristol (1988)
20. Wiggins, S.: Introduction to applied nonlinear dynamical systems and chaos. In: Texts in Applied Mathematics, vol. 2. Springer, New York (1990)

Diffusive Limits of the Master Equation in Inhomogeneous Media

Luca Salasnich, Andrea Bonato and Fabio Sattin

Abstract Diffusion is the macroscopic manifestation of disordered molecular motion. Mathematically, diffusion equations are partial differential equations describing the fluid-like large-scale dynamics of parcels of molecules. Spatially inhomogeneous systems affect in a position-dependent way the average motion of molecules; thus, diffusion equations have to reflect somehow this fact within their structure. It is known since long that in this case an ambiguity arises: there are several ways of writing down diffusion equations containing space dependence within their parameters. These ways are all potentially valid but not necessarily equivalent, meaning that the different diffusion equations yield different solutions for the same data. The ambiguity can only be resolved at the microscopic level: a model for the stochastic dynamics of the individual molecules must be provided, and a well-defined diffusion equation then arises as the long-wavelength limit of this dynamics. In this work we introduce and employ the integro-differential Master Equation (ME) as a tool for describing the microscopic dynamics. We show that is possible to provide a parameterized version of the ME, in terms of a single numerical parameter (α), such that the different expressions for the diffusive fluxes are recovered for different values of α. This work aims to fill a gap in the literature, where the ME was shown to deliver just one diffusive limit. In the second part of the paper some numerical computer models are introduced, both to support analytical considerations, and to extend the scope of the ME to more sophisticated scenarios, beyond the simplest α-parameterization.

Keywords Diffusion · Master equation

L. Salasnich (✉) · A. Bonato
Dipartimento di Fisica e Astronomia Galileo Galilei, Universitá di Padova, Via Marzolo 8, 35131 Padova, Italy
e-mail: luca.salasnich@unipd.it

F. Sattin
Consorzio RFX (CNR, ENEA, INFN, Universitá di Padova, Acciaierie Venete SPA), Corso Stati Uniti 4, 35127 Padova, Italy
e-mail: fabio.sattin@igi.cnr.it

V. Carmona et al. (eds.), *Nonlinear Systems, Vol. 1*,
Understanding Complex Systems, https://doi.org/10.1007/978-3-319-66766-9_10

1 Introduction

Let us attempt to describe the dynamics of some fluid, quantified by its concentration $n(x, t)$ evolving in time and space. For simplicity, throughout this work we will restrict ourselves to one space dimension. In the absence of sink and sources, n fulfils the conservation law

$$\frac{\partial n}{\partial t} = -\frac{\partial j}{\partial x}$$

The kind of dynamics involved depends upon the analytical expression of the flux j. In the case of a collective motion where all the fluid parcels drift with the same velocity V, $j = nV$. Superimposed to this motion, the single fluid particles may experience disordered independent movements: Diffusion is the name given to the macroscopic realization of the random movement of large numbers of particles in space. The first experimental investigations on diffusion date back to the first half of nineteenth century, prominently with Thomas Graham's studies on the mixing of gases and of salts in water [12, 27, 28, 30, 32], and with Robert Brown and his studies on the movement of small pollen particles in aqueous suspension, which were to be explained later by Einstein [7, 30]. Another fundamental advancement was brought forth around the middle of the same century by Adolf Fick, who unified the diffusion in fluids with the conduction of heat in solids, studied earlier by Joseph Fourier [11, 30]. The Fick-Fourier's (FF) law relates the flux j with the gradient of the diffusing density n through:

$$j = -D\frac{\partial n}{\partial x} \tag{1}$$

The coefficient D compactly summarizes the effect of the interaction between the individual particles and the surrounding milieu. In principle, if inter-particle interactions are not negligible, it may depend from n, too, but we will not consider this possibility here. In thermodynamics, Eq. (1) is an instance of a linear relationship between the flux and a thermodynamic force (the gradient of the free energy).

Equation (1) was derived by Fick as a purely empirical relation. Alternatively, one can tackle a formal approach: the motion of each individual particle is modelled as a sequence of uncorrelated jumps, an instance of a Markovian stochastic process. This mathematical formalization leads to describing the trajectory of each individual particle through a Langevin equation. Then, the average over a large ensemble of such particles leads to the Fokker-Planck (FP) Equation [17, 31]:

$$\frac{\partial n}{\partial t} = -\frac{\partial j}{\partial x} \equiv \frac{\partial}{\partial x}\left[\frac{\partial (nD)}{\partial x} - nV\right] \tag{2}$$

In this expression, we may identify the overall flux as done by two contributions, that we will label as diffusive ($-\partial(nD)/\partial x$) and convective (nV). Analogously, we may generalize Eq. (1) by adding a convective contribution:

$$j \equiv -D\frac{\partial n}{\partial x} + Vn \tag{3}$$

So far, no consideration has been made of the homogeneity of the background. As a matter of fact, the law (1) was worked out in contexts where it was impossible to discriminate experimentally any consequence of the inhomogeneity of the medium upon the dynamics of particles; implicit in it is therefore the postulate of the homogeneity of the medium. In homogeneous systems, D, V, must be constant, since they are postulated to be dependent by the properties of the medium alone. It is straightforward to acknowledge that, in this case: (I) The fluxes j appearing in Eqs. (2) and (3) are identical; (II) There is a clear-cut unambiguous definition of convective and diffusive fluxes.

Just like (1)–(3) were got by experiment and by theory, the coefficients D, V appearing therein may be regarded either as unknown quantities to be fixed by matching with experiments, or as known from some more fundamental theory. Real systems, however, are often inhomogeneous. This prompts to consider the possibility that the parameters quantifying the strength of the interaction with the background, D, V, become position-dependent: $D = D(x)$, $V = V(x)$. In this case, the fluxes in (2) and (3) are no longer identical although they are still closely related:

$$\begin{aligned}\frac{\partial n}{\partial t} &= \frac{\partial}{\partial x}\left[\frac{\partial(nD)}{\partial x} - nV\right]\\ &= \frac{\partial}{\partial x}\left[D\frac{\partial n}{\partial x} + n\frac{\partial n}{\partial D} - nV\right] = \frac{\partial}{\partial x}\left[D\frac{\partial n}{\partial x} - nV'\right]\end{aligned}$$

Thus, the flux initially written, e.g., in the Fokker-Planck form may ultimately be written in the Fick-Fourier one through a redefinition of the convective term:

$$V' = V - \frac{dD}{dx}. \tag{4}$$

This interchangeability may lead to speculate that the question of which flux to use is actually devoid of relevance: it has been argued in literature that only the total flux must be given physical meaning, not the diffusive and convective contributions separately [23]. This is trivially true as long as D and V are seen as pure fitting coefficients. However, the terms entering Eq. (4) come from distinct sources. Following van Kampen, we divide them in geometric and thermal terms [3, 18]. The latter expression originally implied that diffusion was caused by thermal agitation of the molecules. Since the gas we are considering is not necessarily made of real molecules, we will be employing it here in a broader acceptation, meaning any mechanism that acts on the microscopic disordered motion of the particles and thus affect the diffusivity D. Geometric terms, conversely, refer to the biasing on particle motion caused by background geometry such as external forcing, ratchet effects ... ordinarily on large scales. Therefore, although formally, on the basis of Eq. (4) a varying diffusivity is indistinguishable from a genuine convection, we may expect to be potentially able to discriminate between the two on the basis of the physical mechanisms acting on the system under consideration. In this work we will postulate that geometric effects are not involved. This removes the previous ambiguity. For clarity we rewrite here down the versions of Eqs. (2), (3) that we will be considering from now on:

$$\frac{\partial n}{\partial t} = \frac{\partial}{\partial x}\left[D\frac{\partial n}{\partial x}\right] \tag{5}$$

$$\frac{\partial n}{\partial t} = \frac{\partial}{\partial x}\left[\frac{\partial (nD)}{\partial x}\right] \tag{6}$$

At this stage, Eqs. (5), (6) are no longer equivalent: when one and the same D is inserted in each of them, the resulting solutions $n(x, t)$ are different (for the same boundary and initial conditions, of course). Furthermore, some classes of solutions are specific to the one or the other of these equations. One example is provided by uphill transport, sometimes observed in high-temperature magnetized plasmas, where the (matter or heat) flux goes along the same direction of the local gradient. This phenomenon is unexplainable within Eq. (5), since therein the flux flows only opposite to the gradient. Hence, in order to cope with this evidence, one has to enlarge the scope of Eq. (5) by allowing for some convection to exist, thereby reverting to Eq. (3), or equivalently to (6) [9, 10] (For completeness we mention that it has been speculated also that other kind of transport might be at work here, producing Lévy flights, and therefore somewhat outside of the diffusive-convective paradigm).

The question: which expression between (5) and (6) is to be used, acquires therefore importance in this framework. The recent paper [2] highlights its relevance in the context of the modelling of ecological systems, assessing to which extent different choices for the diffusive fluxes do affect the results of several predefined test problems. Thus, it would be very convenient to have some guidelines telling us in advance of our analysis whether, in the system under consideration, it is more appropriate to use the one rather than the other equation. These guidelines have to be extracted from some knowledge of the microscopic physics driving the systems, and are accordingly expected to potentially be largely system-specific and non-universal. It is instead possible to answer on quite general grounds a weaker question: it is possible to establish a simple parametrization such that a single numerical parameter uniquely identifies whether a generic system is driven by the one or the other Eqs. (5), (6). This numerical parameter has a close connection with the microscopic stochastic dynamics, hence it is possible to assess its value only after a thorough knowledge of the system's microphysics.

There are several ways of introducing this formalism. One way is from the Lagrangian point of view, as adopted, e.g., by Lançon et al. [21, 22]. They modeled the motion of each individual particle as a Brownian motion with space dependent diffusion coefficient. An ambiguity arises in this case. Namely, the rule for updating the walker position writes:

$$x(t + dt) = x(t) + \eta\sqrt{2Ddt} \tag{7}$$

where η is a white noise of unit amplitude. At this stage the question arises: at which location $D(x)$ should be evaluated? At the initial position, $x(t)$; at the final one, $x(t+dt)$, or somewhere else? The need of making such a choice goes under the name of Itô-Stratonovich convention. Adhering to the one or the other choice ultimately leads to Eqs. (5) or (6) when one goes from the single particle to the fluid-like picture of the motion. Details can be found in Lançon et al.

An Eulerian viewpoint is possible as well. This time the location of space is held fixed, and accounting is done of fluid elements entering and leaving it according to given stochastic rules. The resulting balance equation is named Master Equation as introduced by van Kampen [16]. The Master Equation (ME) formalism and the Brownian walker one are, of course, different ways of expressing the same physics (although the former one is more flexible) and therefore are expected to produce the same results. To the best of our knowledge, however, the exercise of extracting diffusion equations (5), (6) from the ME has been fulfilled so far only partially, by van Milligen et al., who recovered Eq. (6) [40]. No analogous result for (5) has been produced. This led van Milligen et al. to argue that perhaps Eq. (5) does lack a true microscopic justification. In this work, instead, we will complete this exercise by showing how Eq. (5) can be extracted from the ME in analogy with the Brownian walker result.

Section 2 provides a brief introduction to the ME. Sections 3 and 4 are devoted to show how Eqs. (6), (5) respectively do arise as suitable limits of the ME. Section 5 supplements the previous analytical results with numerical exercises: lattice models are designed and implemented as computer codes and we will show that, on the basis of the stochastic rules imposed, the numerical simulations do conform to the predictions done in the previous sections. Indeed, we have remarked earlier that the ME is fairly a flexible tool for the investigation of stochastic system. Still in Sect. 5 we will validate this claim, by producing an instance of a system whose dynamics is modelled by the ME and that cannot be reduced to either Eqs. (5) or (6). Finally, Sect. 6 provides a brief recap of the results.

2 The Master Equation

The ME is an integro-differential equation expressing the rate of change for the local scalar density n in terms of transition probabilities. In one spatial dimension and with consideration of just one transported quantity it writes

$$\frac{\partial n(x,t)}{\partial t} = -\frac{n(x,t)}{\tau} + \int dz\, n(z,t)\,\frac{P(x,z)}{\tau} \tag{8}$$

In (8), $P(x,z)$ represents the probability for the lump of matter $dz\, n(z)$ to be moved from z to x, and τ sets the corresponding temporal scale, which may depend upon x as well.

Thus the first term in the r.h.s. of (8) represents the rate of particles leaving the location x, whereas the integral stands for the total rate of particles that, started elsewhere, land on x. Equation (8) does not account for the presence of any source and/or sink; we will not consider them in this work although they can be added straightforwardly. We cursorily note that P may be considered as a transfer function without necessarily a connection with probability. A deterministic version of Eq. (8) is used, e.g., in biophysics, where it is known as Amari's equation [1].

The limits of integration in Eq. (8) are left unspecified. The reader may assume that they range from $-\infty$ to $+\infty$; we will not address here such issues as finite-size boundary effects (some related analysis can be found in [34]).

Equation (8) may contain virtually all the information available about the system, except for the velocity degrees of freedom of the particles. It is fairly flexible, too: depending on the functional form chosen for P, τ, it can accommodate a wide range of dynamics, from sub- to super-diffusive transport, including possibly nonlinearities. Indeed, in the form (8) it has already been circumscribed from a more general expression (Generalized Master Equation) [19] that involves a convolution not only over space, but over time as well: $P = P(x, z, t, t')$. Explicit consideration of a finite temporal memory may lead to the emergence of a vastly more complicated dynamics, varying from diffusive to wave-like, as argued by Maxwell, Cattaneo and Vernotte [4, 13–15, 20, 29, 38, 41].

3 From the Master Equation to the Fokker-Planck Flux

In most practical applications, one has available just coarse-grained information; i.e., not knowledge of the full P, τ, but just averages $<P>$, $<\tau>$ over finite regions of the system. Mathematically, this amounts to saying that only the long-wavelength limit of Eq. (8) is actually relevant. In this and the next section we will derive the result that by removing small-scale details-i.e. taking its long-wavelength limit-Eq. (8) reduces a parabolic partial differential equation, the diffusion equation, and that can be either in the form (5) or (6).

For simplicity, throughout this work, we will drop any dependence of τ from space or time: it will just play the role of a constant, characteristic time scale of the process. Accordingly, the whole physical content of the theory is brought by P. Since we are dealing with position-dependent systems, one could legitimately wonder whether dropping any spatial dependence from τ does impose too severe constraints upon the dynamics we may model. Setting $\tau = \tau(x)$ does-of course-affect quantitatively the shape of the resulting equations, since additional terms proportional to $\tau'(x)$, $\tau''(x)$ do appear. However, conclusions drawn in the next sections will remain qualitatively unaffected.

In order to carry out any further analysis one needs first to make explicit the dependence of P upon its arguments. In spatially homogeneous systems, the only dependence can enter through the jump length $(x - z)$ since P is invariant under translations. In inhomogeneous systems, instead, this invariance must be broken, and explicit dependence of P upon x, z separately must appear. The most intuitive choice is to make P dependent from the starting location as well:

$$P(x, z) \equiv P(x - z; z) \tag{9}$$

Heuristically, this choice is close to the standard way of looking at problems in dynamics-or in computer programming: the evolution of a system is determined (although, in this case, only on a statistical basis) by the law of motion (P) and by the initial conditions (z).

It is convenient switching to the variable $\Delta = z - x$ in Eq. (8):

$$\frac{\partial n(x,t)}{\partial t} = -\frac{n(x,t)}{\tau} + \int d\Delta\, n(x+\Delta, t)\, \frac{P(-\Delta, x+\Delta)}{\tau} \tag{10}$$

The long-wavelength limit is taken by postulating that the jump probability P is virtually zero beyond some maximum jump length Δ_{max}, and that both n and P vary slowly over lengths lesser than Δ_{max}. Therefore, the argument of the integral in (10) is expanded around x in powers of Δ, and the Taylor expansion is stopped to some finite order. Pawula theorem [31] warrants that, by stopping to the second order, no such unphysical artefacts as locally negative densities may occur. The true justification for the truncation to the second order, however, has been provided only quite recently by Ryskin [33]. Ryskin's proof is essentially a consequence of the Central Limit Theorem, and states that for analytic P's and over time scales just moderately longer than τ, the dynamics always converges to be diffusive. Since Ryskin's result is essential for this paper, we will provide in the Appendix a sketch of his proof.

Armed with these results we eventually carry out the expansion of (10):

$$\frac{\partial n(x,t)}{\partial t} = -\frac{n(x,t)}{\tau} + \frac{1}{\tau}\int d\Delta\, n\, P + \frac{1}{\tau}\int d\Delta\, \Delta\, \frac{\partial (nP)}{\partial x} + \frac{1}{\tau}\int d\Delta\, \frac{\Delta^2}{2}\, \frac{\partial^2 (nP)}{\partial x^2} \tag{11}$$

In Eq. (11), for brevity, we have not written explicitly the arguments of n and P: $n = n(x,t)$, $P = P(-\Delta, x)$.

The two first terms in the r.h.s. of (13) mutually cancel by virtue of the normalization

$$\int d\Delta\, P(-\Delta; x) = 1$$

Equation (11) takes thus the form of a conservation equation:

$$\frac{\partial n}{\partial t} = \frac{\partial^2 (D(x) n)}{\partial x^2} - \frac{\partial (U(x) n)}{\partial x}$$

with

$$U(x) = \frac{1}{\tau}\int d\Delta\, \Delta\, P(-\Delta; x); \quad D(x) = \frac{1}{\tau}\int d\Delta\, \frac{\Delta^2}{2}\, P(-\Delta; x) \tag{12}$$

In accordance with the guidelines set up in the Introduction, we will restrict to scenarios without convective flux. This is achieved by imposing unbiased jump probability: $P(\Delta; x) = P(-\Delta; x)$, thus $U(x) = 0$, and

$$\frac{\partial n}{\partial t} = \frac{\partial^2 (D(x)n)}{\partial x^2} \tag{13}$$

Equation (13) is thus a continuity equation for n, the flux being written in the FP form (6); It comes fairly straightforwardly from the postulate (9), which looks like quite natural. We add a comment in order to avoid potential confusion about this regard. The complete transition probability $P(x - z; x; z)$ may be-and usually is in inhomogeneous environments-biased: $\int d\Delta\, \Delta\, P \neq 0$, but the average defining U in Eq. (12) involves the reduced probability $P(x - z'; x; z = x)$ in which z is set equal to x in all arguments but in the step length. It is this reduced probability that has to be unbiased.

This result thus may suggest that the FP is the natural long-wavelength limit of the ME-within the above constraint of unbiased P. Van Milligen et al. [40] supported this conclusion both by recalling earlier numerical results-namely, the computer simulations of particles hopping on a lattice by Collins et al. [6]-as well as carrying out and modelling some experiments of viscous fluids dynamics [39].

We mention that this kind of question (which choice is physically motivated) has often occurred in the literature [31, 37]. We have repeatedly stated throughout this work that the question has no answer, since both choices may be valid. Actually, unambiguous evidence for Fickian transport in inhomogeneous media with unbiased P does exist in literature. For instance, one may mention (i) generic one-dimensional Hamiltonian systems [8, 24]; (ii) the experiments on the dynamics of colloidal particles [21, 22]; (iii) the analytical and computer models of particles scattering against a Lorentz background made by Bringuier [3], as well as (iv) the extensive study by Schnitzer [35]. Several of these results are based upon microscopic models of the stochastic dynamics, thus there is no doubt that the Fick's flux is a valid limit for some classes of systems.

4 From the Master Equation to the Fick-Fourier Flux

It is therefore clear that we have not yet exhausted all the physics potentially embedded into Eq. (8). The purpose of this section is to make explicit the constraints imposed for the derivation of Eq. (13) and see how they have to be relaxed to allow for wider classes of solutions. To this purpose, we will outline a recipe patterning the classical Itô-Stratonovich [5, 17, 31] one within the framework of the Master Equation.

Let us start by looking at the structure of the arguments of the transition probability P within a particle-hopping picture: a generic particle is bound to location z for a duration τ, and then is propelled away. The initial conditions of the particle are completely fixed (although only in a statistical sense) by the local background at the starting point z (whence the appearance of z as second argument of P in Eq. 9), whereas during its travel it suffers an interaction with its environment that damps its motion up to a complete stop at point x, whence the meaning of the first argument of P as the total length travelled: xz. This picture is fairly intuitive, and for some classes of systems it provides a close approximation of the real dynamics, but we should remember that Eq. (8) is meant generically to give just a coarse description for the actual dynamics, which is much more complicated since involves kinetic degrees of freedom, too: we are neglecting, for instance, all effects related to the finite inertia of the moving particles. It is the same kind of ignorance about the true path travelled by the particle that, in Brownian motion (Eq. 7), leads to choosing between the different Itô or Stratonovich recipes. Therefore we should expect that P depends explicitly not only on the starting location z but potentially all the points between z and the final one, x. It is intuitive that such a detailed accounting would likely lead to a fairly complicated expression when inserted into Eq. (8). However, since only the initial (z) and the final (x) locations are well defined, not the exact trajectory between the two a more reasonable argument is that just them should appear as arguments of P. The simplest ansatz is to allow for a linear combination of the two:

$$P = P(x - z; \bar{x}), \quad \bar{x} = (1 - \alpha)z + \alpha x, \quad 0 \le \alpha \le 1 \tag{14}$$

Thus, with $\Delta = z - x$ and $G = P/\tau$ the ME writes

$$\frac{\partial n(x,t)}{\partial t} = -\frac{n(x,t)}{\tau} + \int d\Delta\, n(x+\Delta,t)\, G(-\Delta, x + (1-\alpha)\Delta)$$

Now we expand the terms inside the integral in power series around x dropping as customary all terms involving odd powers of Δ. In order to keep expressions short, in the next lines we label the spatial derivative with the apex sign: $' \equiv \partial/\partial x$ and the time derivative as $\partial_t \equiv \partial/\partial t$. We get

$$\partial_t n = \int d\Delta \frac{\Delta^2}{2} \left\{2(1-\alpha)G'n' + Gn'' + (1-\alpha)^2 G''n\right\} + \int d\Delta \{nG\} - \frac{n}{\tau} \tag{15}$$

By replacing the integrals in (15) using D given in (12) we get

$$\partial_t n = \left\{Dn' + D'(1-2\alpha)n\right\}' + \left[\alpha^2 D''n + \int d\Delta \{nG\} - \frac{n}{\tau}\right] \tag{16}$$

The normalization condition for particles jumping from the fixed location z to an arbitrary one x reads

$$\int dx\, P(z; x) = 1 \tag{17}$$

After inserting Eq. (14) for P:

$$1 = \int dx P(x - z; (1 - \alpha)z + \alpha x) \tag{18}$$

We rewrite in this expression the arguments of P as $P = P(-\Delta; z - \alpha\Delta)$, and expand (18) up to the second order in Δ:

$$1 = \int d\Delta\, P(-\Delta; z) - \alpha \frac{d}{dz} \int d\Delta\, \Delta\, P(-\Delta; z) + \alpha^2 \frac{d^2}{dz^2} \int d\Delta\, \frac{\Delta^2}{2} P(-\Delta; z) + \ldots \tag{19}$$

The first-order term in α vanishes by virtue of the symmetry of P. After the multiplication by $n(z)/\tau$ and the replacement of the second integral with D we get

$$\frac{n}{\tau} = \int d\Delta\, \{nG\} + \alpha^2 D'' n.$$

Therefore the term inside square brackets in Eq. (16) vanishes and we are left with

$$\partial_t n = \left\{Dn' + D'(1 - 2\alpha)n\right\}' \tag{20}$$

Let us now specialize Eq. (20): first consider the case $\alpha = 0$. We get accordingly

$$\partial_t n = (nD)'' \tag{21}$$

which is just the FPE once again, as expected since Eq. (14) reduces now to (9). The choice $\alpha = 1/2$ yields instead:

$$\partial_t n = (Dn')'$$

i.e, the FF diffusion. We can thus conclude that the ME may actually produce Fickian transport as long as its diffusive limit makes sense.

For completeness, let us add some brief comments about the third special case $\alpha = 1$.

$$\partial_t n = (Dn' - D'n)', \tag{22}$$

It is interesting to perform a comparison between (21) and (22). Both are instances of flux without a gradient, i.e., fluxes can be sustained even in the absence of a gradient in the concentration n. By setting $n' = 0$ in both we get respectively

$$\partial_t n = nD'', \quad \partial_t n = -nD''.$$

Hence, the two cases ($\alpha = 0, 1$) are related by the mirror symmetry $D'' \leftrightarrow -D''$. We conclude this section with a minor comment about the parallels between our procedure and that of Lançon et al. [21, 22] using the Brownian walk formalism. We point out how the ME approach grants a small advantage: namely, their derivation could retain terms containing D, D' but not higher derivatives (unlike ours) since it would turn the Brownian walker algorithm updating the position $(x(t) \to x(t+dt))$ into an algebraic equation for $x(t+dt)$ of order larger than one, with related issues of multiple roots.

5 Numerical Experiments: Lattice Models of Diffusion

The ansatz (14) suggests that final conditions are to be involved in order to recover the FF flux. One might wonder which systems do fulfil it. In this section we design two different models for particles hopping between nodes of a one-dimensional lattice, hence variants of the Collins' model [6]. Differences are in the statistical rules obeyed by the particles. Model 1 provides a flexible knob easily interpolating between different dynamics. Model 2 is somehow less amenable to direct inspection of its emergent dynamics, but implements perhaps a more realistic mechanism.

5.1 Model 1

At each time step, particles perform a random jump from their starting node to some neighbouring one. The width of the jump, i.e. the number of nodes the particle can travel, is picked randomly from a uniform distribution $[-\ell(i), \ell(i)]$, where i is the starting node, and ℓ a maximum jump length that depends on the starting location. Furthermore, each node j is assigned an acceptance rate $0 \le P_a(j) \le 1$. Once a particle has been chosen to hop from node i to node j, a second statistical test is performed based upon the acceptance rate: it succeeds with probability $P_a(j)$, and the particle moves from i to j, whereas with probability $1 - P_a(j)$ the test fails, and in this case the particle does not move. The hopping probability thus writes

$$P(j|i) = P_j(i-j|i) \times P_a(j).$$

Provided that ℓ is small with respect to the length of the lattice, we can consider a continuous version of the ME (where we set $\tau = 1$)

$$\frac{\partial n(x,t)}{\partial t} = -n(x,t) + \int d\Delta\, n(x+\Delta,t)\, P_j(-\Delta, x+\Delta) P_a(x)$$

which, after the usual Taylor expansion, writes

$$\frac{\partial n(x,t)}{\partial t} = -\,n(x,t) + \int d\Delta\, n(x,t)\, P_j(-\Delta,x) P_a(x) \tag{23}$$
$$+ \int d\Delta\, \Delta\, P_a(x) \frac{\partial}{\partial x} \left(n(x,t) P_j(-\Delta,x) \right)$$
$$+ \frac{1}{2} \int d\Delta\, \Delta^2\, P_a(x) \frac{\partial^2}{\partial x^2} \left(n(x,t) P_j(-\Delta,x) \right)$$
$$= P_a(x) \left[-\frac{\partial}{\partial x} \left(n(x,t) U_j \right) + \frac{\partial^2}{\partial x^2} \left(n(x,t) D_j \right) \right] \tag{24}$$

The odd moment U_j vanishes because P_j has mirror symmetry and (24) reduces to

$$\frac{\partial n(x,t)}{\partial t} = P_a(x) \frac{\partial^2}{\partial x^2} \left(n(x,t) D_j \right) \tag{25}$$

Eq. (25) is obviously not in the form of a Fokker-Planck Equation unless P_a is a constant—but can be cast into it by defining the auxiliary functions $\nu = n/P_a$, $D = P_a D_j$:

$$\frac{\partial \nu}{\partial t} = \frac{\partial^2 (\nu D)}{\partial x^2} \tag{26}$$

We check now that the presence of the P_a function outside the derivative term in (25) does not cause problems with the particle conservation. We postulate $P_a(x)$ to be a smooth function. To lowest order, we can get P_a constant, which obviously turns Eq. (25) into a continuity equation for n. Then, we go to the next order by linearly expanding P_a around some fixed point, conventionally $x = 0 : P_a(x) = P_a(0) + P_a'(0)x$. This allows rewriting Eq. (25):

$$\frac{\partial n(x,t)}{\partial t} = \frac{\partial}{\partial x} \left[P_a(x) \frac{\partial (n D_j)}{\partial x} - \frac{d P_a}{dx} n D_j \right] = \frac{\partial}{\partial x} \left[\frac{\partial (n D)}{\partial x} - 2 n D \frac{\ln d P_a}{dx} \right] \tag{27}$$

which is still in conservative form.

The convective flux is still geometric in nature, since is proportional to space derivatives of the jumping probabilities, but it has a more elaborated dependence than just from dD/dx, hence cannot be derived from within the Itô-Stratonovich α-parametrization.

Equation (27) reduces to the FPE when $dP_a/dx = 0$. Conversely, if P_a is not constant, the probability for a particle to jump between two nodes depends on the arrival site as well as the departure one, and we expect non-FP features to arise. Fick's flux arises in connection with the condition:

$$\frac{d \ln D}{dx} = 2 \frac{d \ln P_a}{dx} \rightarrow D_j \propto P_a \tag{28}$$

This, together with $D_j(x) = 1/2 < \ell^2 >= \ell^2(x)/6$, that comes from our choice of the statistical distribution, establishes that Fick's diffusion arises for just a specific functional form of the jumping length:

$$\ell \propto \sqrt{P_a}. \tag{29}$$

In the following we show the result of a numerical simulation done tracking the evolution of $N_p = 10^6$ particles over a lattice with $N = 2048$ nodes: a Monte Carlo implementation of solution of the Master Equation. All particles are initially located at $i = N/2$ and reflecting boundaries are imposed at both sides. As far as the acceptance rate is concerned, we consider two scenarios:

(A) variable acceptance rate: $P_a = 0.1 + 0.85(i/N)$ (B) constant acceptance rate: $P_a \equiv 1\ \forall i$ The jump length is instead taken in both cases as $\ell = \left[12\sqrt{0.1 + 0.85(i/N)}\right]$ (where [...] means the integer part); thus it fulfils constraint (29) in the scenario (A), whereas yields Fokker-Planck flux in scenario (B). There is nothing special about these numerical values; they have simply been chosen on the basis of the following fairly trivial considerations: (i) the larger $\Delta\ell/\Delta i$, the more any effect related to inhomogeneity shows up. However, $\ell(i)$ must be small enough in order for D to make sense as a local quantity: hops must not be too large. With the choice above we get that $D_{max} > 100 D_{min}$ and $\ell_{max}/N \approx 1/200$, which fulfil both these constraints. (ii) The same rationale applies to P_a as well: the larger dP_a/dx, the more clearly the departure from FP flux appears, but P_a must stay in the range $0 < P_a \leq 1$, and values too close to zero make the numerical treatment cumbersome, since particles take very long times to sample accurately these regions.

We compare the particle simulation with the numerical solution of the diffusion equation for both the choices Fokker-Planck (Eq. 6) and Fick-Fourier (Eq. 5). In the Fig. 1 the dots are the number of particles found at the different nodes after $t = 10^4$ time steps, the black (red) curves the FP and FF solutions respectively. There is a clear agreement between expectations and numerical results in both cases.

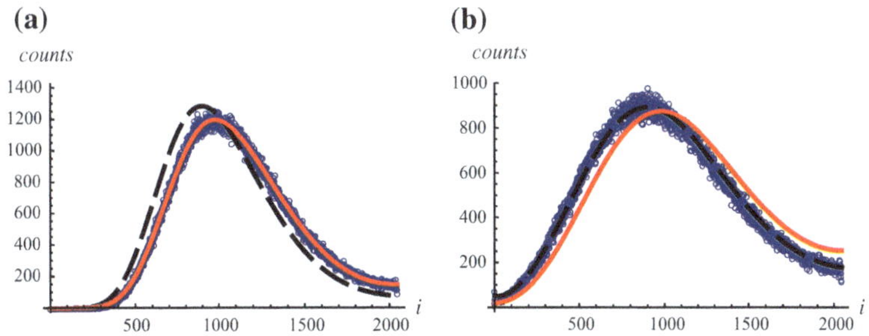

Fig. 1 Left plot, particle distribution after 10^4 steps for scenario (**a**) with variable acceptance rate; right plot, the same for scenario (**b**) with constant acceptance rate. Dots are from the numerical simulation of the particle model; red solid curve is the solution of fluid diffusion equation using FF's flux (Eq. 5); black dashed curve, the same using FP flux (Eq. 6)

5.2 *Model 2*

In model 1 the particle is required to pass a test relative to the destination node before it actually reaches the node itself. This time, we restore the locality in the model: all the tests are taken at the node occupied by the particle.

We consider a lattice version of mono-energetic particle motion: at each time clock particles are moved of exactly one node: say $(j-1 \to j)$. However, *after the displacement is done* a test is carried out: with probability $1-q(j)$ the particle conserves its direction during the next time clock, whereas with probability $q(j)$ it will reverse its direction. The next step, in the two cases, will be either $(j \to j+1)$ or $(j \to j-1)$. Thus, the total jump of the particle is defined by summing the number n of nodes travelled between two successive changes of direction, its probability being $P = q(j)\prod_{n-1}(1-q(l))$, where the index l stands for the nodes' indices between the starting and the final one; therefore P will depend on all the nodes visited as well. This way, the appearance of the final location between the arguments of P appears clearer: it is not necessary that the particle collects information about the arrival site in advance of its hopping, like in model 1. The model built this way is a version of random walk, thus its large-scale dynamics has to be diffusive. Unlike model 1, in this case, we were not able to guess an explicit expression for the diffusivity; however, since P is position-dependent, D must be as well, $D = D(x)$.

We have carried out a numerical Monte Carlo simulation for this system, which yields the solution of the Master Equation. Starting from a collection of particles all placed exactly in the middle of the lattice, we have let them to move randomly over times so long that the final stationary equilibrium is reached. The equilibrium density is shown in Fig. 2, and is spatially constant. Furthermore, at equilibrium, the flux must be null. Although we do not know explicitly the diffusivity D, we know that the stationary solution using the FP flux is $-d(Dn)/dx = 0 \to n \propto 1/D \neq$ constant. Hence, regardless of the value of $D(x)$, the flux cannot be in the Fokker-Planck form. Conversely, the FF flux is $-Ddn/dx$, and is consistent with the stationary solution $n =$ constant.

6 Summary

In this paper we have provided a simple procedure to derive diffusion equations as several different limits from within the single framework of the ME. We have worked out an analytical expression, Eq. (20), that can interpolate between both Fick-Fourier and Fokker-Planck limits in dependence of a single numerical parameter α. We have highlighted that the value of α should arise from the knowledge of the microscopic physics of the system examined. In most actual situations, it is likely difficult to extract it, but we have already provided several instances where this exercise was carried out [3, 6]. Other examples include [36] and-fairly recently-[42], although these results have been latter questioned [25, 26, 43].

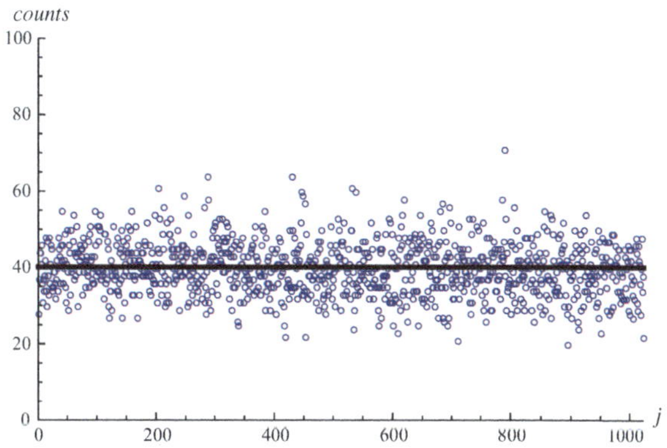

Fig. 2 Dots: particle distribution for the model 2 after 2.5×10^5 time steps. Parameters of the simulation are: number of nodes $N = 1024$; number of particles $N_p = 40 \times 1024$; particles initialized at $j = N/2$; $q(j) = 0.1 + 0.4 * j/N$. Reflecting boundaries are used. The resulting final distribution is uniform within the statistical noise: for reference, the black line is the perfectly constant density. This result is at a variance with expectations from Fokker-Planck flux, whereas it is consistent with Fick-Fourier flux

Regardless of the fact that one is able to determine a priori which is the best expression for the diffusive flux to be employed, the main conclusion stressed in this work is that both choices are legitimate long-wavelength limits of some underlying microscopic model. Furthermore, we have shown that the ME formalism is much more comprehensive than the Langevin equation one.

The analytical conclusions have further been supported by implementing and solving numerically simple intuitive models.

Acknowledgements FS wishes to thank Prof. E. Bringuier and Prof. M. Baiesi for providing him with some of the literature quoted, Dr. D.F. Escande, Dr. S. Cappello, Dr. I. Predebon for reading drafts of this paper and Prof. G. Ryskin for interesting discussions about his proof. This project has received funding from the European Union's Horizon 2020 research and innovation programme under grant agreement number 633053. The views and opinions expressed herein do not necessarily reflect those of the European Commission. LS thanks MIUR for partial support (PRIN Project 2010LLKJBX).

Appendix

The rationale of Ryskin's result is based upon the Central Limit Theorem (CLT). The total displacement of a particle is approximated as the sum of several uncorrelated jumps. The single jumps are not necessarily identical, but picked up from some statistical distribution. Regardless of the details of the distribution, provided that the variance of the jumps remains finite, the CLT warrants that the total displacement distributes according to a statistical distribution that quickly approaches a Gaussian distribution after even a moderate numbers of steps.

For brevity we will sketch Ryskin's proof for homogeneous systems only. Its generalization to inhomogeneous systems adds some mathematical labour but does not differ conceptually. In homogeneous systems P depends just from the difference of its arguments: $P = P(x - z)$, as explained in Sect. 3. This allows for a dramatic simplification after the Fourier transform of Eq. (8) is taken:

$$\frac{\partial \tilde{n}(k,t)}{\partial t} = -\frac{\tilde{n}(k,t)}{\tau} + \frac{\tilde{n}(k,t)\tilde{P}(k)}{\tau} \tag{30}$$

Equation (30) can be solved analytically:

$$\tilde{n}(k,t) = \tilde{n}(k,t=0) \times \exp\left(\int_0^t \frac{dt}{\tau}(\tilde{P}(k)-1)\right) = \exp\left(\frac{t}{\tau}(\tilde{P}(k)-1)\right) \tag{31}$$

Formally, $n(x,t)$ comes from the inverse transform of (31) and it does not appear analytically computable for generic P. However, we can write

$$\tilde{P}(k) = \int dz\, e^{ikz} P(z) = \sum_{m=0}^{\infty} \frac{(ik)^m}{m!} \int dz\, z^m\, P(z) \tag{32}$$

We will be considering just specularly symmetric transitions, as customary: $P(x - z) = P(z - x)$; the moments in Eq. (32) become

$$\int dz\, P = 1 = \tilde{P}(k=0) \tag{33}$$

$$\int dz\, z\, P = 0 \tag{34}$$

$$\int dz\, z^2 P = \sigma^2 \tag{35}$$

$$\int dz\, z^m P = < z^m > \equiv \mu_m, \quad m \geq 3 \tag{36}$$

and $\mu_m = 0$ for all odd m's.

Let us now rewrite Eq. (32) using the trigonometric expression of the exponential:

$$\tilde{P}(k) = \int dz \cos(kz)P(z) + i \int dz \sin(kz)P(z) \tag{37}$$

$$\to |\tilde{P}(k)|^2 = \left(\int dz \cos(kz)P(z)\right)^2 + \left(\int dz \sin(kz)P(z)\right)^2$$

$$= < \cos(kz) >^2 + < \sin(kz) >^2$$

This implies that $|\tilde{P}(k)| \leq 1$ for generic $k \neq 0$. Furthermore,

$$|\tilde{P}(k)| \to 0, k \to \infty \tag{38}$$

To demonstrate this, let us note that $\sin(kz)$, $\cos(kz)$ are periodic with wavelength $\lambda = 2\pi/k \to 0$, $k \to \infty$, while P is a smoothly varying function, hence is almost constant over λ: $P(z) \approx P(z + \lambda) = P_0$. Therefore

$$\int_0^\lambda dz \cos(kz)P(z) \approx P_0 \int_0^\lambda dz \cos(kz) \approx 0 \tag{39}$$

(the same holds for $\sin(kz)$).

We define m, Δt such that $t = j\Delta t$, with j integer and $\Delta t \approx O(\tau)$. Equation (31) becomes

$$\exp\left(\frac{t}{\tau}(\tilde{P}(k) - 1)\right) = \left[\exp\left(\frac{\Delta t}{\tau}(\tilde{P}(k) - 1)\right)\right]^j = \tag{40}$$

$$\left[\exp\left(\frac{\Delta t}{\tau}\left(-\frac{k^2\sigma^2}{2} + \sum_{m=3}^{\infty} \frac{(ik)^m}{m!}\mu_m\right)\right)\right]^j$$

In the last line of (41) we have taken advantage of (33)–(36).

Let us define $\xi = j^{1/2}k$. Equation (41) becomes

$$\frac{\tilde{n}(k,t)}{\tilde{n}(k,t=0)} = \exp\left[\frac{\Delta t}{\tau}\left(-\frac{\xi^2\sigma^2}{2} + \sum_{m=3}^{\infty} \frac{1}{j^{\frac{m}{2}-1}} \frac{(i\xi)^m}{m!}\mu_m\right)\right] \tag{41}$$

Then, we consider separately the two limits $\xi > 1$ and $\xi \leq 1$ (Notice that the boundary between $\xi > 1$ and $\xi \leq 1$ is a dynamical one: it varies with time, *i.e.* with j). The former limit corresponds, for any fixed time, to taking $k \to \infty$ and therefore the result (38) holds: there is not contribution to the density from features at these wavelengths. Conversely, when $\xi \leq 1$ the first term inside the exponent in Eq. (41) dominates over the others hence we can retain just it and, reverting to the original variables

$$\tilde{n}(k,t) = \tilde{n}(k,t) \times \exp\left[-\frac{t}{\tau}\frac{(k\sigma)^2}{2}\right]$$

which is the propagator of the diffusion equation, with diffusivity = $\sigma^2/2\tau$. This concludes the proof.

References

1. Amari, S.: Dynamics of pattern formation in lateral-inhibition type neural fields. Biol. Cybern. **27**, 77 (1977)
2. Bengfort, M., Malchow, H., Hilker, F.: The Fokker-Planck law of diffusion and pattern formation in heterogeneous environments. J. Math. Biol. https://doi.org/10.1007/s00285-016-0966-8 (2016)
3. Bringuier, E.: Kinetic theory of inhomogeneous diffusion. Phys. A **388**, 2588 (2009)
4. Cattaneo, C.: Sulla conduzione del calore. Atti Sem. Mat. Fis. Univ. Modena **3**, 83 (1948)
5. Coffey, W., Kalmykov, Y.P., Valdron, J.: The Langevin Equation. World Scientific (1998)
6. Collins, R., Carson, S., Matthew, J.: Diffusion equation for one-dimensional unbiased hopping. Am. J. Phys. **65**, 230–7 (1997)
7. Einstein, A.: Investigations on the Theory of the Brownian Movement. Dover publications, New-York (1956)
8. Elskens, Y., Escande, D.: Microscopic Dynamics of Plasmas and Chaos. Institute of Physics Publishing (2003)
9. Escande, D., Sattin, F.: When can the Fokker-Planck equation describe anomalous or chaotic transport? Phys. Rev. Lett. **99**, 185005 (2007)
10. Escande, D., Sattin, F.: When can the Fokker-Planck equation describe anomalous or chaotic transport? intuitive aspects. Plasma Phys. Control. Fusion **50**, 124023 (2008)
11. Fick, A.: On liquid diffusion. J. Membr. Sci. **100**, 33–38 (1995)
12. Graham, T.: On the law of the diffusion of gases. J. Membr. Sci. **100**, 17–21 (1995)
13. Green, A., Naghdi, P.: A re-examination of the basic postulates of thermomechanics. Proc. R. Soc., Math. Phys. Sci. **432**, 171 (1991)
14. Guyer, R., Krumhansl, J.: Solution of the linearized phonon boltzmann equation. Phys. Rev. **148**, 766 (1966)
15. Joseph, D., Preziosi, L.: Heat waves. Rev. Mod. Phys. **61**, 41 (1989)
16. van Kampen, N.: The expansion of the master equation. Adv. Chem. Phys. **34**, 245 (1976)
17. van Kampen, N.: Itô versus Stratonovich. J. Stat. Phys. **24**, 175–187 (1981)
18. van Kampen, N.: Diffusion in inhomogeneous media. J. Phys. Chem. Solids **49**, 673–677 (1988)
19. Kenkre, V., Montroll, E., Shlesinger, M.: Generalized master equations for continuous-time random walks. J. Stat. Phys. **9**, 45 (1973)
20. Kovács, R., Ván, P.: Generalized heat conduction in heat pulse experiments. Int. J. Heat Mass Transf. **83**, 613 (2015)
21. Lançon, P., Batrouni, G., Lobry, L., Ostrowsky, N.: Drift without flux: Brownian walker with a space-dependent diffusion coefficient. EPL **54**, 28 (2001)
22. Lançon, P., Batrouni, G., Lobry, L., Ostrowsky, N.: Brownian walker in a confined geometry leading to a space-dependent diffusion coefficient. Phys. A **304**, 65–76 (2002)
23. Landsberg, P.: D grad v or grad (D v)? J. Appl. Phys. **56**, 1119 (1984)
24. Lichtenberg, A., Lieberman, M.: Regular and Stochastic Motion. Springer (1983)
25. Mannella, R., McClintock, P.: Comment on influence of noise on force measurements. Phys. Rev. Lett. **107**, 078901 (2011)
26. Mannella, R., McClintock, P.: Itô versus Stratonovich: 30 years later. Fluct. Noise Lett. **11**, 1240010 (2012)
27. Mason, E., Evans, R.: Graham's laws: simple demonstrations of gases in motion: part i, theory. J. Chem. Educ. **46**, 359–423 (1969)
28. Mason, E., Kronstadt, B.: Graham's laws of diffusion and effusion. J. Chem. Educ. **44**, 740 (1967)
29. Maxwell, J.: On the dynamical theory of gases. Philos. Trans. R. Soc. London **157**, 49–88 (1867)
30. Philibert, J.: One and a half century of diffusion: Fick, Einstein, before and beyond. Diffus. Fundam. **4**, 6 (2006)
31. Risken, H.: The Fokker-Planck Equation. Springer (1996)

32. Ruckstuhl, A.: Thomas Graham's study of the diffusion of gases. J. Chem. Educ. **28**, 594 (1951)
33. Ryskin, G.: Simple procedure for correcting equations of evolution: application to Markov processes. Phys. Rev. E **56**, 5123–5127 (1997)
34. Sattin, F.: Fick's law and Fokker-Planck equation in inhomogeneous environments. Phys. Lett. A **372**, 3941 (2008)
35. Schnitzer, M.: Theory of continuum random walks and application to chemotaxis. Phys. Rev. E **48**, 2553 (1993)
36. Smythe, J., Moss, F., McClintock, P., Clarkson, D.: Itô versus Stratonovich revisited. Phys. Lett. A **97**, 95–98 (1983)
37. Sokolov, I.: Itô, Stratonovich, Hänggi and all the rest: the thermodynamics of interpretation. Chem. Phys. **375**, 359–363 (2010)
38. Ván, P., Fülöp, T.: Universality in heat conduction theory: weakly nonlocal thermodynamics. Ann. Phys. (Berlin) **524**, 470–478 (2012)
39. Van Milligen, B.P., Bons, P., Carreras, B., Sánchez, R.: On the applicability of Fick's law to diffusion in inhomogeneous systems. Eur. J. Phys. **26**, 913 (2005)
40. Van Milligen, B.P., Carreras, B., Sánchez, R.: Pulse propagation in a simple probabilistic transport model. Plasma Phys. Control. Fusion **47**, B743 (2005)
41. Vernotte, M.: Les paradoxes de la theorie continue de i'equation de la chaleur. C. R. Acad. Sci. **246**, 3154–3155 (1958)
42. Volpe, G., Helden, L., Brettschneider, T., Wehr, J., Bechinger, C.: Influence of noise on force measurements. Phys. Rev. Lett. **104**, 170602 (2010)
43. Volpe, G., Helden, L., Brettschneider, T., Wehr, J., Bechinger, C.: Influence of noise on force measurements. Phys. Rev. Lett. **107**, 078902 (2011)

Part IV
Computational Methods

Anticipating Abrupt Changes in Complex Networks: Significant Falls in the Price of a Stock Index

Antonio Cordoba, Christian Castillejo, Juan J. García-Machado and Ana M. Lara

Abstract Early prediction of abrupt changes in complex systems is of great interest in preventing unwanted effects. This has recently led to the establishment of indicators whose evolution may be indicative of some of such changes. Here we present a criterion to predict the sharp fall in the prices of a stock market index. We have studied the moving networks constituted by the companies included in several indexes (IBEX35, CAC40, DAX30 and Euro Stoxx50), constructing the corresponding "Minimal Spanning Tree (MST)". When the number of leading nodes in the network decreases in a substantial manner, the network has few leaders, and if those suffer any fall, the index might fall as well. By means of this hypothesis, we are looking for a rotation direction beforehand, a downward rotation. Using daily closing price series from 2007 to 2017 for these indexes, we can point out that when the number of leading nodes is small, and the average correlation of companies forming an index decreases, placing itself below 0.4–0.5, depending on the index, and this decrease is accompanied by a significant increase in the correlation deviation, the price tends to fall at around 70% of reliability.

Keywords Complex networks · Phase transitions · Stock markets · Price predictability · Financial analysis · Econophysics

A. Cordoba (✉)
Departamento de Física de la Materia Condensada, Universidad de Sevilla,
Avda Reina Mercedes s/n, 41012 Sevilla, Spain
e-mail: cordoba@us.es

C. Castillejo · A. M. Lara
Instituto IBT, Parque Empresarial Nuevo Torneo, C/ Astronomía, 1, torre 2,
planta 10, 41001 Sevilla, Spain
e-mail: automatasibt@institutoibt.com

A. M. Lara
e-mail: formacion@institutoibt.com

J. J. García-Machado
Departamento de Economía Financiera, Contabilidad y Dirección de Operaciones,
Universidad de Huelva, Plaza de la Merced, 11, 21002 Huelva, Spain
e-mail: machado@uhu.com

V. Carmona et al. (eds.), *Nonlinear Systems, Vol. 1*,
Understanding Complex Systems, https://doi.org/10.1007/978-3-319-66766-9_11

1 Introduction

Dynamic systems and, in particular, dynamic networks that evolve over time can undergo abrupt changes in their behavior for a given value of a set of parameters that at a given instant characterize the behavior of the system. This affects both natural and social systems, with well-defined evolutionary rules or with empirical values recorded and represented by time series. These critical thresholds—also called tipping points—often point to the onset of catastrophic situations such as earthquakes, species extinction, climate change, etc., and in particular sudden and significant falls in share prices in stock markets.

It is very important to be able to predict these changes so that one can avoid or minimize their damage or even take advantage of them (for example, by trading profitably in the stock market). In cases of systems with well defined rules of evolution, embodied in a set of differential equations, this task, without being easy in general, can be carried out in a systematic way. However, in cases where only empirical time series are available, without knowledge of the mechanisms that originate them, the problem of making predictions is a huge challenge [5, 8, 31]. With this objective, a set of indicators, of a statistical nature, applied in isolation or in concurrence with others, has been defined for different systems in order to try to predict with certain probability of reliability the occurrence of one of these critical changes.

The list of indicators applied to complex systems of different nature is broad: recovery rate or return time of the system state when is subjected to a disturbance, dominant eigenvalue, rate of change of amplitude, autocorrelation at lag 1, spatial correlation, analysis of fluctuations, variance, skewness, spectral analysis, spectral reddening, Fisher information, Shannon index … A catalog of these indexes and their applicability to different systems to anticipate critical transitions has been published by Scheffer et al. [32]. Another approach is to establish a measure of the resilience of the system and, in particular, of a complex network [16] with the idea that if at any given time the resilience is low the system is more fragile when it has to react to perturbations of nature unknown or unforeseen, and thus, it is more likely that in these situations there will be a collapse of the whole complex network. Bardoscia et al. [3], using a parallelism with ecological systems, have recently analyzed the emergence of instability in financial systems.

In any case, as it has been indicated, there is no method that anticipates with certainty the advent of these abrupt changes, there is only partial success of a probabilistic nature for certain systems, so that continual attempts are made to design new indicators that can serve as early warnings of the occurrence of unwanted collapses. This is of great importance, both for policy makers and traders, to anticipate significant falls in the financial markets due to their great economic impact. One aspect that has often been addressed in network theory and which is relevant to our study is how we can identify subsets of the network formed by elements more closely connected to each other than to the other elements of the network, forming functional groups or cliques that are more or less compact or linked through a hub that plays the leading role of the group and establishes a certain hierarchy [15, 39]. Heiberger [17] has analyzed S&P100 index and concludes that collective shifts precede structural changes in stock market networks and that this connection is mostly carried by companies that already dominate the development of the S&P100. A metric between nodes can be defined, which allows the clustering of the closest ones by different grouping techniques, by a relevance index introduced by Roli et al. [29] or by the method used by Bonanno and Mantegna [6] to draw a minimal spanning tree (MST), which we will apply here.

Since the 1960s, many theories and models of investment have been developed, some of them more successful and operative than others, or more sophisticated, with greater amount of data needed, and so on. Starting from the well-known models of Markowitz and Sharpe, through CAPM (Capital Asset Pricing Model), APT (Arbitrage Pricing Theory) and OPT (Option Pricing Theory), to the Technical Analysis and Chartism (also known as "behavioral finance"), there have been many academics, researchers, statisticians, economists, professionals and investors who have tried to develop a model or methods in order to explain the markets' behavior and to adopt some trading strategy, allowing them to take a competitive advantage and obtain benefits from the stock market investment.

The models based on the behavior of the investor, in the "Big Data", etc., are among the latest advances that try to seek advantage of the incorporation of psychologists, mathematicians, physicists and engineers to the teams of Research in Financial Markets, to find their patterns of behavior. Recent advances found a method, in order to observe the most suitable values of the same index to obtain diversified portfolios. In addition, this breakthough could predict the collapse point of a particular market. One of the most interesting aspects of these theories is the possibility of predicting sudden changes in the system, forcing it to develop a modified structure.

The different financial markets are interconnected through the network. In the "theory of networks", stock values are connected to each other, according to their degree of correlation. This theory's main goal is to identify market leaders and predict the values that will set behavioral patterns in other markets. The study is based on a series of mathematical filters used to detect who are the leaders and followers of a stock market, according to the quotation of each value. Moreover, it is verified that the market consist of united organizations, meaning that the collapse of a market directly affects the development of others. The "backbone" of these network (the main nodes), and for different markets, would be the banking sector, as well as the "mid and small cap" in their ramifications.

What does it consist of? A good example would be the comparison between our system and Facebook. In this social networks, the users relate to each other through links or request for friendship. Some of these individuals are influencers, that is, their opinion has more impact and reaches a greater number of users, and therefore they are considered "leaders", who lead the way in the systematic behavior of an interconnected group. Something similar happens in the financial markets. In this area, stock exchanges are connected with each other, according to their degree of correlation. Here, Big Data plays a very important role. Its constant development produces more accesible information to improve the model and results offered by network theory.

So, how can this theory improve the optimization of Trading? When applying this method, professionals can detect not only the network structure generated in a particular financial system, but also it allows them to study the risk of a possible collapse in a stock exchange network. Thank to this advance, a strategy can be created and adapted to the most optimal values, depending on the client' portfolio and the degree of exposure to risk.

In this chapter, we consider the structure of different financial systems as moving networks, their nodes, edges, relationships and their behaviour in the previous peaks to a fall in the prices of their market indexes. This paper is organized as follows. We start reviewing the conceptual framework and related literature as theoretical background in Sect. 2. Next, in Sect. 3, we describe the methodology used in this study. After carrying out a general description of the model, we provide a more detailed description of each kind of information and how it is obtained (extraction of index's data) and processed in order to create the model in the two following subsections: "construction of the networks" and "use of the number of leading nodes in the network as a possible indicator of a fall in index prices". Finally, Sect. 4 present empirical results, and a summary and conclusions are provided in Sect. 5, respectively.

2 Conceptual Framework

In the last decades, many studies have been carried out on financial markets, and many diverse techniques and hypotheses have been applied in an attempt to explain the behavior of the price of diverse financial products. At the same time, many professionals outside the finance world have carried out many studies in an attempt to apply advances within their own fields. These professionals range from economists (who are naturally the most common ones), mathematicians, statisticians, or, as in this case, physicists.

The irruption of these professionals has allowed us study financial markets as groups of elements that are interwined and influence each other. From the 70s on, we can find studies (Levine [21]) in which the behavioral patterns applied had only been applied to physical systems, until then. This opened the doors for a new way of studying financial markets.

In recents years, much progress has been made in this sense, and studies (such as Fiedor [12–14], for instance) make reference to how companies in a sector or stock market index behave in a nolineal manner, which dynamically causes them to affect each other. This has been proven through the application of the hierarchical network theory to the system. The appearance of artificial intelligence, alongside the possibility to teach systems to take decisions according to a historical basis was also brought to the studies of financial markets by neural network theories (Siripurapu [33]).

Another contribution with which Physics has contributed to the study of financial markets has been that of the phase transitions and critical phenomena, which in this case can be applied to prevent possible changes in the trend of asset prices. This allows us to interpret them as the end of a trend or market phase by using model that are at the forefront in physical studies, such the Ising model, percolation theory and other studies (Sornette [36]).

Barrio et al. [4] analyze how buyers and sellers relate to each other, and which allows us to draw some general conclusions about how different commercial strategies could affect the distribution of wealth in different types of societies. Lemieux et al. [20] have explored the application of three different portfolio formation rules using standard clustering techniques. Musmeci et al. [26] analyze the best methods of creating a portfolio and the best hierarchical structures, as well as the different ways in which sectors react to events, such as a financial crisis. Ren et al. [28] has also recently proposed a new dynamics portfolio strategy based on the time-varying structure de networks in Chinese stock markets.

Other studies (Alkan and Khashanah [1]) refer to the temporary development of modelled indexes, such as networks, along with their change over time when they are affected by external facts. In addition, not only the influence of companies constituting an index and the index itself have been studied, but also the influence of different indexes have upon each other (Sandoval et al. [30]). Fiedor [13] has introduce a way to incorporate nonlinear dynamics and dependencies into hierarchical networks to study financial market using the concepts mutual information and mutual information rate and and applying them to two stock indexes.

Carlsson and Memoli [7] have studied hierarchical clustering schemes under an axiomatic view. Huang et al. [18] has done a study of MST comparing the algorithms of Prim and Kruskal to define a metric, expressing that Kruskal algorithm is more suitable for sparse edged networks. Song et al. [35] have introduced a graph-theoretic approach to extract clusters and hierarchies in complex data-sets in an unsupervised and deterministic manner, without the use of any prior information, and have constructed networks containing the subset of most significant links and analyzed the network structure to differentiate meaningful clusters and hierarchies in a variety of real data-sets. Mantegna et al. have addressed the study of different aspects of financial markets to extract relevant information from a broad set of stock indices based on the construction and analysis of correlation graphs [23, 34, 38]. Tse et al. [37] have constructed a network using US stock prices, price returns and trading volumes and established that the variations of the stock prices are strongly influenced by a relatively small number of stocks and report that all network based on connecting highly correlated stock prices display a scalefree degree distribution, in a similar way that other self-organized systems [2]. Cimini et al. [9] have designed a method to reconstruct financial networks from partial information through statistical mechanics methods to improve the possibility of correctly estimating the resilience of these systems to events such as financial shocks, crises and cascade failures. Other recent studies on this matter are Lima Dias [22], Donnat et al. [10], and Marti et al. [25].

In this chapter we have focused on how the network behaves in the moments right before a fall in the prices of the index. In order to do this, we have focused on the use of the network as our starting point obtaining both the correlations of the companies and the existing nodes in each moment in the network.

Regarding the structuring of financial systems as network, Peralta [27] also offers information on the correlations and behaviors of the companies that form the network.

3 Methodology

On the basis of diverse studies and articles on financial markets, we have combined characteristics and conclusions from a new model to obtain an indicator that will allow us to predict, with enough reliability and anticipation, the maximums that occurs in an index price by studying the companies that form that index.

Firstly, we have carried out a study on the network constituted by the companies of an index in such a way that the correlated behavior between all these companies can be observed. Consequently, we can obtain the average correlation between them and the average deviation of said correlation. After that, we have studied the disposition of the companies from the point of view of a network. Thus, we can affirm that there are situations in which the network has more or less leading nodes. We call leading nodes those that in the MST have more than two links. In the following, for brevity, we will refer to the leading nodes simply as nodes in the network (the total of nodes in the network is always the total of companies that compose the index and does not change). This variation in the number of nodes in the network for differentes situations of the market provides extra information, as will be discussed later.

Once the information about both strands has been obtained, tests have been carried out on various stock indexes around the world. These results have been analyzed, providing us with relevant conclusions. In addition to numerical results, the possibility of choosing a securities that is close to the optimal is also obtained through the network's graphic layout. This can be accomplished by choosing companies that are as minimally correlated as possible, something that in this case can be attained by selecting the companies located as far as possible from the next graph.

As can be observed in Fig. 1, leading companies, in the form of nodes, are the ones located at the center, and from which several edges sprout up. At the same time, nodes located at the ends represent follower companies. This distribution, which unites the more correlated companies, causes companies to unite themselves by sectors. This creates various sectors in the graph. Some studies (Sornette [36]) address the importance of sectors in the development of business prices, thus causing companies of the same sector to be placed in a very correlated way. Therefore, BEI.DE is less related to DBK.DE than to CON.DE, which implies that if we want to obtain an optimal portfolio, we would better choose BEI.DE and DBK.DE rather BEI.DE and CON.DE, since risk would be more diversified this way.

The network study is based fundamentally on two sets of data. One of them is an index (such as IBEX35 or NASDAQ), and the other is a "moving network" or "moving candlestick chart" built up from data from the companies that this index takes into consideration. This network is a moving network in the same sense as a moving average, for example. By using this moving network, the characteristics of different networks structures (observed at different instants), may be observed and analyzed. Such analysis shows that, over time, this structure undergoes significant changes. The basic idea behind the model is that certain changes in this structure may constitute signals that anticipate the future behavior of the market. The goal is to combine information from both the index and the changing structure, to try to find some clues which will hopefully allow us to predict abrupt changes in the closing price, all this with a certain probability of success.

One important decision to make is, therefore, the "length" of the moving network, which will be later explained in detail. Basically, this governs the size of the data set used to calculate the moving network in each instant of time. Previous research from one model considering the "minimal spanning trees of stock portfolios at different time horizons", shows that many parameters that can be calculated from a network change when the length of the network (called "time horizon" in the model) is altered. With regard to this fact, we have considered that the length chosen should be large enough so that it does not suffer too much the "Epps effect" (the Epps effect, named after T. W. Epps (Epps [11]), is the phenomenon that the empirical correlation between the returns of two different stocks decreases as the sampling frequency of data increases), in order to give useful information, but also small enough so that it provides sufficiently frequent changes in the structure of the moving network. We believe that a seemingly working value for this purpose is comprised between approximately ten to twenty days.

After this general description of the model, we provide a more detailed description of each kind of information, how it is obtained, and how it is processed in order to create the model. This is explained in the following subsections: "construction of the networks", and "use of the number of leading nodes in the network as a possible indicator of a fall in index prices".

3.1 Construction of the Networks

In order to build the network, historic candlestick data are used. This data comes from Yahoo Finance®, thanks to an API (Application Programming Interface) provided by MATLAB. The candlestick contain the closing, open, maximum and minimum price for a given day. The volume will also be considered. Let's suppose, for example, that candlesticks from 01-Jan-2008 to 01-Jan-2009 are downloaded.

It is important to bear in mind that the network is a moving network. That mean that only a limited selection of this data are used to create a given network. The exact amount is defined by a parameter that we will call "length". For example, a length of 15 days would mean that the first network would have to be built up to data comprising the dates from 01-Jan-2008 to 16-Jan-2008, that is, the starting and ending days of the network. Since it is a moving network, the next network will comprise dates from 02-Jan-2008 to 17-Jan-2008, the next one from 03-Jan-2008 to 18-Jan-2008, and so on, until the limit of the data 01-Jan-2009 is reached. Obviously, only those days in which the corresponding market is open are considered. Days in

Table 1 Example of the correlation matrix (R) for an index with three companies

	Company A	Company B	Company C
Company A	1.00	0.54	−0.48
Company B	0.54	1.00	0.58
Company C	−0.48	0.58	1.00

which the market is closed are simply skipped. The purpose of this moving network is to study the market as it evolves in the time.

Now, let's examine the process of creating the network for one of the given time ranges. A matrix of correlation coefficients is calculated from each company's closing price every day of the time range. This correlation matrix is not calculated directly from prices, but from logarithmic rate of return instead:

$$l = \ln(\frac{c(d)}{c(d-1)})$$

where l is the logarithmic rate of return (in the analyzed sesion), and $c(d)$ the closing price for a given day. This expression was already used by King [19] to analyze the observed covariance matrix of a large set of series of monthly changes in closing prices and to see the kind of association that is present in the movement over time of a cross-section of security price changes. As shown in Table 1, each row or column of this correlation matrix R corresponds to a certain company, and the correlation of two companies, for example, A and B, is given by the element located at row A and column B or conversely row B and columna A, since the matrix is symmetric.

The "meaning" of the network, that is, the data that is going to be somehow encoded in the distance matrix, is going to be a correlation that was previously computed. So, from this correlation matrix, we define an intermediate matrix (I) of size n × n, in which the element in the i-th row and j-th column in the distance matrix is defined as follow:

$$I_{i,j} = \sqrt{2(1 - R_{i,j})}$$

Then, we define the elements of the distance matrix as:

$$d_{i,j}(x) = \begin{cases} I_{i,j} \text{ if } i < \text{j} \\ 0 \quad \text{if } i \geq \text{j} \end{cases}$$

That means that the elements of the distance matrix are the upper triangular matrix of the intermediate matrix. From this distance matrix, with the help of MATLAB, a minimum spanning tree is made from this distance matrix, in a similar way as the one shown in Mantegna and Stanley [24]. The algorithm implemented in MATLAB starts from the daily quotes of companies that compose an index arranged in a matrix in which in each column corresponds to the quotes of a company ordered temporarily. This matrix is transformed into another one in which in each column the variations of the logarithms of the quotes of each company are represented (a matrix m × n,

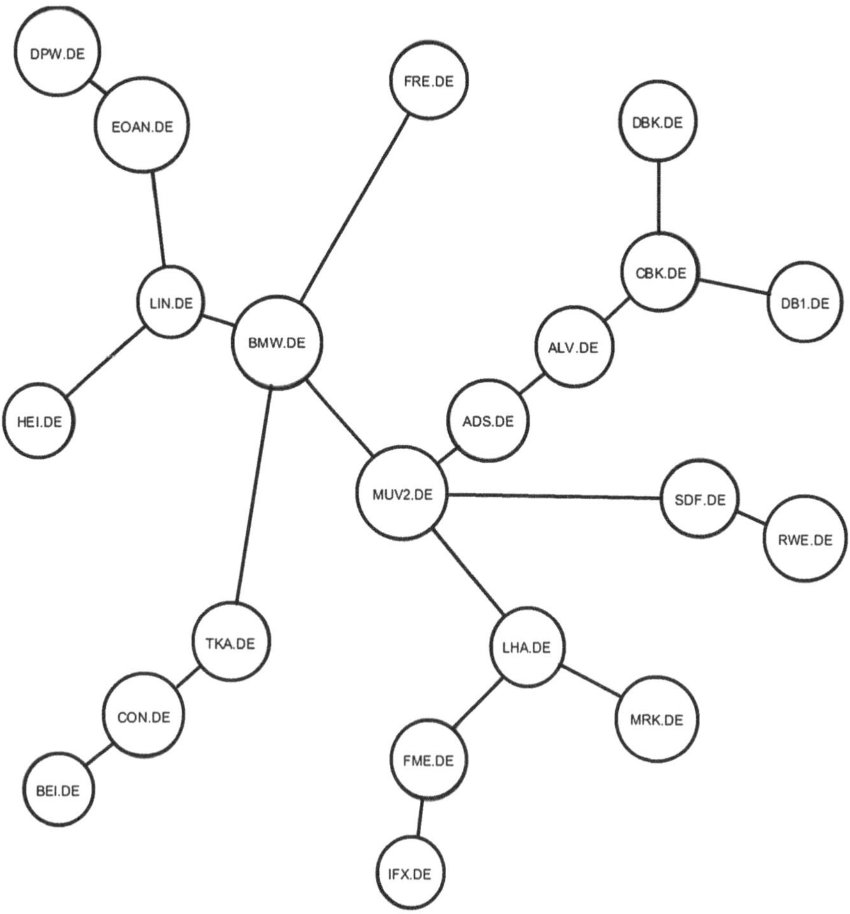

Fig. 1 Example of a minimal spanning tree (MST) for the DAX index from which centrality, number of edges and number of leading nodes will be extracted

where m is the number of days quoted and n the number of companies that compose the network). Then the Kruskal theorem is applied to this matrix to eliminate the noise and later its correlation matrix is obtained. This new matrix is represented graphically. An example of such a tree is shown in Fig. 1.

From both the correlation matrix and the minimum spanning tree, some parameters are extracted to be used in the study. Beginning with the correlation matrix, the following parameters are calculated:

- Maximum distance: It is simply the maximum of all elements in the distance matrix.
- Minimum distance: As before, it is simply the minimum of all elements in the distance matrix.

- Mean distance: It is given by the mean of all the non-zero elements in the distance matrix.
- Standard deviation of the distance: It is the standard deviation of all the non-zero elements in the distance matrix.
- Maximum correlation: It is the maximum of all the elements in the correlation matrix.
- Minimum correlation: It is the minimum of all the elements in the correlation matrix.
- Mean correlation: It is the mean of all the elements in the upper triangular matrix of the correlation matrix.
- Standard deviation of the correlation: It is the standard deviation of all the elements in the upper triangular matrix of the correlation matrix.

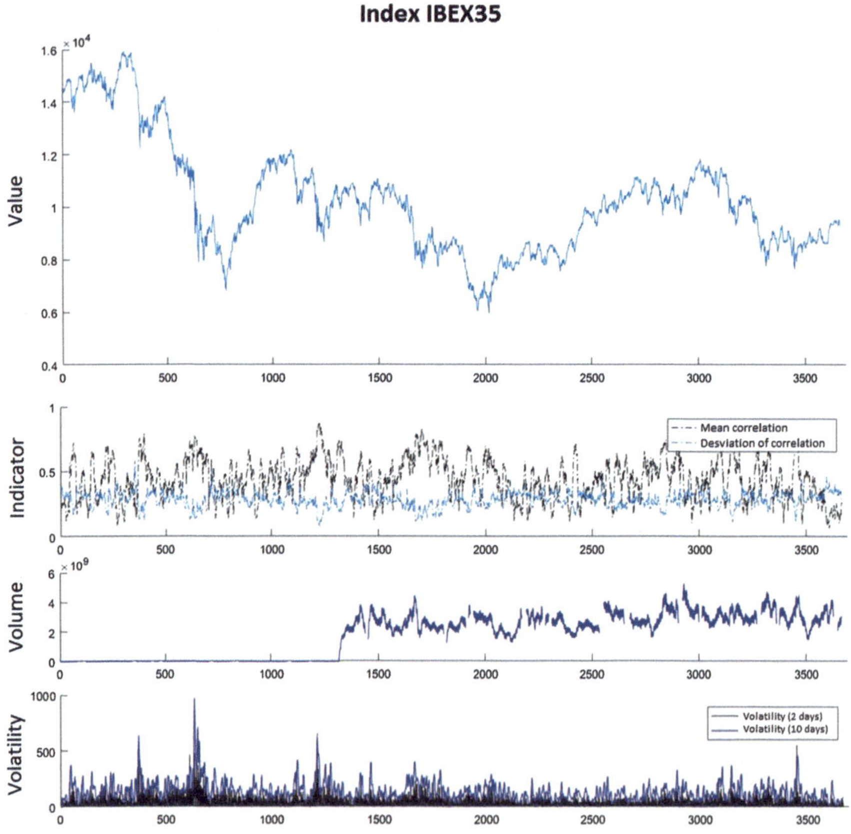

Fig. 2 Example of graphical results obtained when running the app. IBEX35 index between 2007 and 2017

From the minimum spanning tree, the following parameters are calculated:

- Centrality.
- Number of edges
- Number of nodes

Also, by an application implemented in MATLAB the quantities appearing in the example of Fig. 2 are calculated and represented graphically.

3.2 Use of the Number of Nodes in the Network as a Possible Indicator of a Fall in Index Prices

By building a network with the companies that make up an index, we can obtain information about the companies arrangement. As result, we would adquire information about which companies are leading companies and which are follower companies. Depending on the moment, the same network can have a different number of leading companies. Moreover, companies regarded as leaders in a specific moment can cease to be so in the next minute. This fact is obtained through the network nodes. One must remember that when we are talking about nodes we refer to leading nodes. If a network has various nodes, it means that it has various leaders. When a network is made up of a few nodes, it means that the index leadership falls on a few number of companies, and this implies that if there is a few number of leading companies and they happen to suffer a fall, the whole index can be affected, since the other companies are following the leaders (Peralta [27]). However, when the index leadership is distributed among many companies, the remaining leading companies can sustain a possible fall of the whole index if one of the leaders suffers a fall. To measure the fall in prices, the volatility of the index is taken as a reference and measured by establishing a ratio of n times said volatility, adjusting this ratio in each index.

In this study, we try to unite the directions provided by both strands so that we will be able to predict when a fall in the index prices will occur, that is, when the index will achieve its maximum.

4 Results

Results obtained in these tests have proven that when the average correlation of companies forming an index suffers a fall, placing itself below 0.4 or 0.5 (depending on the index), and this fall is accompanied by a significant increase of the correlation deviation, the price tends to range at around 70% reliability.

We could benefit from this significant reliability through, for example, the use of financial options that would allow us to use strategies such as "long straddle" or "long butterfly", which will result in benefits of substantial volatility. These strategies consists of buying call and put options in order to benefit from price changes in spite of not knowing their direction. Through the use of the number of nodes in the network, the rotation direction can be established. This would establish the hypothesis mentioned above, that is, when the number of nodes decreases in a substantial manner, the networks has few leaders, and if those suffer any fall, the index might fall as well. By mean of this hypothesis, we are looking for a rotation direction beforehand, a downward rotation. Thus, combining the reduction of the number of nodes in the network with a decrease in the mean correlation below 0.4 or 0.5, according to the index, accompanied by a significant increase of the correlation deviation, we can establish a criterion that anticipates a significant fall in the price of the index with a high reliability. In this sense, this is the main contribution of this study.

Start Date	End Date	Mean Distance	Deviation of Distance	Max Correlation	Mean Correlation	Deviation of correlations	Nodes	Mean Centrality	Radio	Diameter	Index Price	Index Volume
13/03/2007	28/03/2007	0.90	0.31	0.97	0.65	0.21	5	4.80	0.23	1.65	5552.69	1731229200
11/08/2008	22/08/2008	0.87	0.38	0.98	0.65	0.30	5	4.40	0.21	1.83	4400.45	1216451900
30/03/2010	14/04/2010	1.04	0.30	0.97	0.53	0.25	5	4.80	0.23	1.64	4057.70	
15/10/2010	29/10/2010	1.21	0.29	0.95	0.27	0.34	5	4.60	0.30	1.85	3833.50	1281326600
28/10/2011	11/11/2011	0.76	0.35	1.00	0.77	0.16	5	6.60	0.00	1.49	3149.38	2178432600
31/10/2011	14/11/2011	0.78	0.35	1.00	0.75	0.17	5	6.40	0.00	1.46	3108.95	2122820200
20/02/2014	07/03/2014	1.07	0.28	0.92	0.49	0.24	5	4.40	0.40	1.61	4366.42	1503118600
24/02/2015	11/03/2015	1.09	0.28	0.96	0.45	0.28	5	4.40	0.30	1.79	4997.75	716054000
09/07/2007	23/07/2007	0.96	0.32	0.96	0.58	0.25	6	4.33	0.27	1.70	6009.16	1359213000
16/07/2007	31/07/2007	0.90	0.30	0.97	0.65	0.19	6	4.00	0.24	1.53	5751.08	1815822800
15/08/2007	30/08/2007	0.90	0.29	0.97	0.65	0.18	6	4.50	0.23	1.42	5592.53	1770437700
13/09/2007	28/09/2007	1.01	0.31	0.95	0.52	0.28	6	4.00	0.30	1.82	5715.69	2084068000
23/06/2008	04/07/2008	1.09	0.33	0.97	0.42	0.35	6	4.50	0.23	1.78	4266.00	1843443500
30/06/2008	14/07/2008	0.98	0.30	0.98	0.56	0.24	6	3.83	0.18	1.73	4142.53	1992200800
09/07/2008	24/07/2008	1.02	0.32	0.95	0.51	0.31	6	4.00	0.32	1.81	4347.99	2469601500
08/08/2008	22/08/2008	0.89	0.38	0.97	0.63	0.32	6	3.83	0.25	1.83	4400.45	1375416900
15/06/2009	30/06/2009	1.04	0.31	0.97	0.52	0.26	6	4.00	0.26	1.66	3140.44	1646311900
17/06/2009	02/07/2009	0.92	0.33	0.97	0.66	0.20	6	4.33	0.24	1.63	3116.41	1672818400
27/07/2009	11/08/2009	1.20	0.25	0.92	0.31	0.29	6	4.17	0.40	1.67	3456.18	1635297200
08/02/2010	22/02/2010	1.20	0.30	0.93	0.30	0.35	6	4.33	0.38	1.88	3756.70	1777480900
16/06/2010	01/07/2010	0.83	0.34	0.97	0.75	0.14	6	4.50	0.25	1.41	3339.90	1853220000
23/06/2010	08/07/2010	0.75	0.36	1.00	0.77	0.19	6	5.50	0.00	1.99	3538.25	1756330300
02/08/2010	13/08/2010	1.00	0.31	0.96	0.54	0.26	6	4.17	0.27	1.61	3610.91	1197627000
12/08/2010	27/08/2010	0.92	0.30	0.96	0.62	0.21	6	4.33	0.27	1.49	3507.44	1364284000
12/10/2010	27/10/2010	1.19	0.29	0.94	0.29	0.34	6	4.33	0.36	1.88	3815.77	1460896100
17/01/2011	28/01/2011	1.21	0.29	0.92	0.27	0.35	6	3.50	0.40	1.86	4002.32	1468088800
08/02/2011	23/02/2011	1.27	0.24	0.92	0.20	0.30	6	3.83	0.40	1.86	4013.12	2104246600
24/10/2011	04/11/2011	0.72	0.35	1.00	0.81	0.13	6	4.17	0.09	1.41	3123.55	1955005200
24/10/2011	08/11/2011	0.76	0.33	0.97	0.78	0.13	6	4.50	0.26	1.41	3143.30	2339387400
28/11/2011	13/12/2011	0.87	0.35	0.97	0.66	0.26	6	3.33	0.26	1.58	3078.72	1893805400
07/12/2011	22/12/2011	0.82	0.35	0.97	0.71	0.22	6	4.67	0.26	1.49	3071.80	1761057800
09/05/2012	24/05/2012	1.00	0.35	0.97	0.49	0.36	6	3.67	0.26	1.82	3038.25	2201108600
16/08/2012	31/08/2012	1.16	0.29	0.95	0.32	0.33	6	4.17	0.31	1.91	3413.07	1341289600
15/10/2012	26/10/2012	1.16	0.32	0.95	0.33	0.38	6	3.67	0.30	1.92	3435.09	1616097400
25/10/2012	09/11/2012	1.14	0.30	0.94	0.36	0.34	6	3.83	0.35	1.87	3423.57	1630735000
21/11/2012	06/12/2012	1.20	0.24	0.90	0.30	0.28	6	4.17	0.45	1.88	3601.65	1712701400
27/12/2012	11/01/2013	1.08	0.29	0.97	0.45	0.29	6	3.50	0.26	1.80	3706.02	
16/05/2013	31/05/2013	1.08	0.28	0.92	0.48	0.25	6	3.83	0.41	1.78	3948.59	1120847400
02/12/2013	16/12/2013	1.02	0.34	0.97	0.52	0.31	6	4.33	0.25	1.84	4119.88	989606800
24/02/2014	10/03/2014	1.05	0.30	0.97	0.51	0.25	6	4.00	0.25	1.63	4370.84	1407298300
23/06/2014	04/07/2014	1.22	0.29	0.97	0.27	0.35	6	3.83	0.25	1.80	4468.98	1044701200
21/05/2015	05/06/2015	0.97	0.31	0.95	0.60	0.24	6	4.33	0.32	1.76	4920.74	1249282900
04/08/2015	19/08/2015	0.86	0.33	0.97	0.68	0.20	6	4.33	0.23	1.51	4884.10	1056702700
01/09/2015	16/09/2015	0.81	0.33	0.98	0.73	0.17	6	3.83	0.21	1.41	4645.84	1201989800
11/09/2015	25/09/2015	0.75	0.40	0.98	0.75	0.32	6	3.83	0.20	1.83	4480.66	1355045600
09/11/2015	24/11/2015	0.97	0.34	1.00	0.53	0.34	6	5.33	0.07	1.97	4820.28	1218475400
15/02/2016	01/03/2016	1.00	0.28	0.96	0.51	0.25	6	4.17	0.28	1.70	4406.84	1347311700
27/06/2016	11/07/2016	0.84	0.30	0.97	0.68	0.20	6	4.33	0.23	1.48	4264.53	1270292000
28/06/2016	13/07/2016	0.91	0.29	0.96	0.61	0.22	6	4.00	0.28	1.52	4335.26	1385686700
31/08/2016	15/09/2016	1.01	0.28	0.93	0.51	0.26	6	4.17	0.37	1.74	4373.22	1057363300

Fig. 3 Extract of data IBEX35 index between 2007 and 2017

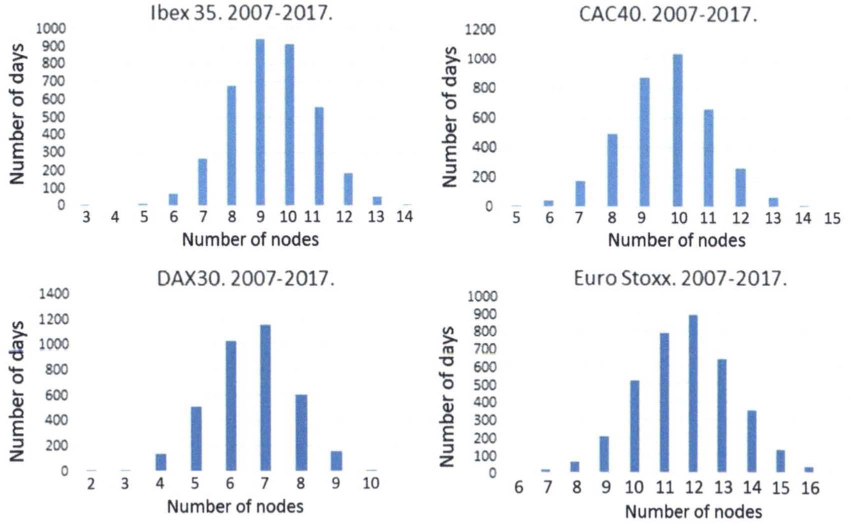

Fig. 4 Distribution of the number of nodes between 2007 and 2017 in IBEX35, CAC40, DAX30 and Euro Stoxx50

In order to carry out several tests, the following stock indexes have been studied: IBEX35, DAX30, CAC40 and Euro Stoxx50, between years 2007 and 2017.

Once a study about a specific index and its constituent companies has been carried out, the results are presented in three different ways so that each one of them provides a different kind of information.

- Graph of the disposition of the companies: in the same way that can be seen in the previous Fig. 1.
- Graph showing the price of the index, the average correlation of its constituting companies, the average deviation of the correlation, the index's traded volume, and its volatility. It can be seen in an example in Fig. 2.
- Spreadsheet with the following information: average distance, distance deviation, maximum correlation, average correlation, correlation deviation, correlation variation, number of nodes, average centrality, radius, diameter, closing price, total volume, price variation and volume variation. It can be seen in an example in Fig. 3.

After carrying out the study about the different indexes, it is clear that the number of nodes belonging to the network in each moment is similar to a normal distribution. Therefore, it can be observed how those are distributed in IBEX35, CAC40, DAX30 and Euro Stoxx50 (Fig. 4).

In order to carry out this tests, the days when the number of nodes is at minimum have been analyzed. The way in which the average correlation and the correlation deviation behave in each case has also been studied.

In the case of IBEX35, we have studied the days when the number of nodes is 3, 5 and 6. In the case of CAC40, we have studied the days in which the number of nodes is 5. In the case of DAX30 we have studied the days in which the number of nodes is 2 and 3, and in the case of Euro Stoxx50 we have studied the days in which the number of nodes is 6 and 7. In total, 145 cases have been analyzed.

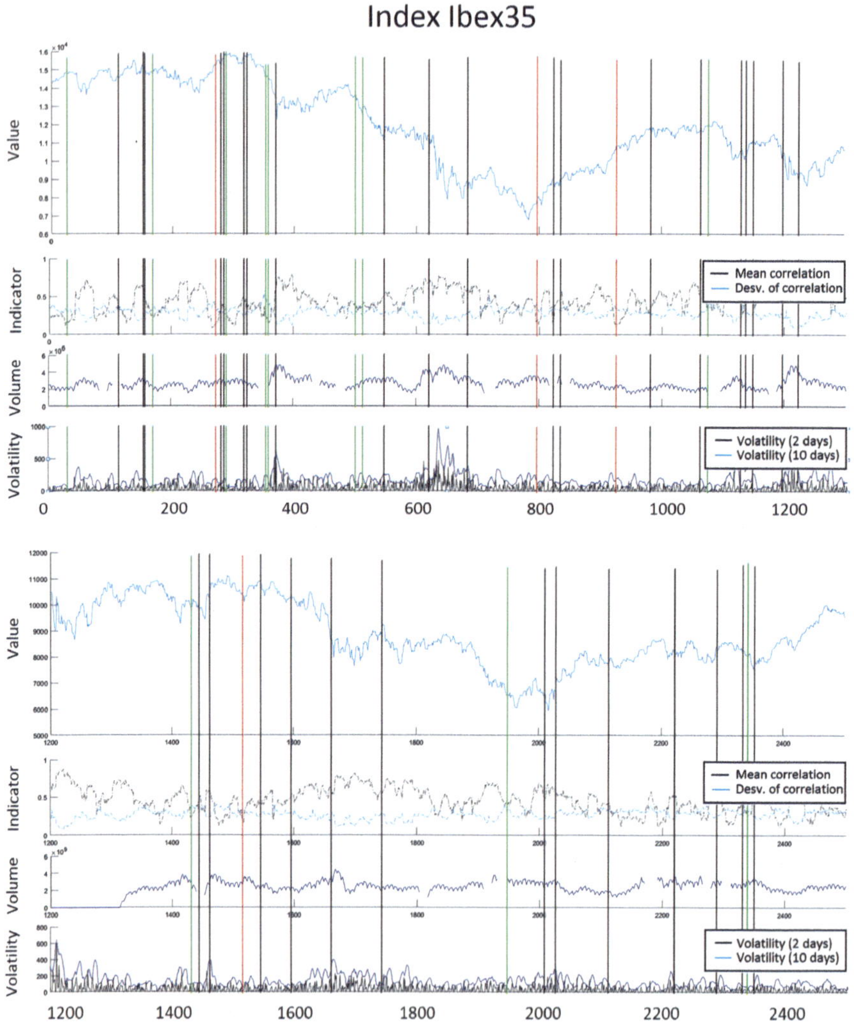

Fig. 5 Results for the index. Time interval is split into two parts to improve display

Table 2 Details of the index between 2007 and 2017 according to the number of nodes in the networks

IBEX35					
Number of nodes	Mean correlation	Mean correlation deviation	Mean distance	Mean distance deviation	Mean maximum correlation
3	0.548	0.272	1.272	0.334	1.000
5	0.408	0.319	1.124	0.313	0.960
6	0.421	0.298	1.149	0.294	0.953

IBEX35
In the IBEX35, the average number of nodes is 9, but it decreases to the point of reaching 3 in the extreme cases.

We have analyzed the characteristics of the index network, observing how the average correlation and the average correlation deviation varies these days. Our conclusion is that out of 64 days in which the number of nodes was 3, 4 or 5, in 16 occasions the average correlation decreased significantly, coinciding with an increase in the deviation. And in those 16 occasions, the price fell on 12 of them. This translates into 75% reliability.

In Fig. 5 and Table 2, we can observe that each vertical line corresponds to a day in which the number of nodes was at its minimum, out of which the ones in black are the ones that did not coincide with a fall in the average correlation and an increase in its deviation. The green lines are the ones in which, both facts coinciding, a fall in the price happened, and the red ones are the ones in which no fall in the prices happened depite both facts coinciding.

CAC40
In the CAC40, the average number of nodes is 10, varying between 5 and 15. For the purposes of this study the days in which it decreases down to 5 and 6 have been selected. Out of the 48 cases studied, 13 occasions show a decrease in the correlation and an increase in the correlation deviation. Prices fell in 8 occasions, which translates into a 62% reliability. In Fig. 6 and Table 3, we can observe a part of the result obtained.

DAX30
DAX30 has an average of 7 nodes, varying between 2 and 11. In this case, the days with 2 and 3 nodes have been studied, a fact that has been repeated in 11 occasions, out of which 6 have coincided with a fall in the average correlation and an increase in its deviation. Out of those 6, 4 have seen a fall in the prices, so we can obtain

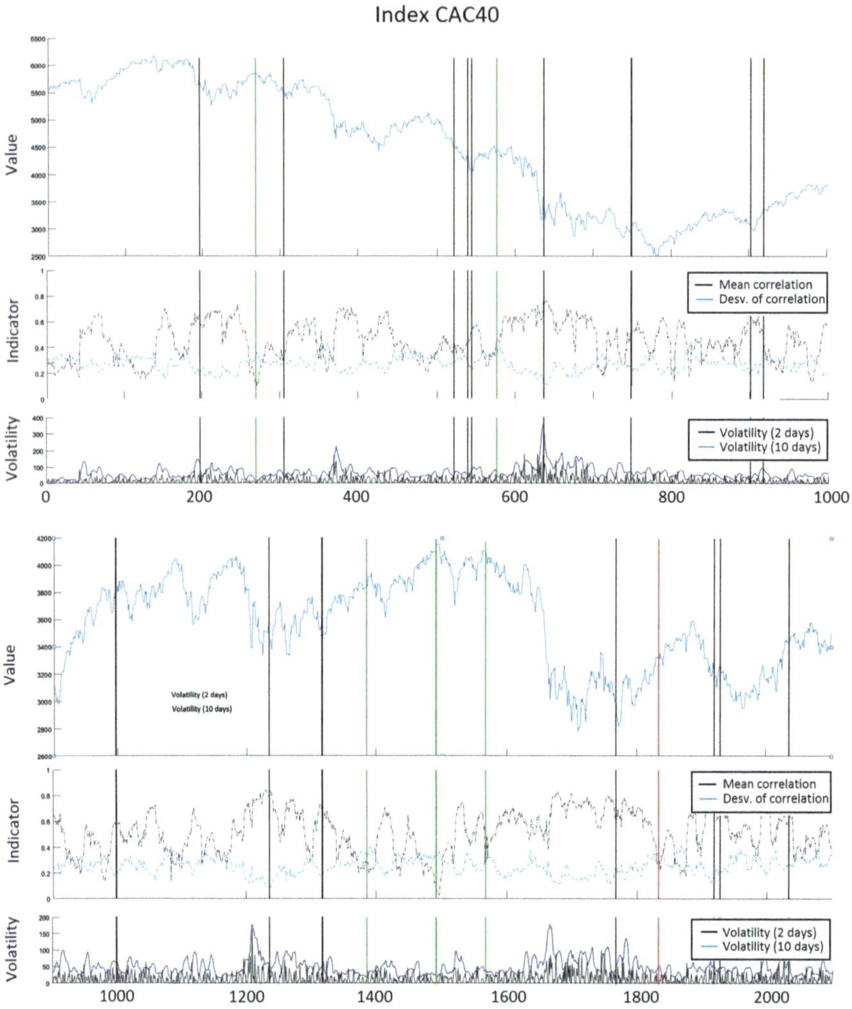

Fig. 6 Results for the index. Time interval is split into two parts to improve display

Table 3 Details of the CAC40 index between 2007 and 2017 according to the number of nodes in the networks

CAC40					
Number of nodes	Mean correlation	Mean correlation deviation	Mean distance	Mean distance deviation	Mean maximum correlation
5	0.570	0.244	0.963	0.318	0.969
6	0.522	0.267	1.998	0.312	0.959

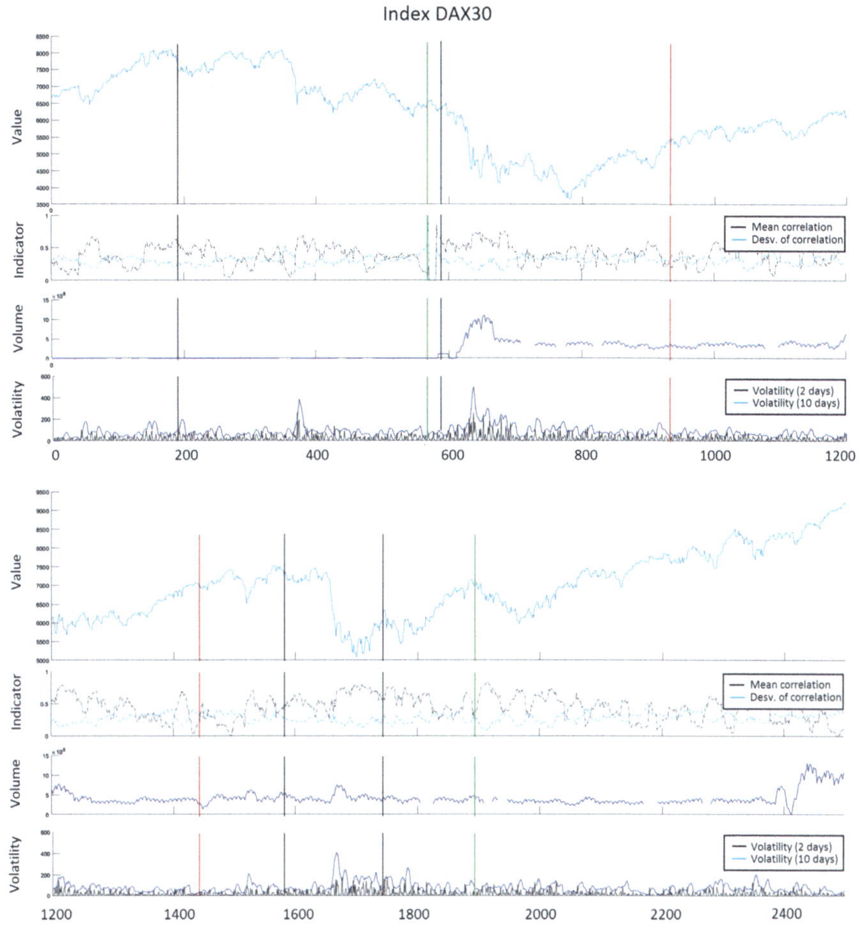

Fig. 7 Results for the DAX30 index. Time interval is split into two parts to improve display

a 67% reliability. In this index we have noticed a delay of a few days between the signal of the nodes fall and the decrease in the average correlation (Fig. 7 and Table 4).

Euro Stoxx50

In order to prove this study's consistency, a test on the Euro Stoxx50 has been carried out. Due to the fact that it is constituted by companies located in different countries, it has some characteristics that are different from the indexes discussed above. In this case, we have analyzed 21 days in which the number of nodes decreased to 6 or 7. Out of those, 7 times correspond to a fall in the average correlation and an increase in its deviation. A fall in the prices happened in 5 occasions, which translates into a 71% reliability (Fig. 8 and Table 5).

Table 4 Details of the DAX30 index between 2007 and 2017 according to the number of nodes in the networks

DAX30					
Number of nodes	Mean correlation	Mean correlation deviation	Mean distance	Mean distance deviation	Mean maximum correlation
2	0.380	0.559	0.917	0.554	1.000
3	0.472	0.273	1.983	0.271	0.948

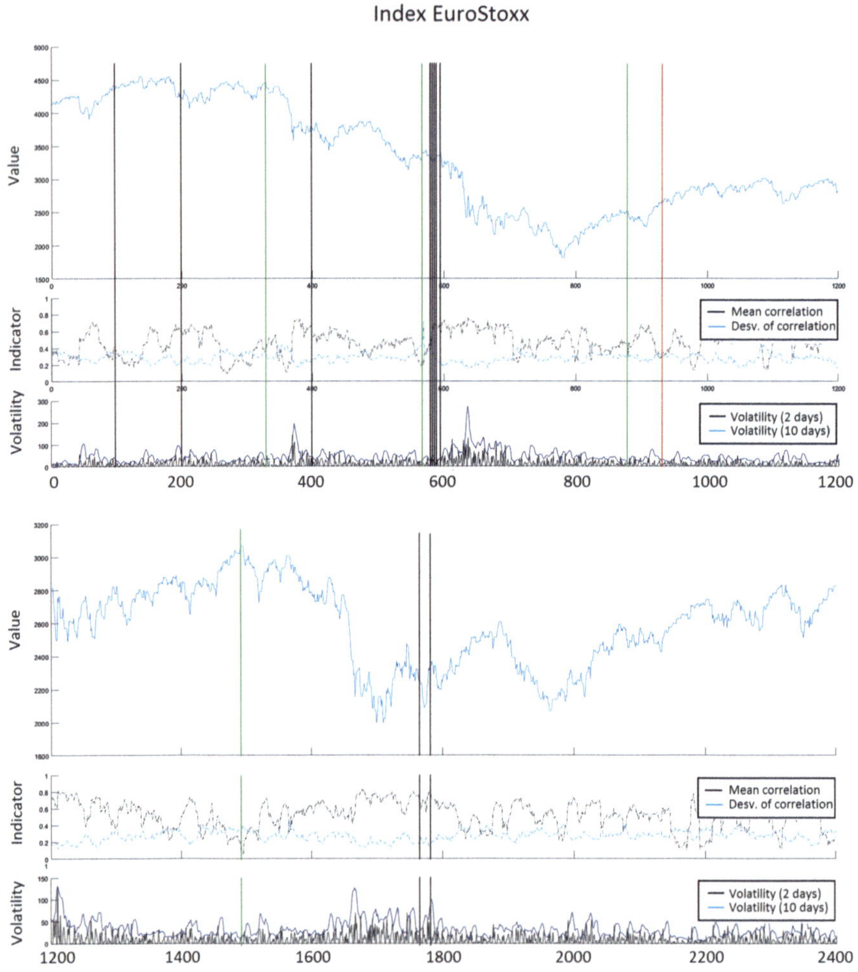

Fig. 8 Results for the Euro Stoxx50 index. Time interval is split into two parts to improve display

Table 5 Details of the Euro Stoxx50 index between 2007 and 2017 according to the number of nodes in the networks

Euro Stoxx50					
Number of nodes	Mean correlation	Mean correlation deviation	Mean distance	Mean distance deviation	Mean maximum correlation
6	0.480	0.381	0.906	0.383	0.987
7	0.547	0.289	1.884	0.307	0.977

Table 6 Summary of results

Summary of results				
Stock index	Cases in the reduction of nodes	Cases in which the average correlation decreased and the correlation deviation increased	Fall in the prices	Reliability (%)
IBEX35	64	16	12	75
CAC40	48	13	8	62
DAX30	11	6	4	67
Euro Stoxx50	21	7	5	71

From the four indexes analyzed, we have obtained that 42 out of the 144 cases that have been studied, the decrease in the number of nodes coincided with a decrease in the average correlation of the companies constituting the index, and in 29 occasions a fall in the prices of the corresponding index happened, which translates into a 70% reliability (Table 6).

5 Discussion and Conclusions

In this paper, we study the structure of different financial systems as moving networks offering information on the correlations and behaviors of the companies that form that network.

Physics has contributed to the study of financial markets in different ways, such as the phase transitions, which in this case can be applied to prevent possible changes in the trend of asset prices. This allows us to interpret them as the end of a trend or market phase by using models that are at the forefront in physical studies.

The different financial markets are interconnected through a network. In the theory of networks, stock values are connected to each other, according to their degree of correlation. This theory's main goal is to identify market leaders and predict the values that will set behavioral patterns in other markets. The study is based on a series of mathematical filters used to detect who are the leaders and followers of a stock market, according to the quotation of each value. In this research, we have focused on how the network behaves in the moment right before a fall in the prices of the index. In order to do this, we have focused on the use of the network as our starting point obtaining both the correlations of the companies and the existing nodes in each moment in the network.

We have studied the networks constituted by the companies included in several indexes (IBEX35, CAC40, DAX30 and Euro Stoxx50), in such a way that the Minimal Spanning Tree (MST) and the correlated behavior between all companies that make up them, have been able to be observed. Therefore, we can obtain the average correlation between them and the average deviation of said correlation. After that, we have studied the disposition of the companies from the point of view of a network. If this network is made up of few leading nodes (those having more than two links), it means that the index leadership fall on a few number of companies, and this implies that if there is a few number of leading companies and they happen to suffer a fall, the whole index can be affected, since the other companies are following the leaders.

Using daily closing price series from 2007 to 2017 for these main European indexes, we have obtained that in 42 of the 144 cases that have been studied, the decrease in the number of nodes coincided with a decrease in the average correlation of the companies constituting the index, and that in 29 occasions a fall in the prices of the corresponding index happened. Hence, we can point out that when the number of leading nodes is reduced, and the average correlation of companies forming an index suffers a fall, placing itself below 0.4–0.5 (depending of the index), and this fall is accompanied by a significant increase of the correlation deviation, then the price tends to fall at around 70% of reliability.

We think that our findings provide empirical evidence on the determinants to understand the behavior of the financial market through their indexes moving networks and have relevant trading implication for investors, hedgers and speculators. On the other hand, this criterion could be complemented by other indicators or applied in other systems than financial markets, described by empirical time series. We believe that this is an interesting line of research that is still in its beginnings and we are sure that it will produce surprising results and conclusions in the near future.

Acknowledgements AC acknowledges Junta de Andalucía (Spain) by partially funding to his research group (FQM-122).

References

1. Alkan, S., Khashanah, K.: Structural evolution of the stock networks. In: 11th International Conference on Signal-Image Technology and Internet-Based Systems, pp. 406–412. IEEE (2015)
2. Bak, P.: How Nature Works: The Science Ofself-organized Criticality. Springer, New York, USA (1999)
3. Bardoscia, M., Battiston, S., Caccioli, F., Caldarelli, G.: Pathways towards instability in financial networks. Nat. Commun. **8**, 14416 (2017)

4. Barrio, R.A., Govezensky, T., Ruiz-Gutierrez, E., Kaski, K.: Modelling trading networks and the role of trust. Phys. A **471**, 68–79 (2016)
5. Bauch, C.T., Sigdel, R., Pharaon, J., Anand, M.: Early warning signals of regime shifts in coupled human-environment systems. Proc. Natl. Acad. Sci. USA **113**, 14560–14567 (2016)
6. Bonanno, G., Mantegna, R.N.: Networks of equities in financial markets. Eur. Phys. J. B **38**, 363–371 (2004)
7. Carlsson, G., Memoli, F.: Characterization, stability and convergence of hierarchical clustering methods. J. Mach. Learn. Res. **11**, 1425–1470 (2010)
8. Channgam, S., Sae-Tang, A., Termsaithong, T.: A prediction method for large-size event occurrences in the sandpile model. Int. J. Math. Comp. Phys. Elec. Comp. Eng. **10**, 255–258 (2016)
9. Cimini, G., Tiziano, S., Garlaschelli, D., Gabrielli, A.: Systemic risk analysis on reconstructed economic and financial networks. Sci. Rep. **5**, 15758 (2015)
10. Donnat, P., Marti, G., Very, P.: Toward a generic representation of random variables for machine learning. Pattern Recognit. Lett. **70**, 24–31 (2016)
11. Epps, T.W.: Comovements in stock prices in the very short run. J. Am. Stat. Assoc. **74**, 291–298 (1979)
12. Fiedor, P.: Information-theoretic approach to lead-lag effect on financial markets. Eur. Phys. J. B **87**(8), 168 (2014)
13. Fiedor, P.: Networks in financial markets based on the mutual information rate. Phys. Rev. E **89**, 052801 (2014)
14. Fiedor, P.: Sector strength and efficiency on developed and emerging financial markets. Phys. A **413**, 180–188 (2014)
15. Filisetti, A., Villani, M., Roli, A., Fiorucci, M., Serra, R.: Exploring the organization of complex systems through the dynamical interaction among their relevant subsets. In: Proceedings of the European Conference on Artificial Life 2015—ECAL2015, pp. 286–293. MIT Press eBooks, Cambridge, MA, USA (2015)
16. Gao, J., Barzel, B., Barabasi, A.L.: Universal resilience patterns in complex networks. Nature **530**, 307–312 (2016)
17. Heiberger, R.H.: Shifts in collective attention and stock networks. In: Thai, T., Nam, P., Nguyen-Huawei, S. (eds.) International Conference on Computational Social Networks, LNCS, vol. 9197, pp. 296–306. Springer (2015)
18. Huang, F., Gao, P., Wang, Y.: Comparison of prim and kruskal on shangai and shenzhen 300 index hierarchical structure tree. In: Thai, T., Nam, P., NguyenHuawei, S. (eds.) International Conference on Systems and Mining, pp. 139–190. WISM IEEE, Shangai, China (2009)
19. King, B.F.: Market and industry factors in stock price behavior. J. Bus. **39**, 139–190 (1966)
20. Lemieux, V., Rahmdel, P.S., Rick Walker, R., Wong, B., Flood, M.: Clustering techniques and their effect on portfolio formation and risk analysis. In: Proceedings of the International Workshop on Data Science for Macro-Modeling, pp. 1–6. ACM, New York, NY, USA (2014)
21. Levine, J.H.: The sphere of influence. Am. Sociol. Rev. **37**(1), 14–27 (1972)
22. Lima Dias, R.F.: Monitoring Evolving Stock Networks. https://repositorio-aberto.up.pt/bitstream/10216/80783/2/36789.pdf (2015)
23. Mantegna, R.N.: Hierarchical structure in financial markets. Eur. Phys. J. B **11**, 193 (1999)
24. Mantegna, R.N., Stanley, H.E.: Introduction to Econophysics: Correlation and Complexity in Finance. Cambridge University Press, Cambridge, UK (1999)
25. Marti, G., Binkowski, M., Donnat, P.: A review of two decades of correlations, hierarchies, networks and clustering in financial markets. http://arxiv.org/pdf/1703.00485.pdf (2017)
26. Musmeci, N., Aste, T., Di Matteo, T.: Relation between financial market structure and the real economy: comparison between clustering methods. PLOS One **10**(4), e0126998 (2015)
27. Peralta, G.: Three essays on network theory applied to capital markets. Ph.D. thesis, Universidad Carlos III de Madrid (2016)
28. Ren, F., Lu, Y.N., Li, S.P., Jiang, X.F., Zhong, L.X., Qiu, T.: Dynamics portfolio strategy using clustering approach. PLoS ONE **12**, e0169299 (2017)
29. Roli, A., Villani, M., Caprari, R., Serra, R.: Identifyng critical states through the relevance index. Entropy **19**, 73 (2017)

30. Sandoval, L., Mullokandov, A., Kenett, D.Y.: Dependency relation among international stock market indices. J. Risk Financ. Manag. **8**, 227–265 (2015)
31. Scheffer, M., Bascompte, J., Brock, W.A., Brovkin, V., Carpenter, S.R., Dakos, V., Held, H., van Nes, E.H., Rietkerk, M., Sugihara, G.: Early-warning signals for critical transitions. Nature **461**, 53–59 (2009)
32. Scheffer, M., Carpenter, S.R., Lenton, T.M., Bascompte, J., Brock, W., Dakos, V.: Anticipating critical transitions. Science **338**, 344–348 (2012)
33. Siripurapu, A.: Convolutional networks for stock trading. Stanford University Department of Computer Science. Technical Report (2015)
34. Song, D.M., Tumminello, M., Zhou, W.X., Mantegna, R.N.: Evolution of worldwide stock markets, correlation structure, and correlation-based graphs. Phys. Rev. E **84**, 026108 (2011)
35. Song, W.M., Di Matteo, T., Aste, T.: Hierarchical information clustering by means of topologically embedded graphs. PLOS One **7**, e31929 (2012)
36. Sornette, D.: Physics and financial economics (1976–2014): puzzles, ising and agent-based models. Rep. Prog. Phys. **77**, 062001 (2014)
37. Tse, C.K., Liu, J., Lau, F.C.M.: A network persperctive of the stock market. J. Empir. Financ. **17**, 659–667 (2010)
38. Tumminello, M., Aste, T., Di Matteo, T., Mantegna, R.N.: A tool for filtering information in complex systems. Proc. Natl. Acad. Sci. USA **102**, 10421–10426 (2005)
39. Villani, M., Roli, A., Filisetti, A., Fiorucci, M., Poli, I., Serra, R.: The search for candidate relevant subsets of variables in complex systems. Artif. Life **21**, 395–397 (2015)

On the Numerical Approximation to Generalized Ostrovsky Equations: I

A Numerical Method and Computation of Solitary-Wave Solutions

Ángel Durán

Abstract In the present chapter, two numerical procedures to simulate the dynamics of generalized versions of the Ostrovsky equation are presented. First, a numerical method to approximate the corresponding periodic initial-value problem is introduced. The scheme consists of a spatial discretization based on Fourier collocation methods, which is justified by the presence of nonlocal terms. Due to the stiff character of the semidiscretization in space, the time integration is performed with a fourth-order, diagonally implicit Runge-Kutta method, which provides additional theoretical and computational properties. The second point treated in this chapter concerns the solitary-wave solutions of the equations. Their numerical generation is carried out by using a Petviashvili-type method, along with acceleration techniques. The resulting procedure is able to compute both classical and generalized solitary waves in an efficient way. The speed-amplitude relation and the asymptotic behaviour of the waves are studied from the computed profiles.

Keywords Generalized Ostrovsky equation · Fourier collocation
Petviashvili-type methods · Solitary waves

1 Introduction

There are many references about the Ostrovsky equation (also called the Rotation-Modified Korteweg-de Vries equation or RMKdV)

$$(u_t - \beta u_{xxx} + uu_x)_x = \gamma u, \quad x \in \mathbb{R}, \tag{1}$$

from the original paper by Ostrovsky [45]. In (1) β and γ are constant parameters. Equation (1) was proposed as a one-dimensional model for the propagation of gravity waves of small amplitude in a rotating fluid in a horizontal channel. If the variables x

Á. Durán (✉)
Applied Mathematics Department, University of Valladolid, 47011 Valladolid, Spain
e-mail: angel@mac.uva.es

V. Carmona et al. (eds.), *Nonlinear Systems, Vol. 1*,
Understanding Complex Systems, https://doi.org/10.1007/978-3-319-66766-9_12

and t are proportional to distance along the channel and time respectively, then $u(x,t)$ represents the fluid velocity in the x-direction. A way to understand (1), adopted by several references, [15, 25, 26, 29] (see also the introduction in [44]) is starting from the KdV approximation (the KdV equation appears in (1) when $\gamma = 0$) that includes the balance between the small-scale Boussinesq dispersion, proportional to β, and the nonlinear effects, represented by the quadratic term. Then, Eq. (1) incorporates in the balance the dispersion due to a rotation (large-scale Coriolis dispersion), proportional to γ and comparable to the previous effects. We assume for simplicity that γ, the physical parameter that measures the effect of Earth's rotation, is positive. Then the type of dispersion is determined by the sign of the parameter β. With the minus sign in the corresponding term in (1), the equation with $\beta < 0$ (negative dispersion) models surface and internal waves in the ocean and surface waves in a shallow channel with uneven bottom, [5, 23, 30, 45]. When $\beta > 0$ (positive dispersion), then (1) is applied to model capillary waves on the surface of liquid or oblique magneto-acoustic waves in plasma, [20–22]. Some other references for (1) are [4, 24, 27, 46]. The list is far from being complete.

The present chapter is concerned with a generalized version of (1) of the form

$$(u_t - \beta u_{xxx} + f(u)_x)_x = \gamma u, \quad x \in \mathbb{R} \tag{2}$$

where f is a twice continuously differentiable, real-valued and homogeneous function of degree $p \geq 2$ in the sense that

$$f(\lambda s) = \lambda^p f(s), \quad \lambda > 0, \quad s \in \mathbb{R}. \tag{3}$$

According to [39], condition (3) implies that f can be written in the form

$$f(u) = a_e |u|^p + a_o |u|^{p-1} u, \quad a_e, a_o \in \mathbb{R}. \tag{4}$$

The particular case $f(s) = \pm|s|^p$ is emphasized since for $p = 2$ this corresponds to the Ostrovsky equation (1). The relevance of (2), (4) as mathematical model includes the propagation of internal waves in the ocean with cubic nonlinearities, see [20] (and some references therein, like [37, 52, 53]); see also [15].

We first summarize some mathematical properties of (2). Local in time well-posedness of (1) is proved (see [60]) in the spaces

$$X_s = \{f \in H^s(\mathbb{R}) / \mathscr{F}^{-1}\left(\frac{\mathscr{F}f(\xi)}{\xi}\right) \in H^s(\mathbb{R})\},$$

with norm $||f||_{X_s} = ||f||_s + \left|\left|\mathscr{F}^{-1}\left(\frac{\mathscr{F}(f)(\xi)}{\xi}\right)\right|\right|_s$, where $s > 3/2$; the norm in the Sobolev space $H^s = H^s(\mathbb{R})$ is given by

$$||f||_s = \left(\int_{-\infty}^{\infty} (1+\xi^2)^s |\mathscr{F}(f)(\xi)|^2 d\xi\right)^{1/2},$$

where $\mathscr{F}$ stands for the Fourier transform, defined on $H^0 = L^2(\mathbb{R})$,

$$\mathscr{F}(f)(\xi) = \int_{-\infty}^{\infty} e^{-i\xi x} f(x)dx, \quad f \in H^0,$$

and with $\mathscr{F}^{-1}$ as the inverse Fourier transform.

Related results of existence and well-posedness can be seen in [33, 41, 59]. In [41], Linares and Milanés study the initial-value problem for (1) obtaining local well-posedness in $\{f \in H^s(\mathbb{R})/\partial_x^{-1} f \in L^2(\mathbb{R})\}$ for $s > 3/4$ and a global result for the case $s = 1$, $\beta\gamma > 0$. The operator ∂_x^{-1} is defined by using the Fourier symbol as

$$\mathscr{F}(\partial_x^{-1} f)(\xi) = (i\xi)^{-1}\mathscr{F}(f)(\xi), \quad \xi \in \mathbb{R}\backslash\{0\}, \quad \mathscr{F}(\partial_x^{-1} f)(0) = 0, \tag{5}$$

Isaza and Mejía ([33] and references therein) prove local well-posedness in $H^s(\mathbb{R})$ for (1) with $\beta = 1$, with $s > -1/2$ for $\gamma > 0$ and $s > -3/4$ for $\gamma < 0$, along with global well-posedness in $H^s(\mathbb{R})$, $s > -3/10$ for both cases. Finally, in [59], well-posedness is proved in X_s, $s > -3/4$.

On the other hand, Levandosky and Liu [39], refer the study of the associated Cauchy problem to a generalization of the results presented in the literature concerning (1), whose main references may be [33, 41, 60]. Thus, the derivation in [60] of some conservation laws for (1) can be generalized to obtain the zero mass conservation

$$I(u) = \int_{-\infty}^{\infty} u(x,t)dx = 0, \tag{6}$$

along with the preservation of the momentum and the total energy

$$V(u(t)) = \int_{-\infty}^{\infty} u(x,t)^2 dx, \tag{7}$$

$$E(u(t)) = \int_{-\infty}^{\infty} \left(\frac{\beta}{2} u_x(x,t)^2 + \frac{\gamma}{2}(\partial_x^{-1} u(x,t))^2 + F(u(x,t)) \right) dx, \tag{8}$$

for sufficiently smooth and decaying at infinity solutions of (2). In (8), $F' = f$, $F(0) = 0$. Equation (2) can also be written in the Hamiltonian form (see [14] for the case of (1))

$$u_t = \frac{\partial}{\partial x}\frac{\delta E_g}{\delta u},$$

where $\frac{\delta}{\delta u}$ denotes variational derivative. Choudhury and collaborators in [14] prove also the nonintegrable character of (1).

In this chapter particular attention will be paid to solitary-wave solutions of (2). They are solutions u of the form $u(x,t) = \phi_c(x - ct)$ for some profile ϕ_c and speed c. This function ϕ_c must satisfy

$$\left(-c\phi_c' - \beta\phi_c''' + f(\phi_c)'\right)' - \gamma\phi_c = 0, \tag{9}$$

which can be written as a fourth-order ordinary differential equation

$$\phi_c^{(iv)} - Q\phi_c'' + P\phi_c = -\frac{1}{\beta} f(\phi_c)'', \tag{10}$$

where $P = \gamma/\beta$, $Q = -c/\beta$.

Some results on the existence of these waves are now reviewed. Equation (10) is analyzed in [13, 14] for the case of the classical Ostrovsky equation ($f(u) = u^2$) by using normal form theory, [32, 43]. This leads to the existence of several types of traveling wave solutions of (1). This paper will be focused on classical and generalized solitary waves. Note that Eq. (10) can be written as a first-order differential system

$$U' = V(U, c, \gamma, \beta) = LU + R(U), \tag{11}$$

where $U = (\phi_c, \phi_c', \phi_c'', \phi_c''')^T$ and

$$L := L(c, \gamma, \beta) = \begin{pmatrix} 0 & 1 & 0 & 0 \\ 0 & 0 & 1 & 0 \\ 0 & 0 & 0 & 1 \\ -P & 0 & Q & 0 \end{pmatrix}, \quad R := R(U, c, \beta) = \begin{pmatrix} 0 \\ 0 \\ 0 \\ -\frac{1}{\beta} f(U)'' \end{pmatrix}.$$

The normal form theory establishes that under certain conditions the qualitative behaviour of the transformed system is determined by the analogous behaviour of the associated linearized system, [6, 32]. In this case, the linearization of (11) at the equilibrium point $U = 0$ leads to the linear system $U' = LU$. The eigenvalues λ of L will satisfy

$$\lambda^4 - Q\lambda^2 + P = 0.$$

Four regions in the (Q, P) plane where L has different spectrum can be derived, [13]. They are separated by four curves, see Fig. 1:

- $C_0 = \{(Q, P)/P = 0, Q > 0\}$, where $\lambda = 0$ is a double eigenvalue and additionally there are two simple, real eigenvalues.
- $C_1 = \{(Q, P)/P = 0, Q < 0\}$, where $\lambda = 0$ is a double eigenvalue and additionally there are two simple, pure imaginary eigenvalues.
- $C_2 = \{(Q, P)/P > 0, Q = -2\sqrt{P}\}$, where there are two double, pure imaginary eigenvalues.
- $C_3 = \{(Q, P)/P > 0, Q = 2\sqrt{P}\}$, where there are two double, real eigenvalues.

(At the origin, L has a quadruple eigenvalue $\lambda = 0$.) See [13] to relate the properties of $V(U, c, \gamma, \beta)$ with the question of existence of different types of solitary waves. In particular, the existence of classical solitary wave solutions when $\gamma, \beta > 0$ and $c <$

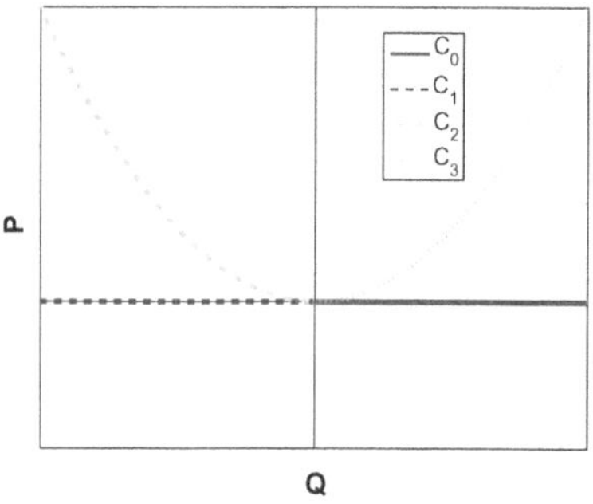

Fig. 1 Regions in the (Q, P) plane where L in (11) has different spectrum, [13, 14]

$2\sqrt{\beta\gamma}$, previously derived in [42] and which also holds in the generalized case (2), [39], corresponds to the region above C_2 and C_3. (The result in [42] on nonexistence of classical solitary waves when $\beta < 0, \gamma > 0$ and $c < \sqrt{140\gamma|\beta|}$ is also included here, as the region below C_0.) On the other hand, Boyd and Chen [8, 12], consider the RMKdV equation

$$(u_t + uu_x + u_{xxx})_x - \varepsilon^2 u = 0, \tag{12}$$

along with the equation satisfied by the 'steadily-translating' solutions $u = u(X)$,

$$-cu_{XX} + u_X^2 + uu_{XX} + u_{XXXX} - \varepsilon^2 u = 0, \tag{13}$$

also called stationary RMKdV equation, where $X = x - ct$. Note that (12) corresponds to (1) for $\gamma = \varepsilon^2 > 0$, $\beta = -1$ and uu_x instead of $2uu_x$. Therefore, for positive speeds c, the solutions of (13) are associated to the region in the (Q, P) plane with $P < 0$, $Q > 0$ (below C_0). Boyd and Chen mention two types of solutions: weakly nonlocal solitary waves (these are the generalized solitary waves, in the sense described e. g. by [43]; they are homoclinic to periodic orbits as $|X| \to \infty$, in contrast with the classical solitary waves of region above C_2 and C_3, that decay to zero at infinity) and spatially periodic solutions. (They call them 'microterons' and 'microteroidal' waves, respectively.) The first have only a single large peak on $(-\infty, \infty)$. They are observed and generated by Hunter [31], of an asymptotic form (for $c = 1$)

$$u \approx 3\text{sech}^2(X/2) - 6\varepsilon\left(\sin(\varepsilon|X|) + \chi\cos(\varepsilon X)\right), \tag{14}$$

with arbitrary χ (which is zero for the microteron wave of minimum amplitude). In [8, 12], the authors extend to matched asymptotic expansions of Hunter to third order and describe the periodic traveling waves in terms of the parameters of the equation. They use a Fourier Galerkin algorithm, thought as both a numerical method and an explicit analytic approximation, see also [7].

We additionally emphasize that the nonexistence of classical solitary wave solutions of the Ostrovsky equation (1) when $\gamma > 0$, $\beta < 0$ (actually, when $\gamma\beta < 0$) was previously obtained in [20, 37]; the first one extends the proof to the generalized case (2) with $f(u) = u^p/p$, $p > 1$. (The proof contains basically the arguments of the theory explained, in a more general way, in [13] and previews the existence of generalized solitary waves.) Finally, concerning the asymptotic decay of the classical solitary waves (when $\beta, \gamma > 0$, $c < 2\sqrt{\beta\gamma}$), Liu and Varlamov [42], for (1) and Levandosky and Liu [39], for (2), (4), prove that the profiles are in fact in $H^\infty(\mathbb{R})$.

In order to study, by computational means, some aspects of the dynamics of (2), the present chapter develops two numerical tools. The first one is a scheme to approximate the corresponding periodic initial-value problem for (2). The method makes use of a Fourier pseudospectral discretization in the spatial variable, which is a natural choice because of the nonlocal terms. Since the semidiscretization leads to a stiff differential system, numerical approximation in time should be carried out with implicit methods. Thus, the spatial discretization is coupled with an implicit, fourth-order accurate time-stepping integrator, based on the Implicit Midpoint Rule. The resulting method is diagonally implicit, satisfies suitable stability and conservation properties and provides the computational advantage of the iterative resolution of the discrete systems for the intermediate stages with the fixed-point algorithm, [19]. Since no explicit solutions of the Ostrovsky-type equations are known, the accuracy of the method is first checked in two classical equations: the generalized KdV equation and the Benjamin-Ono equation. The inclusion of the last one is justified by the aim of studying the influence of the presence of nonlocal terms in the performance of the method. The construction of the fully discrete scheme, some of its properties and tests for accuracy are explained in Sect. 2.

The second issue is concerned with the numerical generation of traveling-wave solutions of (2) and is treated in Sect. 3. The lack of exact formulas for the waves motivates the design of efficient numerical techniques to approximate the corresponding profiles. First, the one considered in this chapter is introduced; it consists of the so-called Petviashvili's method, already used in the literature to this end, see e. g. [20], combined with an acceleration technique based on minimal polynomial extrapolation. The inclusion of the acceleration technique is justified in terms of efficiency: there are some cases where the Petviashvili's method by itself is too slow or directly does not converge. The use of acceleration techniques for these situations was discussed in [3] as an alternative to other procedures presented in the literature, like shooting methods, [40]. The performance of the accelerated Petviashvili's scheme is checked by studying the numerical generation of different types of traveling waves: classical and generalized solitary waves as well as multi-pulse solitary waves (also both classical and generalized). From this accuracy, some properties of the waves, like the speed-amplitude relation or some additional information on the asymptotic behaviour, are suggested from several numerical experiments.

2 The Numerical Method

For simplicity, the 2π periodic initial-valiue problem for (2)

$$(u_t - \beta u_{xxx} + f(u)_x)_x = \gamma u, \quad x \in [-\pi, \pi], t > 0, \tag{15}$$
$$u(x, 0) = u_0(x), \quad x \in [-\pi, \pi],$$

where f is of the form (4) and u_0 is a real-valued, 2π—periodic function, will be considered. For a general period $P = 2l$, (15) can be posed on $(-l, l)$ with the usual change of variable $x = (\pi/l)y$, $y \in [-l, l]$ and the corresponding scaling. We will assume that (15) is well-posed, for some $s > 0$, in

$$X_s(-\pi, \pi) = \{f \in H^s_{per}(-\pi, \pi)/\partial_x^{-1} f \in H^s_{per}(-\pi, \pi)\},$$

where

$$H^s_{per}(-\pi, \pi) = \{f : [-\pi, \pi] \to \mathbb{R}/ \sum_{k\in\mathbb{Z}} (1 + k^2)^s |(\widehat{f})(k)|^2 < \infty\}.$$

For $k \in \mathbb{Z}$, $\widehat{f}(k)$ denotes the k-th Fourier coefficient of f

$$\widehat{f}(k) = \int_{-\pi}^{\pi} e^{-ikx} f(x)dx, \quad k \in \mathbb{Z},$$

and the operator ∂_x^{-1} is defined by using the Fourier coefficients, cf. (5)

$$\widehat{\partial_x^{-1} f}(k) = (ik)^{-1}\widehat{f}(k), \quad k \in \mathbb{Z}\backslash\{0\}, \quad \widehat{\partial_x^{-1} f}(0) = 0.$$

The norm in $H^s_{per}(-\pi, \pi)$ is

$$||f||_{H^s_{per}(-\pi,\pi)} = \left(\sum_{k\in\mathbb{Z}} (1 + k^2)^s |\widehat{f}(k)|^2 \right)^{1/2},$$

while the norm in $X_s(-\pi, \pi)$ is given by

$$||f||_{X_s(-\pi,\pi)} = ||f||_{H^s_{per}(-\pi,\pi)} + ||\partial_x^{-1} f||_{H^s_{per}(-\pi,\pi)}, \quad f \in X_s(-\pi, \pi).$$

Similarly, the quantities

$$I_p(u(t)) = \int_{-\pi}^{\pi} u(x, t)dx = 0, \tag{16}$$

$$V_p(u(t)) = \int_{-\pi}^{\pi} u(x, t)^2 dx, \tag{17}$$

$$E_p(u(t)) = \int_{-\pi}^{\pi} \left(\frac{\beta}{2} u_x(x,t)^2 + \frac{\gamma}{2} (\partial_x^{-1} u(x,t))^2 + F(u(x,t)) \right) dx, \quad (18)$$

are conserved by smooth enough solutions of (15), cf. (6)–(8). In (18), $F' = f$, $F(0) = 0$. The Hamiltonian formulation of (15) also holds. Additionally, $(\cdot, \cdot)$ will denote the usual L^2 inner product in $(-\pi, \pi)$ with $||\cdot||$ standing for the corresponding norm.

2.1 Space Semidiscretization

The description and analysis of the spatial discretization will require, for N a positive integer, to consider the spaces

$$S_N = span\{e^{ikx}, -N \leq k \leq N\},$$

and

$$S_N^0 = \{\phi \in S_N / \widehat{\phi}(0) = \int_{-\pi}^{\pi} \phi(x)dx = 0\} = span\{e^{ikx}, -N \leq k \leq N, k \neq 0\}.$$

Note that, due to periodicity, if $\phi \in S_N$ then $\partial_x \phi \in S_N^0$. Thus, the operator $\partial_x : S_N^0 \to S_N^0$ is invertible.

The semidiscrete Fourier-Galerkin approximation to (15) is defined as a map $u^N : [0, \infty) \to S_N$ such that, for all $\chi \in S_N$,

$$\begin{aligned} &(u_{tx}^N, \chi) + ((-\beta u_{xxx}^N + f(u^N)_x)_x - \gamma u^N, \chi) = 0, \quad t > 0, \\ &u^N(x, 0) = P_N u_0(x), \end{aligned} \quad (19)$$

or, equivalently (due to periodicity)

$$\begin{aligned} &(u_t^N, \chi_x) + ((-\beta u_{xxx}^N + f(u^N)_x), \chi_x) + (\gamma u^N, \chi) = 0, \quad t > 0, \\ &u^N(x, 0) = P_N u_0(x), \end{aligned} \quad (20)$$

where P_N is the orthogonal projection of L^2 onto S_N.

If we choose $\chi = e^{ikx}$, $k = -N, \ldots, N$, $k \neq 0$ then (19) becomes an initial value problem of a differential system for the Fourier coefficients of u^N,

$$\begin{aligned} &(ik)\widehat{u^N}_t(k,t) - (\beta k^4 + \gamma)\widehat{u^N}(k,t) - k^2 \widehat{f(u^N)}(k,t) = 0, \quad t > 0, \\ &\widehat{u^N}(k, 0) = \widehat{u_0}(k). \end{aligned} \quad (21)$$

On the other hand, if we take $\chi = 1$ in (20) we obtain $\widehat{u^N}(0,t) = 0$. Therefore $u^N(\cdot, t) \in S_N^0$, $t > 0$ and an alternative definition of the semidiscrete approximation would be to consider $u^N : (0,\infty) \to S_N^0$ satisfying (20) for all $\chi \in S_N^0$. An equivalent form to (21) is also

$$\begin{aligned}
&\frac{d}{dt}\widehat{u^N}(k,t) = \frac{-i}{k}\left((\beta k^4+\gamma)\widehat{u^N}(k,t) + k^2\widehat{f(u^N)}(k,t)\right), \qquad (22)\\
&t > 0, \quad k = -N, \ldots, N, k \neq 0,\\
&\widehat{u^N}(k,0) = \widehat{u_0}(k), \quad k \in \mathbb{Z},\\
&\widehat{u^N}(0,t) = 0, \quad t \geq 0,
\end{aligned}$$

The form (4) of f implies that the right hand side of (22) is at least locally Lipschitz continuous with respect to the L^2 norm in S_N. Then using standard theory for differential equations we obtain the existence of a unique, local in time solution of (22). Also, standard arguments, [48, 49], prove the existence of a global in time solution if the semidiscretization preserves the L^2 norm. In our case, this is a consequence of the following result.

Lemma 1 *The solution u^N of (19) satisfies, for $t > 0$,*

$$(i) \quad \frac{d}{dt}\int_{-\pi}^{\pi}(u^N(x,t))^2 dx = 0,$$

$$(ii) \quad \frac{d}{dt}\int_{-\pi}^{\pi}\left(\beta(u_x^N(x,t))^2 + F(u^N(x,t)) + \gamma(\partial_x^{-1}u^N(x,t))^2\right)dx = 0,$$

where $F' = f$, $F(0) = 0$.

Remark 1 Since $u^N(t) \in S_N^0$, $t \geq 0$, then u^N satisfies the zero mass conservation property (16). Now Lemma 1 means that the semidiscretization preserves the other two invariants of the periodic problem (17) and (18). Note also that (i) means that $||u^N(\cdot,t)|| = ||P_N u_0||$.

Remark 2 The statement (ii) requires the definition of $v^N = \partial_x^{-1}u^N$. As mentioned above, this has sense because $u^N(t) \in S_N^0$, $t \geq 0$ and $\partial_x : S_N^0 \to S_N^0$ is invertible; in practice v^N is defined via the Fourier coefficients as expected: If

$$u^N(x,t) = \sum_{-N\leq k\leq N}\widehat{u^N}(k,t)e^{ikx},$$

then

$$v^N(x,t) = \sum_{-N\leq k\leq N}\widehat{v^N}(k,t)e^{ikx}, \quad \widehat{v^N}(k,t) = \frac{\widehat{u^N}(k,t)}{ik}, \quad k \neq 0; \quad \widehat{v^N}(0,t) = 0.$$

Proof of Lemma 1 If we take $\chi = v^N$ in (19) and use the periodic boundary conditions, we have

$$(u_{tx}^N, v^N) = -(u_t^N, v_x^N) = -\frac{1}{2}\frac{d}{dt}||u^N(\cdot, t)||^2,$$

and

$$((-\beta u_{xxx}^N + f(u^N)_x)_x, v^N) - (\gamma u^N, v^N) = -\beta(u_{xx}^N, u_x^N) - (f(u^N)_x, u^N) - (\gamma u^N, v^N).$$

Note that integration by parts and again periodicity imply

$$\begin{aligned} (u_{xx}^N, u_x^N) &= 0, \\ (f(u^N)_x, u^N) &= F(u^N(\pi, t)) - F(u^N(-\pi, t)) = 0, \\ (u^N, v^N) &= \frac{1}{2}\left((v^N(\pi, t))^2 - (v^N(-\pi, t))^2\right) = 0. \end{aligned}$$

Therefore $\frac{d}{dt}||u^N(\cdot, t)||^2 = 0$ and this proves (i).

Now we consider P_N^0 the orthogonal projection of L^2 onto S_N^0 and take χ in (19) such that

$$\partial_x \chi = \beta u_{xx}^N - P_N^0 f(u^N) + \gamma \partial_x^{-1} v^N.$$

Then

$$\begin{aligned} (u_{tx}^N, \chi) &= -(u_t^N, \chi_x) \\ &= -\int_{-\pi}^{\pi} \left(\beta u_{xx}^N - P_N^0 f(u^N) + \gamma \partial_x^{-1} v^N\right) u_t^N dx \\ &= \int_{-\pi}^{\pi} \left(\beta u_x^N u_{xt}^N + P_N^0 f(u^N) u_t^N + \gamma v^N v_t^N\right) dx \\ &= \frac{d}{dt}\int_{-\pi}^{\pi} \left(\frac{\beta}{2}(u_x^N)^2 + F(u^N) + \frac{\gamma}{2}(\partial_x^{-1} u^N)^2\right) dx, \end{aligned}$$

(where $F' = f$, $F(0) = 0$). On the other hand, using the property

$$(P_N^0 u, v) = (u, v), \quad u, v \in S_N^0,$$

and that P_N^0 commutes with ∂_x and, consequently, with ∂_x^{-1}, then we have

$$
\begin{aligned}
((-\beta u^N_{xxx} + f(u^N)_x)_x - \gamma u^N, \chi) &= -((-\beta u^N_{xxx} + f(u^N)_x), \chi_x) + (\gamma v^N, \chi_x) \\
&= \int_{-\pi}^{\pi} \left(\beta u^N_{xxx}\right)\left(\beta u^N_{xx} - P_N^0 f(u^N) + \gamma \partial_x^{-1} v^N\right) dx \\
&\quad + \int_{-\pi}^{\pi} f(u^N)_x \left(\beta u^N_{xx} - P_N^0 f(u^N) + \gamma \partial_x^{-1} v^N\right) dx \\
&\quad + \int_{-\pi}^{\pi} \gamma v^N \left(\beta u^N_{xx} - P_N^0 f(u^N) + \gamma \partial_x^{-1} v^N\right) dx \\
&= \beta \int_{-\pi}^{\pi} u_{xxx} \partial_x^{-1} v^N - \gamma \int_{-\pi}^{\pi} f(u^N)_x \partial_x^{-1} v^N dx \\
&\quad - \gamma \int_{-\pi}^{\pi} f(u^N)_x v^N dx = 0.
\end{aligned}
$$

and this proves (ii). □

2.2 *Time Discretization*

The system (22) is stiff and then considering implicit time stepping integration is recommended. Our choice consists of a composition of three steps of the one-stage, second-order accurate Gauss-Legendre implicit Runge-Kutta method, also called implicit midpoint rule (IMR), to integrate from t_n to t_{n+1}. A brief description is made in this section; see [19] for more details.

2.2.1 The Implicit Midpoint Rule (IMR)

Given $0 < t^* < \infty$, a step size Δt and M such that $t^* = M\Delta t$, we consider a discretization of the interval $[0, t^*]$ with the points $t_m = m\Delta t$, $m = 0, \ldots, M$. The fully discrete solution given by the IMR is defined as the sequence $\{U^m\}_{m=0}^M$ of elements of S_N satisfying, for every $\chi \in S_N$ and $m = 1, \ldots, M$

$$
\begin{aligned}
(\left(\frac{U^{m+1} - U^m}{\Delta t}\right)_x, \chi) + ((-\beta (U^{m+1/2})^N_{xxx} + f(U^{m+1/2})_x)_x \\
- \gamma U^{m+1/2}, \chi) = 0, \qquad (23)
\end{aligned}
$$

where $U^{m+1/2} = \dfrac{U^{m+1} + U^m}{2}$. Note that if

$$
U^m(x) = \sum_{k=-N}^{N} \widehat{U^m}(k) e^{ikx},
$$

then the system for the Fourier coefficients $\widehat{U^m}(k)$ has the form $(m = 0, 1, \ldots, M)$

$$\frac{\widehat{U^{m+1}}(k)-\widehat{U^m}(k)}{\Delta t}=\frac{i}{k}\left((\gamma+\beta k^4)\widehat{U^{m+1/2}}(k)+k^2\widehat{f(U^{m+1/2})}(k)\right),$$
$$-N\leq k\leq N, k\neq 0. \qquad (24)$$

System (24) can be written in a fixed point type formulation for $Z=U^{m+1/2}$

$$\widehat{Z}(k)=\alpha_1(k)\widehat{U^m}(k)+\frac{\Delta t}{2}\alpha_2(k)\widehat{f(Z)}(k),\quad -N\leq k\leq N, \qquad (25)$$

where

$$\alpha_1(k)=\frac{k}{k-\frac{i\Delta t}{2}(\gamma+\beta k^4)},\quad \alpha_2(k)=-\frac{ik^2}{\left(k-\frac{i\Delta t}{2}(\gamma+\beta k^4)\right)}, \qquad (26)$$

and if it is possible to solve (25), (26) for $\widehat{Z}(k)$, we recover $\widehat{U^{m+1}}(k)=2\widehat{Z}(k)-\widehat{U^m}(k)$. Note that taking $\chi=1$ in (23) leads to $\widehat{Z}(0)=0$. Therefore, the hypothesis $\widehat{U^0}(0)=0$ on the initial condition and the resolution of (25) imply

$$\widehat{U^m}(0)=\int_{-\pi}^{\pi}U^m(x)dx=0,\quad m\geq 1, \qquad (27)$$

for all $m=1,2\ldots,M$. In addition, if we take $\chi=\partial_x^{-1}U^{m+1/2}$ in (23) then

$$-\frac{1}{2\Delta t}\left(||U^{m+1}||^2-||U^m||^2\right)-((f\left(U^{m+1/2}\right))_x,U^{m+1/2})=0,$$

and by periodicity, if $F'=f$, $F(0)=0$, then

$$((f\left(U^{m+1/2}\right))_x,U^{m+1/2})=-(f\left(U^{m+1/2}\right),U_x^{m+1/2})=-F\left(U^{m+1/2}\right)\Big|_{-\pi}^{\pi}=0.$$

Therefore, we obtain the preservation of the quadratic invariant (17), that is

$$||U^{m+1}||=||U^m||,\quad m=0,1,\ldots,M. \qquad (28)$$

2.2.2 Fourth-Order Composition Method

The method considered here consists, for the step $m\mapsto m+1$, of a consecutive computation of (23) with step lengths $\beta_i\Delta t$, $i=1,2,3$, where, [19],

$$\beta_1=(2+2^{1/3}+2^{-1/3})/3=\frac{1}{2-2^{1/3}};\quad \beta_2=1-2\beta_1,\quad \beta_3=\beta_1.$$

This leads to the fourth-order, diagonally implicit Runge-Kutta composition method with tableau

$$\begin{array}{|ccc}\beta_1/2 & & \\ \beta_1 & \beta_2/2 & \\ \beta_1 & \beta_2 & \beta_3/2 \\ \hline \beta_1 & \beta_2 & \beta_3\end{array}$$

The method inherits several properties of the IMR, such as the conservation properties (27), (28) and simplecticity, see [51] and references therein.

2.3 Numerical Experiments

In this section the convergence of the fully discrete scheme is checked and its performance is explored. The fixed-point algorithm for the iterative resolution of (25)

$$\begin{aligned}\widehat{Z^{[\nu+1]}}(k) &= \alpha_1(k)\widehat{U^m}(k) + \Delta t\alpha_2(k)\widehat{f(Z^{[\nu]})}(k),\\ &-N \leq k \leq N, \quad \nu = 0, 1, \ldots,\end{aligned} \tag{29}$$

will be used. Although other alternatives can indeed be implemented, fixed point iteration converges in all the cases studied and provides a simple way to solve the iteration. Thus, the method will require three iterative steps of the form (29), with the corresponding step sizes $\beta_i \Delta t$, $i = 1, 2, 3$, instead of Δt, see [19] for implementation details.

Since no explicit formulas for the solitary waves of the Eq. (2) are known, the order of the method will be first checked by simulating solitary wave solutions of the generalized Korteweg-de Vries (gKdV) equation, corresponding to $\gamma = 0$, $\beta = -1$, $f(u) = u^p$ in (2). In this case, the solitary wave solution with speed $c_s > 0$ has the form

$$\begin{aligned}\eta(x,t) &= A\text{sech}(K(x - c_s t - x_0))^{2/p},\\ A &= \left(\frac{c_s}{2}(p+1)(p+2)\right)^{1/p}, \quad K = \frac{p}{2}\sqrt{c_s},\end{aligned} \tag{30}$$

with a free parameter $x_0 \in \mathbb{R}$. For $x_0 = 0$, $c_s = 4$ and (30) as initial condition with several values of p, the error in Euclidean norm

$$E(t_n) = \frac{||U^n - \eta_{h,n}||}{||\eta_{h,0}||}, \tag{31}$$

between (30) and the corresponding numerical approximation U^n (where $\eta_{h,n}$ stands for the vector-valued function with entries given by the solution (30) at the collocation

Table 1 Euclidean error (31) at $t^* = 10$ for the gKdV equation and with respect to (29) with $c_s = 4, x_0 = 0$

p	Δt	$E(t^*)$	Rate
2	$2.5E-02$	$1.066976E-03$	
	$1.25E-02$	$6.805057E-05$	3.971
	$6.25E-03$	$4.268835E-06$	3.995
	$3.125E-03$	$2.675602E-07$	3.996
	$1.562525E-03$	$1.681784E-08$	3.992
3	$2.5E-02$	$1.925805E-02$	
	$1.25E-02$	$1.310532E-03$	3.878
	$6.25E-03$	$8.136197E-05$	4.010
	$3.125E-03$	$4.984519E-06$	4.030
	$1.562525E-03$	$3.112465E-07$	4.002

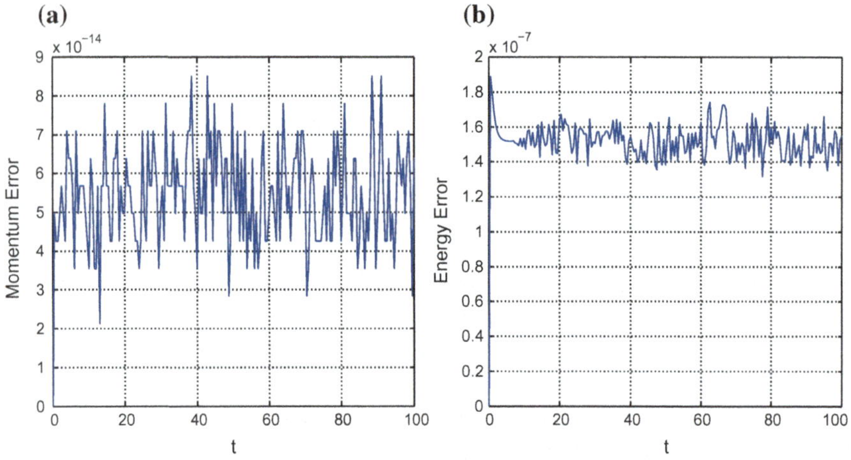

Fig. 2 **a** Energy ($|E_h(U^n) - E_h(U^0)|$) and **b** momentum ($|I_h(U^n) - I_h(U^0)|$) errors versus time, where I_h, E_h are given by (32), (33) resp. and the initial condition U^0 is the exact (29) at $t = 0$ with $x_0 = 0, c_s = 2, p = 2$

points and at time t_n) at $t^* = 10$ and different time steps is shown in Table 1. (The step size in space is $h = 128/1024$.) The results confirm the fourth order of convergence.

The conservation properties of the method are illustrated in Fig. 2a, b. They show, respectively, the behaviour in time of the error in the discrete versions of the momentum and energy invariants of the gKdV equation, defined as

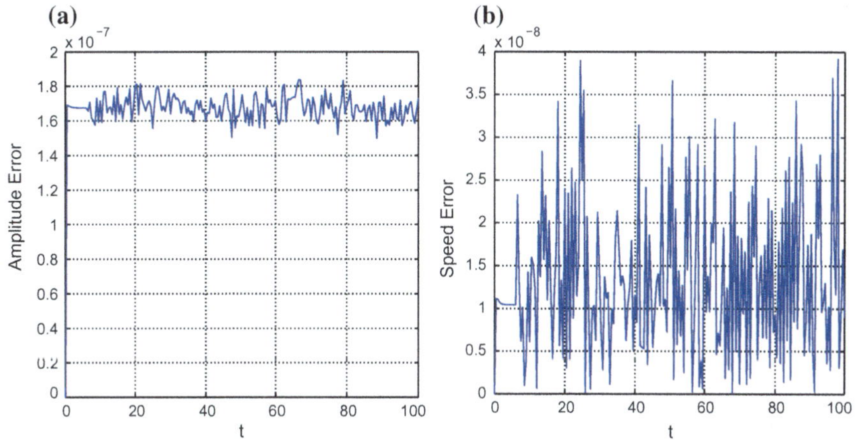

Fig. 3 **a** Amplitude and **b** speed errors versus time, for the numerical integration described in Fig. 2

$$I_h(U) = h \sum_{j=0}^{N-1} U_j^2, \quad U = (U_0, \ldots, U_{N-1})^T, \tag{32}$$

$$E_h(U) = h \sum_{j=0}^{N-1} \left(\frac{1}{2}(DU)_j^2 + \frac{1}{(p+1)(p+2)} U_j^{p+2} \right), \tag{33}$$

(where D stands for the pseudospectral differentiation matrix, [11]). Note that since the method is based on the IMR, its behaviour with respect to the invariants of the problem can be analyzed as was made in [10] for the nonlinear wave and nonlinear Schrödinger equations. In particular, in order to avoid the growth with time of the error in the invariants, the conservation properties of the method have been strengthened by including a projection technique, [28], with respect to the first quantity (32). Then, the behaviour observed in Fig. 2a, b is justified by this additional projection onto the level set of I_h determined by its value at the exact solitary wave (30), the symplectic character of the method and, finally, by the fact that the solitary waves are generated as critical points of the energy subject to fixed value of the momentum, see [39]. The time behaviour of the error is related to the simulation of the parameters of the solitary wave, [18]. Figure 3a, b display, respectively, the errors in the amplitude and speed between the exact solution and the numerical approximation as function of time and for $\Delta t = 6.25E{-}03$ (other time steps give similar results). The parameters are computed in a standard way, see e.g. [16]. Observe that the errors in the first two parameters are small and do not grow with time (at least up to moderate times, like that of this case $t^* = 100$).

The Ostrovsky equations contain nonlocal terms and it may be worth including a short experiment to check if such property affects the performance of the numerical method, [17]. To this end and since the gKdV equation contains only local terms,

Table 2 Normalized Euclidean error $E_2(t^*)$ and error in maximum norm $E_\infty(t^*)$ at $t^* = 50$ for the BO equation (34) and the simulation of (36) with $l = 16, c = 1$, using $N = 1024$ collocation points and exact initial condition

Δt	$E_2(t^*)$	RATE	$E_\infty(t^*)$	Rate
$5E{-}02$	$2.2536E{-}06$		$2.1422E{-}06$	
$2.5E{-}02$	$1.4092E{-}07$	3.999	$1.3278E{-}07$	4.012
$1.25E{-}02$	$8.8095E{-}09$	3.999	$8.2618E{-}09$	4.006
$6.25E{-}03$	$5.5082E{-}10$	3.999	$5.1348E{-}10$	4.008
$3.125E{-}03$	$3.3451E{-}11$	4.042	$3.1199E{-}11$	4.041

some tests with the $2l$**-periodic initial-value problem of the Benjamin-Ono (BO) equation, used in** [57]

$$u_t + uu_x - \mathcal{G}u_{xx} = 0, \quad x \in \mathbb{R}, \quad t > 0, \tag{34}$$

$$u(x, 0) = u_0(x), \quad x \in \mathbb{R}, \tag{35}$$

have been made. In (35), u_0 is $2l$-periodic and $\mathcal{G}$ is the periodic version of the Hilbert transform

$$\mathcal{G}v(x) := PV\frac{1}{2l}\int_{-l}^{l} \cot\left(\frac{\pi y}{2l}\right) v(x - y)dy.$$

The periodic problem for (34) admits periodic traveling wave solutions of the explicit form

$$u_l(x, t) = \frac{2c\delta^2}{1 - \sqrt{1 - \delta^2}\cos(c\delta(x - ct))}, \tag{36}$$

where $\delta = \pi/(cl)$, c arbitrary. For $c = 1, l = 16$, the profile (36) at $t = 0$ has been taken as initial condition and the fully discretization above has been adapted to integrate (34). At $t^* = 50$ the normalized errors, cf. (31), in the Euclidean and maximum norms and with $N = 1024$, have been measured with several time steps. The results are displayed in Table 2. They confirm the fourth-order accurate of the method and suggest that this is not affected by the presence of the nonlocal term in (34) given by $\mathcal{G}$, at least when approximating traveling-wave solutions.

3 Numerical Generation of Solitary Waves

This section deals with the numerical generation of several types of traveling-wave solutions of (2).

3.1 A Numerical Technique to Generate Solitary Waves

Described here is the numerical method considered in this paper to generate approximations to traveling wave solutions of the generalized Ostrovsky equation (2) with homogeneous nonlinearities of the form (4).

3.1.1 The Petviashvili's Method

To our knowledge, the first application of the Petviashvili's method, [50], in equations of the form (2), was made in [20] to generate approximate generalized solitary waves. The Petviashvili's method is also used in [38] to compute approximations to classical solitary waves with speeds c far from the limit $c^* = 2\sqrt{\beta\gamma}$. Due to the highly oscillatory character of the solitary wave profile when c is close to c^*, this algorithm fails in those cases and the authors consider a shooting method as an alternative. Here we will overcome this difficulty in a different way, by using acceleration techniques.

First we make a brief description of the Petviashvili's method, see e.g. [1, 35, 36, 47, 50] for details. Note that (9) can be written in the form

$$\mathscr{L}\phi = \mathscr{N}(\phi), \quad \mathscr{L} = \gamma + c\partial_{xx} + \beta\partial_{xxxx}, \quad \mathscr{N}(\phi) = \partial_{xx} f(\phi). \tag{37}$$

Due to the homogeneous character of f, the operator $\mathscr{N}$ in (37) is also homogeneous of degree p. For $\gamma \neq 0$, the linear operator $\mathscr{L}$ is invertible and if ϕ_c is a solution of (37) then

$$\mathscr{L}^{-1}\mathscr{N}'(\phi_c)\phi_c = p\mathscr{N}(\phi_c) = p\phi_c, \tag{38}$$

which means that ϕ_c is an eigenfunction of the iteration operator with eigenvalue $p > 1$. Therefore, the classical fixed-point algorithm, applied to (37),

$$\mathscr{L}\phi_{n+1} = \mathscr{N}(\phi_n), \quad n = 0, 1, \ldots, \tag{39}$$

will not converge in general. In these situations, the Petviashvili's iterative method is sometimes an alternative. Its formulation for this case is

$$m(\phi_n) = \frac{(\mathscr{L}\phi_n, \phi_n)}{(\mathscr{N}(\phi_n), \phi_n)}, \tag{40}$$

$$\mathscr{L}\phi_{n+1} = m(\phi_n)^{\alpha}\mathscr{N}(\phi_n), \quad n = 0, 1, \ldots, \tag{41}$$

for a given initial iteration ϕ_0 and some parameter α to be specified later. As mentioned in [2], the inclusion of the so-called stabilizing factor (40) in the fixed-point type formula (41) has the goal of modifying the spectrum of the iteration operator $S = \mathscr{L}^{-1}\mathscr{N}'(\phi_c)$, in such a way that the iteration operator of the new iterative procedure (41) shares the spectrum of S except the eigenvalue $\lambda = p$, which for some suitable values of α is converted into an eigenvalue with modulus less than one. In particular, the choice $\alpha = p/(p-1)$ (that will be considered in the experiments below) transforms the eigenvalue $\lambda = p$ of S to $\lambda = 0$ as eigenvalue of the iteration operator of (41) at a solution profile ϕ_c. Thus, if $\lambda = p$ is the only eigenvalue of S with magnitude above one, then thie resulting method is locally convergent method; see [2] for details.

In practice, formulas (40), (41) are implemented in Fourier space for the periodic problem of (9) on a sufficiently long interval $(-l, l)$. Thus if $\widehat{\phi_n}(k)$ denotes the k-th Fourier coefficient of ϕ_n, then (40), (41) become

$$m(\phi_n) = -\frac{\sum_k (\gamma - c\widetilde{k}^2 + \beta\widetilde{k}^4)|\widehat{\phi_n}(k)|^2}{\sum_k \widetilde{k}^2 \widehat{f(\phi_n)}(k)\overline{\widehat{\phi_n}(k)}}, \tag{42}$$

$$\widehat{\phi_{n+1}}(k) = -m(\phi_n)^{\alpha} \frac{\widetilde{k}^2 \widehat{f(\phi_n)}(k)}{\gamma - c\widetilde{k}^2 + \beta\widetilde{k}^4}, \tag{43}$$

where $\widetilde{k} = \dfrac{\pi}{l}k$, $k \in \mathbb{Z}$. The computation of (42), (43) with Discrete Fourier Transform (DFT), [58], will be used to generate approximations to both classical and generalized solitary waves. In all the experiments below, the performance has been measured according to three quantities:

- The discrepancy between one and the stabilizing factor $m_n := m(\phi_n)$. (Note that, in the case of convergence, this factor (40) must tend to one.)
- The Euclidean error between two consecutive iterations

$$ERROR_c(n) = ||\phi_n - \phi_{n-1}||.$$

- The residual error (also in Euclidean norm) at the n−th iteration

$$RES(n) = ||L_0(\phi_n)||, \quad L_0(\phi) = \mathscr{L}\phi - \mathscr{N}(\phi).$$

The discrete version of (42), (43) is carried out by using Fourier pseudospectral approximation, in such a way that the corresponding approximation operators are

$$\mathscr{L}_h = \gamma I_N + cD_N^2 + \beta D_N^4, \quad \mathscr{N}_h(\phi_h) = D_N^2 f(\phi_h),$$

where ϕ_h is a N-vector approximation to the profile ϕ_c at the collocation points $x_j = -l + jh$, $j = 0, \ldots, N-1$, $h = 2l/N$, I_N is the $N \times N$ identity matrix and D_N stands for the pseudospectral differentiation matrix of size N, [11, 58]. For the computations below, three parameters are emphasized: the speed c, the amplitude of the resulting wave and the degree of homogeneity p.

3.1.2 Acceleration Techniques

As mentioned before, the difficulties of the Petviashvili's method to generate classical solitary waves when the speed c is close to the limit c^* (see [39]) have been overcome here by incorporating some acceleration techniques to the iterative procedure. Some of the most widely used methods in the literature to this end are the so-called Vector Extrapolation Methods (VEM). They introduce the extrapolation as a procedure to transform the original sequence of the iterative process ϕ_n from (40), (41) by some strategy. For a more detailed analysis and implementation of the methods see e. g. [9, 34, 54–56] and references therein. The application of acceleration techniques for traveling wave computations can be seen in [3]. Briefly described here are the main stages of the procedure. This is carried out in a cycling mode. A cycle of the iteration is performed by the following steps: Given a width of extrapolation $mw \geq 1$, for $k = 0, 1, \ldots$ and for the advance $k \mapsto k + 1$:

1. Set $\psi_0 = \phi_k$ and compute mw steps of the fixed-point algorithm:

$$L\psi_{j+1} = m(\psi_j)^{\alpha} N(\psi_j),\ j = 0, \ldots mw - 1.$$

2. Compute the extrapolation steps:

$$\psi_{n,k} = \sum_{j=0}^{n} \beta_{j,k}\psi_j,\ n = 0, \ldots, mw,$$

   for some $\beta_{j,k}$.
3. Set $\phi_{k+1} = \psi_{mw,k}$, $\psi_0 = \phi_{k+1}$ and go to step 1.

The cycle 1-2-3 is repeated until the error (residual or between two consecutive iterations) is below a prefixed tolerance or a maximum number of iterations is attained.

On the other hand, in these methods, the coefficients $\beta_{j,k}$ of the extrapolation steps are (linear or nonlinear) functions of some previous steps of the iteration. The derivation is usually established for linear systems and follows different criteria. Some examples are the minimal polynomial extrapolation (MPE), the reduced rank extrapolation (RRE) and the modified minimal polynomial extrapolation (MMPE); they belong to the so-called polynomial methods, [34, 54–56]. These calculate the extrapolation steps as weighted average of the iterations and the weights are determined by setting orthogonality conditions on the generalized residual, [34]. In practice, since the width of extrapolation is not known, the methods are implemented with relatively small values of mw and the one with the best performance is taken.This width of extrapolation was experimentally set as $mw = 3$ or $mw = 4$ in the examples below, where the MPE method was used.

3.2 Numerical Results

Displayed here are numerical experiments to analyze the performance of the method and some properties of the waves suggested by the results.

3.2.1 Classical Solitary Waves

Recall that classical solitary waves exist when $\gamma, \beta > 0$ and speeds $c < c^* = 2\sqrt{\beta\gamma}$. Here $\gamma = 1/4$, $\beta = 1$ (thus $c^* = 1$) have been taken to illustrate the plethora of computations made with different parameters values. In [38], Levandosky shows that without loss of generality, the homogeneous nonlinearity f can be alternatively written in the form

$$
\begin{aligned}
f_\theta &= \cos(\theta) f_e + \sin(\theta) f_o, \quad -\frac{\pi}{2} \leq \theta \leq \frac{\pi}{4}, \\
f_e(s) &= |s|^p, \quad f_o(s) = |s|^{p-1}s, \quad p > 1. \qquad (44)
\end{aligned}
$$

Although experiments with several values of θ were performed, for simplicity only those with $\theta = 0$ will be shown here. This corresponds to taking $a_e = 1, a_o = 0$ in (4).

The first results study the case of negative speed c. By way of illustration, we consider the parameter values $p = 2$ with $c = -0.75$. (Other values of p give similar results.) The performance of the method is shown in Fig. 4a–d. As displayed in Fig. 4a, the solitary wave profile is even, with maximum negative excursion at $x = 0$ and two maximum lobes. As p increases, the maximum negative excursion increases (Fig. 5) and the amplitude of the lobes decreases. The profiles also contain negative excursions to the left and to the right, see Fig. 4b. These oscillations are more intense as the speed tends to c^* (suggesting an oscillatory decay of the profiles) and their formation occurs for any $p > 1$. The convergence of the method is verified by Fig. 4c, d. The first one displays the residual error as function of the number of iterations and in semi-log scale, while Fig. 4d calibrates the computational effort by displaying the residual error (in semi-log scale) as function of the CPU time (in seconds). (The behaviour of the other two quantities that control the iteration, that is, the discrepancy in the error for the stabilizing factor and the error between two consecutive iterations, is similar and will not be shown here.)

In the case of negative c, acceleration techniques were not necessary (although they improved the efficiency of the iteration). This also happens when $c > 0$ but still far from c^*. Since here the behaviour is similar to that of the previous experiments for negative c, this case will be used to illustrate the asymptotic decay of the waves.

Figure 6 displays the phase portraits of the computed solitary waves for two values of p with $\gamma = 0.25$, $\beta = 1$ and $c = 0.5$. (The derivative is computed by using spectral differentiation.) The results suggest, as mentioned above for the case of negative c, and for all the values of p considered (including those of experiments made but not shown here), an exponential, damped decay of the waves at infinity.

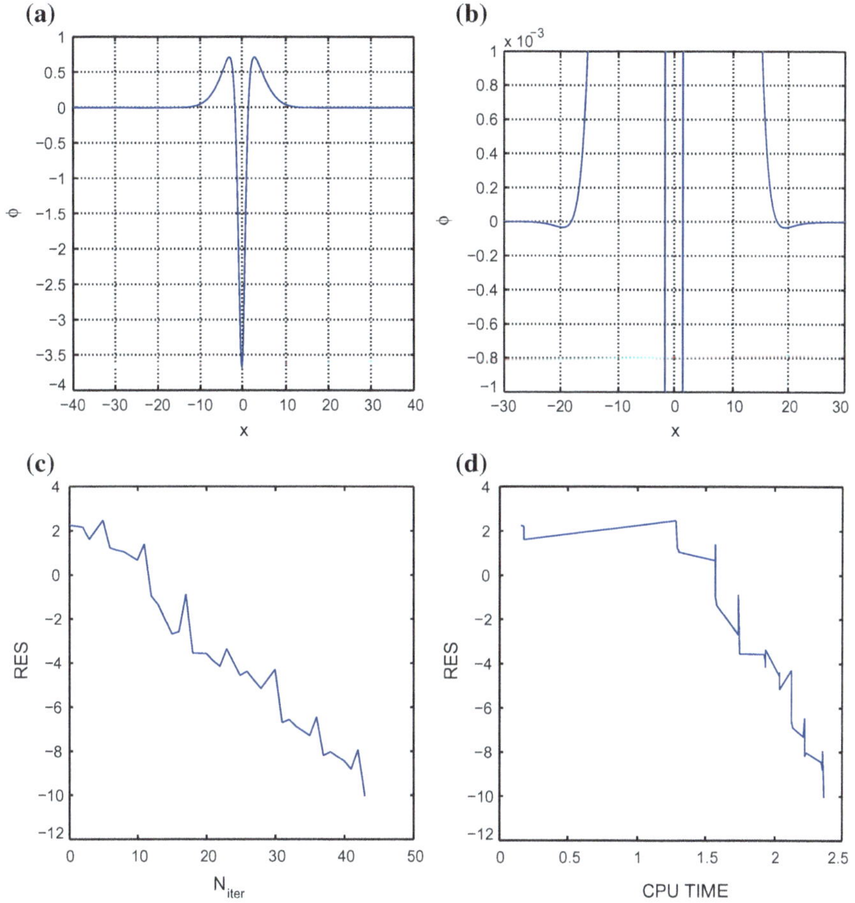

Fig. 4 Generation of classical solitary waves of (2), (4) with $\gamma = 1/4$, $\beta = 1$, $p = 2$ with $a_e = 1$, $a_o = 0$ and $c = -0.75$. **a** Computed solitary-wave profile. **b** Magnification of (**a**). **c** Residual error versus number of iterations and in semi-log scale. **d** Residual error versus cpu time (in seconds)

When c is closer to c^*, the solitary wave profiles contain more and more oscillations and the Petviashvili's method starts to be much slower or to fail. Some reasons for that can be seen in Tables 3 and 4.

For $\gamma = 0.25$, $\beta = 1 (c^* = 1)$ and $p = 3, 4$ (respectively) Tables 3 and 4 show the six largest magnitude eigenvalues of the iteration matrix $S = \mathscr{L}_h^{-1}\mathscr{N}_h'(\phi_h)$ of the classical fixed-point iteration (39) evaluated at the last computed iterate ϕ_h, for several values of the speed c close to c^*. We observe that as $c \rightarrow c^*$ the spectrum of S changes and, besides the eigenvalue which is given by the degree p of homogeneity, see (38) (and which becomes zero in the spectrum of the iteration matrix for the Petviashvili's method) and the eigenvalue $\lambda = 1$ (which corresponds to the translational invariance of the equation and that affects the convergence only in the orbital sense, see [2]),

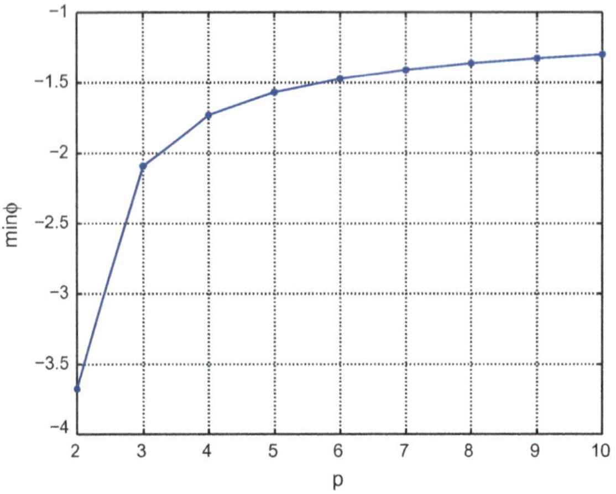

Fig. 5 Minimum (maximum negative excursion) of the computed solitary-wave profile of (2), (4) with $\gamma = 1/4$, $\beta = 1$, $a_e = 1$, $a_o = 0$ and $c = -0.75$ as function of p

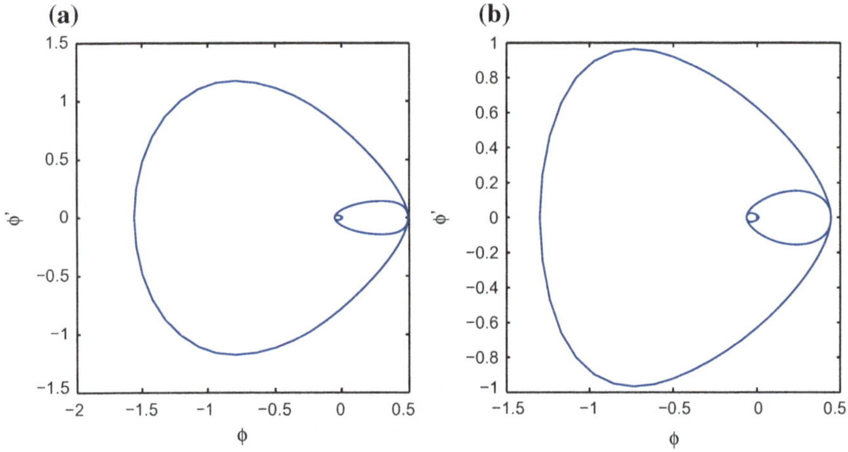

Fig. 6 Generation of classical solitary waves of (2), (4) with $\gamma = 1/4$, $\beta = 1$ with $a_e = 1$, $a_o = 0$ and $c = 0.5$. Phase portraits of the computed profiles: **a** $p = 2$; **b** $p = 3$

the other eigenvalues grow in magnitude, being eventually above one and therefore making the iteration much slower or fail. This behaviour is attenuated as the degree of homogeneity p is increasing. For example, when $p = 2$ and $c = 0.8$, there is already an eigenvalue $\lambda^* = -1.1329E{+}00$ above one in magnitude while when $p = 3$ and $c = 0.95$, the third largest (in magnitude) eigenvalue is still below one. (For $p = 4$ this holds even when $c = 0.995$.) Then, in order to improve the performance of the method, the numerical experiments have been implemented with the MPE acceleration procedure, described above. As an example, with this technique, the

Table 3 Six largest magnitude eigenvalues of the matrix S at the last computed iterate: $p = 3$, $\gamma = 0.25$, $\beta = 1$ ($c^* = 1$), u_m is the maximum negative excursion of the resulting profile

c	0.7	0.8	0.9	0.95
	3.0000E+00	2.9999E+00	2.9999E+00	2.9999E+00
	9.9999E−01	9.9999E−01	9.9999E−01	1.0000E+00
	−6.0465E−01	−7.0854E−01	−8.6278E−01	−9.8560E−01
	−3.9358E−01	−4.6988E−01	−6.0089E−01	−7.2570E−01
	−3.2451E−01	−3.8129E−01	−4.7673E−01	−5.6446E−01
	−3.0712E−01	−3.4696E−01	−4.0090E−01	−4.3818E−01
u_m	−1.1071	−9.8725E−01	−8.3355E−01	−7.2277E−01

Table 4 Six largest magnitude eigenvalues of the matrix S at the last computed iterate: $p = 4$, $\gamma = 0.25$, $\beta = 1$ ($c^* = 1$), u_m is the maximum negative excursion of the resulting profile

c	0.9	0.95	0.99	0.995
	3.9999E+00	3.9999E+00	4.0000E+00	4.0000E+00
	9.9999E−01	9.9999E−01	1.0000E+00	1.0000E+00
	−6.4641E−01	−7.8456E−01	−9.4861E−01	−9.9262E−01
	−4.5054E−01	−5.9275E−01	−8.7971E−01	−9.6064E−01
	−2.9135E−01	−3.8085E−01	5.8402E−01	8.8998E−01
	−2.4150E−01	−2.8448E−01	−5.7376E−01	7.0979E−01
u_m	−8.9107E−01	−8.0239E−01	−6.5046E−01	−5.7459E−01

resulting iteration is able to generate solitary wave profile approximations for the case $p = 4$, $c = 0.995$ (with $c^* = 1$), see Fig. 7a, b.

Another advantage of the application of the acceleration techniques is the reduction of the computational cost. This can be illustrated by Fig. 7c, d, which correspond to the same computation as in Fig. 7a, b but without acceleration.

A final experiment concerning classical solitary waves illustrates the behaviour of the amplitude of the computed profiles with respect to the speed. This is shown in Fig. 8 for different values of p. In all the cases the amplitude is decreasing in magnitude as c tends to c^*. The approximation to zero amplitude as $c \to c^*$, $c < c^*$ looks faster when $p = 2$.

3.2.2 Generalized Solitary Waves

As mentioned above, the classical Ostrovsky equation (1) admits generalized solitary wave solutions. According to [14], these traveling waves, homoclinic to periodic orbits as $|X| \to \infty$, are associated, in the (Q, P) plane, to region below curve C_0, see the Introduction.

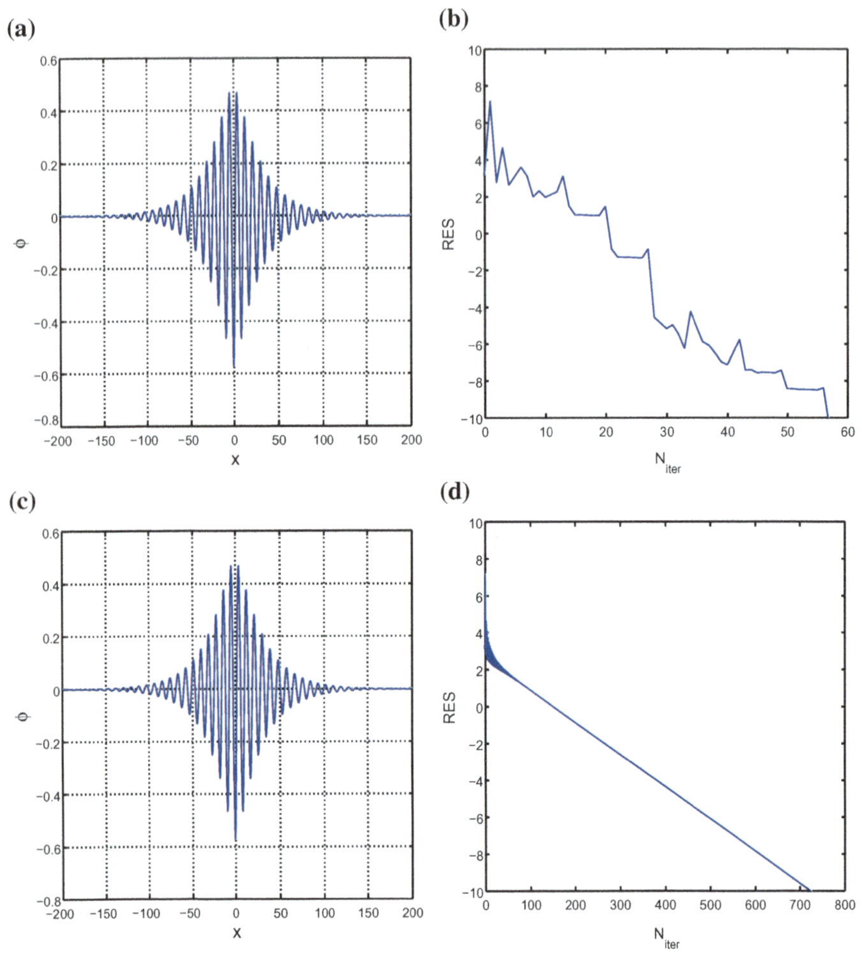

Fig. 7 Generation of classical solitary waves of (2), (4) with $\gamma = 1/4, \beta = 1, p = 4$ with $a_e = 1, a_o = 0, c = 0.995$. In **a**, **b** the (MPE) has been used as acceleration technique, while in **c**, **d**, no acceleration technique is implemented. **a**, **c** Computed solitary-wave profile. **b**, **d** Residual error, displayed as function of the number of iterations and in semi-log scale

In this case, the Petviashvili's method does not need the help of acceleration techniques to generate numerically these profiles, although the MPE scheme has been used in order to improve the computational cost.

Some figures will illustrate the generation of approximate profiles with an initial iteration of the form (14). The experiments also give numerical evidence of existence of such waves for higher values of p; since the resulting profiles are similar, only those corresponding to $p = 3$ will be shown here.

The first group of results (Fig. 9a, b) corresponds to the case $P < 0$, $Q > 0$. Note that Fig. 9b, which shows a wave of elevation, is obtained by considering opposite

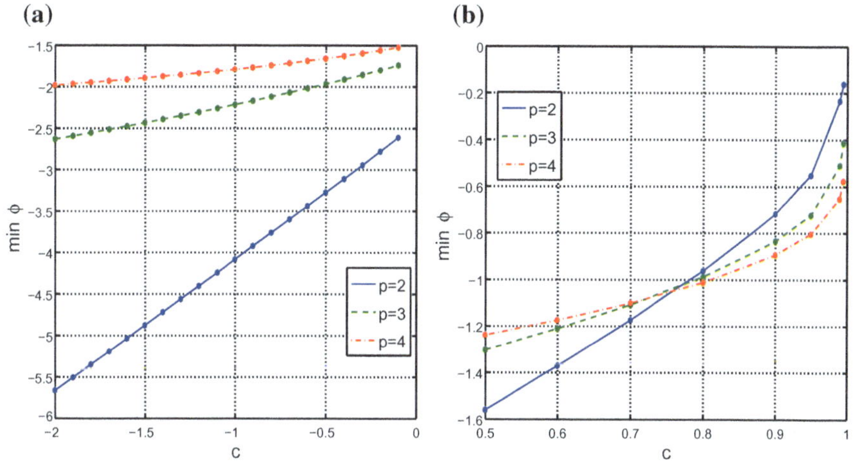

Fig. 8 Maximum negative excursion of the computed solitary-wave profile of (2), (4) as function of the speed c and several values of p. The parameters are $\gamma = 0.25$, $\beta = 1$ (thus the limit is $c^* = 1$) and $a_e = 1$, $a_o = 0$. **a** Case $c < 0$. **b** Case $0 < c < c^*$

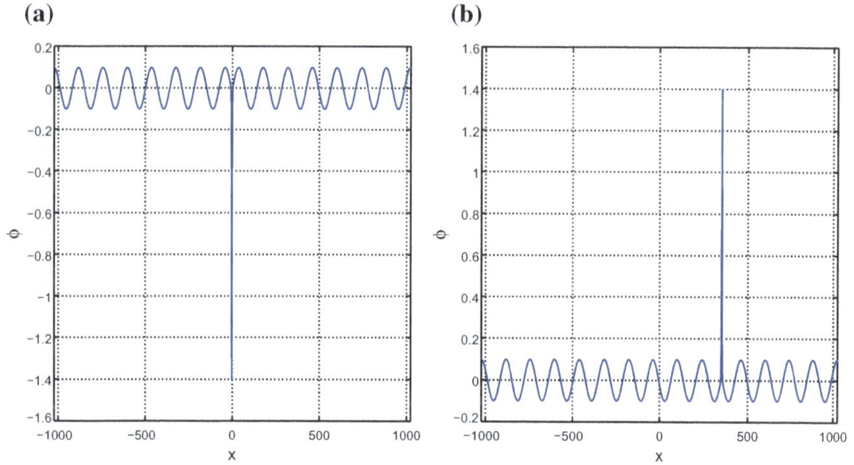

Fig. 9 Generation of generalized solitary waves of (2), (4). Computed profiles with $p = 3$, $a_e = 1$, $a_o = 0$ and: **a** $\gamma = -0.002$, $\beta = 1$, $c = -1$. **b** $\gamma = 0.002$, $\beta = -1$, $c = 1$

values of the parameters used for Fig. 9a (a wave of depression) but both are in the same region. We also note that in the region with $P, Q < 0$ (below curve C_1) only periodic traveling waves were generated.

The corresponding phase portraits confirm the asymptotic behaviour of the profiles, homoclinic to small amplitude periodic waves, see Fig. 10.

Additional experiments can give more information on this behaviour. By way of illustration, the part of the oscillations displayed in Fig. 9b from $x = -1024$ to

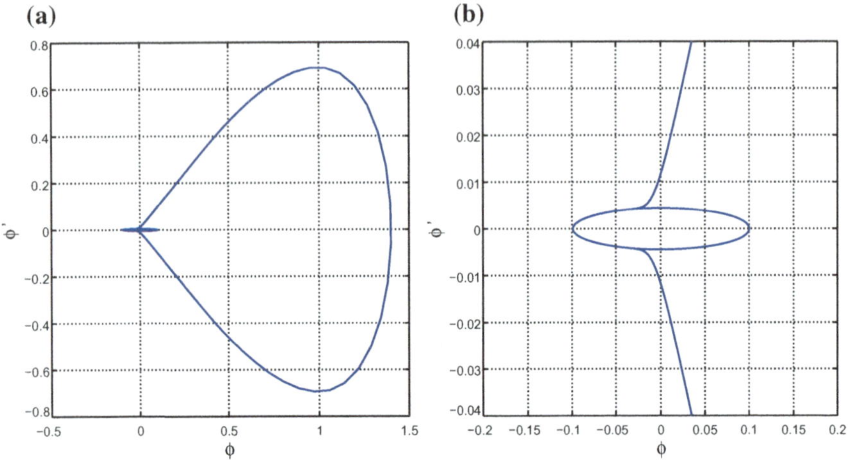

Fig. 10 Generation of generalized solitary waves of (2), (4) with $\gamma = 0.002$, $\beta = -1$, $p = 3$ with $a_e = 1$, $a_o = 0$, $c = 1$. **a** Phase portrait. **b** Magnification of (**a**)

Table 5 Coefficients and goodness of fit for the fitting curves of Fig. 9(b): $f(x) = \sum_{j=1}^{3} a_j \sin(b_j x + c_j)$. SSE and RMSE denote, respectively, the statistical parameters of the sum of squares due to error and the root mean squared error

Coefficients	Coefficients	Coefficients	g.o.f.
$a_1 = 0.1715$	$b_1 = 0.04517$	$c_1 = -2.339$	SSE$= 1.983 \times 10^{-4}$
$a_2 = 0.0202$	$b_2 = 0.09116$	$c_2 = 0.405$	R-squared$= 1$
$a_3 = 0.0039$	$b_3 = 0.1392$	$c_3 = 4.131$	RMSE$= 6.01 \times 10^{-4}$

$x = 0$ has been fitted to a sinusoidal sum. The one that gave the best goodness of fit corresponds to the data shown in Table 5. This represents the oscillations as a sum of three sinusoidal functions containing several frequencies; there exists a fundamental one associated to b_1, while b_2 and b_3 approximate $2b_1$ and $3b_1$, respectively.

3.2.3 Multi-pulse Traveling Waves

Some multi-pulse classical and generalized solitary waves can also been generated. The existence of these waves was derived in [13] for the classical Ostrovsky equation using the reversibility of (1), which forces to admit multi-humped classical and generalized solitary waves in regions above curves C_2 and C_3 and below C_0 respectively. The numerical generation is represented by Figs. 11a, b and 12a, b, both for $p = 2$ and where the superposition of two single-pulse profiles (classical and generalized, respectively) generated by the method, has been taken as initial iteration.

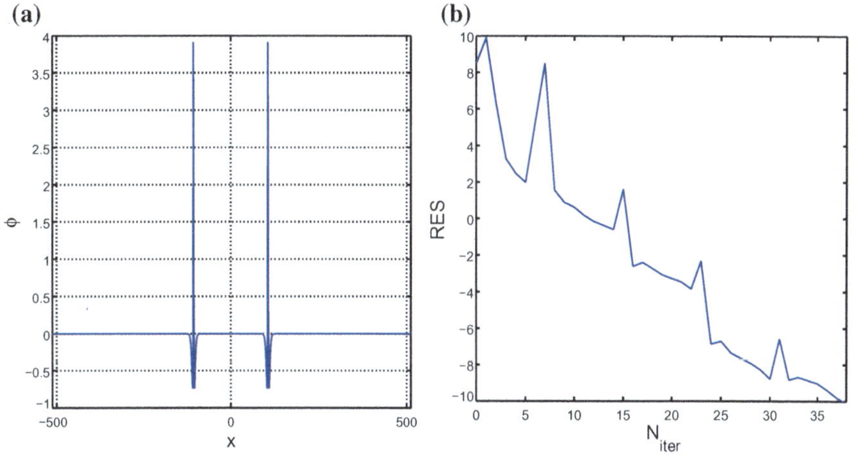

Fig. 11 Generation of multi-pulse solitary waves. Two-pulse classical solitary wave of (2), (4) with $\gamma = -0.25$, $\beta = -1$, $c = 0.9$, $p = 2$ with $a_e = 1$, $a_o = 0$. **a** Computed solitary-wave profile. **b** Residual error, displayed as function of the number of iterations and in semi-log scale

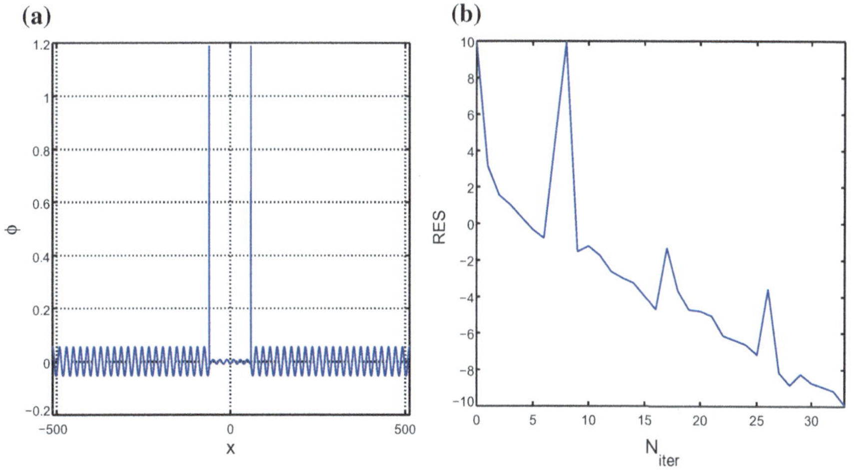

Fig. 12 Generation of multi-pulse solitary waves. Two-pulse generalized solitary wave of (2), (4) with $\gamma = 0.1$, $\beta = -0.001$, $c = 1$, $p = 2$ with $a_e = 1$, $a_o = 0$. **a** Computed solitary-wave profile. **b** Residual error, displayed as function of the number of iterations and in semi-log scale

Acknowledgements This work was supported by Spanish Ministerio de Economía y Competitividad under the Research Grant MTM2014-54710-P. The author would like to thank Professors V. Dougalis, D. Dutykh and D. Mitsotakis for fruitful discussions and so important suggestions.

References

1. Ablowitz, M.J., Musslimani, Z.H.: Spectral renormalization method for computing self-localized solutions to nonlinear systems. Opt. Lett. **30**, 2140–2142 (2005)
2. Álvarez, J., Durán, A.: Petviashvili type methods for traveling wave computations: I. Analysis of convergence. J. Comput. Appl. Math. **266**, 39–51 (2014)
3. Álvarez, J., Durán, A.: Petviashvili type methods for traveling wave computations: II. Acceleration techniques. Math. Comput. Simul. **123**, 19–36 (2016)
4. Apel, J.R., Ostrovsky, L.A., Stepanyants, Y.A., Lynch, J.F.: Internal solitons in the ocean. WHOI Tech. Rep. (2006)
5. Benilov, E.S.: On the surface waves in a shallow channel with an uneven bottom. Stud. Appl. Math. **87**, 1–14 (1992)
6. Bona, J.L., Dougalis, V.A., Mitsotakis, D.E.: Numerical solution of KdV-KdV systems of Boussinesq equations I. The numerical scheme and generalized solitary waves. Math. Comput. Simul. **74**, 214–228 (2007)
7. Boyd, J.P.: Weakly nonlocal solitary waves and beyond-all-orders asymptotics: generalized solitons and hyperasymptotic perturbation theory. In: Mathematics and Its Applications, vol. 442. Kluwer, Amsterdam (1998)
8. Boyd, J.P., Chen, G.Y.: Five regimes of the quasi-cnoidal, steadily translating waves of the rotation-modified Korteweg-de Vries ("Ostrovsky") equation. Wave Motion **35**, 141–155 (2002)
9. Brezinski, C.: Convergence acceleration during the 20th century. J. Comput. Appl. Math. **122**, 1–21 (1975)
10. Cano, B.: Conserved quantities of some Hamiltonian wave equations after full discretizations. Numer. Math. **103**, 197–223 (2006)
11. Canuto, C., Hussaini, M.Y., Quarteroni, A., Zang, T.A.: Spectral Methods in Fluid Dynamics. Springer, New York, Heidelberg, Berlin (1988)
12. Chen, G.Y., Boyd, J.P.: Analytical and numerical studies of weakly nonlocal solitary waves of the rotation-modified Korteweg-de Vries equation. Physica D **155**, 201–222 (2002)
13. Choudhury, S.R.: Solitary-wave families of the Ostrovsky equation: an approach via reversible systems theory and normal forms. Chaos, Solitons Fract. **33**, 1468–1479 (2007)
14. Choudhury, S.R., Ivanov, R.I., Liu, Y.: Hamiltonian formulation, nonintegrability and local bifurcations for the Ostrovsky equation. Chaos, Solitons Fract. **34**, 544–550 (2007)
15. Costanzino, N., Manukian, V., Jones, C.K.R.T.: Solitary waves of the regularized short pulse and Ostrovsky equations. SIAM J. Math. Anal. **41**(5), 2088–2106 (2009)
16. Dougalis, V.A., Durán, A., López-Marcos, M.A., Mitsotakis, D.E.: A numerical study of the stability of solitary waves of the Bona-Smith family of Boussinesq systems. J. Nonlinear Sci. **17**, 569–607 (2007)
17. Dougalis, V.A., Durán, A., Mitsotakis, D.E.: Numerical solution of the Benjamin equation. Wave Motion **52**, 194–215 (2015)
18. Durán, A., Sanz-Serna, J.M.: The numerical integration of relative equilibrium solutions. The nonlinear Schrödinger equation. IMA J. Numer. Anal. **20**, 235–261 (2000)
19. de Frutos, J., Sanz-Serna, J.M.: An easily implementable fourth-order method for the time integration of wave problems. J. Comput. Phys. **103**, 160–168 (1992)
20. Galkin, V.N., Stepanyants, Y.A.: On the existence of stationary solitary waves in a rotating fluid. J. Appl. Math. Mech. **55**, 939–943 (1991)

21. Gilman, O.A., Grimshaw, R., Stepanyants, Y.A.: Approximate analytical and numerical solutions of the stationary Ostrovsky equation. Stud. Appl. Math. **95**, 115–126 (1995)
22. Gilman, O.A., Grimshaw, R., Stepanyants, Y.A.: Dynamics of internal solitary waves in a rotating fluid. Dyn. Atm. Ocean **23**(1), 403–411 (1995)
23. Grimshaw, R.H.: Evolution equations for weakly nonlinear, long internal waves in a rotating fluid. Stud. Appl. Math. **73**, 1–33 (1985)
24. Grimshaw, R.H.: Internal solitary waves. In: Liu (ed.) Advances in Coastal and Ocean Engineering, pp. 1–30. World Scientific, Singapore (1997)
25. Grimshaw, R.H., He, J.M., Ostrovsky, L.A.: Terminal damping of a solitary wave due to radiation in rotational systems. Stud. Appl. Math. **10**, 197–210 (1998)
26. Grimshaw, R.H., Helfrich, K.R., Johnson, E.R.: Experimental study of the effect of rotation on nonlinear internal waves. Phys. Fluids **25**, 0566,021–05660,223 (2013)
27. Grimshaw, R.H., Ostrovsky, L.A., Shira, V.I., Stepanyants, Y.A.: Long nonlinear surface and internal gravity waves in a rotating ocean. Surv. Gheophys. **19**, 289 338 (1998)
28. Hairer, E., Lubich, C., Wanner, G.: Geometric numerical integration. In: Structure-Preserving Algorithms for Ordinary Differential Equations. Springer, New York, Heidelberg, Berlin (2004)
29. Helfrich, K.R.: Decay and return of internal solitary waves with rotation. Phys. Fluids **19**, O26,601 (2007)
30. Helfrich, K.R., Melville, W.K.: Long nonlinear internal waves. Ann. Rev. Fluid Mech. **38**, 395–425 (2006)
31. Hunter, J.K.: Numerical solutions of some nonlinear dispersive wave equations. In: E.L. Allgower (ed.), K.G.E. Computational Solutions of Nonlinear Systems of Equations. Lectures in Applied Mathematics, vol. 26, pp. 301–316. AMS, Providence (1990)
32. Iooss, G., Adelmeyer, M.: Topics in Bifurcation Theory and Applications. World Scientific, Singapore (1998)
33. Isaza, P., Mejía, J.: Global Cauchy problem for the Ostrovsky equation. Nonl. Anal. **67**, 1482–1503 (2007)
34. Jbilous, K., Sadok, H.: Vector extrapolation methods. applications and numerical comparisons. J. Comput. Appl. Math. **122**, 149–165 (2000)
35. Lakoba, T., Yang, Y.: A generalized Petviashvili method for scalar and vector Hamiltonian equations with arbitrary form of nonlinearity. J. Comput. Phys. **226**, 1668–1692 (2007)
36. Lakoba, T., Yang, Y.: A mode elimination technique to improve convergence of iteration methods for finding solitary waves. J. Comput. Phys. **226**, 1693–1709 (2007)
37. Leonov, A.I.: The effect of earth rotation on the propagation of weak nonlinear surface and internal long oceanic waves. Annal. New York Acad. Sci. **373**, 150–159 (1981)
38. Levandosky, S.: On the stability of solitary waves of a generalized Ostrovsky equation. Technical Report. (2006)
39. Levandosky, S., Liu, Y.: Stability of solitary waves of a generalized Ostrovsky equation. SIAM J. Math. Anal. **38**, 985–1011 (2006)
40. Levandosky, S., Liu, Y.: Stability and weak rotation limit of solitary waves of the Ostrovsky equation. Disc. Cont. Dyn. Syst. Ser. B **7**, 793–806 (2007)
41. Linares, F., Milanés, A.: Local and global well-posedness for the Ostrovsky equation. J. Diff. Eq. **222**, 325–340 (2006)
42. Liu, Y., Varlamov, V.: Stability of solitary waves and weak rotation limit for the Ostrovsky equation. J. Diff. Eq. **203**, 159–183 (2004)
43. Lombardi, E.: Topics in Bifurcation Theory and ApplicationsOscillatory Integral and Phenomena Beyond all Algebraic Orders. Springer, Berlin (2000)
44. Obregon, M.A., Stepanyants, Y.A.: On numerical solution of the Gardner-Ostrovsky equation. Math. Model Nat. Phenom. **7**(2), 113–130 (2012)
45. Ostrovsky, L.A.: Nonlinear internal waves in a rotating ocean. Okeanologia **18**, 181–191 (1978)
46. Ostrovsky, L.A., Stepanyants, Y.A.: Nonlinear surface and internal waves in rotating fluids. In: Nonlinear Waves 3, pp. 106–128. Springer, New York (1990)
47. Pelinovsky, D.E., Stepanyants, Y.A.: Convergence of Petviashvili's iteration method for numerical approximation of stationary solutions of nonlinear wave equations. SIAM J. Numer. Anal. **42**, 1110–1127 (2004)

48. Pelloni, B., Dougalis, V.A.: Numerical solution of some nonlocal nonlinear dispersive wave equations. J. Nonlinear Sci. **10**, 1–22 (2000)
49. Pelloni, B., Dougalis, V.A.: Error estimates for a fully discrete spectral scheme for a class of nonlinear, nonlocal dispersive wave equations. Appl. Numer. Math. **37**, 95–107 (2001)
50. Petviashvili, V.I.: Equation of an extraordinary soliton. Soviet J. Plasma Phys. **2**, 257–258 (1976)
51. Sanz-Serna, M., J., Calvo, M.P.: Numerical Hamiltonian Problems. Chapman & Hall, London (1994)
52. Shira, V.: Propagation of long nonlinear waves in a layer of a rotating fluid. Iza. Akad. Nauk SSSR, Fiz Atmosfery i Okeana **17**, 76–81 (1981)
53. Shira, V.: On long essentially nonlinear waves in a rotating ocean. Iza. Akad. Nauk SSSR, Fiz Atmosfery i Okeana **22**, 395–405 (1986)
54. Sidi, A.: Convergence and stability of minimal polynomial and reduced rank extrapolation algorothms. SIAM J. Numer. Anal. **23**, 197–209 (1986)
55. Sidi, A., Ford, W.F., Smith, D.A.: Acceleration of convergence of vector sequences. SIAM J. Numer. Anal. **23**, 178–196 (1986)
56. Smith, D.A., Ford, W.F., Sidi, A.: Extrapolation methods for vector sequences. SIAM Rev. **29**, 199–233 (1987)
57. Thomée, V., Vasudeva Murthy, A.S.: A numerical method for the Benjamin-Ono equation. BIT **38**, 597–611 (1998)
58. Trefethen, L.N.: Spectral Methods in MATLAB. SIAM, Philadelphia (2000)
59. Tsuwaga, K.: Well-posedness and weak rotation limit for the Ostrovsky equation. J. Diff. Eq. **247**, 3163–3180 (2009)
60. Varlamov, V., Liu, Y.: Cauchy problem for the Ostrovsky equation. Disc. Dyn. Syst. **10**, 731–751 (2004)

On the Numerical Approximation to Generalized Ostrovsky Equations: II

Dynamics of Solitary-Wave Solutions

Ángel Durán

Abstract In this chapter generalized versions of the Ostrovsky equations are considered. These were shown to admit classical and generalized solitary wave solutions. The periodic initial-value problem for the equations is numerically solved with a fully discrete scheme based on pseudospectral discretization in space and a fourth-order composition Runge-Kutta method as time integrator. The resulting scheme is checked and applied to study numerically the dynamics of the solitary wave solutions. Specifically, we analyze the stability of classical and generalized solitary waves under small perturbations, the resolution of initial data into several solitary pulses (the so-called resolution property) and various aspects of the interaction of the solitary waves.

Keywords Generalized Ostrovsky equation · Fourier collocation method
Petviashvili-type methods · Solitary waves · Stability · Resolution property

1 Introduction

In a previous chapter of this volume [20], the following generalized Ostrovsky equation

$$(u_t - \beta u_{xxx} + f(u)_x)_x = \gamma u, \quad x \in \mathbb{R} \tag{1}$$

was introduced. Equation (1) is used as mathematical model for the propagation of internal waves in a rotating fluid in a horizontal channel. The variables x and t represent, respectively, the distance along the channel and the time, $u(x, t)$ is the fluid velocity in the horizontal direction at position x and time t, β and γ are constant

Á. Durán (✉)
Applied Mathematics Department, University of Valladolid, 47011 Valladolid, Spain
e-mail: angel@mac.uva.es

V. Carmona et al. (eds.), *Nonlinear Systems, Vol. 1*,
Understanding Complex Systems, https://doi.org/10.1007/978-3-319-66766-9_13

parameters (where $\gamma > 0$ is assumed) which govern, respectively, the small-scale Boussinesq dispersion and the dispersion due to rotation. Finally f, standing for the nonlinear effects in the model, is a twice continuously differentiable, real-valued, homogeneous function of degree $p \geq 2$, in the sense that

$$f(\lambda s) = \lambda^p f(s), \quad \lambda > 0, \quad s \in \mathbb{R}. \tag{2}$$

Condition (2) implies that f can be written in the form [38]

$$f(u) = a_e|u|^p + a_o|u|^{p-1}u, \quad a_e, a_o \in \mathbb{R}. \tag{3}$$

In particular, the choice $f(s) = \pm s^2$ corresponds to the classical Ostrovsky equation [5, 27, 33, 47].

Equation (1) in its general form was introduced by Levandosky and Liu [38], where the study of the corresponding Cauchy problem is referred to that of the classical Ostrovsky equation [34, 40, 57, 58]. Thus, Eq. (1) is well-posed in suitable Sobolev spaces (see [20] and references therein), satisfies the zero-mass condition

$$I(u) = \int_{-\infty}^{\infty} u(x,t)dx = 0,$$

and for smooth and decaying enough solutions, the momentum and energy

$$V(u(t)) = \int_{-\infty}^{\infty} u(x,t)^2 dx, \tag{4}$$

$$E(u(t)) = \int_{-\infty}^{\infty} \left(\frac{\beta}{2}u_x(x,t)^2 + \frac{\gamma}{2}(\partial_x^{-1}u(x,t))^2 + F(u(x,t))\right) dx, \tag{5}$$

are conserved. In (5), $F' = f$, $F(0) = 0$ and the operator ∂_x^{-1} is defined by using the Fourier symbol as

$$\mathcal{F}(\partial_x^{-1} f)(\xi) = (i\xi)^{-1}\mathcal{F}(f)(\xi), \quad \xi \in \mathbb{R}\backslash\{0\}, \quad \mathcal{F}(\partial_x^{-1} f)(0) = 0,$$

where $\mathcal{F}$ stands for the Fourier transform

$$\mathcal{F}(f)(\xi) = \int_{-\infty}^{\infty} e^{-i\xi x} f(x)dx.$$

The last quantity (5) provides a Hamiltonian structure for (1), see [20] for details.

Another relevant property of (1) is the existence of solitary wave solutions. They are solutions u of the form $u(x,t) = \phi_c(x-ct)$ for some function $\phi_c(X)$, $X = x - ct$ and speed c. The profile ϕ_c must satisfy

$$\left(-c\phi_c' - \beta\phi_c''' + f(\phi_c)'\right)' - \gamma\phi_c = 0. \tag{6}$$

Existence results on solitary waves are mainly concerned with the classical Ostrovsky equation, with [13, 15–17, 38, 42] as main references. In particular, normal form theory is applied in [16, 17] to prove the existence of classical solitary waves (CSW), for which $\phi_c \to 0$ as $|X| \to \infty$, when $\gamma, \beta > 0$ and $c < c^* = 2\sqrt{\beta\gamma}$. This was previously derived in [42] and also holds in the generalized case (1). On the other hand, generalized solitary waves (GSW), for which ϕ_c is homoclinic to small oscillations at infinity, are discussed in [15] as microterons, see also [16]. Their existence is suggested in [23] for the case of $f(u) = u^p/p$, $p > 1$.

An additional point on the solitary waves concerns their stability. Here a brief discussion between two types, orbital and asymptotic stability, is made. Orbital stability concerns stability of the waves modulo the symmetry groups associated to the equations, in case they exist. (In our case, we have the group of spatial translations; this essentially means that if $u(x, t)$ is a solution, then any spatial translation $u(x + x_0, t)$, $x_0 \in \mathbb{R}$ is also a solution. This is sometimes related to the preservation of invariants [46].) The definition for the case of the Ostrovsky equations (1), (3) can be seen in [38]. It is therefore a concept of stability for the orbits by the symmetry group. Essentially, it says that a solitary wave is orbitally stable if, given small perturbations of the initial profile, the corresponding solution of (1) is, in some norm, close to any element of the orbit of the solitary wave (that is, in our case, close to some translation of the solitary wave profile) at any time $t > 0$.

The rigorous theory of orbital stability of the solitary waves for the Korteweg-de Vries (KdV) and the Benjamin-Bona-Mahony (BBM) equations appears in [6, 7], by using variational theory and the characterization of the solitary waves as extremal of some invariant of the problem constrained to a fixed value of a second conserved quantity and with the speed of the solitary wave as a Lagrange multiplier. The method was extended to many other specific and more general situations in many papers, see e.g. [1, 12, 26, 59, 60]. This theory was applied to the classical Ostrovsky equation in [42] (see also [39, 41]) and to (1), (3) in [38].

When the solitary waves are generated as solutions of a constrained variational problem of a functional (energy), constrained to some fixed value of another one (momentum), then this theory claims that orbital stability can be measured in terms of the speed as Lagrange multiplier [26]. Specifically, in the case of (1), solitary waves $\varphi(x - ct)$ are critical points of

$$L(u) = E(u) - cV(u),$$

where V and E are given by (4) and (5), respectively, and c is the speed of the wave. Then, defining the function

$$d(c) = d(\beta, c, \gamma) = E(\varphi) - cV(\varphi), \tag{7}$$

in [39] it is proved that, for $\beta, \gamma > 0$ and $c < c^*$, φ is orbitally stable when

$$\frac{\partial^2}{\partial c^2} d(\beta, c, \gamma) > 0,$$

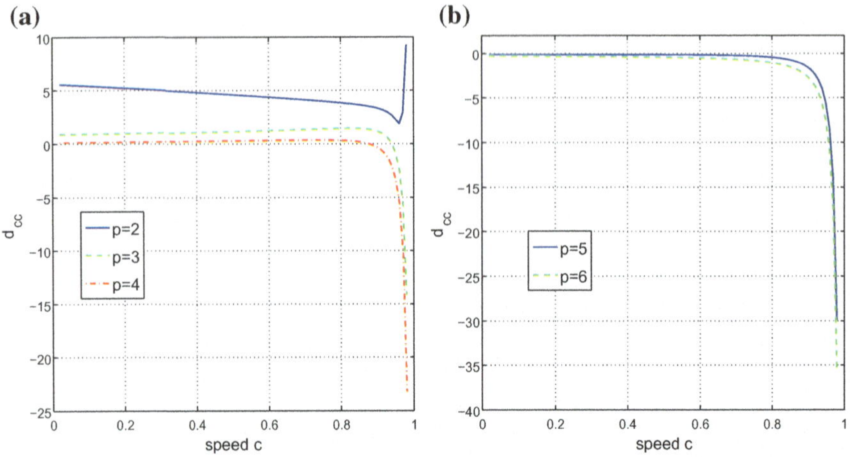

Fig. 1 Behaviour of $\frac{\partial^2 d}{\partial c^2}$ as function of the speed c with $\gamma = 0.25$, $\beta = 1$ and $f(u) = |u|^p$

and unstable when

$$\frac{\partial^2}{\partial c^2} d(\beta, c, \gamma) < 0.$$

The quantity (7) can be computed by using the definition or any of the alternatives provided in [38] along with, (see [38], Theorem 3.1)

$$\frac{\partial d}{\partial c}(\beta, c, \gamma) = -V(\varphi),$$

and then approximating $\frac{\partial^2 d}{\partial c^2}$ by a numerical differentiation formula of second order (central differences). The discrete versions of V and E will be used (see Sect. 3). The results corresponding to the nonlinearity $f(u) = |u|^p$ ($a_e = 1, a_o = 0$ in (3)) are shown in Fig. 1. The main results of orbital stability are summarized in the Introduction of [37], points (i)–(v). They establish stability for CSWs of speed c close to c^*, γ small and $2 \leq p < 5$, while for $p > 5$ (and γ small), the waves are unstable. In the same paper, there are two other conjectures (Conjecture 7.1), as a consequence of the numerical experiments. The numerical results shown in Fig. 1 are in accordance with Table 1 in [37], fifth column (that corresponds to the chosen nonlinearity).

The second concept of stability, asymptotic stability, gives more information about the long time behaviour of solutions u of (1) from initial small perturbations of the solitary wave profiles. The theory assumes that for long times, u behaves like

$$u(x, t) = \varphi_{s_\infty}(x - c_\infty t + x_{0,\infty}) + z(x, t)$$

where φ_{s_∞} is a solitary wave with some speed c_∞ (close to the speed c_s of the original solitary wave profile) and $x_{0,\infty}$ is a phase shift. The second term $z(x,t)$ represents a remainder consisting of small amplitude dispersive oscillatory waves and probably smaller solitary waves. Thus the solitary wave is said to be asymptotically stable when $z \to 0$ as $t \to \infty$ in a suitable sense. We refer to the Introduction of [19] for a more detailed explanation and many references. Essentially, in the case of asymptotically stable solitary waves, what one would typically see, from initial small perturbations of the profile, is an emerging wave that evolves to a solitary wave profile plus ripples or tails and probably other smaller nonlinear waves; the asymptotic behaviour should tend to the persistence of the main wave, evolving to a real asymptotic solitary wave profile, and the disappearance of the remaining waves. The theory requires an exhaustive analysis of the linearization of the equation at the solitary wave, along with estimates of the decay of solutions of the equations satisfied by z, see e.g. [45, 48, 49] (and [21, 43] for a different approach, more related to the use of modified equations satisfied by the asymptotic profile).

To our knowledge, this type of stability has not been studied in the case of classical solitary waves of the Ostrovsky equations (1). In the particular case of GSWs, computational or theoretical studies on the dynamics are less common in the literature, see e.g. the experimental results in [9, 10], for a type of Boussinesq system.

Remark 1 We also mention that the evolution of a KdV solitary-wave profile as initial condition under the classical Ostrovsky equation has been studied, numerically and analytically, in many references [24, 25, 28–30, 32], as a way to analyze the effect of rotation, governed by the parameter γ.

In the first chapter of this study [20], two numerical tools to analyze the dynamics of (1) were introduced. The first one is a scheme of approximation to the solutions, based on a pseudospectral discretization in the spatial variable and a fourth-order, simply diagonally composition method of Runge-Kutta type to integrate in time the resulting semidiscrete system. These choices are justified by the nonlocal character of (1) and the suitability of the time integrator for nonlinear wave problems [22]. The second tool is a numerical technique to generate approximations to classical and generalized solitary wave profiles ϕ satisfying (6). The procedure is based on the Petviashvili's method [51], combined with acceleration techniques to improve the performance and to approximate efficiently highly oscillatory classical solitary waves. Properties concerning the speed-amplitude relation and the asymptotic behaviour of the waves are suggested by the numerical simulations.

In this second chapter, the two procedures will be used to study by numerical means the dynamics of the classical and generalized solitary waves of (1). The study will be focused on the effect of small and large perturbations as well as collisions of the waves. The experiments are mainly devoted to shed some light on the stability of the waves and the dependence of the dynamics on the speed and the degree of homogeneity p.

From the plethora of experiments performed (some of them are shown in this chapter) we derive the following conclusions.

- Both classical and generalized solitary waves are asymptotically stable under small perturbations of the amplitude (and speed) parameters. Perturbed initial data evolve to a new solitary pulse. Small ripples and tails that tend to disperse as time goes by are observed behind and in front of the main pulse (in the case of CSWs) and just behind (in the case of GSWs).
- For larger perturbations, the so-called resolution property is observed in both the CSW and GSW cases: the profiles resolve into a train of solitary waves plus dispersive tails to the left and to the right, along with apparently nonlinear ripples which may monitor the generation of new pulses.
- Interactions of CSWs and of GSWs are inelastic, as expected. (The classical Ostrovsky equation is proved to be nonintegrable in [17].) In the case of overtaking collisions of CSWs (that is, the two waves travel in the same direction), the degree of inelasticity depends on the speed (or amplitude) of the waves to be collided, being low when both pulses are slow, a bit higher when they are both faster and clearly strong when the difference in speeds grows. In the first two cases, some not dispersive ripples are apparently formed, leading to an increment of the energy of the perturbed wave. This is not observed in the fast-slow interaction. On the other hand, head-on collisions of CSWs (when the two waves travel in opposite directions) show, according to the speed-amplitude relation for that case, the same behaviour.

The chapter is structured as follows. Section 2 is devoted to summarize the description of the numerical techniques implemented for the computations, explained in more detail in [20]. This is complemented by some additional experiments of validation. The computational study of the dynamics of the solitary waves makes up Sect. 3. To this end the behaviour of small solutions of (1) is first estimated by using the corresponding linearized equation. Then small and large perturbations of classical and generalized solitary waves are illustrated and analyzed. Small perturbations of the parameters suggest stability of the waves for long times, while the case of large perturbations include experiments leading to the resolution property and experiments of weak and strong interactions. We refer to [20] for the notation used.

The numerical study was perfomed for values of the nonlinearity parameter up to the limit $p = 5$, obtained by Levandosky and Liu [38], for the orbital stability of CSWs. This excludes from this computational work, for instance, the analysis of the possible singularity formation for larger values of p, obtained in generalized versions of other classical nonlinear dispersive wave equations [8, 11]. This approach would take part (along with a more exhaustive analysis of some phenomena observed in this numerical study) of a future research.

2 The Numerical Procedures

In this section we give a brief review of the fully discrete pseudospectral scheme that is used to discretize the periodic initial-value problem for (1) as well as of the accelerated Petviashvili's iteration to generate approximate solitary wave profile solutions of (6). See [20] for details.

2.1 Discretization of the Periodic Initial-Value Problem

Let N be a positive integer, $l > 0$ and consider the space of trigonometric polynomials of degree N

$$S_N = span\{e^{\frac{i\pi k}{l}x}, -N \leq k \leq N\},$$

and the subspace of S_N consisting of zero-mean polynomials

$$S_N^0 = \{\phi \in S_N / \widehat{\phi}(0) = \int_{-l}^{l} \phi(x)dx = 0\} = span\{e^{\frac{i\pi k}{l}x}, -N \leq k \leq N, k \neq 0\}.$$

The semidiscrete Fourier-Galerkin approximation of the $2l-$periodic initial-value problem for (1) with initial condition given by a $2l-$periodic function u_0 is defined as the map $u^N : [0, \infty) \to S_N$ such that for all $\chi \in S_N$,

$$\begin{aligned} &(u^N_{tx}, \chi) + ((-\beta u^N_{xxx} + f(u^N)_x)_x - \gamma u^N, \chi) = 0, \quad t > 0, \\ &u^N(x, 0) = P_N u_0(x). \end{aligned} \tag{8}$$

The periodic conditions lead to the equivalent formulation

$$\begin{aligned} &(u^N_t, \chi_x) + ((-\beta u^N_{xxx} + f(u^N)_x), \chi_x) + (\gamma u^N, \chi) = 0, \quad t > 0, \\ &u^N(x, 0) = P_N u_0(x), \end{aligned} \tag{9}$$

where P_N is the orthogonal projection of L^2 onto S_N. By taking $\chi = 1$ in (9) we obtain that $u^N(\cdot, t) \in S_N^0, t > 0$. This means that u^N satisfies the zero-mass condition

$$I_l(u^N(\cdot, t)) = \int_{-l}^{l} u^N(x, t)dx = 0, \quad t > 0.$$

Standard arguments prove the local existence and uniqueness of solutions of (9). The solution is global in time by using the preservation of the L^2 norm of the discretization

$$V_l(u^N(\cdot, t)) = \int_{-l}^{l} u^N(x, t)^2 dx = V_l(u_0), \quad t > 0.$$

The conservation of the energy

$$E_l(u^N(\cdot, t)) = \int_{-l} \left(\frac{\beta}{2} u^N_x(x, t)^2 + \frac{\gamma}{2}(\partial_x^{-1} u^N(x, t))^2 + F(u^N(x, t))\right) dx = E_l(u_0),$$

also holds, where $F' = f$, $F(0) = 0$ and if

$$u^N(x,t) = \sum_{-N\leq k\leq N} \widehat{u^N}(k,t)e^{ikx},$$

then

$$v^N(x,t) = \sum_{-N\leq k\leq N} \widehat{v^N}(k,t)e^{ikx}, \quad \widehat{v^N}(k,t) = \frac{\widehat{u^N}(k,t)}{ik}, \quad k \neq 0; \quad \widehat{v^N}(0,t) = 0.$$

The system of ordinary differential equation derived from (8) is stiff and for a stable time discretization, the use of implicit schemes is required. The fourth-order, diagonally implicit Runge-Kutta method of tableau

$$\begin{array}{|lll} \beta_1/2 & & \\ \beta_1 & \beta_2/2 & \\ \beta_1 & \beta_2 & \beta_1/2 \\ \hline \beta_1 & \beta_2 & \beta_1 \end{array} \quad \beta_1 = \frac{1}{2-2^{1/3}}, \quad \beta_2 = 1 - 2\beta_1, \tag{10}$$

has been shown to be suitable for nonlinear wave problems [22]. The method (10) is a composition of three stages of length $\beta_1\Delta t$, $\beta_2\Delta t$ and $\beta_1\Delta t$, of the second-order Gauss-Legendre implicit Runge-Kutta method (or implicit midpoint rule), with Δt standing for the time step-size. Thus (10) inherits simplecticity and, as a consequence, the preservation of the discrete L^2 norm [52]. Additionally, it satisfies the discrete version of the zero-mass condition and the implicit systems of the intermediate stages can be numerically solved by using the classical fixed point iteration.

2.2 *Numerical Generation of Solitary Waves*

In order to describe the iterative technique to approximate solitary wave profile solutions of (6), let N be a positive integer and define a uniform grid on a long enough interval $(-l, l)$, $l > 0$ by $x_j = -l + jh$, $j = 0, \ldots, N-1$, $h = 2l/N$. For $\mathbf{v} = (v_0, \ldots, v_{N-1})^T \in \mathbb{R}^N$ we consider the operators

$$\mathscr{L}_h\mathbf{v} = (\gamma I_N + cD_N^2 + \beta D_N^4)\mathbf{v}, \quad \mathscr{N}_h(\mathbf{v}) = D_N^2 f(\mathbf{v}),$$

where I_N is the $N \times N$ identity matrix, D_N stands for the pseudospectral differentiation matrix of size N [14, 56], and the evaluations involved in $f(\mathbf{v})$ are understood in the Hadamard sense. We define the approximation of the profile solution ϕ_c at the grid x_j, $j = 0, \ldots, N-1$, as the vector $\phi_h \in \mathbb{R}^N$ satisfying

$$\mathscr{L}\phi_h - \mathscr{N}(\phi_h) = 0. \tag{11}$$

The system of algebraic equations (11) for the components of ϕ_h is solved iteratively by using the Petviashvili's method [36, 50, 51]: if $\phi^{[0]}$ stands for an initial iteration, then the iterative step $\nu \mapsto \nu + 1$, $\nu = 0, 1, \ldots$ is implemented as

$$m(\phi^{[\nu]}) = \frac{(\mathscr{L}_h \phi^{[\nu]}, \phi^{[\nu]})}{(\mathscr{N}_h(\phi^{[\nu]}), \phi^{[\nu]})}, \tag{12}$$

$$\mathscr{L}_h \phi^{[\nu+1]} = m(\phi^{[\nu]})^{\frac{p}{p-1}} \mathscr{N}_h(\phi^{[\nu]}), \quad n = 0, 1, \ldots, \tag{13}$$

where $\langle \cdot, \cdot \rangle$ denotes the Euclidean inner product in $\mathbb{R}^N$. The use of (12), (13) provides a simple iterative technique of fixed-point type that overcomes the general lack of convergence of the classical fixed-point algorithm [3]. The iteration is implemented in Fourier space by using the Discrete Fourier Transform (DFT) [56], and is complemented by the inclusion of an acceleration technique based on extrapolation [35, 53–55]. The acceleration improves in general the performance and in some cases is able to transform initial divergent processes into convergent iterations. This is particularly interesting in solitary wave generation, see [4]. The convergence of (13) was checked in [20] with the generation of approximations to classical and generalized solitary wave solutions of (1) as well as multi-pulse classical and generalized profiles.

2.3 Some Experiments of Validation

The accuracy of both the numerical profiles generated by (13) and the fully discrete method described in Sect. 2.1 is now checked with some numerical experiments.

The first profile considered, by way of illustration, approximates a solitary wave solution of the classical Ostrovsky equation ($p = 2, a_e = 1, a_o = 0$ in (3)) with parameter values $\gamma = 0.25, \beta = 1$ and speed $c = 0.5$. The computed profile has been taken as initial condition of the time stepping code in the (spatial) interval $[-128, 128]$ and evolved up to a final time $t^* = 100$ with three time steps $\Delta t = 6.25E{-}03, 3.125E{-}03, 1.5625E{-}03$. As a result of this evolution, several figures have been generated. Figure 2a shows the numerical approximation at times $t = 0, 20, 40, 60, 80, 100$ and $\Delta t = 6.25E{-}03$, representing the evolution of the initial computed wave. We observe that the numerical solution does not look to develop any relevant backward or forward disturbance, which may be a good sign of the accuracy of both the initial profile (as approximation to the exact solitary wave) and the numerical integration from it (as approximation to the evolution of the exact wave). This apparently good accuracy of the profile and the evolution is somehow confirmed by the following additional experiments. Figure 2b shows, for several values of Δt, the evolution of the error between the amplitude (maximum negative excursion) of the initial profile and that of the numerical approximation at time t, computed as in e.g. [19]. According to the results, we can confirm the preservation of the amplitude (up to the final computed time $t^* = 100$) in an accurate way, which

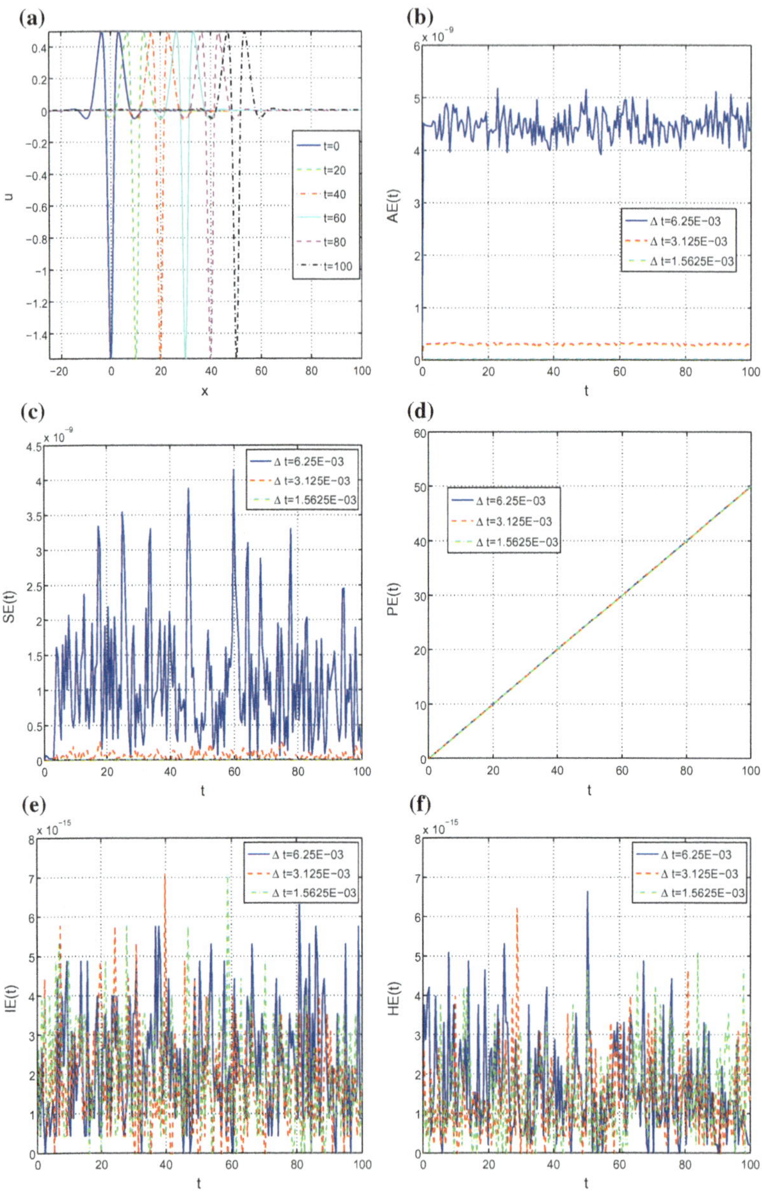

Fig. 2 **a** Numerical approximation of (1) from the profile computed with $\gamma = 0.25$, $\beta = 1$, $c = 0.5$ and $p = 2$, $a_e = 1$, $a_0 = 0$, at several times. **b–d** Amplitude, speed and phase errors versus time, for the numerical approximation of (1) from the profile computed in (**a**). **e**, **f** Momentum ($|V_h(U^n) - V_h(U^0)|$) and Energy ($|E_h(U^n) - E_h(U^0)|$) errors versus time, where V_h, E_h are given by (14), (15) resp. and from the profile computed in (**a**)

in the worst case is of order of $5E-09$. This good behaviour of the numerical integration is also observed in the evolution of the speed error, shown in Fig. 2c. Finally, the error in the phase, displayed in Fig. 2d, suggests a linear growth with time, which is in agreement with the expected behaviour for the global error, observed for the case of the generalized KdV equation in [20] and mentioned in the literature [2].

The last experiments for this case concern the time behaviour of the method with respect to the invariant quantities of the periodic problem. The corresponding discrete versions are

$$V_h(U) = \frac{h}{2}\sum_{j=0}^{N-1} U_j^2, \quad U = (U_0, \ldots, U_{N-1})^T, \tag{14}$$

$$E_h(U) = h\sum_{j=0}^{N-1}\left(\frac{\beta}{2}(D_N U)_j^2 + \frac{\gamma}{2}(D_N^{-1}U)_j^2 + F(U)_j\right), \tag{15}$$

where $V = D_N^{-1}U = (V_0, \ldots, V_{N-1})^T$ represents the solution of the system $D_N V = U$ with

$$\widehat{V}_0 = \sum_{j=0}^{N-1} V_j = 0, \tag{16}$$

where $\widehat{V}$ denotes the DFT of V. (Note that since $\widehat{U}(0) = 0$, then the system $D_N V = U$ is compatible and when the condition (16) is imposed, it also has a unique solution: this corresponds indeed to take the inverse DFT of the vector $\widehat{V}$ with $\widehat{V}_0 = 0$ and $\widehat{V}_k = \widehat{U}_k/(ik)$ when $k \neq 0$.) Figure 2e shows the behaviour in time of the difference between the value of (14) at the initial condition and that of the numerical approximation at times $t_n = n\Delta t$ for three time steps and up to a final time $t^* = 100$. (Formula (14) is computed at each unit of time.) The evolution of the error confirms the virtual preservation of (14) (the error is always below $8E-15$). The time behaviour of the error in the Hamiltonian (15), observed in Fig. 2f, is similar and can be explained by the preservation of (14) and the relation between the two quantities at the solitary wave, see the Introduction. This behaviour would then be an additional sign of the high accuracy in the computation of the numerical profiles. The same experiments have been performed for higher degrees of homogeneity p, leading to the same conclusions.

In an analogous way, we illustrate the simulation of generalized solitary wave solutions with the numerical profile obtained in the case $\gamma = 0.002$, $\beta = -1$, $c = 1$ and $p = 2$, $a_e = 1$, $a_0 = 0$, on an interval $[-1024, 1024]$ with $N = 4096$ collocation points. The time stepping code has been run from this profile as initial condition and up to a final time $t^* = 100$. The results are displayed in Fig. 3. (The phase error is not shown, but its behaviour is similar to that observed in the simulation of classical solitary waves.)

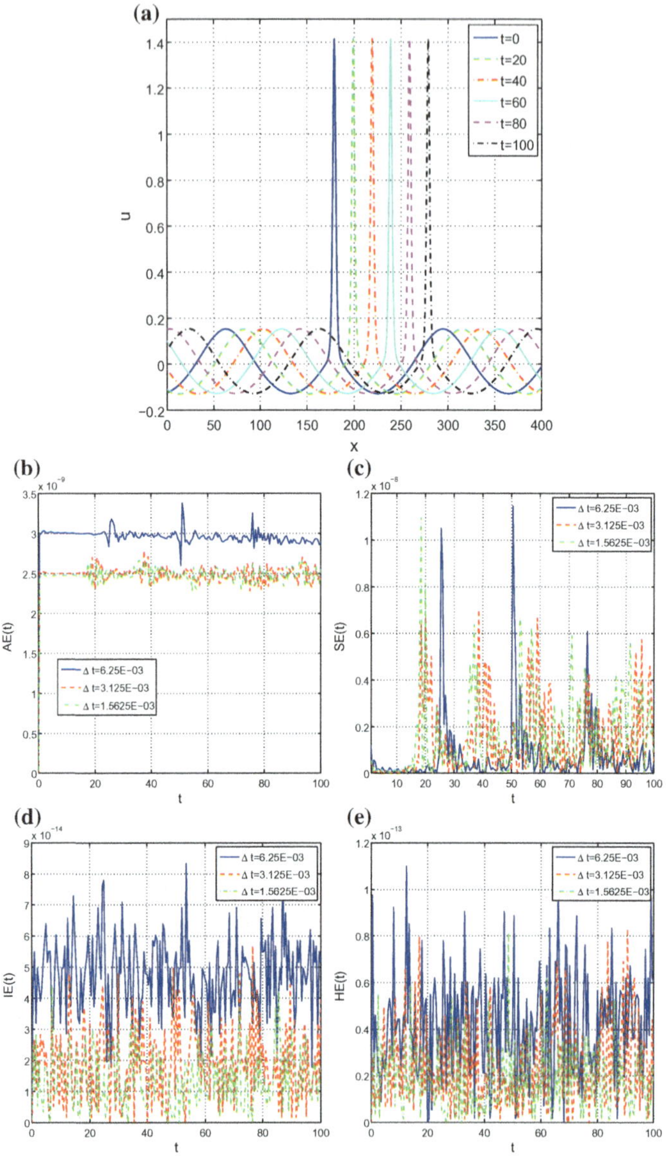

Fig. 3 **a** Numerical approximation of (1) from the profile computed with $\gamma = 0.002$, $\beta = -1$, $c = 1$ and $p = 2$, $a_e = 1$, $a_0 = 0$, at several times. **b**, **c** Amplitude, and speed errors versus time, for the numerical approximation of (1) from the profile computed in (**a**). **d**, **e** Momemtum $|V_h(U^n) - V_h(U^0)|$) and Energy ($|E_h(U^n) - E_h(U^0)|$) errors versus time, where V_h, E_h are given by (14), (15) resp. and from the profile computed in (**a**)

3 Dynamics of Solitary Waves. A Numerical Study

The aim of this section is to study by numerical means the dynamics of solitary wave solutions of (1). The experiments will involve both the classical and generalized solitary waves and will concern the stability of the waves under small perturbations of the parameters, the resolution of initial data into a train of solitary waves (the so-called resolution property) and the interactions of solitary waves (along with properties of the emerging waves after the collisions). The degree p of homogeneity of the nonlinearity used in the computations is limited to $2 \leq p < 5$, the bound established for the orbital stability results of CSWs in [38]. The dynamics for $p > 5$ is expected to be analyzed in the future.

3.1 Study of Small Tails and Ripples

In order to try to explain the results suggested by the experiments below, it may help to describe the behaviour of small amplitude solutions of (1). Assume first that $\beta, \gamma > 0$ and take a (classical) solitary wave of speed $c = c_s < c^* = 2\sqrt{\beta\gamma}$. Then small amplitude solutions of (1) evolve, in a frame moving with the solitary wave, $y = x - c_s t$, according to the linearized equation

$$\partial_y((\partial_t - c_s\partial_y)u - \beta\partial_{yyy}u) - \gamma u = 0.$$

For plane wave solutions $u(x,t) = e^{i(ky-\omega(k)t)}$, $k \in \mathbb{R}$, we have

$$ik\left((-i\omega(k) - ic_s k)u - \beta(ik)^3 u\right) - \gamma u = 0.$$

That is, for $k \neq 0$,

$$\omega(k) = -c_s k + \beta k^3 + \frac{\gamma}{k}.$$

The local phase speed, relative to the speed of the solitary wave, is therefore

$$v(k) = \frac{\omega(k)}{k} = -c_s + \phi(k^2), \quad \phi(x) = \beta x + \frac{\gamma}{x}.$$

The function $\phi(x)$, $x > 0$ has a minimum at $x^* = \sqrt{\gamma/\beta}$, see Fig. 4a. Thus

$$\phi(x) \geq \phi(x^*) = 2\sqrt{\beta\gamma} > 0, \quad x > 0.$$

Then, for $k \neq 0$ and since $c_s < c^*$,

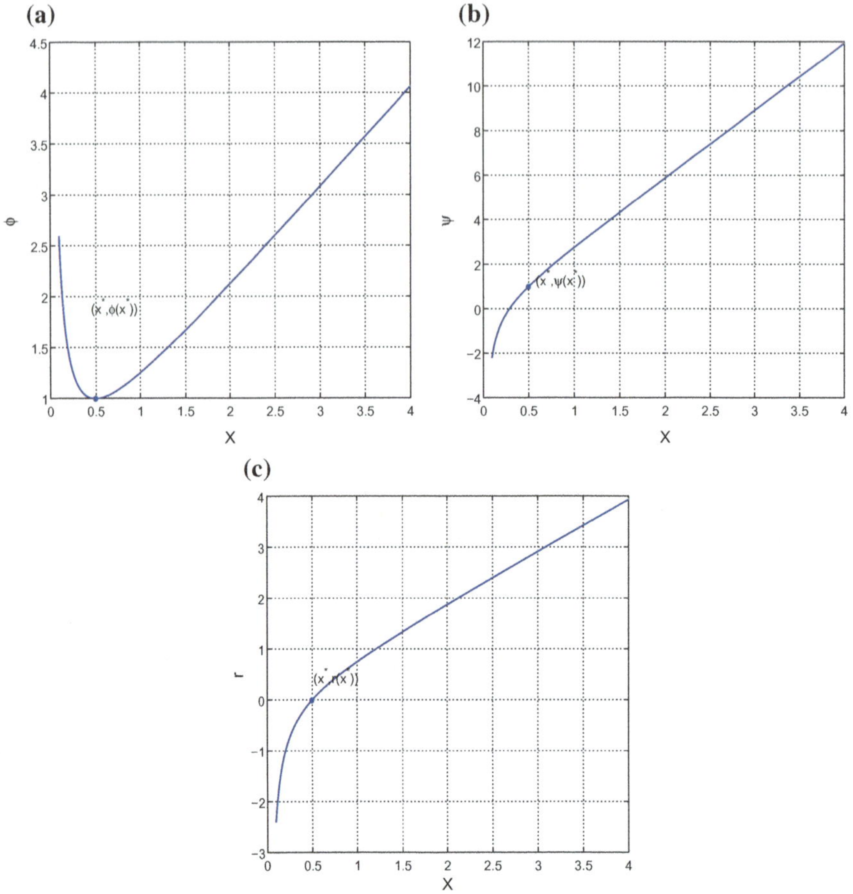

Fig. 4 Case $\gamma = 0.25$, $\beta = 1$ (then $x^* = 1/2$). **a** Function $\phi(x) = \beta x + \frac{\gamma}{x}$, $(\phi(x^*) = 1)$. **b** Function $\psi(x) = 3\beta x - \frac{\gamma}{x}$, $x > 0$ $(\psi(x^*) = 1)$. **c** Function $r(x) = \beta x - \frac{\gamma}{x}$, $x > 0$

$$v(k) = -c_s + \phi(k^2) \geq -c_s + 2\sqrt{\beta\gamma} > 0.$$

Therefore $v(k) > 0$ for all wavenumbers $k \neq 0$ and the solution component $e^{i(ky-\omega(k)t)}$ is leading the solitary pulse. This suggests the existence of propagating ripples in front of the main pulse.

Since ϕ is decreasing for $0 < x < x^*$ and increasing for $x > x^*$ (going to infinity with x) then for the components with wavenumbers $k > x^*$, those corresponding to longer wavelengths (smaller k) are slower than those of shorter wavelength. If x^* is small, then all the components will eventually have wavenumbers $k > x^*$, so the phase speed is fast. This generates a computational drawback, observed in some numerical experiments.

The associated group velocity is, in this frame of reference, of the form

$$\omega'(k) = -c_s + \psi(k^2), \quad \psi(x) = 3\beta x - \frac{\gamma}{x}, \quad x > 0.$$

The function ψ is increasing for all $x > 0$ with

$$\lim_{x\to 0+} \psi(x) = -\infty; \quad \lim_{x\to +\infty} \psi(x) = +\infty.$$

(See Fig. 4b.) Furthermore, $\psi(x) < 0$ if $x < x_c = \sqrt{\gamma/(3\beta)}$ and $\psi(x) > 0$ if $x > x_c$. (Note that $x_c < x^*$.) Note that $\psi(x) > c_s$ is satisfied when

$$P(x) = 3\beta x^2 - c_s x - \gamma > 0, \quad x > 0.$$

The polynomial $P(x)$ has two zeros at some $x = x- < 0$ and $x = x^*$. Thus, if $x > x^*$, then $P(x) > 0$ and $\psi(x) > c_s$, $\psi(x) > 0$ (since $x^* > x_c$). There is a group traveling to the right and in front of the solitary wave. Note that if $x < x_c$ then $\psi(x) < 0$ and for those k with $k^2 < x_c$ we have $\omega'(k) < -c_s$ so the group travels to the left.

If $x < x^*$ then $\psi(x) < c_s$ and those components with k such that $k^2 < x^*$ satisfy $\omega'(k) = -c_s + \psi(k^2) < 0$. They may travel to the left (if $\psi(k^2) < 0$) or to the right (if $\psi(k^2) > 0$) but behind the solitary wave. Note also that the difference between the phase speed and the group velocity is

$$\omega'(k) - v(k) = 2r(k^2), \quad r(x) = \beta x - \frac{\gamma}{x}, \quad x > 0.$$

The function $r(x)$ is increasing, positive when $x > x^*$ and negative when $x < x^*$ (see Fig. 4c). Therefore, the group velocity exceeds the phase speed for those components k with $k^2 > x^*$.

Remark 2 On the other hand, the behaviour for the case of generalized solitary waves has been studied in some references, see e.g. [23, 31] and references therein. By way of illustration, assume that $\beta < 0, \gamma > 0$. In this case, the function ϕ is always decreasing and the phase speed $v(k)$ is positive when $R(k^2) < 0$ being

$$R(x) = x^2 + Qx + P, \quad Q = -\frac{c_s}{\beta} > 0, \quad P = \frac{\gamma}{\beta} < 0.$$

The polynomial R has two real zeros $x_\pm$ with $x_- < 0 < x_+$ and then $R(x) < 0$ for $0 < x < x_+$. Therefore, for some wave number, we will have $k^2 > x_+$ and consequently $R(k^2) > 0$. Then the phase speed $v(k)$ will be negative: the solution component $e^{i(ky-\omega(k)t)}$ will be behind the pulse.

As for the group velocity $\omega'(k)$ note that now $\psi(x)$ is negative for all $x > 0$; thus $\omega'(k) < -c_s$ for all k and the group travels to the left (relative to the generalized solitary pulse).

3.2 Some Aspects of the Dynamics of Classical Solitary Waves

The first part of the computational study is concerned with the dynamics of classical solitary waves.

3.2.1 Small Perturbations of Classical Solitary Waves

In the following numerical experiments a first approach to the stability of the classical solitary waves under small perturbations is made. Among all the possibilities in this sense, see e.g. [19], we will focus on the dynamics of waves from small perturbations of the amplitude. The implementation will consist of computing an approximate profile ϕ_0 with the accelerated Petviashvili's method (13). Then the initial condition for the time stepping code will be of the form

$$u(x, 0) = r\phi_0(x), \tag{17}$$

with a perturbation parameter $r > 0$; thus, u is obtained from ϕ_0 with a perturbation of the amplitude via the factor r. The code is run and the corresponding numerical solution is monitored up to a final time of integration. As mentioned above, we limit our approach to the cases $2 \leq p < 5$ and $c < c^*$ in the case of classical solitary waves [37, 38]. Finally, among the experiments performed with nonlinearities of the form (3), only those with $a_e = 1$, $a_0 = 0$, for which $f(u) = |u|^p$, and several values of p will be, for the sake of clarity, shown here.

In order to study by computational means the dynamics of the waves under small perturbations, several values of r have been used, By way of illustration, we take Eq. (1) with $\gamma = 0.25$, $\beta = 1$ (thus $c^* = 1$). In the first experiment, the corresponding classical solitary wave profile ϕ_0 with speed $c = 0.5$ and $p = 2$ is computed. From the perturbed wave (17) with $r = 1.05$, the numerical method is run with $\Delta t = 1.5625 \times 10^{-3}$ and up to a final time $t = 100$. In Fig. 5a–d the form of the numerical approximation at several times is displayed. Figure 5b is a magnification of Fig. 5a and Fig. 5d is a magnification of Fig. 5c. Confirming the results in Sect. 3.1, some ripples are observed behind and in front of the main emerging wave. Those in front of it travel very fast and this makes the computation harder. (The spatial interval is $[-1024, 1024]$.) The ripples look to disperse as time evolves. Figure 6a shows the evolution of the difference between the amplitude of the initial profile (without perturbation, about $u_m = -1.55818$) and that of the emerging wave. Note that this discrepancy is stabilized in some below 0.1. (The amplitude of the emerging wave at $t^* = 100$ is about $u_m^* = -1.635783$.) This is an argument in favour of the idea that a solitary wave profile, close to the unperturbed one, is generating. This is also confirmed by Fig. 6b, which displays the evolution of the difference between the speed of the unperturbed profile ($c = 0.5$) and that of the main emerging wave. The results suggest that this last one evolves with a constant speed, approximately $c^* = 0.4465$ (then the profile is slower than the unperturbed one). On the other hand, the preservation of the quantities (14) and (15) is under admissible thresholds, as

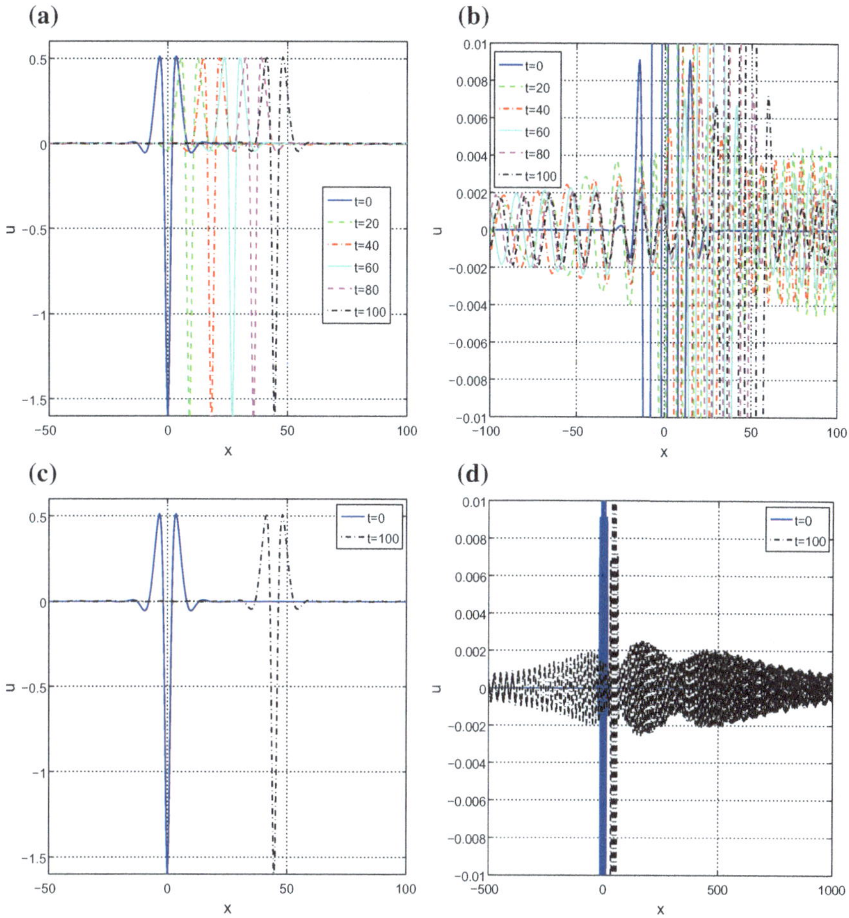

Fig. 5 Evolution of numerical approximation to a perturbed solitary wave for $\gamma = 0.25$, $\beta = 1$, $c = 0.5$, $p = 2$. The initial condition is of the form (17) with $r = 1.05$. **a** Numerical solution at different times; **b** magnification of (**a**); **c** numerical solution at times $t = 0, 100$; **d** magnification of (**c**)

shown in Fig. 6c, d. This behaviour is essentially preserved when p is increasing, $p < 5$. The results corresponding to $\gamma = 0.25, \beta = 1, c = 0.75, p = 3, r = 1.05$ are shown in Figs. 7 and 8. Since the speed is higher, the ripples on the right go faster (they cross the computational window in a shorter time). See also the differences in the behaviour of the evolution of the speed and of the Hamiltonian error. In this case, while the unperturbed wave has amplitude and speed $u_m = -1.049969$, $c = 0.75$, respectively, the emerging wave at $t^* = 100$ has values $u_m^* \approx -1.115819$, $c^* \approx 0.625$.

We observe that for moderate r, the initial perturbed wave evolves to the formation of only one classical solitary wave. The perturbation affects the amplitude (the

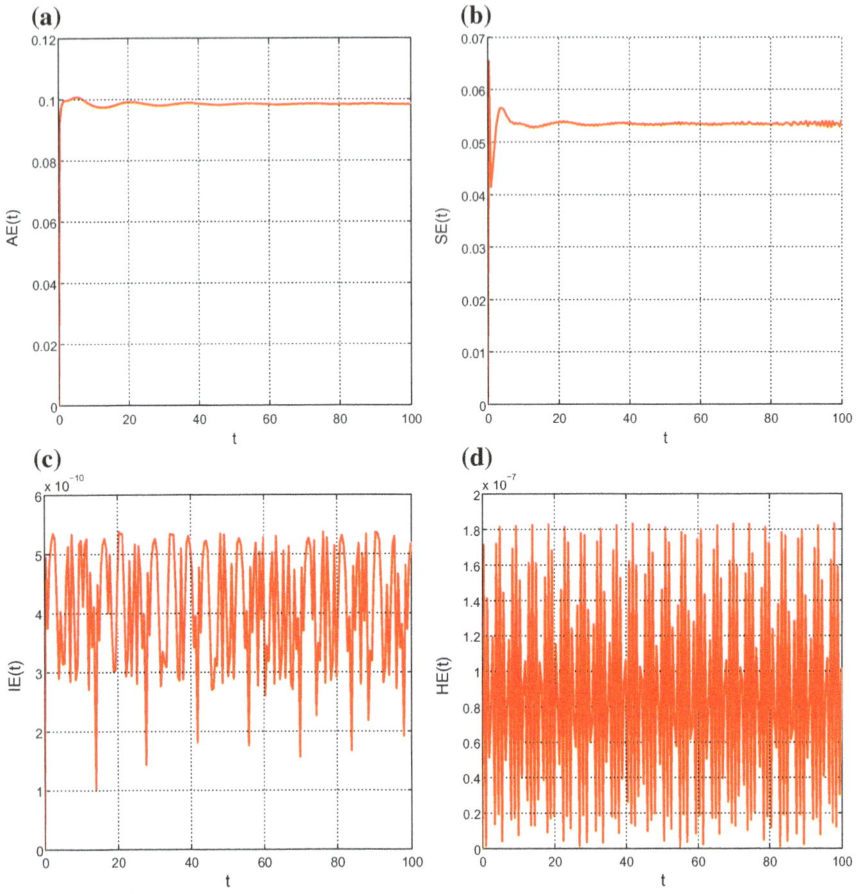

Fig. 6 Numerical approximation to a perturbed solitary wave for $\gamma = 0.25$, $\beta = 1$, $c = 0.5$, $p = 2$. The initial condition is of the form (17) with $r = 1.05$. **a** Amplitude error; **b** speed error; **c** momentum $|V_h(U^n) - V_h(U^0)|$) error; **d** Energy ($|E_h(U^n) - E_h(U^0)|$) error. All are displayed against time, and V_h, E_h are given by (14), (15) respectively

maximum negative excursion increases) and the speed (which decreases, becoming negative in some cases; the emerging wave travels to the left, see Table 1) but does not apparently generate more than one solitary pulse. In all the computed cases, the ripples look to disperse as $t \to \infty$.

3.2.2 Large Perturbations of Classical Solitary Waves

For larger perturbations, other types of behaviour are observed. This section is devoted to illustrate the phenomena of the resolution property and the collisions of classical solitary pulses.

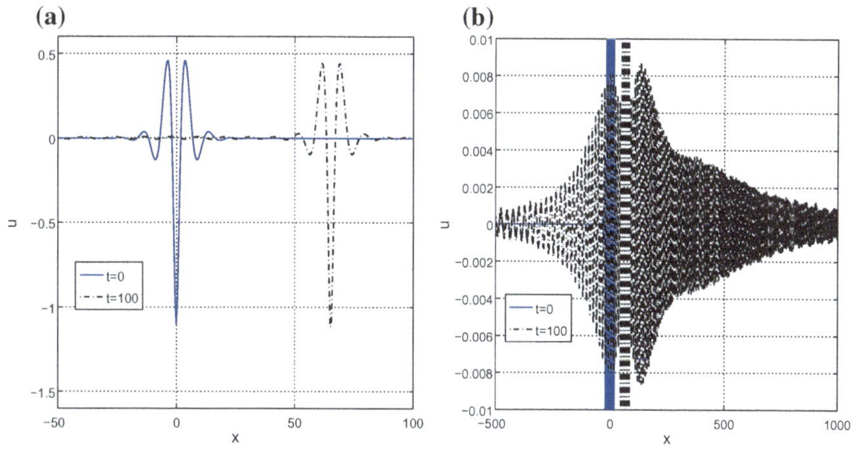

Fig. 7 Evolution of numerical approximation to a perturbed solitary wave for $\gamma = 0.25, \beta = 1, c = 0.75, p = 3$. The initial condition is of the form (17) with $r = 1.05$. **a** Numerical solution at times $t = 0, 100$; **b** magnification of (**a**)

Table 1 Amplitude and speed of the emerging classical solitary wave from (17) with ϕ_0 the computed profile corresponding to $\gamma = 0.25, \beta = 1, p = 2, c = 0.5$. The computed amplitude of ϕ_0 is $u_m = -1.55818$

r	u_m^N	c^N
1.2	−1.95231	0.2805
1.5	−2.528343	−0.072

Resolution property.

Resolution of arbitrary initial data into sequences of solitary waves plus some other tails is a typical property to study in the dynamics of the models and in some sense is related to the stability of solitary waves (see [19] and references therein). A typical ini tial data from which the solution resolves into a train of solitary waves are Gaussian-type functions. In this case, the resolution into more than one pulses is numerically observed when initial conditions of the form (17) with r large are considered. This is illustrated by Figs. 9 and 10a, which show the numerical results corresponding to considering a perturbed initial data (17) for (1) with $p = 2, c = 0.5, \gamma = 0.25, \beta = 1$ and $r = 4$. The evolution displays a clear formation of two solitary wave profiles, one traveling to the left and the other one to the right. Both develop ripples and tails to the right and to the left and the formation of new solitary wave profiles hidden into the ripples for longer times does not look to be dismissed. The maximum negative excursion of the first profile is $u_m^{(1)} \approx -7.518072$ with speed $c^{(1)} \approx -3.45$, while the second profile has amplitude $u_m^{(2)} \approx -1.291566$, see Fig. 10c, d. Some of the ripples might also contain smaller solitary wave profiles, not completely form by the final time of simulation.

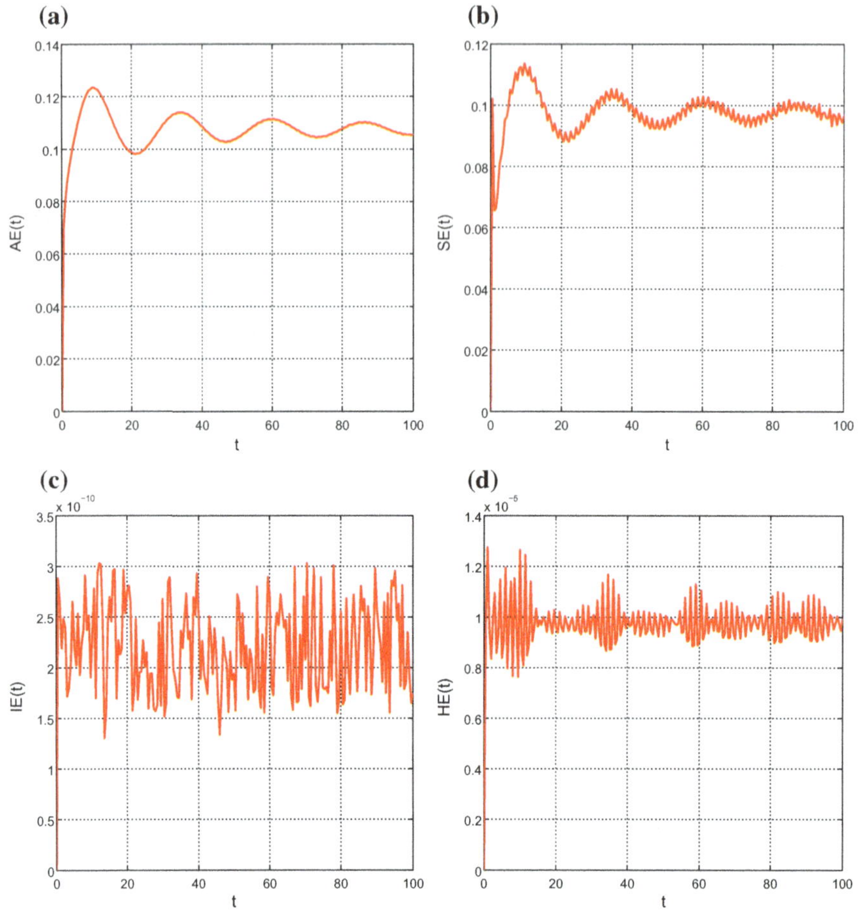

Fig. 8 Numerical approximation to a perturbed solitary wave for $\gamma = 0.25$, $\beta = 1$, $c = 0.75$, $p = 3$. The initial condition is of the form (17) with $r = 1.05$. **a** Amplitude error; **b** speed error; **c** momentum $|V_h(U^n) - V_h(U^0)|$) error; **d** energy ($|E_h(U^n) - E_h(U^0)|$) error. All are displayed against time, and V_h, E_h are given by (14), (15) respectively

Collisions of solitary waves

Interactions between solitary waves are usually studied with two main goals: the integrable/nonintegrable character of the equation and the stability of the solitary waves. In the first case, it is known [17], that the classical Ostrovsky equation is nonintegrable and it is reasonable to think that generalized versions (1), (3) will also be. Thus one can expect that when initial data consist of two solitary wave profiles with different speed and traveling in the same direction then the corresponding solution will evolve with the faster overcoming the slower one in an inelastic way. This means that the emerging waves will have, compared to the original, new amplitudes and speeds.

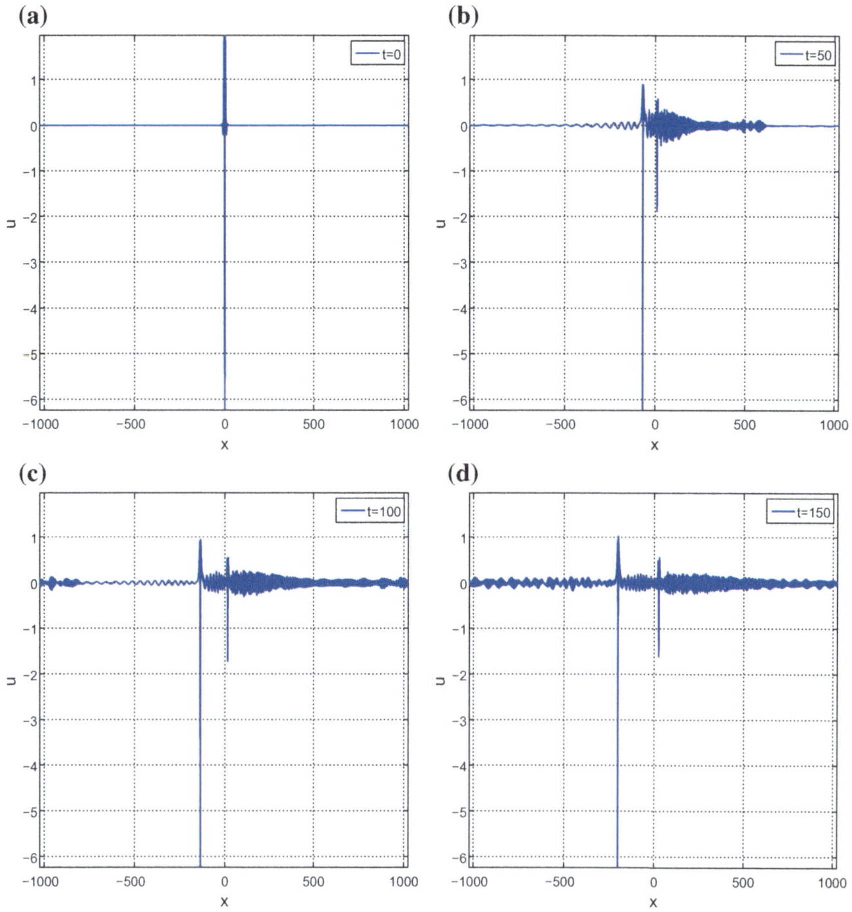

Fig. 9 Resolution into solitary waves. Numerical approximation at several times of (1) with $p = 2$, $\gamma = 0.25$, $\beta = 1$, $c = 0.5$ and initial condition of the form (17) with $r = 4$

According to the computations, the evolution after the interaction resolves in different ways. We observe that this behaviour depends on the speeds of the waves to interact (and therefore the amplitudes) in such a way that three categories, using the terminology *speed* $c^{(2)}$ *of the second wave-speed* $c^{(1)}$ *of the first wave* (with the second overtaking the first one), can be distinguished: slow-slow, fast-slow and fast-fast. The behaviour observed in the computations suggests a different stability of the waves for larger perturbations, always in terms of their speed/amplitude. We recall (see [20]) that as the speed approaches the limiting value c^*, the oscillatory decay of the classical solitary pulse increases and the maximum negative excursion decreases in magnitude. The slow-slow case is illustrated in Fig. 11. They correspond to the interaction of two classical solitary waves of (1) for $p = 3$, $\gamma = 0.25$, $\beta = 1$

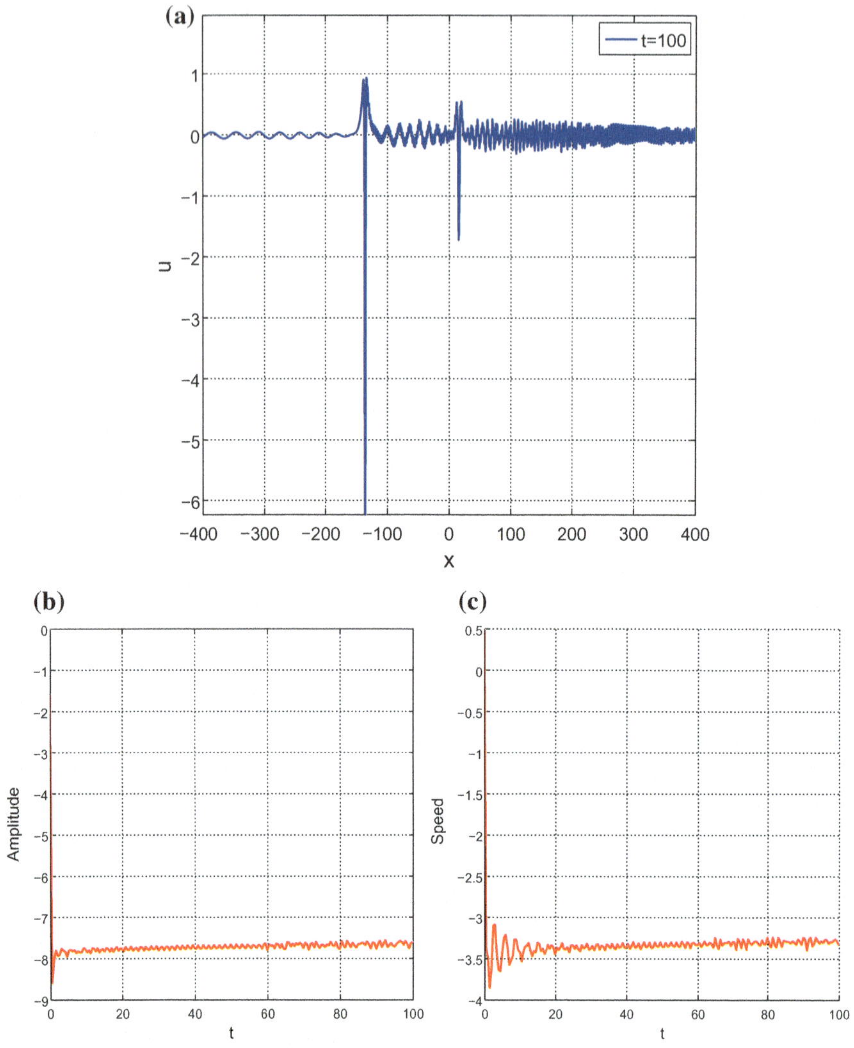

Fig. 10 Resolution into solitary waves. Numerical approximation of (1) with $p = 2$, $\gamma = 0.25$, $\beta = 1$, $c = 0.5$ and initial condition of the form (17) with $r = 4$. **a** Magnification of Fig. 9 at $t = 100$; **b** amplitude of the larger (in magnitude) emerging wave versus time; **c** speed of the larger (in magnitude) emerging wave versus time

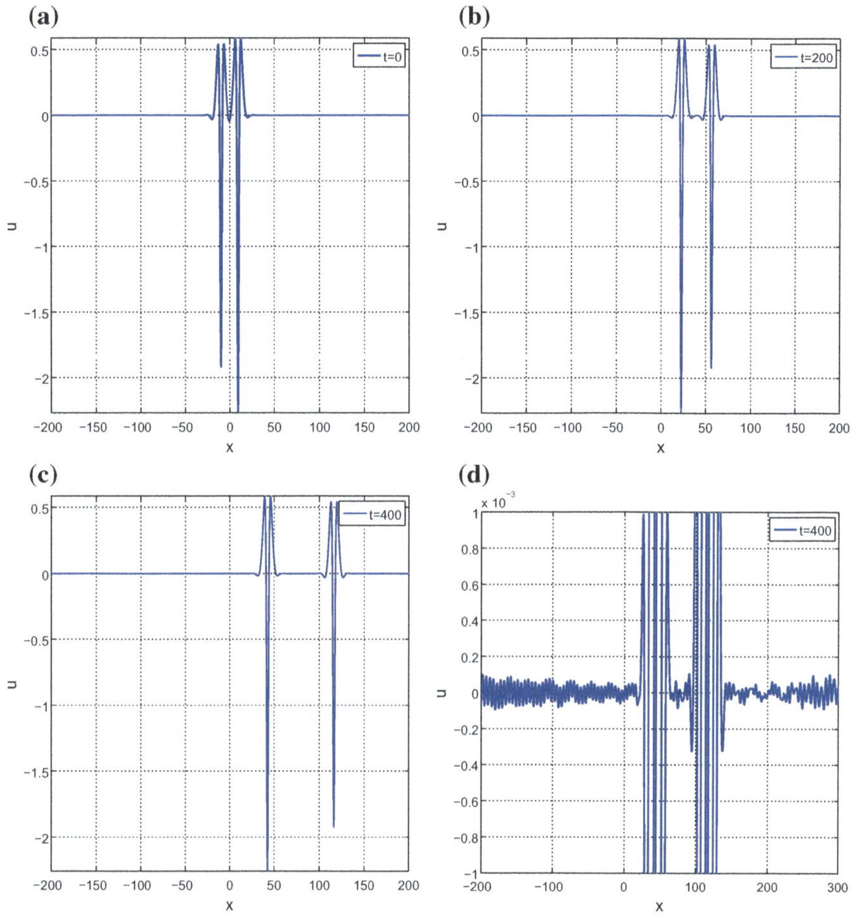

Fig. 11 Overtaking collision of solitary waves (slow-slow case). Numerical approximation of (1) with $p = 3$, $\gamma = 0.25$, $\beta = 1$ and initial condition given by a superposition of two solitary profiles with speeds $c^{(1)} = 0.1$, $c^{(2)} = 0.3$ centered at $x_0 = 10$ and -10 respectively. **a** Initial condition. **b** Numerical solution at $t = 200$. **c** Numerical solution at $t = 400$. **d** Magnification of (**c**)

with speeds $c^{(1)} = 0.1$, $c^{(2)} = 0.3$ centered at $x_0 = 10$ and -10 respectively. The initial maximum negative excursions are, respectively, $u_m^{(1)} \approx -2.2688$ and $u_m^{(2)} \approx -1.9204$. The results suggest the following conclusions:

- After the collision, two solitary pulses emerge. The evolution of the amplitudes (not displayed here) shows that by $t = 400$ they have stabilized at approximate new values $u_m^{(2)} \approx -1.9224$ and $u_m^{(1)} \approx -2.2569$. Therefore there is an exchange with the taller wave (the slower one) decreasing the maximum negative excursion in magnitude while for the smaller emerging wave this minimum slightly increases in magnitude. According to the speed-amplitude relation observed and analyzed

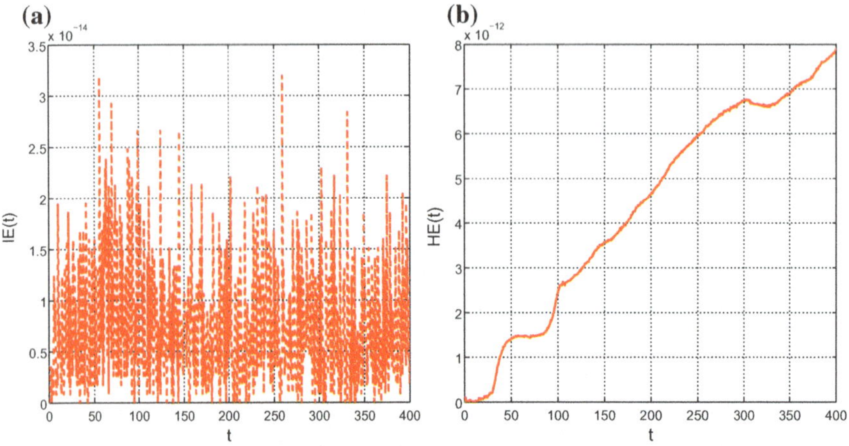

Fig. 12 Overtaking collision of solitary waves (slow-slow case). Numerical approximation of (1) with $p = 3$, $\gamma = 0.25$, $\beta = 1$ and initial condition given by a superposition of two solitary profiles with speeds $c^{(1)} = 0.1$, $c^{(2)} = 0.3$ centered at $x_0 = -10$ and 10 respectively. **a** Momentum $|V_h(U^n) - V_h(U^0)|$) error; **b** Energy ($|E_h(U^n) - E_h(U^0)|$) error. All are displayed against time, and V_h, E_h are given by (14), (15) respectively

in [20], this affects the speeds of the emerging waves with respect to those of the original pulses. For instance, the speed of the taller emerging wave changes to $c^{(1)} \approx 0.1007$.

- By $t = 400$, small tails are formed (see Fig. 11d) in front of and behind the new pulses. Indeed the presence of the tails and the changes in amplitude (and speed) after the collision imply an inelastic interaction, but with a low degree of inelasticity, suggesting that collisions with slow pulses are close to elastic. This was observed and explained in other nonintegrable models [44].
- The behaviour of the conserved quantities (14), (15) before and after the interaction is displayed in Fig. 12. Observe that while the difference in the momentum remains small and bounded in time during the simulation (a virtual preservation) the collision leads to a slight growth with time of the energy functional. Two conjectures for this behaviour are concerned with the tails formed after the collision. One would say that a longer simulation will disperse the ripples and stabilize the energy error small and bounded. The second one would suggest that some of the radiated tails are not dispersive and will generate some growth with time of the energy.

The case fast-slow is illustrated in Fig. 13. Here the initial condition is a superposition of two classical solitary pulses of (1) for $p = 3$, $\gamma = 0.25$, $\beta = 1$ with speeds $c^{(1)} = 0.1$, $c^{(2)} = 0.9$ centered at $x_0 = 10$ and -10 respectively. Compared to the slow-slow case, some differences are observed:

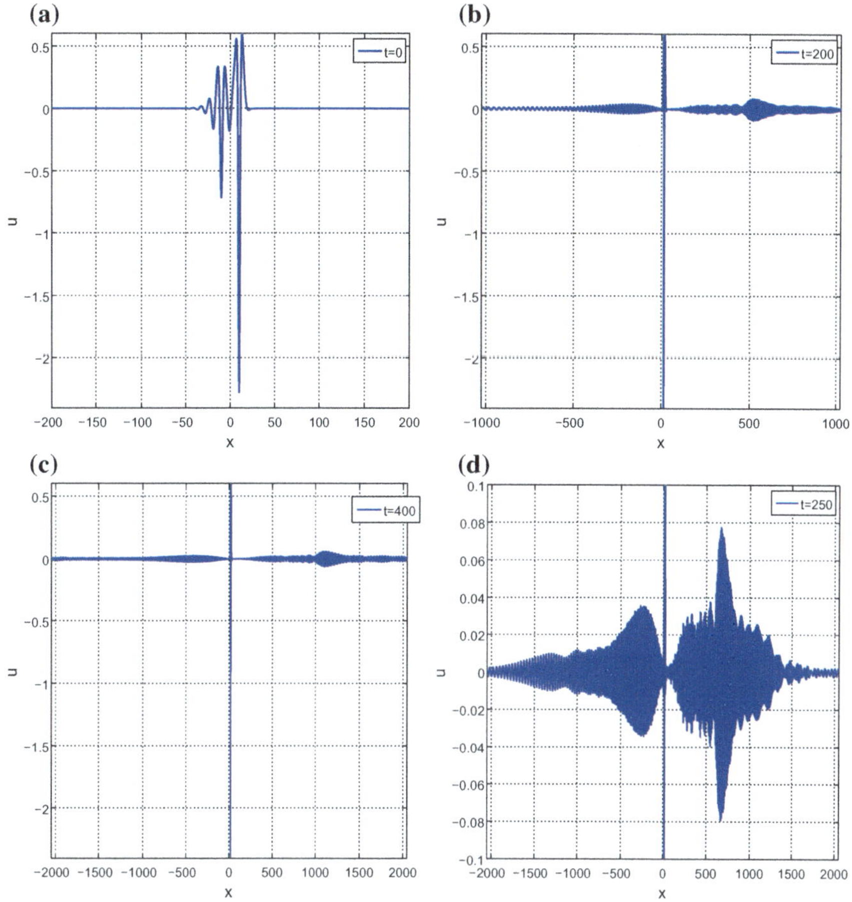

Fig. 13 Overtaking collision of solitary waves (fast-slow case). Numerical approximation of (1) with $p = 3$, $\gamma = 0.25$, $\beta = 1$ and initial condition given by a superposition of two solitary profiles with speeds $c^{(1)} = 0.1$, $c^{(2)} = 0.9$ centered at $x_0 = 10$ and -10 respectively. **a** Initial condition. **b** Numerical solution at $t = 200$. **c** Numerical solution at $t = 400$. **d** Magnification of numerical solution at $t = 250$

- After the interaction the faster wave suffers a strong reduction in amplitude (and, consequently, a strong increment of speed) from $u_m^{(2)} \approx -0.7169$ to $u_m^{(2)} \approx -0.06$ (by $t = 400$) while the maximum negative excursion for the slower pulse changes at $t = 400$ from $u_m^{(1)} \approx -2.2790$ to $u_m^{(1)} \approx -2.4027$, moving very slowly, with a final speed of $c^{(1)} \approx 0.02$. These data suggest a strongly inelastic collision.
- In this case, long dispersive tails, backwards and forwards, emerge and, as displayed in Fig. 14a, the evolution of the difference between the energy (15) of the initial data and that of the numerical approximation shows a bounded behaviour and a virtual preservation by the final time of simulation. The momentum (14)

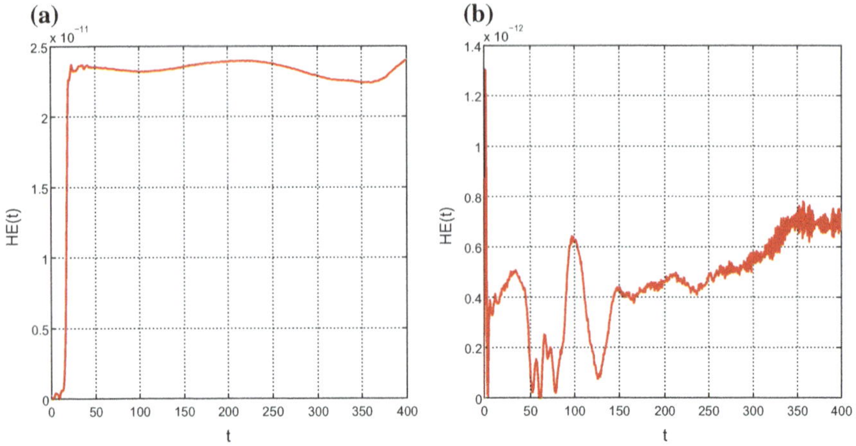

Fig. 14 Overtaking collision of solitary waves. Numerical approximation of (1) with $p = 3$, $\gamma = 0.25$, $\beta = 1$. Energy ($|E_h(U^n) - E_h(U^0)|$) error against time, with E_h given by (15). **a** Fast-slow case. **b** Fast-fast case

is also preserved; the corresponding figure, very similar to Fig. 12a, will not be shown here.

Finally, the fast-fast case corresponds to Fig. 15, where the superposition of two classical solitary pulses of (1) for $p = 3$, $\gamma = 0.25$, $\beta = 1$ with speeds $c^{(1)} = 0.8$ ($u_m^{(1)} \approx -0.9718$) and $c^{(2)} = 0.9$ ($u_m^{(2)} \approx -0.7169$) centered at $x_0 = 10$ and -10 respectively, is taken as initial condition and evolved up to $t = 400$. After the interaction, the two emerging pulses have amplitudes $u_m^{(1)} \approx -1.0460$, $u_m^{(2)} \approx -0.5747$. The degree of inelasticity is then between the other two cases (the speed of the slower emerging wave is reduced to $c^{(1)} = 0.76$. In this case, the energy error also shows a slight growth with time see Fig. 14b), suggesting similar conjectures to those of the slow-slow case. By the final time, however, this error looks to stabilize.

Remark 3 In the case of head-on collisions (that is, when the profiles travel in opposite directions) we notice that the speed-amplitude relation for negative speeds is, compared to the case of positive speeds, the reverse: the taller (in magnitude) the wave the faster it travels. Taking this into account, our experiments of head-on interactions do not reveal a different behaviour from the overtaking collisions above. (For instance, the slow-slow head-on collision is compared to a fast-slow overtaking collision, etc.)

3.3 Generalized Solitary Waves

Our purpose now is to study computationally the stability of generalized solitary waves. In this case the experiments will concern perturbations of initial GSWs, which somehow cover several phenomena: stability under small and large perturbations,

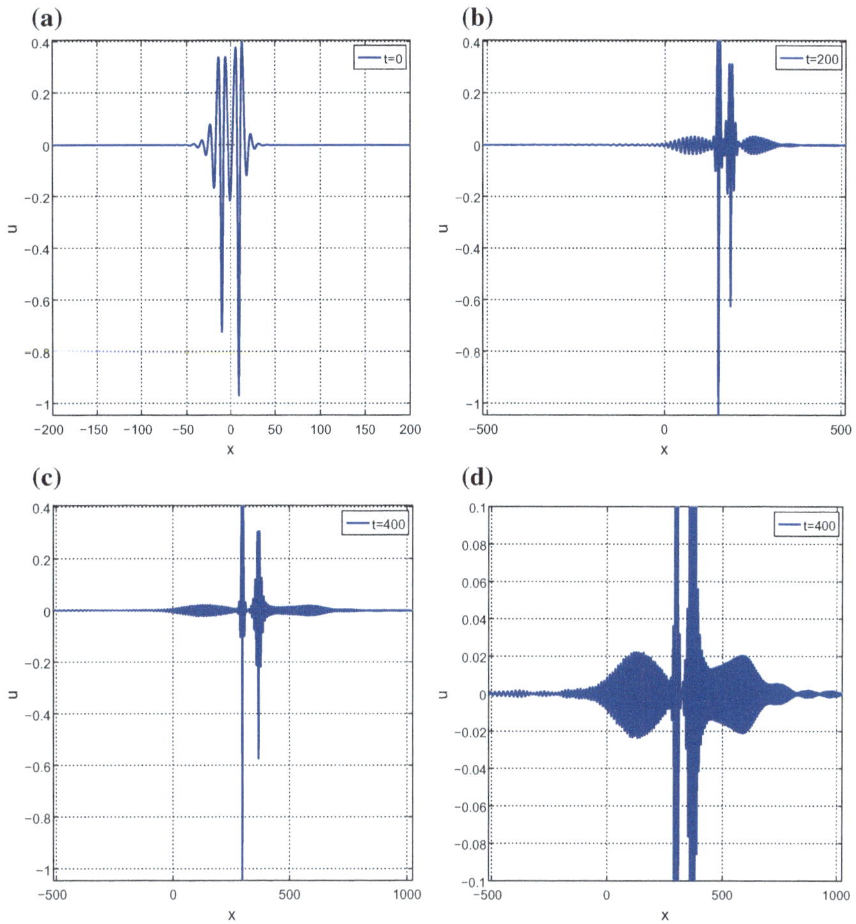

Fig. 15 Overtaking collision of solitary waves (fast-fast case). Numerical approximation of (1) with $p = 3$, $\gamma = 0.25$, $\beta = 1$ and initial condition given by a superposition of two solitary profiles with speeds $c^{(1)} = 0.8$, $c^{(2)} = 0.9$ centered at $x_0 = 10$ and -10 respectively. **a** Initial condition. **b** Numerical solution at $t = 200$. **c** Numerical solution at $t = 400$. **d** Magnification of (**c**)

resolution property and also some of interactions. (Note that the existence theory [16], forces to have in this case overtaking collisions only.) Since the experiments concerning this last phenomenon are not conclusive at all and it is hard to distinguish some correspondence with amplitudes or speeds of the waves (mainly because of the interactions of the oscillating tails) then specific computations of collisions will not be considered below, although they will be observed in some experiments concerning the resolution property.

In some models, see for instance [10], small perturbations of GSW resolve into radiating solitary waves (which are not waves in the classical sense, since they dissipate). A main pulse is formed, with dispersive tails to the left and to the right.

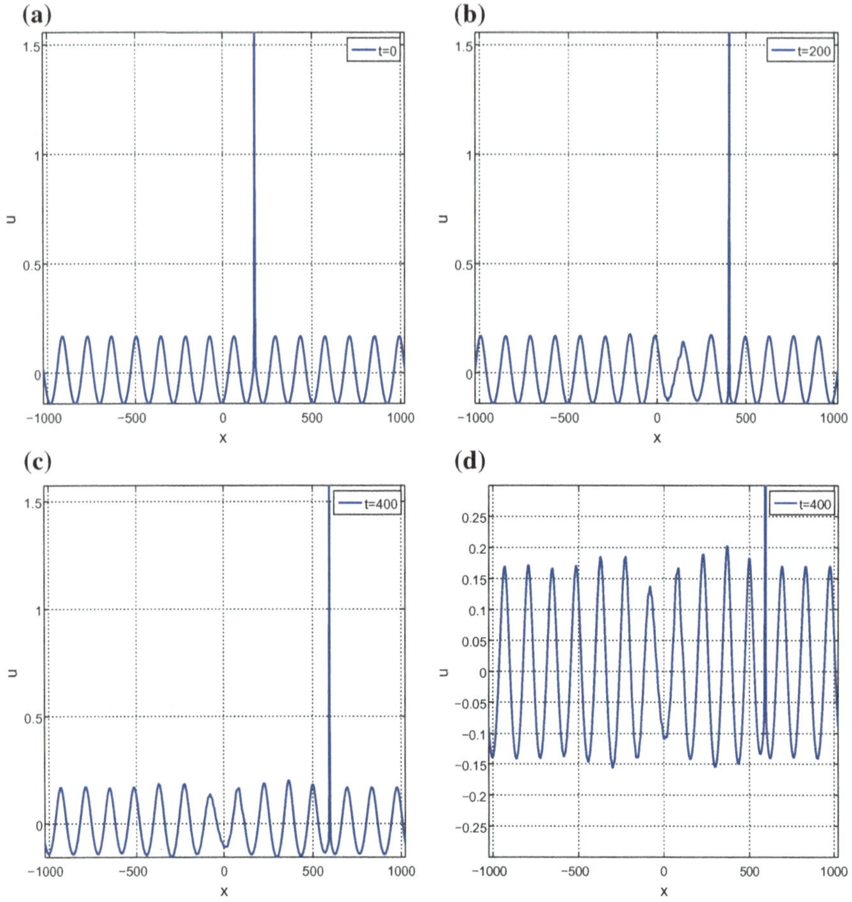

Fig. 16 Perturbations of generalized solitary waves. Numerical approximation of (1) with $p = 2$, $\gamma = 0.002$, $\beta = -1$ and initial condition is given by (17) with ϕ_0 a GSW with speed $c = 1$ and $r = 1.1$. **a** Initial condition. **b** Numerical solution at $t = 200$. **c** Numerical solution at $t = 400$. **d** Magnification of (**c**)

Additionally, ripples of larger magnitude appear mainly in front of the main pulse. The magnitude of the ripples grow with the size of the perturbation.

The evolution of perturbations of generalized solitary wave solutions of (1) is illustrated in the experiments below. In all the cases, a GSW profile ϕ_0, with amplitude of the main pulse of $u_m \approx 1.4152$, for $p = 2$, $\gamma = 0.002$, $\beta = -1$ and $c = 1$. The perturbed initial data were taken of the form (17) with several values of r. As a check of accuracy of the computations, the discrete momentum and energy, (14) and (15) respectively, are preserved to at least eight digits up to the final time ($t = 400$) of simulation. (For an interpretation of these quantities in the context of generalized solitary waves, see e.g. [18].)

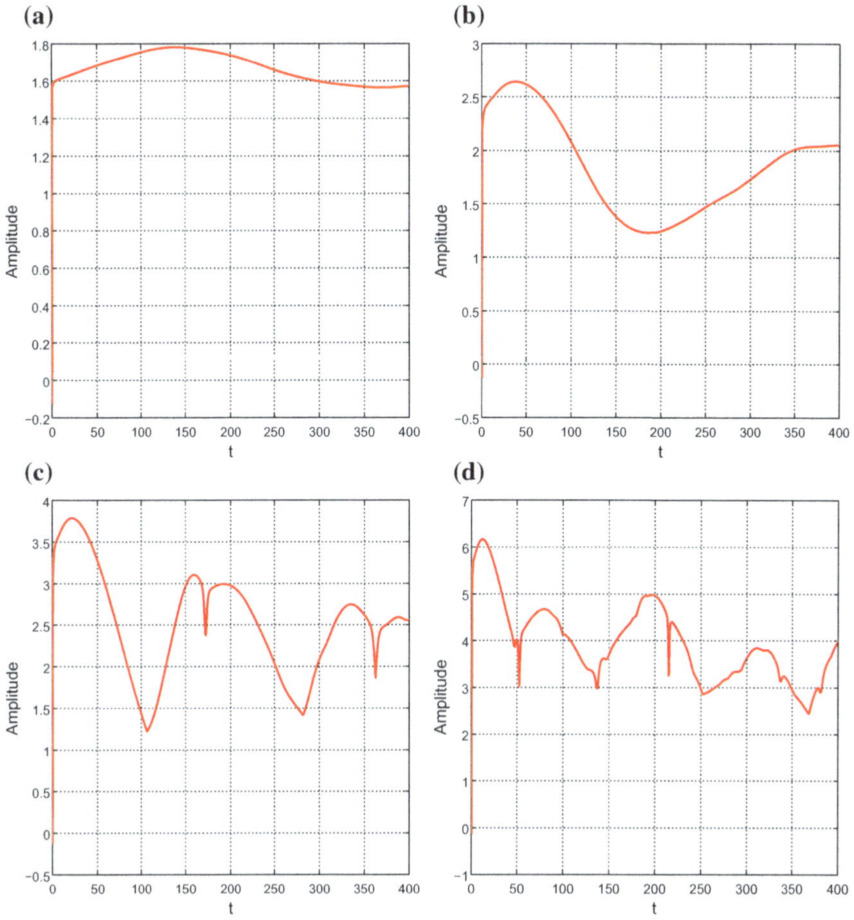

Fig. 17 Perturbations of generalized solitary waves. Numerical approximation of (1) with $p = 2$, $\gamma = 0.002$, $\beta = -1$ and initial condition is given by (17) with ϕ_0 a GSW with speed $c = 1$. Evolution of the amplitude of the perturbed wave. **a** $r = 1.1$. **b** $r = 1.5$. **c** $r = 2$. **d** $r = 3$

3.3.1 Small Perturbations

Figure 16 illustrates the case $r = 1.1$ (the perturbed initial wave has then an amplitude $u_m \approx 1.5567$). The numerical approximation evolves to a new GSW form, with a main pulse of approximate amplitude 1.5737 and a slightly slower speed of about 0.97. The oscillating tail is also slightly perturbed by an apparently small dispersive term traveling to the left (with respect to the GSW) along the oscillations. Figure 17a shows the evolution of the amplitude of the main pulse. When the parameter r grows to $r = 1.5$, the perturbation in the oscillating tail is more intense and the oscillating character of the GSW tail is suggested to be broken somehow, see Fig. 18. This

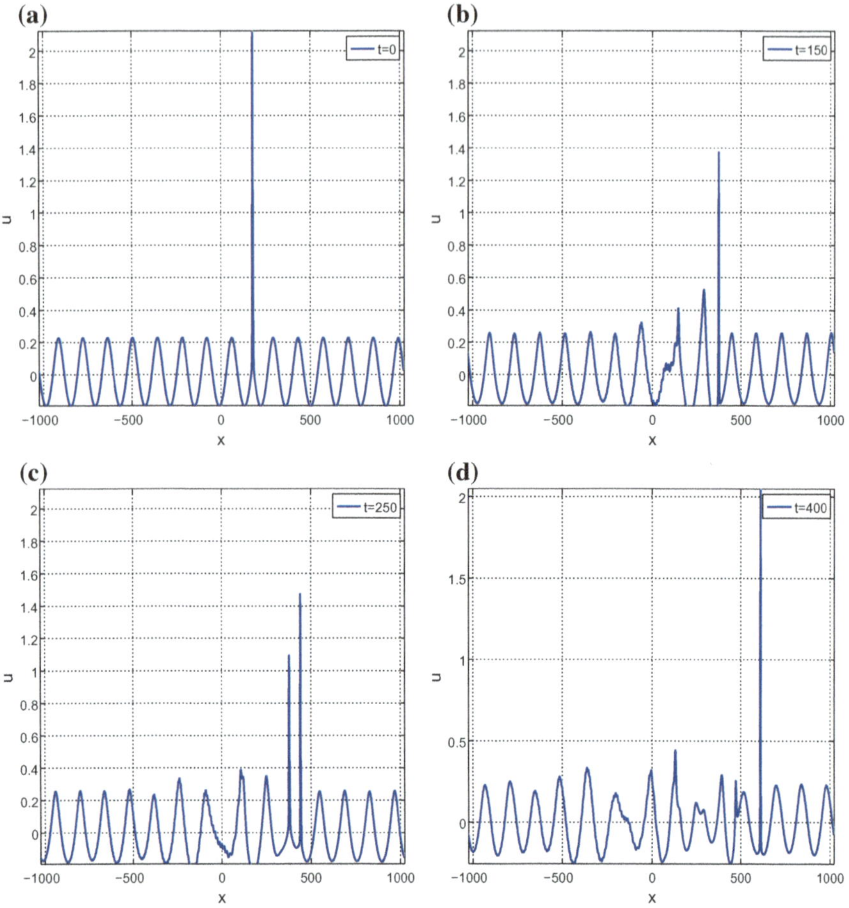

Fig. 18 Perturbations of generalized solitary waves. Numerical approximation of (1) with $p = 2$, $\gamma = 0.002$, $\beta = -1$ and initial condition is given by (17) with ϕ_0 a GSW with speed $c = 1$ and $r = 1.5$. **a** Initial condition. **b** Numerical solution at $t = 150$. **c** Numerical solution at $t = 250$. **d** Numerical solution at $t = 400$

behaviour appears to betoken the formation of additional not dispersive structures, always to the left (cf. Sect. 3.1). The amplitude of the main pulse goes in this case from that of the perturbed initial data (approx. 2.1228) to approx. 2.0480, see the evolution in Fig. 17b, while the speed increases to approximately 1.17. Compared to the previous experiment, we can conclude that increasing the factor r leads, among other properties, to an increment of the speed of the emerging GSW profile.

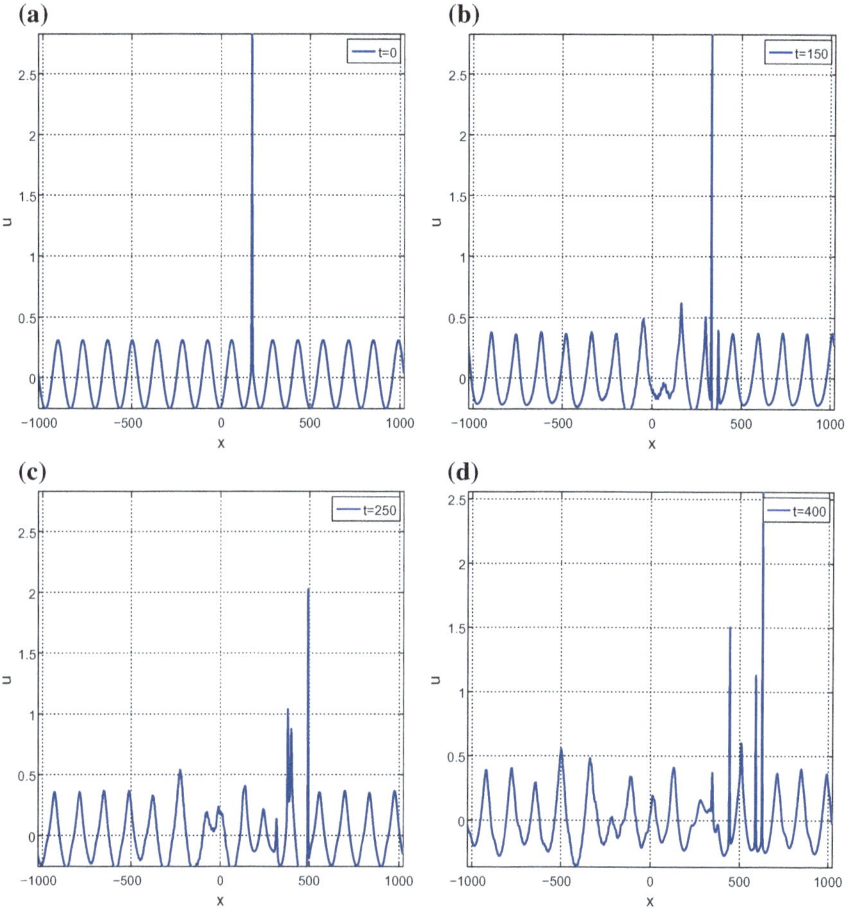

Fig. 19 Perturbations of generalized solitary waves. Numerical approximation of (1) with $p = 2$, $\gamma = 0.002$, $\beta = -1$ and initial condition is given by (17) with ϕ_0 a GSW with speed $c = 1$ and $r = 2$. **a** Initial condition. **b** Numerical solution at $t = 150$. **c** Numerical solution at $t = 250$. **d** Numerical solution at $t = 400$

3.3.2 Larger Perturbations

As in the case of classical solitary waves, experiments with larger values of the perturbation parameter r show new phenomena. The formation of nonlinear structures suggested by the last case $r = 1.5$ is confirmed in Fig. 19, corresponding to $r = 2$. The most relevant feature is the generation of three main pulses, in a sort of resolution property for generalized solitary waves. From a perturbed GSW of amplitude 2.8304, the resulting main pulses have approximate amplitudes 2.5548 (see Fig. 17c), 1.1345 and 1.5072. It is clear that the new generalized solitary wave profiles interact, although it is not easy to observe, from the experiments performed, additional

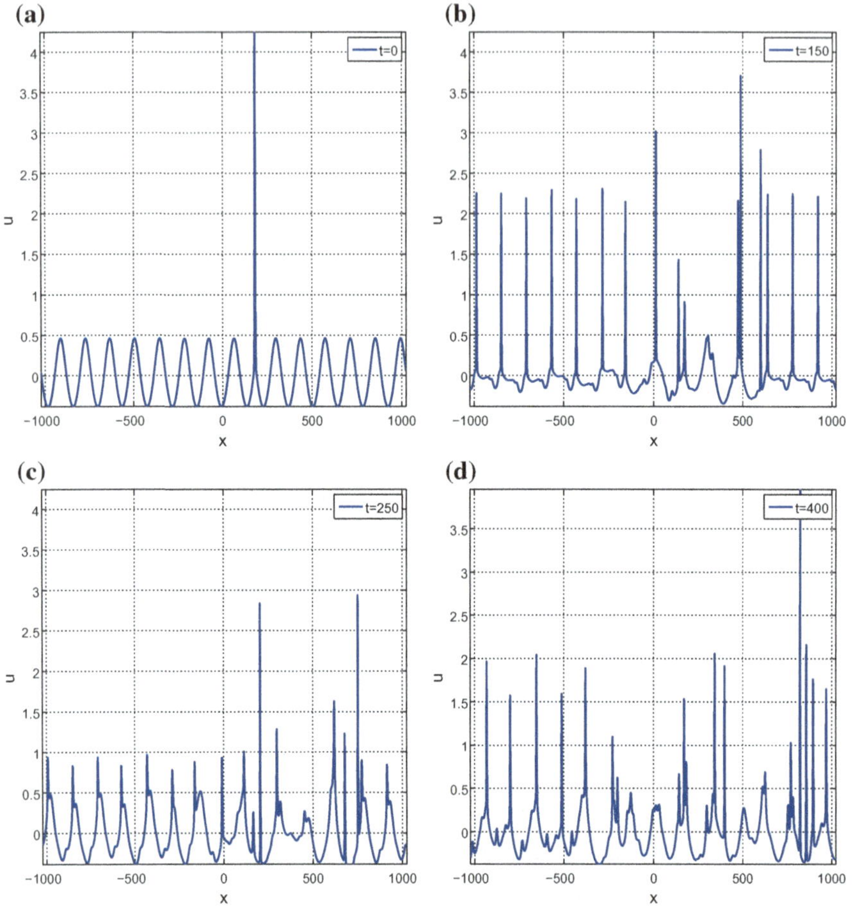

Fig. 20 Perturbations of generalized solitary waves. Numerical approximation of (1) with $p = 2$, $\gamma = 0.002$, $\beta = -1$ and initial condition is given by (17) with ϕ_0 a GSW with speed $c = 1$ and $r = 3$. **a** Initial condition. **b** Numerical solution at $t = 150$. **c** Numerical solution at $t = 250$. **d** Numerical solution at $t = 400$

effects from these collisions. The tendency to this resolution property of GSWs by using large perturbations, is finally confirmed in the last experiment, shown in Fig. 20 and corresponding to $r = 3$. Note that up to a time of simulation of approximately $t = 200$, the approximation evolves, from a perturbed wave of approximate amplitude 4.2455 of the main pulse to a structure of two main pulses with a nice train of periodic forms with small asymmetric humps on the oscillations, see Fig. 21. For longer times, the structure is affected by the unavoidable interactions. The amplitude of the final main pulse (at $t = 400$) is approximately 3.9563.

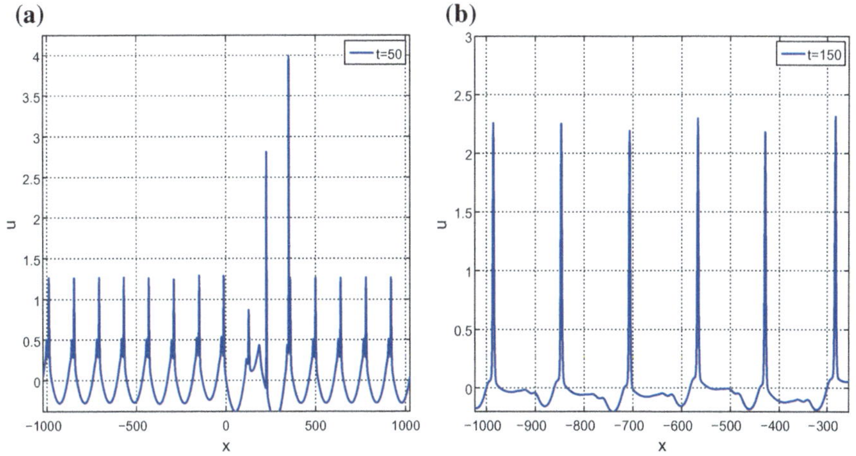

Fig. 21 Perturbations of generalized solitary waves. Numerical approximation of (1) with $p = 2$, $\gamma = 0.002$, $\beta = -1$ and initial condition is given by (17) with ϕ_0 a GSW with speed $c = 1$ and $r = 3$. **a** Numerical solution at $t = 50$. **b** Numerical solution at $t = 150$ (magnification)

Acknowledgements This work was supported by Spanish Ministerio de Economía y Competitividad under the Research Grant MTM2014-54710-P.

References

1. Albert, J.P., Bona, J.L., Henry, D.: Sufficient conditions for stability of solitary-wave solutions of model equations for long waves. Phys. D **24**, 343–366 (1987)
2. Álvarez, J., Durán, A.: Error propagation when approximating multisolitons: the KdV equation as a case study. Appl. Math. Comput. **217**, 1522–1539 (2010)
3. Álvarez, J., Durán, A.: Petviashvili type methods for traveling wave computations: I. Analysis of convergence. J. Comput. Appl. Math. **266**, 39–51 (2014)
4. Álvarez, J., Durán, A.: Petviashvili type methods for traveling wave computations: II. Acceleration techniques. Math. Comput. Simul. **123**, 19–36 (2016)
5. Benilov, E.S.: On the surface waves in a shallow channel with an uneven bottom. Stud. Appl. Math. **87**, 1–14 (1992)
6. Benjamin, T.B.: The stability of solitary waves. Proc. R. Soc. Lond. Ser. A **328**, 153–183 (1972)
7. Bona, J.L.: On the stability theory of solitary waves. Proc. R. Soc. Lond. Ser. A **344**, 363–374 (1975)
8. Bona, J.L., Dougalis, V.A., Karakashian, O.A., McKinney, W.R.: Conservative, high-order numerical schemes for the generalized Korteweg-de Vries equation. Philos. Trans. R. Soc. Lond. A **351**, 107–164 (1995)
9. Bona, J.L., Dougalis, V.A., Mitsotakis, D.E.: Numerical solution of KdV-KdV systems of Boussinesq equations I. The numerical scheme and generalized solitary waves. Math. Comput. Simul. **74**, 214–228 (2007)
10. Bona, J.L., Dougalis, V.A., Mitsotakis, D.E.: Numerical solution of KdV-KdV systems of Boussinesq equations II. Evolution of radiating solitary waves. Nonlinearity **21**, 2825–2848 (2008)

11. Bona, J.L., Kalisch, H.: Singularity formation in the generalized Benjamin-Ono equation. Discret. Contin. Dyn. Syst. **11**, 27–45 (2004)
12. Bona, J.L., Souganidis, P.E., Strauss, W.A.: Stability and instability of solitary waves of KdV type. Proc. R. Soc. Lond. Ser. A **411**, 395–412 (1987)
13. Boyd, J.P., Chen, G.Y.: Five regimes of the quasi-cnoidal, steadily translating waves of the rotation-modified Korteweg-de Vries ("Ostrovsky") equation. Wave Motion **35**, 141–155 (2002)
14. Canuto, C., Hussaini, M.Y., Quarteroni, A., Zang, T.A.: Spectral Methods in Fluid Dynamics. Springer, New York-Heidelberg-Berlin (1988)
15. Chen, G.Y., Boyd, J.P.: Analytical and numerical studies of weakly nonlocal solitary waves of the rotation-modified Korteweg-de Vries equation. Phys. D **155**, 201–222 (2002)
16. Choudhury, S.R.: Solitary-wave families of the Ostrovsky equation: an approach via reversible systems theory and normal forms. Chaos Solitons Fractals **33**, 1468–1479 (2007)
17. Choudhury, S.R., Ivanov, R.I., Liu, Y.: Hamiltonian formulation, nonintegrability and local bifurcations for the Ostrovsky equation. Chaos Solitons Fractals **34**, 544–550 (2007)
18. Clamond, D., Dutykh, D., Durán, A.: A plethora of generalized solitary gravity-capillary water waves. J. Fluid Mech. **784**, 664–680 (2015)
19. Dougalis, V.A., Durán, A., López-Marcos, M.A., Mitsotakis, D.E.: A numerical study of the stability of solitary waves of the Bona-Smith family of Boussinesq systems. J. Nonlinear Sci. **17**, 569–607 (2007)
20. Durán, A.: On the numerical approximation to generalized Ostrovsky equations: I. A numerical method and computation of solitary-wave solutions. Technical Report (2017)
21. El Dika, K.: Asymptotic stability of solitary waves for the Benjamin-Bona-Mahony equation. Discret. Contin. Dyn. Syst. **13**, 583–622 (2005)
22. de Frutos, J., Sanz-Serna, J.M.: An easily implementable fourth-order method for the time integration of wave problems. J. Comput. Phys. **103**, 160–168 (1992)
23. Galkin, V.N., Stepanyants, Y.A.: On the existence of stationary solitary waves in a rotating fluid. J. Appl. Math. Mech. **55**, 939–943 (1991)
24. Gilman, O.A., Grimshaw, R., Stepanyants, Y.A.: Approximate analytical and numerical solutions of the stationary Ostrovsky equation. Stud. Appl. Math. **95**, 115–126 (1995)
25. Gilman, O.A., Grimshaw, R., Stepanyants, Y.A.: Dynamics of internal solitary waves in a rotating fluid. Dyn. Atm. Ocean **23**(1), 403–411 (1995)
26. Grillakis, M., Shatah, J., Strauss, W.A.: Stability of solitary waves in the presence of symmetry: I. J. Funct. Anal. **74**, 170–197 (1987)
27. Grimshaw, R.H.: Evolution equations for weakly nonlinear, long internal waves in a rotating fluid. Stud. Appl. Math. **73**, 1–33 (1985)
28. Grimshaw, R.H., He, J.M., Ostrovsky, L.A.: Terminal damping of a solitary wave due to radiation in rotational systems. Stud. Appl. Math. **10**, 197–210 (1998)
29. Grimshaw, R.H., Helfrich, K.R.: The effect of rotation on internal solitary waves. IMA J. Appl. Math. **10**, 1–14 (2012)
30. Grimshaw, R.H., Helfrich, K.R., Johnson, E.R.: Experimental study of the effect of rotation on nonlinear internal waves. Phys. Fluids **25**, 0566,021–05660,223 (2013)
31. Grimshaw, R.H., Ostrovsky, L.A., Shira, V.I., Stepanyants, Y.A.: Long nonlinear surface and internal gravity waves in a rotating ocean. Surv. Gheophys. **19**, 289–338 (1998)
32. Helfrich, K.R.: Decay and return of internal solitary waves with rotation. Phys. Fluids **19**, O26,601 (2007)
33. Helfrich, K.R., Melville, W.K.: Long nonlinear internal waves. Annu. Rev. Fluid Mech. **38**, 395–425 (2006)
34. Isaza, P., Mejía, J.: Global Cauchy problem for the Ostrovsky equation. Nonlinear Anal. **67**, 1482–1503 (2007)
35. Jbilous, K., Sadok, H.: Vector extrapolation methods, applications and numerical comparisons. J. Comput. Appl. Math. **122**, 149–165 (2000)
36. Lakoba, T., Yang, Y.: A generalized Petviashvili method for scalar and vector Hamiltonian equations with arbitrary form of nonlinearity. J. Comput. Phys. **226**, 1668–1692 (2007)

37. Levandosky, S.: On the stability of solitary waves of a generalized Ostrovsky equation. Technical Report (2006)
38. Levandosky, S., Liu, Y.: Stability of solitary waves of a generalized Ostrovsky equation. SIAM J. Math. Anal. **38**, 985–1011 (2006)
39. Levandosky, S., Liu, Y.: Stability and weak rotation limit of solitary waves of the Ostrovsky equation. Discret. Contin. Dyn. Syst. Ser. B **7**, 793–806 (2007)
40. Linares, F., Milanés, A.: Local and global well-posedness for the Ostrovsky equation. J. Differ. Equ. **222**, 325–340 (2006)
41. Liu, Y., Ohta, M.: Stability of solitary waves for the Ostrovsky equation. Proc. AMS **136**, 511–517 (2008)
42. Liu, Y., Varlamov, V.: Stability of solitary waves and weak rotation limit for the Ostrovsky equation. J. Differ. Equ. **203**, 159–183 (2004)
43. Martel, Y., Merle, F.: Asymptotic stability of solitons for subcritical generalized KdV equations. Arch. Ration. Mech. Anal. **157**, 219–254 (2001)
44. Martel, Y., Merle, F., Mizumachi, T.: Description of the inelastic collision of two solitary waves for the BBM equation. Arch. Ration. Mech. Anal. **196**, 517–574 (2010)
45. Miller, J.R., Weinstein, M.I.: Asymptotic stability of solitary waves for the Regularized Long-Wave equation. Commun. Pure Appl. Math. **49**, 399–441 (1996)
46. Olver, P.J.: Applications of Lie Groups to Differential Equations. Springer, New York (1986)
47. Ostrovsky, L.A.: Nonlinear internal waves in a rotating ocean. Okeanologia **18**, 181–191 (1978)
48. Pego, R.L., Weinstein, M.I.: Asymptotic stability of solitary waves. Commun. Math. Phys. **164**, 305–349 (1994)
49. Pego, R.L., Weinstein, M.I.: Convective linear stability of solitary waves for Boussinesq equations. Stud. Appl. Math. **99**, 311–375 (1997)
50. Pelinovsky, D.E., Stepanyants, Y.A.: Convergence of Petviashvili's iteration method for numerical approximation of stationary solutions of nonlinear wave equations. SIAM J. Numer. Anal. **42**, 1110–1127 (2004)
51. Petviashvili, V.I.: Equation of an extraordinary soliton. Sov. J. Plasma Phys. **2**, 257–258 (1976)
52. Sanz-Serna, M..J., Calvo, M.P.: Numerical Hamiltonian Problems. Chapman & Hall, London (1994)
53. Sidi, A.: Convergence and stability of minimal polynomial and reduced rank extrapolation algorothms. SIAM J. Numer. Anal. **23**, 197–209 (1986)
54. Sidi, A., Ford, W.F., Smith, D.A.: Acceleration of convergence of vector sequences. SIAM J. Numer. Anal. **23**, 178–196 (1986)
55. Smith, D.A., Ford, W.F., Sidi, A.: Extrapolation methods for vector sequences. SIAM Rev. **29**, 199–233 (1987)
56. Trefethen, L.N.: Spectral Methods in MATLAB. SIAM, Philadelphia (2000)
57. Tsuwaga, K.: Well-posedness and weak rotation limit for the Ostrovsky equation. J. Differ. Equ. **247**, 3163–3180 (2009)
58. Varlamov, V., Liu, Y.: Cauchy problem for the Ostrovsky equation. Discret. Dyn. Syst. **10**, 731–751 (2004)
59. Weinstein, M.I.: Lyapunov stability of ground states of nonlinear dispersive evolution equations. Commun. Pure Appl. Math. **39**, 51–68 (1986)
60. Weinstein, M.I.: Existence and dynamic stability of solitary-wave solutions of equations arising in long wave propagation. Commun. Partial Differ. Equ. **12**, 1133–1173 (1987)

Simulating Laser Dynamics with Cellular Automata

Francisco Jiménez-Morales, José Luis Guisado and José Manuel Guerra

Abstract The long-established approach to study laser dynamics uses a set of differential equations known as the laser rate equations. In this work we present an overview of an alternative model based on a cellular automaton (CA). We also present a panorama of different variants of the model: the original one, designed to simulate general laser dynamics; an additional one, that was proposed to simulate pulsed pumped lasers; and finally a new model to simulate lasers that exhibit antiphase dynamics, which is proposed here. Despite its simplicity, the CA model reproduces qualitatively the phenomenology encountered in many real laser systems: (i) the existence of a threshold value of the pumping rate R_t; (ii) the exact dependence of R_t on the life times of the photons and the inversion population; (iii) the two main laser regimes: a steady state and an oscillatory one.

1 Introduction

Laser devices are one of the paradigmatic examples of a complex system, capable of showing many different dynamic behaviors. They are traditionally described macroscopically by the laser rate equations which model the time evolution of aggregate variables like the laser beam intensity and the total population inversion. However, we introduced a new type of model to describe laser dynamics based on cellular automata

F. Jiménez-Morales (✉)
Departamento de Física de la Materia Condensada, Universidad de Sevilla,
Avda. Reina Mercedes s/n, 41012 Sevilla, Spain
e-mail: jimenez@us.es

J. L. Guisado
Departamento de Arquitectura y Tecnología de Computadores,
Universidad de Sevilla, Avda. Reina Mercedes s/n, 41012 Sevilla, Spain
e-mail: jlguisado@us.es

J. M. Guerra
Departamento de Óptica, Facultad de CC. Físicas, Universidad Complutense
de Madrid, 28040 Madrid, Spain
e-mail: jmguerra@fis.ucm.es

V. Carmona et al. (eds.), *Nonlinear Systems, Vol. 1*,
Understanding Complex Systems, https://doi.org/10.1007/978-3-319-66766-9_14

(CA) [9]. The CA approach describes the system as composed of many individual component parts that interact with each other and with their local environment to produce emergent behaviors.

In the following chapter we review a panorama of different specific CA models of particular laser systems and the results obtained by using them. We begin in Sect. 2 by introducing laser devices, their dynamics and their traditional description based on rate equations. The alternative CA model of general laser dynamics and the results that it has generated is presented in Sect. 3. In Sect. 4 we define a variation of the model to simulate a special type of laser devices, pulsed pumped lasers, and present its results. Another version is defined in Sect. 5 in order to simulate antiphase dynamics observed in certain laser devices, and results are presented. Finally, conclusions are drawn in Sect. 6 and future prospects are discussed.

2 Laser Dynamics

Lasers are devices that generate or amplify coherent electromagnetic radiation based on the *stimulated emission* phenomenon, at frequencies that can range from the infra-red to the X-ray region of the spectrum [17, 20].

A laser system is made of three basic components:

1. *Pump source* that provides energy to the laser medium.
2. *Active medium* composed of any kind of particles (atoms or molecules) whose electrons can be excited from the ground level to a higher level.
3. *Optical resonator* that provides feedback of the light.

The principle behind the laser action is the *stimulated emission*. This process is the interaction of a photon with an excited atom. If the energy of the photon is equal to the gap between the two energy levels of the atoms, the atom decays to the lower energy state and a second photon is emitted. The newly produced laser photon can induce a new stimulated emission in another atom of the laser medium, and so on. As a result, a chain reaction can give rise to a coherent laser beam, with characteristic properties of high monochromaticity, uniform polarization and phase (established by the optical cavity design) and high collimation.

In a laser stimulated emission competes with the absorption process, in which atoms in the lower energy state can be excited by absorbing the incoming photon. Stimulated emission and absorption have the same probability and hence, in order to have laser action, some pumping process must produce excited atoms and a population inversion condition must be fulfilled. Figure 1 shows a schematic energy level diagram of a typical four level laser.

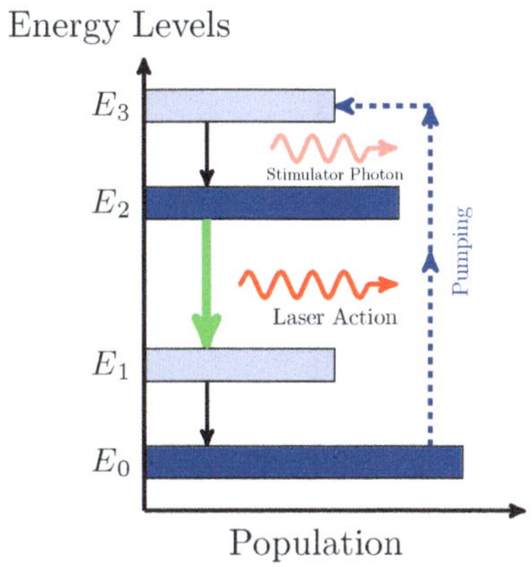

Fig. 1 Four-level laser medium. Atoms in the ground state E_0 are pumped to level E_3, from where they decay by a fast non-radiative transition into level E_2. The lifetime of the laser transition between levels 3–2 is much lower than the corresponding lifetime between levels 2–1. Therefore, excited atoms accumulate in level E_2. From level E_2 the atoms may decay to level E_1 and from there to the ground laser level E_0 by another fast non-radiative transition

2.1 *Laser Rate Equations*

The laser rate equations [17, 20] is a set of two coupled differential equations that describes the temporal evolution of the number of photons $n(t)$ and the population inversion $N(t)$.

$$\frac{dn(t)}{dt} = K\,N(t)\,n(t) - \frac{n(t)}{\tau_c} \tag{1}$$

$$\frac{dN(t)}{dt} = R - \frac{N(t)}{\tau_a} - K\,N(t)\,n(t) \tag{2}$$

The other variables in these equations are: K a gain coefficient, τ_a and τ_c the life time of the atoms and photons respectively and R the pump strength. In the stationary state both the inversion population and the number of photons are constant: $N_s = \frac{1}{K\tau_c}$ and $n_s = \tau_c\left(R - \frac{1}{K\tau_a\tau_c}\right)$. It is clear that to achieve lasing action $n(t)$ must be a positive value and thus the pump strength R should be higher than a threshold value R_t given by:

$$R_t = \frac{1}{K\,\tau_a\,\tau_c} \tag{3}$$

Another important result, obtained by a linear stability analysis of the laser rate equations [9], is that $n(t)$ and $N(t)$ can display relaxation oscillations when the following condition is fulfilled:

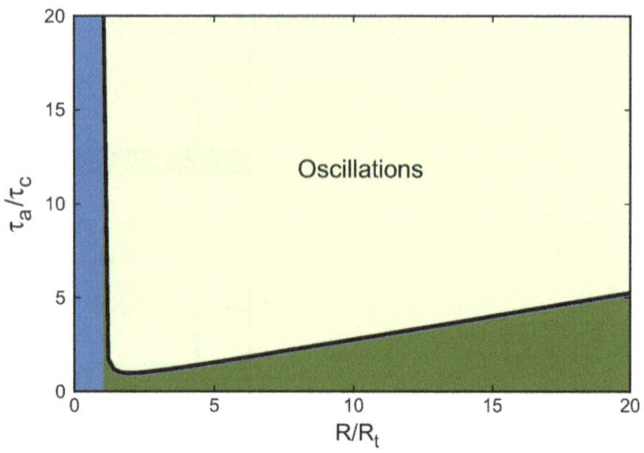

Fig. 2 The parameter space for the laser. Blue area: there is no lasing. Green area: over-damped non-oscillatory behavior. Yellow area: relaxation oscillations. The black line is the stability curve defined by Eq. 4

$$\frac{\tau_a}{\tau_c} > \frac{\left(\frac{R}{R_t}\right)^2}{4\left(\frac{R}{R_t} - 1\right)} \tag{4}$$

On the other hand, if this condition is not satisfied, a non-oscillatory over-damped behavior is found. The different laser outputs can be located in Fig. 2 which plots the fraction between the life times of the atoms and photons versus the pumping strength normalized by its threshold value. In the blue region there is not lasing action (R is lower than R_t). The green region corresponds to the non-oscillatory behaviour whereas the oscillatory state corresponds to the yellow region.

2.2 Limitations of the Laser Rate Equations

The use of the laser rate equations to model laser dynamics has limitations that reduce its applicability, making it very difficult or even impossible to offer satisfactory results in the following cases:

1. Lasers for which the normal numerical methods for solving them are numerically unstable (they magnify approximation errors) [14]. Special procedures must be used and in some cases they are very difficult to integrate [2, 3, 8, 13, 21].
2. Systems with a laser cavity that does not have a simple geometry, or with a non-homogeneous active medium. These conditions are normally assumed in the derivation of the laser rate equations. When they are not fulfilled, the equations are not perfectly valid.
3. Very small size laser sources for which the use of field optics and the Maxwell equations lead to inaccurate conclusions.

To avoid all of those drawbacks of the laser rate equations, an alternative point of view based on a discrete model such as a cellular automaton was proposed in [9], that is free from these limitations.

3 The Cellular Automata Model

Cellular Automata (CA) are computational systems spatially and temporally discrete characterized by local interactions [12, 15, 22]. Formally a CA is given by a tuple $\{\mathcal{G}, \mathcal{S}, \mathcal{N}, \Phi\}$ where $\mathcal{G}$ is the cellular space, $\mathcal{S}$ is the set of states, $\mathcal{N}$ is the neighborhood of a cell and Φ is the local transition function. Despite their simplicity, CA can show complex emergent behaviors that arise from the interactions among their constituents. They have been used in numerous and different fields of science and technology: biomedicine [18], fluid dynamics [5], magnetization in solids [7], reaction-diffusion phenomena [6], telecommunications networks [19], biological population dynamics [1], economics [16], etc.

3.1 Cellular Space

Although the real laser systems are three dimensional we made a simplification and the space of our computational system $\mathcal{G}$ is a two dimensional array of $N_c = L \times L$ cells with periodic boundary conditions.

3.2 Set of States

The state of a cell i at time t is a vector of 2 dimensions: $s_i(t) = \{a, c\}$. Where $a \in \{0, 1\}$ is the electronic state (ground or excited) and $c = \in \{0, \ldots, M\}$ is the number of photons that are present in that cell. Associated with every electron there is a temporal variable $\tilde{a} \in \{0, \ldots, \tau_a\}$ where τ_a is the life time that an electron can stay in the excited state. Each photon has also a life time τ_c and another time variable $\tilde{c} \in \{0, \ldots, \tau_c\}$ is associated with them.

3.3 Neighborhood

The neighborhood specifies with which surrounding cells can each particular cell interact locally. We use the *Moore Neighborhood* which is composed of the cell itself and the eight cells around it.

3.4 Transition Function

The transition function Φ determines the state of each cell at a time $t+1$: $s_i(t+1) = \Phi(s_i(t))$. The transition function is the most important ingredient of any realistic CA model. In our case Φ can be divided in four different rules $\Phi = \Phi_1 \otimes \Phi_2 \otimes \Phi_3 \otimes \Phi_4$ each one representing a particular physical process.

- $\boldsymbol{\Phi}_1$: The Pumping. Each electron in the ground state ($a = 0$) can go to the excited state ($a = 1$) with a probability λ.
- $\boldsymbol{\Phi}_2$: The stimulated emission. If an electron is in the excited state ($a = 1$) and is surrounded in its neighborhood by more than a photon, then the electron decays to the ground state and a new photon is created.
- $\boldsymbol{\Phi}_3$: The decaying of photons. A given photon can only be present during a time τ_c, after that the photon is destroyed.
- $\boldsymbol{\Phi}_4$: The decaying of electrons. Excited electrons can only be in that state during a time τ_a, after that the electron decays to the ground state. In this process no photons are created.

And finally, to make the model more realistic, a very small number of noise photons ($n_{np} \approx 10^{-4} N_c$) are introduced in random positions.

3.5 Results

We have used arrays of 200×200 and 400×400 lattice cells with periodic boundary conditions. The simulations of the CA model are studied varying the parameters τ_a, τ_c and λ.

First of all we have to check if the model reproduces the existence of a threshold pumping probability which is one of the main characteristics of the lasers. After that, we identify the different kinds of behaviors exhibited by the CA in the space of the parameters. And finally we compare the CA outputs with those predicted by the laser rate equations.

3.5.1 Threshold Pumping

An essential feature of lasers is that there is a specific threshold pumping probability value, that will be denoted by λ_t. It is necessary that the pumping probability is higher than this threshold value to have laser action.

In order to compute λ_t, after a transient time t_0, the system is let to evolve during $\Delta t = 200$ time steps and the average number of laser photons $\bar{n}$ is recorded:

$$\bar{n} = \frac{\sum_{\mathbf{r} \in L,\, t \in (t, t_0 + \Delta t]} c_{\mathbf{r}}(t)}{\Delta t} \tag{5}$$

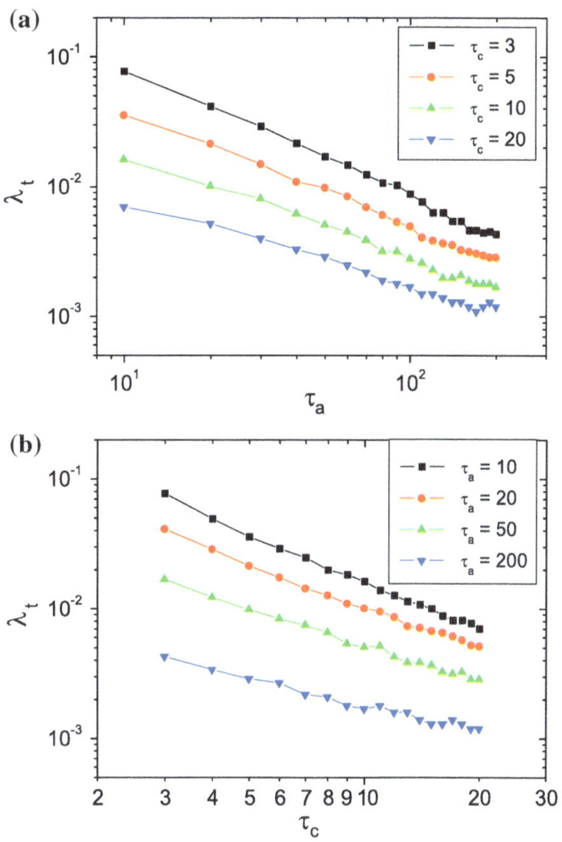

Fig. 3 Threshold pumping probability λ_t from the CA laser model: **a** λ_t dependence on the upper laser level life time (τ_a) for different values of the cavity life time (τ_c). **b** λ_t dependence on τ_c for different values of τ_a. Both Figures are plotted on a logarithmic scale. τ_a and τ_c are measured in time steps

The average number of photons $\overline{n}$ will depend linearly with the pumping probability λ, therefore for small values of λ it is expected that $\overline{n} \approx n_{np}$. We consider that λ_t is the minimum value of the pumping for which the photon average $\overline{n}$ is a 25% higher than the noise level represented by n_{np}.

The results can be seen in Fig. 3, that shows the resulting values of the threshold pumping probability λ_t obtained in this way for different values of τ_a and τ_c. The values of λ_t decrease with τ_a and τ_c as expected from Eq. (3).

In order to investigate if the exact dependence of the threshold pumping probability λ_t with τ_a and τ_c is the same as the dependence of the threshold pumping rate in Eq. (3) we have plotted, in Fig. 4, λ_t versus $\tau_a\,\tau_c$ on a logarithmic scale. Here, all the different curves of Fig. 3 collapse in a unique *straight line* with a slope close to -1 in Fig. 4. This is in good agreement with the behavior predicted by the laser rate equations, i.e. Eq. (3), showing that the CA model reproduces the threshold pumping behavior that is a characteristic of the dynamics of laser.

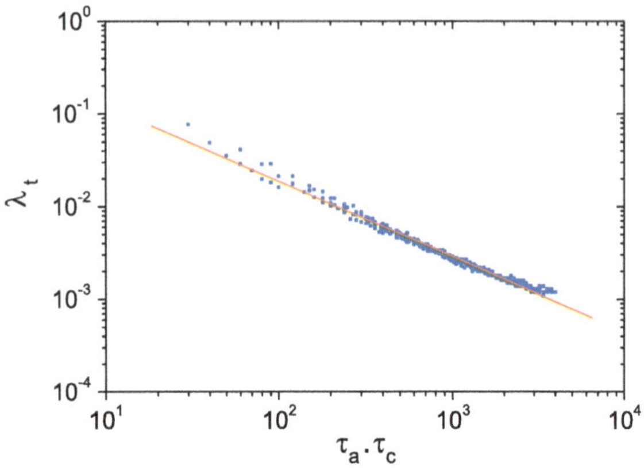

Fig. 4 Log-log plot of the threshold pumping probability λ_t versus τ_a, τ_c. The comparison between the simulations results of the CA model (the dots) and the theoretical result of the laser rate equations (the straight line $\lambda_t \propto \frac{1}{\tau_a \tau_c}$) shows a good agreement

3.5.2 Characterization of the Behavior Using the Entropy

The outcome of the CA simulation is given in terms of the time series of the number of photons, $n(t)$, and the population inversion $N(t)$, values that are dependent on the parameters λ, τ_a and τ_c. To characterize the whole set of possible behaviors one of the most commonly used quantity is the entropy function defined by [11]:

$$S(\lambda, \tau_c, \tau_a) = -\sum_{i=1}^{m} f_i \log_2 f_i \tag{6}$$

where f_i is the frequency of a given value of $n(t)$ and m is the number of non-void bins. Values of $n(t)$ that differ by less than 10^{-3} are taken as the same in order to calculate f_i and so the number of bins is $m = 10^3$.

Figure 5 shows a contour plot of the entropy obtained from the time series of $n(t)$ as a function of the relative pumping $\left(\frac{\lambda}{\lambda_t}\right)$ and the quotient between the life time of the electrons and photons $\left(\frac{\tau_a}{\tau_c}\right)$. For simplicity we take the value of $\tau_c = 10$. Similar pictures of the entropy are found for other values. For a comparison with the results of the laser rate equations the stability curve is also plotted, where it has been taken into account that $\frac{R}{R_t} = \frac{\lambda}{\lambda_t}$, as shown in [9].

For values of the pumping λ less than λ_t—pumping below threshold—the entropy has a very low value, the number of photons $n(t)$ is almost zero and there is not laser action. An increase in λ results in a growth in the number of photons and in the value of entropy. The highest values of S are found in the darkest regions ($\tau_a \gg \tau_c$ and $\lambda \approx 7\lambda_t$) where the most complex laser behaviors are expected to be found.

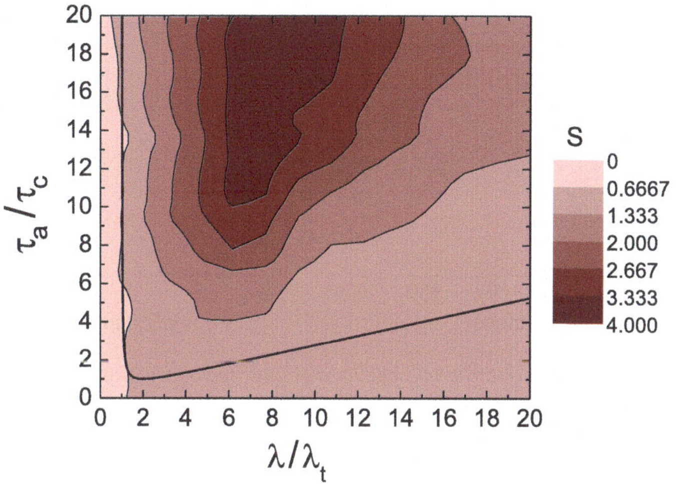

Fig. 5 Contour plot of the entropy as a function of the relative pumping $\left(\frac{\lambda}{\lambda_t}\right)$ and $\left(\frac{\tau_a}{\tau_c}\right)$. The value of τ_c has been fixed to 10. The stability curve obtained by the laser rate equations is plotted as the black line, taking into account that $\frac{R}{R_t} = \frac{\lambda}{\lambda_t}$

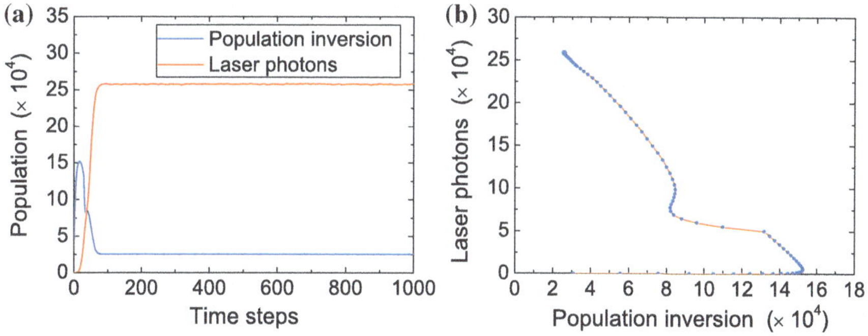

Fig. 6 Steady state behavior in lasers. **a** Time series of the number of photons $n(t)$ and the population inversion $N(t)$. **b** Plot of $n(t)$ versus $N(t)$. After a transient time $n(t)$ and $N(t)$ both show a constant value. Parameters: $\lambda = 0.192$, $\tau_a = 30$, $\tau_c = 10$

3.5.3 Spiking Behavior

It is remarkable that the CA model, besides its simplicity, is able to capture the main results obtained by the laser rate equations which are found in lasers.

Time series of the number of photons $n(t)$ and the population inversion $N(t)$ are shown in Figs. 6 and 7. The first one is an example of the stationary constant regime that is found for low values of S. While the second one corresponds to an oscillatory behavior, known as laser spiking, found for high values of S.

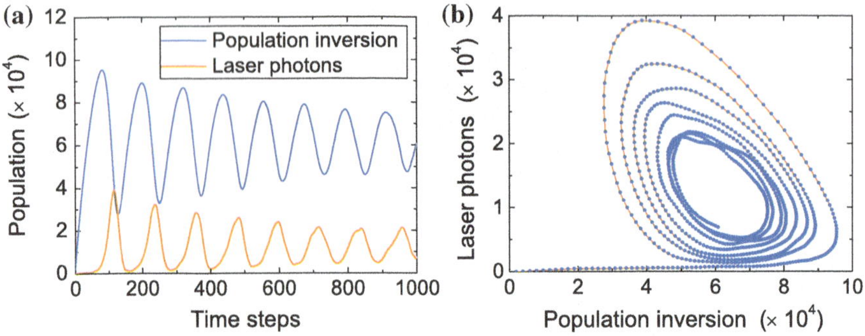

Fig. 7 Oscillatory behavior (*relaxation oscillations* or *laser spiking*). **a** Time series of the number of photons $n(t)$ and the population inversion $N(t)$. **b** Phase space of $n(t)$ versus $N(t)$ showing a cyclic limit with a decreasing amplitude of the oscillation. Parameters: $\lambda = 0.0125$, $\tau_a = 180$, $\tau_c = 10$

4 Pulsed Pumped Laser

In the previous sections the pumping rate R takes a constant value. In this one we describe a modification to study a particular type of laser systems, *pulsed pumped lasers*, in which the pumping rate changes in time according to a pulse function $R(t)$ of width $t_p > \tau_a$. A typical example of this kind of pumping is given by:

$$R(t) = \begin{cases} R_m \left| cos\left(2\pi \frac{t}{t_p}\right)\right| & 0 < t < t_p \\ 0 & t > t_p \end{cases} \tag{7}$$

Unlike in the previous cases, now we also have to contemplate that a small fraction ε of the spontaneous emission process is radiative. Thus the laser rate equations for pumped laser [10] reads:

$$\frac{dn(t)}{dt} = \varepsilon \frac{N(t)}{\tau_a} + K\, N(t)\, n(t) - \frac{n(t)}{\tau_c} \tag{8}$$

$$\frac{dN(t)}{dt} = -\frac{N(t)}{\tau_a} + R(t) - K\, N(t)\, n(t) \tag{9}$$

Figure 8 shows the results of the numerical integration of the Eqs. (8) and (9). The dynamics of the laser intensity $n(t)$ follows the shape of $R(t)$ whereas the population inversion $N(t)$, after a transient time, grows until a constant value is achieved. Once the pumping diminishes below a threshold value both $n(t)$ and $N(t)$ go to zero without any relaxation oscillations. This kind of regime is known as *quasistationary pumped laser*.

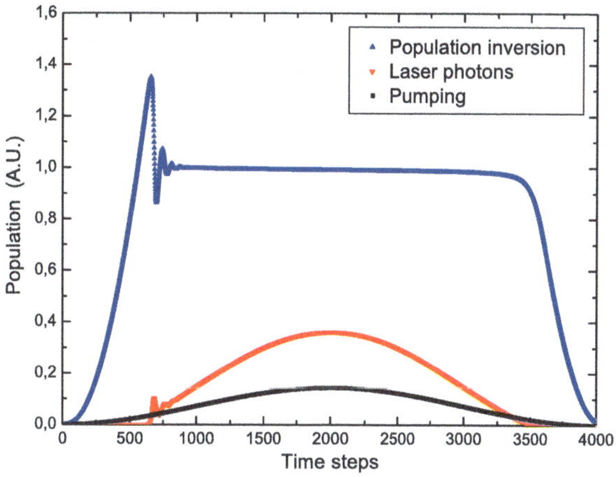

Fig. 8 Time series of the population inversion (in blue), the laser photons (in red) and the pumping rate (in black) in arbitrary units. The set of parameters used to integrate equations (8) and (9) are: $\tau_c = 3$, $\tau_a = 45$, $K = 6 \cdot 10^{-6}$, $\varepsilon = 10^{-5}$, $R_m = 4000$, $t_p = 2000$

4.1 CA Rule

The CA rule Φ that simulates the pulsed laser can be expressed as composed of four different processes: $\Phi = \Phi_1 \otimes \Phi_2 \otimes \Phi_3 \otimes \Phi_4$, where:

- Rule Φ_1—the pumping—considers a time dependent probability $\lambda(t)$ for the electrons to be in state 1:

$$\lambda(t) = \lambda_m \left| cos\left(2\pi \frac{t}{t_p}\right)\right|$$

- Rule Φ_2—the stimulated emission—and rule Φ_3—the temporal evolution of photons—are the same rules of the original CA model.
- Rule Φ_4, the decaying of the electrons, is considered now to be a radiative process with a probability θ for the emission of a new photon.

4.2 Results of Pulsed Pumped Lasers Dynamics Using CA

Figure 9 shows a typical result of the CA simulation, with the parameters $\lambda_m = 0.1$ and $\theta = 0.01$ starting from an initial state in which every electron is in the ground state and there is not any photon. Due to the discrete nature of the model and the probabilistic rules, both time series of $n(t)$ and $N(t)$ have a small noise not seen in the numerical integration.

Qualitatively there is a very good agreement between the analytic calculations shown in Fig. 8 and the CA simulations. The discrete model is able to reproduce the phenomenology observed in real pulsed pumped lasers: (i) The number of electrons in the excited state reaches a maximum value, afterwards decreases and maintains an

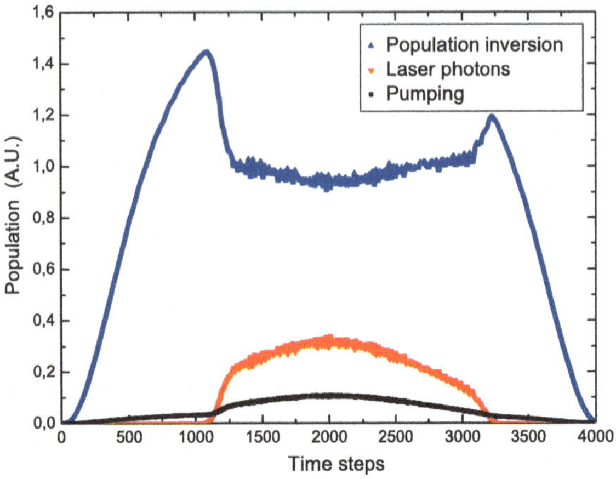

Fig. 9 Time series of the population inversion (in blue), the laser photons (in red) and the pumping rate (in black) in arbitrary units obtained by the CA model. The set of parameters used $\lambda_m = 0.1, \theta = 0.01$. Lattice size of 400×400 cells

almost constant value that will decrease towards zero when the pumping $R(t)$ is low enough; (ii) The evolution of the number of photons, $n(t)$, shows a direct dependence on the pumping $R(t)$, as it was found by the laser rate equations.

5 Antiphase Dynamics in Lasers

5.1 Laser Rate Equations for the Antiphase Dynamics in Lasers

Antiphase phenomena in lasers is an interesting subject of nonlinear dynamics. Experimentally it has been observed in many solid-state lasers, for example, in a class B Nd:YAG laser [4]. In this type of lasers it has been shown that the polarization of the beam plays an important role and thus the theoretical framework used to study this antiphase dynamics considers the laser as composed of two subsystems (v and h). Each subsystem is associated with a different polarization state which are coupled by their intensities and population inversions. Phenomenologically the antiphase dynamics in laser can be described by the following set of equations:

$$\frac{dn_v}{dt} = \kappa(N_v + \beta \cdot N_h) \cdot n_v - \frac{n_v}{\tau_c} \tag{10}$$

$$\frac{dn_h}{dt} = \kappa(N_h + \beta \cdot N_v) \cdot n_h - \frac{n_h}{\tau_c} \tag{11}$$

$$\frac{dN_v}{dt} = \gamma[R - (1 + n_v + \beta \cdot n_h) \cdot N_v] \tag{12}$$

$$\frac{dN_h}{dt} = \gamma[R - (1 + n_h + \beta \cdot n_v) \cdot N_h] \tag{13}$$

where $n_v(n_h)$ and $N_v(N_h)$ are the intensity and the population inversion corresponding to each state of polarization, $\frac{1}{\kappa}$ and $\frac{1}{\gamma}$ are the lifetimes corresponding to the photons and the population inversion, R is the pumping power and $0 \leq \beta \leq 1$ is the parameter that characterizes the coupling between both populations.

The main result obtained from the set of the laser equations is the existence of relaxation oscillations in the total intensity $(n_v + n_h)$ and in the total population inversion $(N_v + N_h)$, both with the same rapid frequency f_R which is given by:

$$f_R = \frac{1}{2\pi}\sqrt{\gamma\kappa[R(1+\beta)-1]} \tag{14}$$

whereas the difference of the intensities $(n_v - n_h)$ and the difference of population inversion $(N_v - N_h)$ between the two polarization states exhibit also relaxation oscillations but with a lower frequency f_L. The relationship between both frequencies depends only on the coupling parameter β and it is given by:

$$f_R = \left(\frac{1+\beta}{1-\beta}\right) f_L \tag{15}$$

5.2 CA Model for the Antiphase Dynamics

For the case of antiphase dynamics in laser we consider a CA model in which there are two different kinds of photons and atoms (v and h) in each lattice node. Also it is considered that the evolution rule Φ is a totalistic rule. For this purpose the following variables are defined:

$$\Gamma_i^v(t) = \sum_{neighbours} c_i^v(t) + \beta c_i^h(t) \tag{16}$$

$$\Omega_i^v(t) = \sum_{neighbours} a_i^v(t) + \beta a_i^h(t) \tag{17}$$

$\Gamma_i^v(t)$ represents the photonic state surrounding cell "i" at time t and $\Omega_i^v(t)$ stands for the atomic state. $0 \leq \beta \leq 1$ is the coupling parameter between both polarization states. For the other polarization state (h) there are two other similar functions $\Gamma_i^h(t)$ and $\Omega_i^h(t)$.

As in the previous sections the CA transition function Φ can be split in four rules $\Phi = \Phi_1 \otimes \Phi_2 \otimes \Phi_3 \otimes \Phi_4$, each one representing a different physical process.

Φ_1: Pumping Process.

In the original CA model if an atom is in the ground state ($a_i^v(t) = 0$) it will be pumped to the activated state ($a_i^v(t+1) = 1$) with a probability p. But now it will happen if the condition $\Omega_i^v(t) \geq \Omega_i^h(t)$ is also fulfilled. For the other polarization state ($a_i^h(t+1) = 1$) with probability p if $\Omega_i^h(t) \geq \Omega_i^v(t)$.

Φ_2. Stimulated Emission:

For each polarization state a photon will be created if there is an excited atom and the photonic state surrounded it is higher than a threshold value $\delta = 2$. The polarization of the photon will be v if $\Gamma_i^v(t) \geq \Gamma_i^h(t)$, otherwise a photon with polarization h will be emitted. While the atom will go to the ground state.

Φ_3 and Φ_4: The decaying of photons and excited atoms:

For both subsystems photons are destroyed after a time τ_c since they were created and the excited atoms can only remain in that state during a time τ_a.

5.3 Simulations of Antiphase Dynamics with the CA Model

A lattice size of $N = 200 \times 200$ cells has been used. The simulations begin with every cells in the null state but a small number of noise photons (0.001%) are introduced at random positions, the CA evolves for 6000 time steps. We found that for low pumping parameters ($p < 0.001$) no laser action is observed. Relaxation oscillations are observed for $p > 0.01$.

Figure 10 shows the time series of the number of photons of each polarization (n_v and n_h) and the inversion population (N_v and N_h) for a pumping of $p = 0.02$ and a coupling parameter $\beta = 0.6$. We have used the parameters $\tau_c = 10$, $\tau_a = 2500$. It can be observed that after a transient time n_v oscillates in antiphase with n_h. Antiphase dynamics is also seen between N_v and N_h.

Figure 10 also shows the total intensity ($n_v + n_h$), the total inversion population ($N_v + N_h$) and the differences ($n_v - n_h$) and ($N_v - N_h$). The total intensity shows rapid oscillations with a frequency f_R whereas the difference ($n_v - n_h$) oscillates with a low frequency f_L. Those two frequencies are clearly seen in Fig. 11 which shows the power spectrum of the time series corresponding to Fig. 10.

To study the dependence of f_R and f_L on the coupling parameter β we focus on the low frequency f_L. Figure 12 shows the power spectrum of the difference ($n_v - n_h$) in the low frequencies range. Our CA model reproduces qualitatively the fact that the low frequency of the oscillation decreases when the coupling parameter β increases ($f_L \propto \frac{1-\beta}{1+\beta}\beta^{1/2}$).

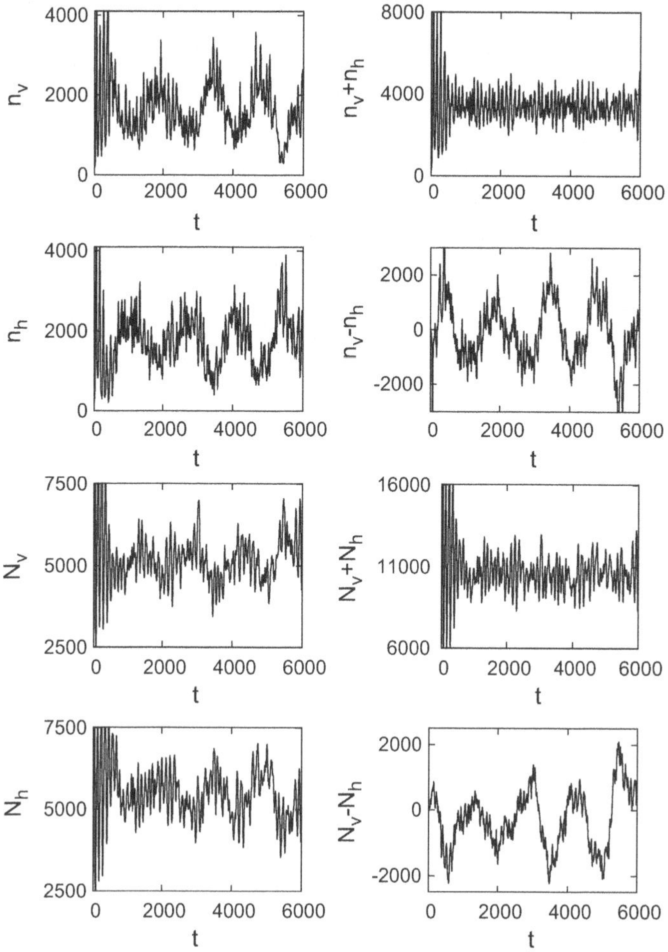

Fig. 10 To model antiphase dynamics in lasers with a CA two laser subsystems are considered. Time series of the intensity of both populations, the total intensity and the difference of intensities. The pumping probability is $p = 0.02$ and the coupling parameter $\beta = 0.6$

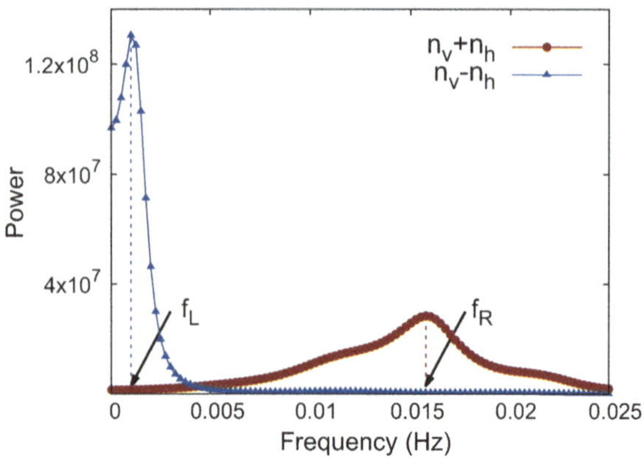

Fig. 11 Power Spectrum of the time series of the total number of photons $n_v + n_h$ (red) and the difference $n_v - n_h$ (blue) for a simulation of the CA. The coupling parameter is $\beta = 0.6$ and the pumping probability $p = 0.02$. The CA simulations show that $n_v + n_h$ oscillates with a frequency f_R which is higher than f_L that is the corresponding to $n_v - n_h$

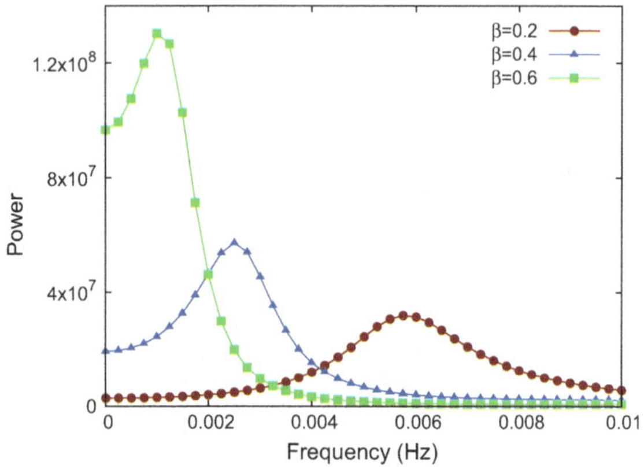

Fig. 12 Power Spectrum for the antiphase CA in the range of low frequencies for different values of the coupling parameter β. Pumping probability $p = 0.02$. Increasing the coupling parameter β between the two polarization states diminishes the low frequency f_L

6 Conclusions and Future Work

In this work we have presented a panorama of variants of a discrete model of a laser system: (i) a CA model of general laser dynamics; (ii) a modification to study pulsed pumped lasers; and (iii) another version to simulate antiphase dynamics observed in certain laser devices.

This type of models does not describe the system from a macroscopic point of view, like the standard modeling approach based on coupled differential equations, but instead it describes the elementary components of the system and their local interactions, specified by simple rules. As it is a characteristic of complex systems, the macroscopic behavior displayed by the system arises from the interaction of the elementary components as an emergent and cooperative phenomena.

Although simplified, the model is able to capture the main features of laser phenomenology, as shown by the simulation results: existence of a threshold pumping rate which depends inversely on the decaying life time of the atoms and the photons; constant or oscillatory behavior with relaxation oscillations (spiking), depending on these life times and on the pumping rate; reproduction of the qualitative population dynamics behavior of pulsed pumped lasers; and reproduction of the antiphase dynamics shown by some class B Nd:YAG lasers. It can also be used to study spatio-temporal pattern formation. The main drawback of the CA model approach is that it is a fully particle-like description. Therefore, it would be interesting to introduce wave-like properties in the CA model as a future work.

The CA modeling approach can be a good complement to the classical analysis that relies on the laser rate equations for situations in which it may be difficult or even impossible to obtain satisfactory results: (i) when there are stiff differential equations and the normal numerical methods for solving them are numerically unstable, so that they magnify approximation errors [2, 8, 14, 21]; (ii) when the equations are not perfectly valid if the assumptions that the laser cavity has a simple geometry and that the active medium is homogeneous are not fulfilled; (iii) for very small-size laser devices for which the use of Maxwell's equations and field optics, which involve many approximations, produce results that are not very accurate.

References

1. Bandini, S., Pavesi, G.: Simulation of vegetable populations dynamics based on cellular automata (2002)
2. Brydon, D., Pearson, J., Marder, M.: Solving stiff differential equations with the method of patches. J. Comput. Phys. **144**, 280–298 (1998)
3. Byrne, G.D., Hindmarsh, A.C.: Stiff ODE solvers: a review of current and coming attractions. J. Comput. Phys. **70**, 1–62 (1987)
4. Cabrera, E., Calderón, O.G., Guerra, J.: Experimental evidence of antiphase population dynamics in lasers. Phys. Rev. A **72**, 043824 (2005)
5. Chopard, B., Droz, M.: Cellular Automata Modeling of Physical Systems. Cambridge University Press (1998)
6. Chopard, B., Luthi, P., Droz, M.: Reaction-diffusion cellular automata model for the formation of liesegang patterns. Phys. Rev. Lett. **72**, 1284–1387 (1994)
7. Creutz, M.: Deterministic Ising dynamics (1986)
8. Dinand, M., Schuette, C.: Theoretical modeling of relaxation oscillations in Er-doped waveguide lasers. J. Lightwave Technol. **13**(1), 14–23 (1995)
9. Guisado, J.L., Jiménez-Morales, F., Guerra, J.M.: Cellular automaton model for the simulation of laser dynamics. Phys. Rev. E **67**(6), 066708 (2003)
10. Guisado, J.L., Jiménez-Morales, F., Guerra, J.M.: Simulation of the Dynamics of Pulsed Pumped. In: Lecture Notes in Computer Science, vol. 3305, pp. 278–285 (2004)
11. Guisado, J.L., Jiménez-Morales, F., Guerra, J.M.: Application of Shannon's entropy to classify emergent behaviors in a simulation of laser dynamics. Math. Comput. Model. **42**(7–8), 847–854 (2005)
12. Ilachinski, A.: Cellular Automata: a discrete universe. World Scientific (2001)
13. Lega, J., Moloney, J.V., Newell, A.C.: Universal description of laser dynamics near threshold. Phys. D **83**(4), 478–498 (1995)
14. Miranker, W.L.: Numerical Methods for Stiff Equations and Singular Perturbation Problems: and singular perturbation problems. D. Reidel—Springer, Dordrecht, The Netherlands (1981)
15. von Neumann, J.: Theory of Self-Reproducing Automata. University of Illinois Press, Urbana (1966)

16. Qiu, G., Kandhai, D., Sloot, P.M.A.: Understanding the complex dynamics of stock markets through cellular automata. Phys. Rev. E—Stat. Nonlinear Soft Matter Phys. **75**(4) (2007)
17. Siegman, A.: Lasers. Unversity Science Books (1986)
18. Sloot, P., Chen, F., Boucher, C.: Cellular Automata Model of Drug Therapy for HIV Infection (2002)
19. Subrata, R., Zomaya, A.Y.: Evolving cellular automata for location management in mobile computing networks. IEEE Trans. Parallel Distrib. Syst. **14**(1), 13–26 (2003)
20. Svelto, O.: Principles of Lasers. Plenum Press (1989)
21. Veasey, D.L., Gary, J.M., Amin, J., Aust, J.A.: Time-dependent modeling of erbium-doped waveguide lasers in lithiumniobate pumped at 980 and 1480 nm. IEEE J. Quantum Electron. **33**(10), 1647–1662 (1997)
22. Wolfram, S.: Cellular Automata and Complexity: collected papers. Addison-Wesley (1994)

Index

V. Carmona et al. (eds.), *Nonlinear Systems, Vol. 1*, Understanding Complex Systems,
https://doi.org/10.1007/978-3-319-66766-9

The manufacturer's authorised representative in the EU is Springer Nature Customer Service Centre GmbH, Europaplatz 3, 69115 Heidelberg, Germany. If you have any concerns regarding our products, please contact ProductSafety@springernature.com

Printed and bound by CPI Group (UK) Ltd, Croydon, CR0 4YY
05/10/2026
02228122-0015